W9-BMH-335

Sensors *for* Mobile Robots

To my parents, who let me build a workshop in my bedroom, and generally tolerated my creative obsessions.

And to my wife, Rachael, who allowed the robots to roam freely in our home, in spite of the hydraulic leaks, battery spills, and incessant chatter from their speech synthesizers.

Sensors *for* Mobile Robots

Theory and Application

H.R. Everett
Naval Command, Control and
Ocean Surveillance Center
San Diego, California

A K Peters, Ltd.
Wellesley, Massachusetts

Editorial, Sales, and Customer Service Office

A K Peters, Ltd.
289 Linden Street
Wellesley, MA 02181

Library of Congress Cataloging-in-Publication Data

Everett, H. R., 1949-
Sensors for mobile robots : theory and application / H.R. Everett.
p. cm.
Includes bibliographical references and index.
ISBN 1-56881-048-2
1. Mobile robots. 2. Robots—Control systems. I. Title.
TJ211.415.E83 1995
629.8 ' 92—dc20 95-17178
CIP

Principle illustrator: Todd Ashley Everett

Printed in the United States of America
99 98 97 96 95 10 9 8 7 6 5 4 3 2 1

Table of Contents

Foreword

A robot's ability to sense its world and change its behavior on that basis is what makes a robot an interesting thing to build and a useful artifact when completed. Without sensors, robots would be nothing more than fixed automation, going through the same repetitive task again and again in a carefully controlled environment. Such devices certainly have their place and are often the right economic solution. But with good sensors, robots have the potential to do so much more. They can operate in unstructured environments and adapt as the environment changes around them. They can work in dirty dangerous places where there are no humans to keep the world safe for them. They can interact with us and with each other to work as parts of teams. They can inspire our imaginations and lead us to build devices that not so long ago were purely in the realms of fiction.

Sensors are what makes it all possible.

When it comes right down to it there are two sorts of sensors. There are visual sensors, or eyes, and there are non-visual sensors. Lots of books have been written about visual sensors and computer vision for robots.

There is exactly one book devoted to non-visual sensors. This one.

We tend to be a little vision-centric in our "view" (there we go again...) of the world, as for humans vision is the most vivid sensor mechanism. But when we look at other animals, and without the impediment of introspection, another picture (hmmm...) begins to emerge. Insects have two eyes, each with at most perhaps 10,000 sensor elements.

Arachnids have eight eyes, many of them vestigial, some with only a few hundred sensor elements, and at most 10,000 again. But insects have lots and lots and lots of other sensors. Cockroaches, for example, have 30,000 wind-sensitive hairs on their legs, and can sense a change in wind direction and alter the direction in which they are scuttling in only 10 milliseconds. That is why you cannot stomp on one unless you have it cornered, and on top of that get lucky. The cockroach can sense your foot coming and change course much faster than you can change where you are aiming. And those 30,000 sensitive hairs represent just one of a myriad of specialized sensors on a cockroach. Plus each different insect has many varied and often uniquely different sensors. Evolution has become a master at producing non-visual sensors.

As robotics engineers we find it hard to create new sensors, but are all aware that in general our robots have a rather impoverished connection to the world. More sensors would let us program our robots in ways that handled more situations, and do better in those situations than they would with fewer sensors. Since we cannot easily create new sensors, the next best thing would be to know what sensors were already available. Up until this point we have all maintained our own little libraries of sensors in our heads. Now Bart Everett has written down all he had in his own private library and more. Bart's robots have always stood out as those with the most sensors, because interactive sensing has always been a priority for Bart. Now he is sharing his accumulated wisdom with us, and robotdom will be a better place for it. Besides providing us with an expanded library, Bart has also done it in a way that everyone interested in robotics can understand. He takes us through the elementary physics of each sensor with an approach that a computer scientist, an electrical engineer, a mechanical engineer, or an industrial engineer can relate to and appreciate. We gain a solid understanding of just what each sensor is measuring, and what its limitations will be.

So let's go build some new robots!

Rodney A. Brooks
MIT AI Lab
Cambridge, MA

Preface

My underlying goal in the preparation of this manuscript was to present some general background on the sensing needs of a mobile system, followed by sufficient theory of operation and illustrative examples such that the overall result is both informative and of practical use. Perhaps the most challenging problem I faced early on in this endeavor was how to arrange reams of information on all the various sensors into some semblance of logical order. One considered possibility was to categorize by class of robot (i.e., airborne, underwater, indoor, exterior, autonomous, teleoperated). Given the emphasis of the book, however, it seemed more appropriate to break down the discussion by sensor type.

In an attempt to bound the problem, I decided to eliminate any treatment of airborne or underwater scenarios and focus instead on interior and exterior land-based applications. Even so, there was still considerable difficulty associated with organizing the flow. For example, at least seven different methods of non-contact ranging techniques are known to exist; one of these methods alone (triangulation) can be implemented in five different ways. Almost all such ranging systems can operate in the acoustical or electromagnetic regions of the energy spectrum; can be active or passive; and may have markedly different assigned functions in actual deployment.

After much weighing of alternative strategies, I chose to present the material in a manner that to some extent parallels the strategy often employed in robotic development. The initial thrust of most early research efforts in which I participated was simply aimed at how to get the robot to move about in a controlled and purposeful fashion. Once this hurdle is surmounted, attention can be turned to collision avoidance, wherein the system learns not to run into things while enroute. The proud builders soon realize the robot can perform admirably for some finite length of time but eventually will get lost, whereupon developmental focus shifts to navigational referencing. Applications are tacked on later, sometimes almost as an afterthought.

Accordingly, following some general background discussions in Chapter 1, we start by taking a look in Chapter 2 at the sensors employed in vehicle dead reckoning, with a careful analysis of potential error sources. Tactile and proximity sensors are introduced next in Chapter 3, providing a rudimentary capability to at least detect potential obstructions in time to stop. Chapters 4 through 7 provide an overview of the various distance measurement techniques available, such as triangulation, time of flight, frequency modulation, phase-shift measurement, and interferometry. Related discussion of implementation in the acoustical, radio frequency, and electro-optical domains is presented in Chapters 8 and 9, with a special emphasis on the various factors affecting performance.

This approach hopefully provides a good foundation for later examining how such non-contact ranging sensors are employed in specific roles, first and foremost being in support of collision avoidance (Chapter 10). Navigational referencing, the subject of Chapters 11 through 16, is addressed in considerable detail as it represents one of the biggest remaining stumbling blocks to successful fielding. A few representative samples of application-specific sensors are treated in closing in Chapter 17.

In retrospect, there is considerably less emphasis than I originally intended on image-based systems, as the subject of machine vision quite obviously could be the focus of a book all in itself. And since a number of distinguished individuals far better qualified than myself have in fact taken that very objective to task, I have purposely limited discussion in this volume, and concentrated instead on various alternative (and often less complex) sensing strategies less documented in the open literature. Reference is made throughout the text to candidate systems, both commercially available and under development, in hopes of complementing theory of operation with some practical lessons in real-world usage. These illustrative examples are called out under separate headings where the discussion becomes rather detailed.

I have very much enjoyed the preparation of this manuscript, both in terms of what I learned in the process and the new contacts I made with other researchers in this exciting field. I hope the results as presented here will be useful in promoting the successful employment of mobile robotic systems through increased awareness of available supporting technologies.

H.R. Everett
San Diego, CA

Acknowledgments

A number of people have assisted me in my educational and research endeavors over the years and collectively contributed to making this book a reality. I would like to express my heart-felt appreciation to:

My uncles, Gene Everett and Joe Hickey, who introduced me to electronics at an early age.

My high school geometry teacher, Mrs. Nell Doar, for providing discipline, inspiration, and the mathematical foundation upon which I was to build.

Professor Robert Newton, my thesis advisor at the Naval Postgraduate School, who made it possible for me to pursue a rather unorthodox topic in the field of mobile robotics.

Vice Admiral Earl B. Fowler, USN (Ret.) for creating a robotics program office within the Naval Sea Systems Command, and giving me a job after graduate school.

Dr. Anita Flynn of MIT for all the late nights and weekends we spent hacking code and building our own sensors in my basement in Virginia.

Gary Gilbreath of the Naval Command Control and Ocean Surveillance Center for transforming ROBART II into a truly intelligent machine.

My son, Todd Everett, for his tremendous help in generating all the graphics used in the figures.

All those people kind enough to review this manuscript in the various stages of its completion, offering helpful insights on how best to present the material: Ron Arkin, Johann Borenstein, Fernando Figueroa, Anita Flynn, Doug Gage, Bob Garwood, Tracy Heath, Susan Hower, Robin Laird, Richard Langley, Richard Lao, Larry Mathies, and Hoa Nguyen.

In addition, portions of the material presented in Chapters 4 through 7 were previously published in *Sensors* and later *Robotics and Autonomous Systems* magazines, and updated in this book with their kind permissions.

1

Introduction

The past several years have brought about a tremendous rise in the envisioned potential of robotic systems, along with a significant increase in the number of proposed applications. Well-touted benefits typically associated with the installation of fixed-location industrial robots are improved effectiveness, higher quality, reductions in manpower, as well as greater efficiency, reliability, and cost savings. Additional drivers include the ability to perform tasks of which humans are incapable, and the removal of humans from demeaning or dangerous scenarios.

The concept of mobility has always suggested an additional range of applications beyond that of the typical factory floor, where free-roaming robots move about with an added versatility fostering even greater returns. Early developmental efforts introduced potential systems for fighting fires, handling ammunition, transporting materials, and patrolling warehouses and storage areas, to name but a few. Most of the resulting prototypes met with unexpected difficulty, primarily due to an insufficient supporting technology base. Even today, after decades of extensive research and development, the successful application of mobile robots remains for the most part an elusive dream, with only a small handful of fielded systems up and running.

While a number of technological hurdles have impeded progress, the three generally regarded as having the greatest impact are: 1) *computational resources*, 2) *communications*, and 3) *sensors*. The first two areas have been addressed for a variety of commercial reasons with remarkable progress. In just a little over 10 years we have transitioned from *6502*- and *Z80*-based personal computers running under *C/PM* with a maximum 64-kilobyte address space, to *Pentium*-based systems running at 90 MHz and addressing up to 32 megabytes of memory. The recent surge in popularity of laptop computers has provided an extra impetus, with special emphasis on reduced power consumption and extended battery life. Wireless local area networks and spread-spectrum technology have likewise advanced in kind, to the point where there are now a number of vendors offering full-duplex Ethernet-compatible high-speed datalinks with ranges of several miles.

The third category of *sensors* now stands somewhat alone as the most significant technical challenge still facing developers, due primarily to a lack of high-volume applications. While there has indeed been some carry-over sensor technology from advances in flexible automation for manufacturing, it has fallen far short of the explosive growth seen in the computer and communications industries. Successful adaptation of what progress has been made is further hampered by the highly unstructured nature of a mobile robot's operating environment. Industrial process-control systems used in repetitive manufacturing scenarios, in contrast, rely on carefully placed sensors that exploit the target characteristics. Background conditions are arranged to provide minimal interference, and often aid in the detection process by purposely increasing the *on-off* differential or contrast. Unfortunately, such optimized configuration control is usually no longer possible once mobility is introduced as a factor in the equation.

Consider for example the issue of collision avoidance: any mobile robot intended for real-world operation must be capable of moving around without running into surrounding obstructions. In practice, however, the nature and orientation of obstacles are not known with any certainty; the system must be capable of detecting a wide variety of target surfaces under varying angles of incidence. Control of background and ambient conditions may not be possible. *A priori* information regarding the relative positions, orientations, and nature of objects within the sensor's field of view becomes very difficult to supply.

The situation only worsens when the operating environment is taken outdoors, for a number of reasons. To begin with, problems of scale introduce a need for additional range capability that significantly adds to system complexity and cost. While an indoor collision avoidance system may need to see only 4 to 6 feet in front of the robot, for example, exterior scenarios typically require effective coverage over a 20- to 30-foot span, sometimes more. In addition, the outdoor environment often poses additional complicating hazards to safe navigation (i.e., terrain traversabilty, oncoming traffic, atmospheric obscurants) that demand appropriate engineering solutions not even addressed on interior systems.

On the positive side, worldwide interest in a rapidly expanding field known as *intelligent vehicle highway systems (IVHS)* has already created a huge potential market for sensors to address many of these problems as faced by the automotive industry (Catling, 1994). Lower-volume autonomous mobile robot applications are sure to benefit from the inevitable spin-off technologies that have already begun to emerge in the form of low-cost laser and millimeter-wave systems, for example. Many of these new and innovative products will be presented as illustrative examples in the following chapters, in hopes of further stimulating this technology-transfer process.

1.1 Design Considerations

The problems confronting most mobile robotic development efforts arise directly from the inherent need to interact with the physical objects and entities in the environment. The platform must be able to navigate from a known position to a desired new location and orientation, avoiding any contact with fixed or moving objects while en route. There has been quite a tendency in early developmental efforts to oversimplify these issues and assume the natural growth of technology would provide the needed answers. While such solutions will ultimately come to pass, it is important to pace the evolution of the platform with a parallel development of the needed collision avoidance and navigation technologies.

Fundamental in this regard are the required sensors with which to acquire high-resolution data describing the robot's physical surroundings in a timely yet practical fashion, and in keeping with the limited onboard energy and computational resources of a mobile vehicle. General considerations for such sensors are summarized below:

- *Field of view* — Should be wide enough with sufficient depth of field to suit the application.
- *Range capability* — The minimum range of detection, as well as the maximum effective range, must be appropriate for the intended use of the sensor.
- *Accuracy and resolution* — Both must be in keeping with the needs of the given task.
- *Ability to detect all objects in environment* — Objects can absorb emitted energy; target surfaces can be specular as opposed to diffuse reflectors; ambient conditions and noise can interfere with the sensing process.
- *Real-time operation* — The update frequency must provide rapid, real-time data at a rate commensurate with the platform's speed of advance (and take into account the velocity of other approaching vehicles).
- *Concise, easy to interpret data* — The output format should be realistic from the standpoint of processing requirements; too much data can be as meaningless as not enough; some degree of preprocessing and analysis is required to provide output only when action is required.
- *Redundancy* — The system should provide graceful degradation and not become incapacitated due to the loss of a sensing element; a multimodal capability would be desirable to ensure detection of all targets, as well as to increase the confidence level of the output.
- *Simplicity* — The system should be low-cost and modular to allow for easy maintenance and evolutionary upgrades, not hardware-specific.
- *Power consumption* — The power requirements should be minimal in keeping with the limited resources on board a mobile vehicle.
- *Size* — The physical size and weight of the system should be practical with regard to the intended vehicle.

The various issues associated with sensor design, selection, and/or integration are complex and interwoven, and not easily conveyed from a purely theoretical perspective only. Actual device characterization in the form of performance validation is invaluable in matching the capabilities and limitations of a particular sensor technology to the application at hand. Most manufacturers of established product lines provide excellent background information and experienced applications engineers to assist in this regard, but some of the more recently introduced devices are understandably a bit behind the power curve in terms of their documented performance results. In addition to the general theory of sensor operation, therefore, this book attempts to provide the reader with some important exposure to the practical experiences and insights of system developers involved in this rapidly evolving field.

1.2 The Robots

I consider myself very fortunate to have been personally associated with the development of a number of mobile systems over the past 30 years and will refer to several of these throughout this text for purposes of illustration. The following introductory sections are intended to provide a brief overview of these robots for those interested in the background. It is somewhat amusing to note the advancements made in the supporting technologies over this time span. The bottom line, however, is that the most sophisticated mobile robots in existence today still fall orders of magnitude short in achieving the utility and perception of their most inept human counterparts. While we have come a long way as developers, there is still much left to be done.

1.2.1 WALTER (1965-1967)

WALTER (Figure 1-1) was a 5-foot-tall anthropomorphic robot I constructed my sophomore year in high school as a science fair entry. Strictly a teleoperated system with no onboard intelligence, WALTER was capable of forward or reverse travel, using two 8-inch rear drive wheels made of ¾-inch plywood and a pair of 2-inch roller-skate wheels in front for steering. The steering mechanism was solenoid-actuated under *bang-bang* control, with a spring-loaded center default position. A 20-foot umbilical tether supplied 117-volt AC power from the control station shown on the left side of the photo.

The right arm was capable of two-degree-of-freedom movement (elbow and shoulder) driven by linear actuators constructed from ¼-inch threaded rod, powered by a sewing machine motor and a kitchen mixer, respectively. The left arm had only a single degree of freedom at the elbow (I ran out of motors), its associated linear actuator being coupled to the prime mover from an old movie projector. All the motors were single-speed series-wound universal type controlled (by onboard relays) from the remote operator console. The linear actuators were coupled to their respective joints by tendons made from bicycle hand-brake cables.

Figure 1-1. *WALTER* (circa 1964) was a teleoperated anthropomorphic robot constructed as a high school science fair entry.

The left and right grippers were also different (it's no fun building the same thing twice...), but similar in that they both lacked wrist movement. The right gripper was fashioned from a 10-inch fuse puller, aligned for grasping objects oriented in the horizontal plane. The left gripper was somewhat more complex, constructed from ¼-inch hardwood with two degrees of freedom as illustrated in Figure 1-2, and oriented to grasp vertical objects. All gripper joints were tendon-driven by cables spring-coupled to powerful solenoids removed from innumerable washing machines.

WALTER's head could pan left or right approximately 45 degrees either side of center, driven through tendons by a linear actuator mounted in the base to keep the center of gravity low. Load-bearing joints (head pan axis, shoulder, elbows) were fashioned from ball-bearing roller-skate wheels. There was a photocell mounted on top of the head to monitor ambient light conditions, and, of course, the obligatory flashing lamps for eyes and nose. Two microphone ears and a speaker behind the mouth opening provided for remote communications via the telephone handset shown in Figure 1-1. (After all, 20 feet is a long way to yell when we have the technology.)

Figure 1-2. WALTER'S left gripper was tendon actuated with two degrees of freedom.

The electronics for both the robot and the control console were vacuum-tube based. One interesting submodule was a capacity-operated relay (see Chapter 3) coupled to a touch sensor in the right gripper. The sole purpose of this circuitry was to discourage pulling and prodding by curious onlookers; any stray finger that poked its way into the open claw would be met by a startling and decidedly effective warning snip. The resounding thump of the actuating solenoid only served to accentuate the message.

WALTER met his demise one day in 1967 at the hands of our cleaning lady (bless her heart). I had been experimenting with some Japanese six-transistor portable radios that sold at the time for around five dollars apiece, trying to come up with a low-cost radio control scheme. The idea was to tune each of the four receivers to a blank spot on the AM dial, and use a continuous-tone RF transmitter that could be switched to any one of these four frequencies. Half-wave rectifiers attached to the audio outputs of the individual radios activated sensitive meter relays that controlled the forward, reverse, left, and right power relays in the drive circuitry.

As fate would have it, the unsuspecting maid bravely entered the confines of my bedroom workshop one day when I was not at home and turned on the ancient pre-World-War-II *Lewyt* vacuum cleaner my dad had rebuilt six times just in my brief lifetime. The motor brushes had long since worn down to their springs, which arced across the pitted commutator segments with such intensity that all TV and radio reception for two blocks was blanked out whenever the machine was running. WALTER's radios responded instantly to this rich broad-band interference, randomly applying power in a mindless fashion to drive motors and steering solenoids alike. The robot lurched forward, twisting and turning, motors whining and solenoids clacking, only to be immediately decapitated with one mighty swing of a *Lewyt* rug sweeper. When I got home the vacuum was still running, WALTER was a total loss, the front door was swinging on its hinges, and the maid had vanished, never to return.

1.2.2 CRAWLER I (1966-1968)

I had been bitten by the bug, it seemed, and was now fascinated with the idea of building a free-roaming robot unencumbered by any sort of tether. There was little point in trying to refurbish WALTER; structural damage notwithstanding, all the electrical components were rated for 117 volts AC. My next creation had to be battery powered. And so I began to amass an impressive collection of DC motors, relays, and other diverse components while sorting out the design in my head. The end result was CRAWLER I (Figure 1-3), intended to be my junior-year science fair project. (The eagerly anticipated event was unfortunately canceled due to faculty indifference.)

Figure 1-3. Photo of *CRAWLER I* (circa 1966) in early stages of development.

I had also decided to build a tracked vehicle for improved maneuverability. Two 24-volt DC gearmotors from a aircraft surplus catalog were mounted on a 18- by 13-inch plywood base (Figure 1-4), driving left and right tracks fashioned from 1.5-inch rubber timing belts turned inside out. Control was again provided by relays, but the motors each had centrifugal speed-limiting switches that could be adjusted to achieve straight-line travel. By adding an override circuit on the stationary side of the slip rings that fed the centrifugal governor, it was possible to momentarily boost the motor rpm to maximum. *Skid steering* was achieved by providing differential speed commands in this fashion or stopping one motor altogether. The vehicle could also turn in place by reversing one track.

The tough part in building an autonomous vehicle, of course, lies in how to control its motion, made even tougher still in an era that predated microprocessors and low-cost sensors. I had in mind a platform that would drive around until it encountered an object, then alter course in an intelligent fashion. I also wanted it to automatically recharge the onboard lead-acid motorcycle batteries when they ran low. Like most engineers, I tackled the tougher issue first: automatic recharging. I settled on a beacon homing scheme and elected to use an ordinary light bulb as the source. (It would take me some time, and several follow-on robots, to shake this mind set.) Details of this tracking and homing design are presented later in Chapter 15.

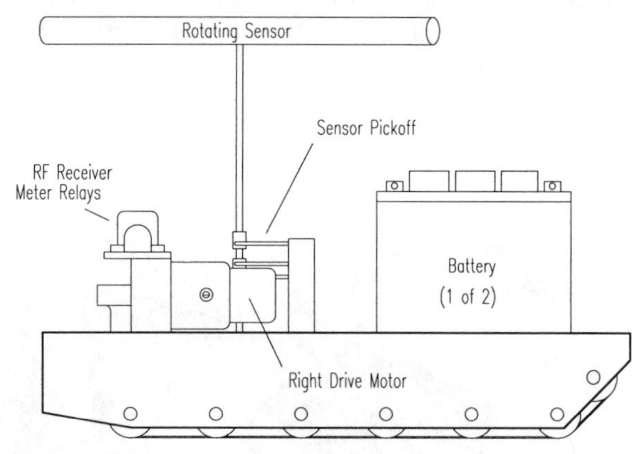

Figure 1-4. A rotating photocell sensor was used on *CRAWLER I* to locate and track a homing beacon for automatic recharging.

Providing for truly autonomous operation meant adding some type of collision avoidance sensor and implementing a scheme of intelligent reaction. Tactile sensors made from guitar strings were subsequently installed on the four corners of the platform to support this task and are described in Chapter 3. Intelligent response was another matter; single-chip microcontrollers were not yet even a figment of anyone's imagination in those days. My Hollywood-inspired image of a computer centered around a lot of flashing lights and punched cards. I had already wired dozens of very impressive indicator lamps in parallel with the relay coils of the CRAWLER's logic and control circuitry (for diagnostic purposes, of course). Operating the CRAWLER with the four-channel radio control developed on WALTER had quickly become boring, so it seemed the appropriate thing to do was build a punched-card reader.

The robot's environment could be simplistically described by four state variables associated with the tactile sensors situated at each of the four corners of the platform. By comparing these sensor input states to a 4-bit address field punched into each card, the correct response to any particular scenario could be read from the output section of the one card with an address code matching the specified input conditions. The robot would simply stop whenever input conditions changed state and cycle the cards until finding a match. The preprogrammed response (i.e., drive and steering commands) to the new conditions would be punched into the 4-bit output field of the correct card.

I was really excited about the prospect of building this card reader and made pretty fair progress using modified index cards with eight photocells to detect ¼-inch holes made by a standard office hole punch. An actual 3.5- by 8-inch card is shown in Figure 1-5; the top row of holes represented the inputs, while the bottom row controlled the outputs. The individual illumination sources for the eight opposing photocells were 3-volt pilot lamps, wired in series to ensure the entire string would extinguish to prevent faulty readings if any single bulb burned out. The lamps were powered by the 12-volt battery at half their rated filament voltage to ensure extended life, and the reduced light output prevented premature activation of the photodetectors through the thin index-card paper. But the mechanics associated with reliably recycling the stack of cards (once all had been read) proved too much for my limited shop facilities, so I resorted to using a 12-inch circular disk of poster paper traced from a 33-rpm record album.

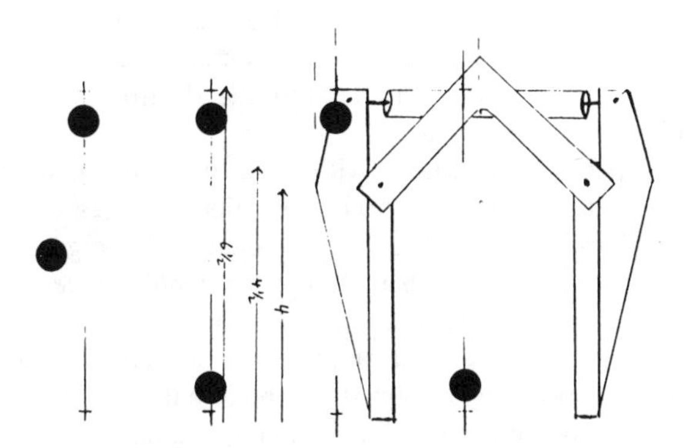

Figure 1-5. An actual 3- by 5-inch card used on *CRAWLER I* showing the two rows of punched holes representing input and output data. The sketch on the back is a preliminary gripper design that was abandoned in favor of the vise-grip implementation shown later in Figure 1-7

This approach greatly simplified matters. The address and output fields were aligned along the radial axis of the disk with 16 possible states as shown in Figure 1-6, with the most significant bit towards the outer periphery. The disk would rotate at 6 rpm while the photocells looked for a hole pattern corresponding to the sensor input states. When a match was found, the disk drive motor was disabled and the output field would be read, thus determining the desired control relay states for left and right track drive and direction. The output holes were punched in radial columns offset exactly 78.75 degrees from their associated input columns to allow sufficient room for the two photocell arrays. The circular card was secured to a rubber-covered drive capstan with a ¼-inch wingbolt and washer.

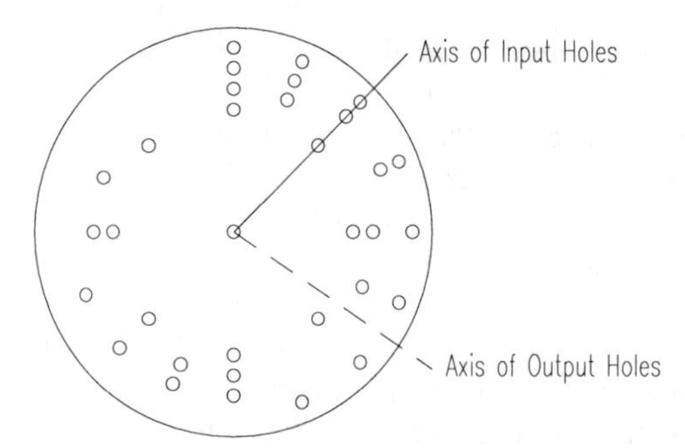

Figure 1-6. Mechanical problems with the stacked-card transport mechanism forced a switch to the circular card format shown above. Punched output holes (not shown) were inserted between the input address fields, offset approximately 90 degrees.

1.2.3 CRAWLER II (1968-1971)

All the added features (particularly the 12-inch disk reader) necessitated a complete repackaging of the *CRAWLER's* mechanical layout, so I elected to scrap the plywood prototype altogether and build an aluminum chassis. The result was *CRAWLER II*, basically the same size, but with the electronics implemented in a layered approach as shown in Figure 1-7.

I had begun experimenting earlier with some miniature hydraulic cylinders fashioned by encapsulating 30- and 50-cc irrigation syringes inside of copper-tubing sheaths with epoxy glue. Considerable force could be generated with one of these devices when operated at about 100 psi; solenoid valves from discarded washing machines were modified to provide control. A surplus chemical-injection pump was used to pressurize an accumulator made from a 4-inch length of 3-inch-diameter copper pipe capped on both ends. CRAWLER II was eventually modified and equipped with a hydraulic arm and gripper configuration as illustrated in Figure 1-7. The gripper force was quite powerful. While attempting to explore the limits of remote-operator dexterity, I once squeezed the locomotive of my brother's train set just a wee bit too hard, rendering it no longer compatible with H-O gauge track.

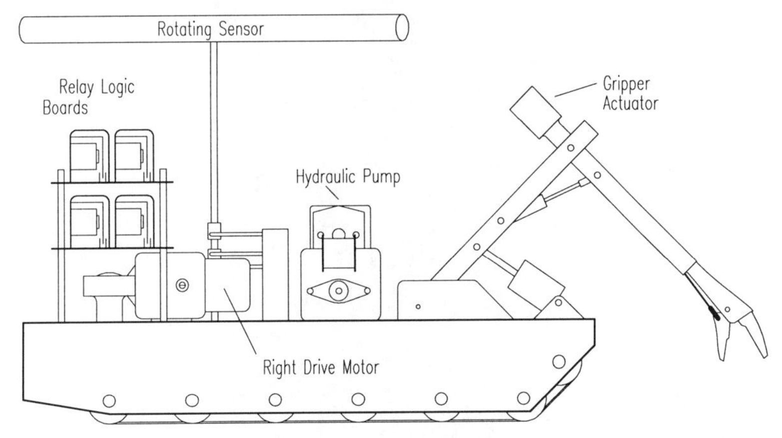

Figure 1-7. *CRAWLER II* (shown here without the circular disk reader) was a teleoperated platform equipped with a 2-DOF hydraulic gripper.

Unfortunately, the bulky disk reader and the manipulator would not both fit on the platform at the same time, and the modified hydraulic components were all rated for 117 volts AC. In addition, there was essentially no way to control the new manipulator in an autonomous fashion, so CRAWLER II had to revert back to tethered control. The few photographs I have of CRAWLER I were taken by one of my high school buddies who owned a Polaroid camera; since most of the CRAWLER II development was performed while I was away at college, I regrettably don't have any pictures. Work on CRAWLER II ceased my junior year, when I "borrowed" the onboard logic control unit to automate our (very) mechanized homecoming display at Georgia Tech.

1.2.4 ROBART I (1980-1985)

ROBART I (Figure 1-8) was my thesis project at the Naval Postgraduate School in Monterey, CA (Everett, 1982a; 1982b). Its assigned function was to patrol a normal home environment, following either a random or set pattern from room to room, checking for unwanted conditions such as fire, smoke, intrusion, etc. The security application was chosen because it demonstrated performance of a useful function and did not require an end-effector or vision system, significantly reducing the required system complexity. Provision was made for locating and connecting with a free-standing recharging station when battery voltage began running low (Figure 1-9). Patrols were made at random intervals, with the majority of time spent immobile in a passive intrusion-detection mode to conserve power.

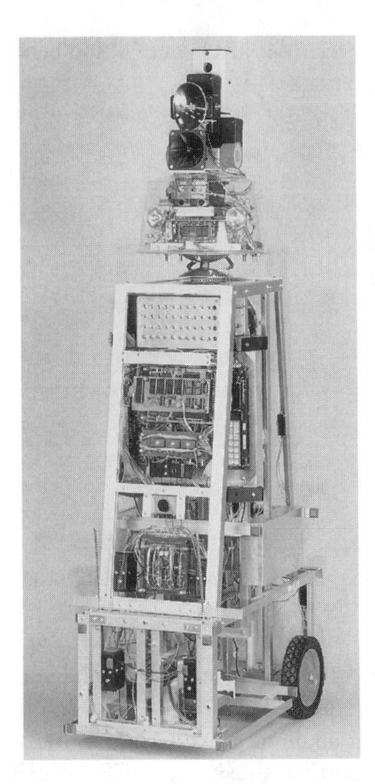

Figure 1-8. *ROBART I* was a fully autonomous interior security robot (courtesy Naval Surface Weapons Center).

A Synertek *SYM-1* single-board computer formed the heart of the onboard electronics. Speech synthesis (to allow the system to announce any unwanted conditions detected in the course of a random patrol) was implemented through National Semiconductor's *Digitalker DT1050* synthesizer chip. Two sets of vocabulary instructions were stored on EPROMs for a total vocabulary of 280 words. A fixed vocabulary was chosen over an unlimited vocabulary created through use of phonemes in light of the greatly decreased demand on the onboard microprocessor in terms of execution time and memory space.

The software maintained the robot in one of two modes of operation: *Alert Mode* or *Passive Mode*. In the *Passive Mode*, the majority of sensors were enabled, but a good deal of the interface and drive control circuitry was powered down to conserve the battery. The robot relied on optical motion detection, ultrasonic motion detection, and hearing to detect an intruder, while at the same time monitoring for vibration (earthquake), fire, smoke, toxic gas, and flooding (Everett, 1982a). Some of these inputs were hard-wired to cause an alert (switch from *Passive Mode* to *Alert Mode*), whereas others had to be evaluated first by software that could then trigger an alert if required. Either mode could be in effect while recharging, and recharging could be temporarily suspended if conditions so warranted.

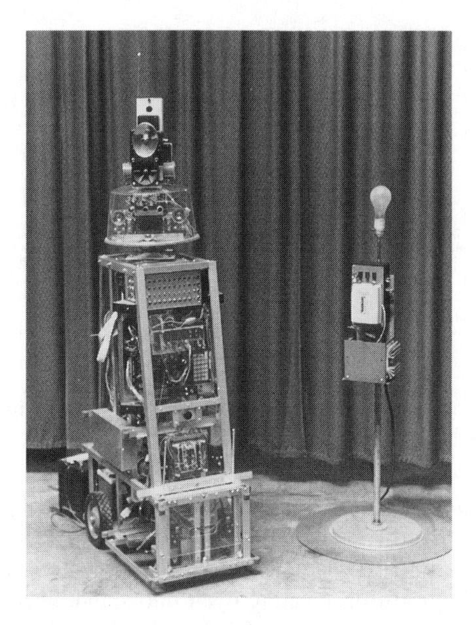

Figure 1-9. An optical homing beacon mounted on top of the recharging station was used to guide *ROBART I* to the charger when a low-battery condition was detected.

Recharging was handled automatically. The 12-volt 20-amphour lead-acid battery gave about six hours of continuous service and then required 12 hours of charge. Roughly one hour of power was available to locate the charging station (by means of a visual homing beacon) after the battery monitor circuits detected a low condition. The homing beacon was activated by a coded signal sent out from an RF transmitter located atop the robot's head, and the recharging supply was activated only when a demand was sensed after connection (Figure 1-10). The robot could elect to seek out the recharging station before a low battery condition actually arose, such as between patrols.

The software employed in homing on the recharger and effecting a connection was able to deal with a multitude of problems that could arise to hinder the process. Provision was made to skirt around obstacles between the robot and the recharging station. If, as a result of a collision avoidance maneuver, the robot were oriented with respect to the charger so as to preclude a successful docking, the vehicle would back up and realign itself before continuing. The robot could also tell when a return from a forward-looking proximity detector was due to the presence of the recharging station, so the software would not try to steer the platform away. (The collision-avoidance strategy will be discussed in more detail later in Chapter 10.)

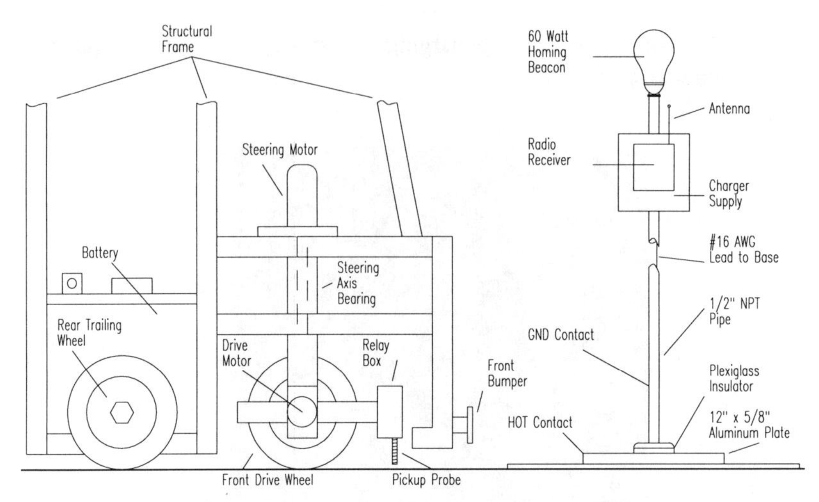

Figure 1-10. Diagram of the optical homing beacon used by *ROBART I* for automatic battery charging (adapted from Everett, 1982a).

A special near-infrared proximity sensor mounted on the head provided reliable detection of diffuse wall surfaces for ranges out to 6 feet. This sensor could be positioned at any angle up to 100 degrees either side of centerline by panning the head and was extremely useful in locating open doors and clear paths for travel. Excellent bearing information could be obtained, allowing this sensor to establish the location of the edge of a doorway, for example, to within 1 inch at a distance of 5 feet.

The hallway navigation scheme employed on ROBART I was based in part on the concept of beacon tracking. The recharging station optical beacon was suitably positioned in a known location to assist the robot in entering the hallway. Once in the hallway, the robot would move parallel to the walls in a reflexive fashion, guided by numerous near-infrared proximity sensors. General orientation in the hallway could be determined by knowing which direction afforded a view of the beacon. With *a priori* knowledge of where the rooms were situated with respect to this hallway, the robot could proceed in a semi-intelligent fashion to any given room, simply by counting off the correct number of open doorways on the appropriate side of the hall.

ROBART I was purposely intended to be a crude and simplistic demonstration of technical feasibility and was built on an extremely limited budget using oversimplified approaches. This philosophy assumed that if the concept could be successfully demonstrated under such primitive conditions of implementation, a reasonable extrapolation would show promise indeed for a more sophisticated second-generation version. (I had actually started work on this follow-on prototype just before leaving the Naval Postgraduate School in 1982.) As my interests shifted more in this direction, ROBART I was loaned to the Naval Surface Weapons Center in White Oak, MD, entrusted to the watchful care of an

MIT co-op student by the name of Anita Flynn (now a famous pioneer in the field of microrobotics). All work with ROBART I ended in 1985, when the prototype was shipped to Vancouver, BC, for display in the *Design 2000* exhibit at *EXPO '86.*

1.2.5 ROBART II (1982-)

ROBART II (Figure 1-11) became the center of focus for the next several years in my basement workshop in Springfield, VA. The system basically performed the same functions as its predecessor but employed a multiprocessor architecture to enable parallel real-time operations. Optimization of performance was addressed through significantly increased sensor capability, distributed processing, and precise vehicle motion control. Upon my transfer in 1986 to the Naval Command Control and Ocean Surveillance Center (NCCOSC) in San Diego, CA (then Naval Ocean Systems Center), the prototype was made available to the Navy for use as a testbed in support of mobile robotics research. The initial development effort focused on two specific technology areas.

Figure 1-11. *ROBART II* was constructed in my basement in Springfield, VA between 1982 and 1986.

The first of these addressed the navigational concerns that were hindering successful implementation of a number of robotic applications requiring mobility (Gilbreath & Everett, 1988). Simply put, an autonomous vehicle must be able to determine its position and orientation in the workspace, plan a path to its intended destination, and then execute that path without running into any obstructions. Numerous proximity and ranging sensors were incorporated on the robot to support map generation, position estimation, collision avoidance, navigational planning, and terrain assessment, enabling successful traversal of congested environments with no human intervention.

The second thrust was aimed at producing a robust automated security system exhibiting a high probability of detection with the ability to distinguish between actual and nuisance alarms. ROBART II was therefore also equipped with a multitude of intrusion and environmental sensors in support of its role as an intelligent sentry. These sensors monitor both system and room temperature, relative humidity, barometric pressure, ambient light and noise levels, toxic gas, smoke, and fire. Intrusion detection is addressed through the use of infrared, optical, ultrasonic, microwave, and video motion detection, as well as vibration monitoring and discriminatory hearing.

All high-level planning and assessment software runs on a desktop IBM-PC/AT computer connected to the robot via a 1200-baud Repco RF modem as shown in Figure 1-12 (Everett, et al., 1990). Robot position as well as sensor monitoring are represented graphically for the operator. The security assessment software package (Smurlo & Everett, 1993) displays time-stamped sensor status as well as environmental conditions, and can be overlaid on live video transmitted from a camera on-board the robot.

The scope of involvement was broadened in 1988 to include enhancements to the world modeling scheme to incorporate fixed installation security sensors (thereby allowing a mobile robot to operate in a secure area already protected by installed motion sensors) and inventory monitoring capability (allowing the robot to detect missing objects). In addition, a *reflexive teleoperated control* capability was added in 1989 to free the operator from the lower-level concerns associated with direct teleoperation. Speed of the vehicle and direction of motion are servo-controlled by an onboard processor in response to local sensor inputs, but under the high-level supervisory control of the remote operator (Laird & Everett, 1990).

In spite of having been built at home from hobbyist-grade components, ROBART II has proven to be an amazingly reliable piece of equipment, with only four documented cases of hardware failure since officially coming to life in early 1983. These included:

- A cold solder joint on a drive-motor shaft encoder.
- A defective power transistor in a drive-motor H-bridge amplifier.
- An oxidized variable capacitor in the CPU clock circuit for the sonar controller.
- An intermittent optical motion detector in the security sensor suite.

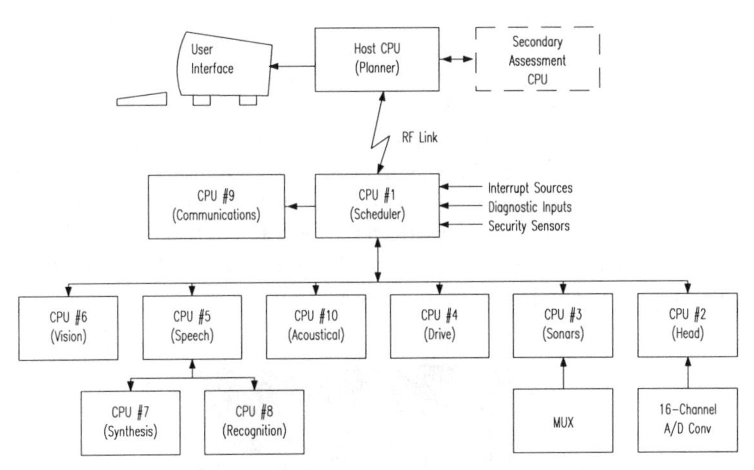

Figure 1-12. Block diagram of the computer architecture employed on *ROBART II.*

This record is somewhat noteworthy, considering the workout given the system over its 13-year lifetime to date; records indicate the robot performed in 53 live demonstrations for visiting students, faculty, scientists, and government officials in 1987 alone. ROBART II has been continuously on line now without a power interruption since sometime back in 1988.

1.2.6 MODBOT (1990-)

The *Modular Robotic Architecture* was developed by NCCOSC as a generic platform control system offering developers a standard set of software and hardware tools that could be used to quickly design modular robotic prototypes with minimum start-up overhead (Smurlo & Laird, 1990). The concept facilitates customization of a testbed system by providing sensor, actuator, and processing modules that can be configured on demand as required by the particular needs of the application being addressed. The ability to later accept newer modules of increasing sophistication provides for evolutionary growth potential, ensuring maximum effective service life before the hardware becomes obsolete.

The *ModBot* (Figure 1-13) is an example of a mobile robot implemented under this modular concept, employing several independent modules of varying intelligence and sophistication connected together in a generalized distributed network. The platform is made up of a detachable base with accompanying power source and various sensor, actuator, and processing modules. Each of these modules enables the robot to obtain and process different information about its surroundings.

The *Collision Avoidance Sonar Module* is active whenever the robot is in motion. It continuously looks for obstacles within a predefined distance and reports back to the *High-Level Processing Module* for appropriate action if an object is detected. The *Near-Infrared Proximity Sensor Module* is another means of determining if objects are in close proximity to the robot. This ring contains 11 Banner diffuse-mode optical proximity sensors (see Chapter 3) facing the forward 180 degrees, each one having a range of approximately 3 feet. This module is used to complement data obtained by the *Collision Avoidance Sonar Module*. The *High-Level Processing Module*, housing a WinSystems AT286 computer mounted in a card cage, receives commands from the remote control station. This module uses its internal map representation, as well as information from other modules, to plan and execute a path to the desired destination.

Figure 1-13. The *ModBot* is an autonomous robotic testbed that can be quickly reconfigured as needed to support a variety of potential research issues (courtesy Naval Command Control and Ocean Surveillance Center).

During data transfers, the *Control Station Module* communicates with the *ModBot* via the *Communications Module*. An RS-232 umbilical cable was initially used during the early stages of development and later replaced by an OCI *LAWN* spread-spectrum RF link. Some exploratory work was also performed using a full-duplex near-infrared datalink made by STI. The modular nature of the robot allowed the *Communications Module* to be upgraded without any other reconfiguration necessary to the rest of the *ModBot* systems.

The flexibility and extendibility of the *ModBot* architecture have made it a valuable testbed for the pursuit of new ideas and applications involving robot mobility. One of the first was a significantly upgraded version of the robotic security concept carried over from ROBART II. The *Intrusion Detection Module* is used to detect intruders in the vicinity of the robot and reports the bearing back to a remotely located *Control Station Module*. The *Intrusion Detection Module* consists of ultrasonic, passive-infrared, and microwave motion detectors which cover the full 360-degree surrounding area. A video motion detector in this module also receives information from the acoustic and video sensors on the

Stereoscopic Pan-and-Tilt Module to determine if an intruder is present. Audio and composite video signals are transmitted back to the operator via two separate analog RF links.

1.2.7 USMC TeleOperated Vehicle (1985-1989)

The *TeleOperated Vehicle (TOV)* was developed for the US Marine Corps by NCCOSC as part of the *Ground Air TeleRobotic Systems (GATERS)* program, and continued under the Unmanned Ground Vehicle Joint Program Office (UGV/JPO) *Ground-Launched Hellfire* program (Metz, et al., 1992). I served as Chief Engineer on the latter effort from July 1988 until October of the following year, during which time we designed and built a hardened second-generation version of the vehicle to support a major milestone demonstration in September 1989. During this series of live-fire exercises at Camp Pendelton, CA, the *TOV* system achieved a perfect record of eight direct hits with *Hellfire* missiles and four direct hits with laser-guided *Copperhead* projectiles.

Figure 1-14. One of three remotely driven reconnaissance, surveillance, and target acquisition (RSTA) vehicles developed by NCCOSC for the USMC *TeleOperated Vehicle (TOV)* program (courtesy Naval Command Control and Ocean Surveillance Center).

Three distinct modules for mobility, surveillance, and weapons firing allow the remote *TOV* platforms to be configured for various tactical missions (Aviles, et al., 1990; Metz, et al., 1992). The first, the *Mobility Module*, encompasses the necessary video cameras and actuation hardware to enable remote driving of the HMMWV. Figure 1-14 shows *TOV-2 (TeleOperated Vehicle 2)*, one of three platforms operated from a control van several kilometers away. A robot in the driver's seat of the HMMWV was slaved to the operator's helmet back in the control van so as to mimic his head movements (Martin & Hutchinson, 1989). If the helmet turned to the left and down, so did the slave robot in the remote vehicle. The two cameras on the robot that look like eyes feed two miniature video monitors on the operator's helmet, so that the operator would see in the van whatever the robot was viewing out in the field.

Two microphones on either side of the head served as the robot's ears, providing the operator with stereo hearing to heighten the *remote-telepresence* effect. Electric and hydraulic actuators for the accelerator, brakes, steering, and gearshift were all coupled via a fiber-optic telemetry link to identical components at the driver's station inside the control van (Figure 1-15). Actual HMMWV controls were replicated in form, function, and relative position to minimize required operator training (Metz, et al., 1992). After a few minutes of remote driving, one would actually begin to feel like one was sitting in the vehicle itself. A low-tension 30-kilometer cable payout system dispensed the fiber-optic tether onto the ground as the vehicle moved, avoiding the damage and hampered mobility that would otherwise arise from dragging the cable (Aviles, et al., 1990).

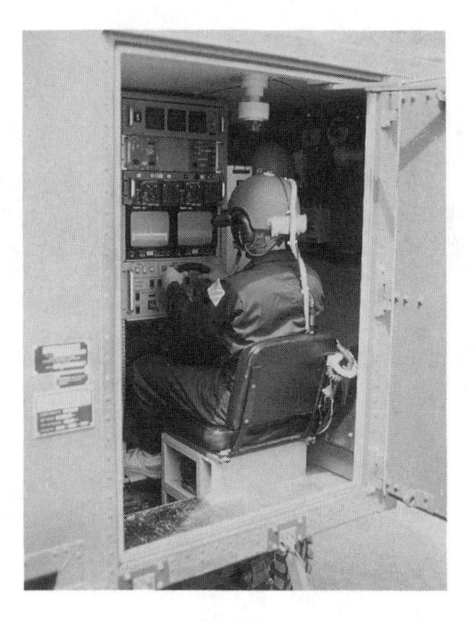

Figure 1-15. The *TOV Control Van* consists of a HMMWV-mounted environmental shelter containing three operator control stations and a fourth supervisor station (courtesy Naval Command Control and Ocean Surveillance Center).

Probably the most bizarre feeling I had driving one of these things remotely was operating the gearshift. You naturally want to look down at the shift lever when you grab it, which of course causes the slave robot at the other end to look down also (Figure 1-16). Your eyes see the shift lever on the remote vehicle, while your hand feels the shift knob in the control van. The problem is your hand doesn't appear in the video that your eyes see. When you move the lever, you feel it move and see it move in the video, but there's no hand there doing the moving. The human brain automatically fuses sensory inputs from two different sources, several kilometers apart, back into one composite image.

Figure 1-16. The remote slave robot is situated in the HMMWV driver's seat just behind the hydraulically actuated steering wheel (courtesy Naval Command Control and Ocean Surveillance Center).

The *Surveillance Module* was basically a glorified pan-and-tilt unit transporting a high-resolution sensor package, all mounted on a scissors-lift mechanism that could raise it 12 feet into the air. The sensor suite weighed approximately 300 pounds and consisted of a low-light-level zoom camera, an *AN/TAS-4A* infrared imager (FLIR), and an *AN/PAQ-3 MULE* laser designator. The remote operator would look for a tank or some other target with the camera or the FLIR, then switch over to the designator to light it up for a laser-guided *Hellfire* missile or *Copperhead* artillery round.

The *Weapons Module* provided each of the designating vehicles with a remotely-actuated .50-caliber machine gun for self defense. In addition to pan-and-tilt motion, electric actuators were provided to charge the weapon, release the safety, and depress the trigger. A fixed-focus CCD camera was mounted just above the gun barrel for safety purposes. The weapon could be manually controlled with the joystick in response to video from this camera, or slaved to the more sophisticated electro-optical sensors of the *Surveillance Module*. One of the remote HMMWVs had a *Hellfire* missile launcher (Figure 1-17) instead of a *Surveillance Module*, the idea being that one platform looked and designated while the other did the shooting. Meanwhile, all the humans could be up to 15 kilometers away, which is important in chemical or biological warfare scenarios.

Figure 1-17. A Rockwell *Hellfire* missile comes off the rail in response to a remote command from the *TOV* operator located in the *Control Van* several kilometers away during demonstrations for a high-level Department of Defense audience at Camp Pendelton, CA, in September 1989 (courtesy Naval Command Control and Ocean Surveillance Center).

1.2.8 MDARS Interior (1989-)

The *Mobile Detection Assessment and Response System (MDARS)* program is a joint Army-Navy effort to develop a robotic security and automated inventory assessment capability for use in Department of Defense warehouses and storage sites. The program is managed by the US Army Physical Security Equipment Management Office, Ft. Belvoir, VA, with NCCOSC providing all technical direction and systems integration functions. Near-term objectives are improved effectiveness (with less risk) to a smaller guard force, and significant reduction in the intensive manpower requirements associated with accounting for critical and high-dollar assets. The initial *Interior* implementation involves eight Cybermotion *K2A Navmaster* robots (Figure 1-18) configured as remote security platforms (Laird, et al., 1993).

From a technical perspective, the objective is to field a *supervised robotic security system* which basically runs itself until an unusual condition is encountered that necessitates human intervention. This requirement implies the MDARS host architecture must be able to respond to exceptional events from several robots simultaneously. Distributed processing allows the problem to be split among multiple resources and facilitates later expansion through connection of additional processors. The individual processors are connected via an *Ethernet* LAN (Figure 1-19) that supports peer-to-peer communications protocol. Distribution of function enables human supervision and interaction at several levels, while the hierarchical design facilitates delegation and assignment of limited human resources to prioritized needs as they arise.

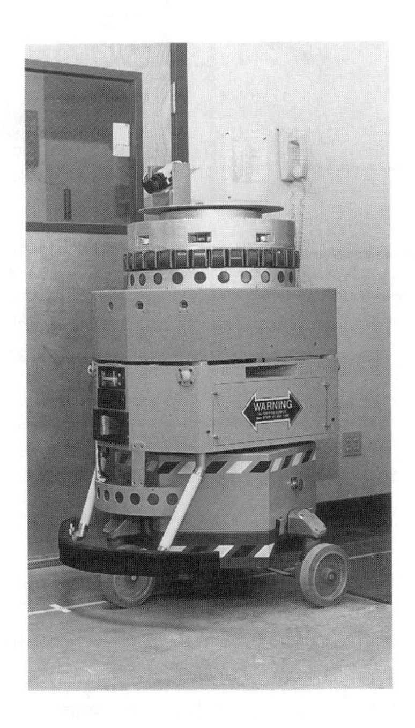

Figure 1-18. The early *MDARS Interior* feasibility prototype developed by the government employed the same modular-ring design used on the *ModBot* (courtesy Naval Command Control and Ocean Surveillance Center).

The *Supervisor* computer sits at the top of the hierarchy, responsible for overall system coordination and graphical display of the "big picture." The *Supervisor* has at its disposal a number of computing resources, such as one or more *Operator Displays*, two or more *Planner/Dispatchers*, a *Product Database* computer, and a *Link Server*. The *Supervisor* and *Operator Displays* have been similarly configured to provide the guard with consistent user-friendly visual displays. Both modules support a point-and-choose menu interface for guard-selectable options, commands, and navigational waypoints. The *Operator Display* allows a security guard to directly influence the actions of an individual platform, with hands-on control of destination, mode of operation, and camera functions. An optional *Virtual Reality Display* can be connected to the network if desired to provide a realistic three-dimensional model of the operating environment (Everett, et al., 1993).

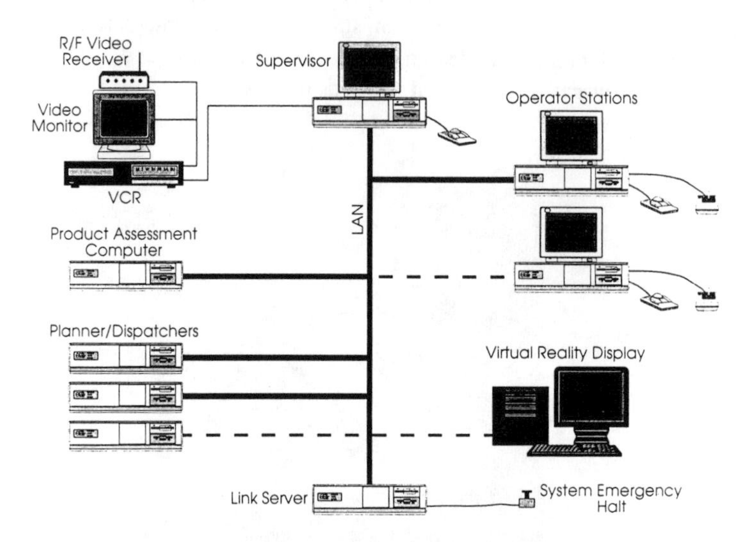

Figure 1-19. Block diagram of the *Multiple Robot Host Architecture* developed at NCCOSC for the coordinated control of multiple platforms.

The *Planner/Dispatcher* computers (an integration of the Cybermotion *Dispatcher* and the NCCOSC *Planner*) are responsible for navigation and collision avoidance. The *Product Database* computer maintains a listing of high-value inventory as verified by an RF tag reading system on board the robot, correlated to geographical location within the warehouse. The *Link Server* provides an interface to a spread-spectrum RF link between the host and the various robots, and maintains a blackboard data structure of robot status information for immediate retrieval by other computers on the LAN.

In October 1993 the MDARS Interior system began extensive test and evaluation in an actual semi-structured warehouse environment at Camp Elliott in San Diego, CA (Laird, et al., 1993). The original staring-array security sensor suite was replaced in December 1993 with the more optimal Cybermotion *SPI (Security Patrol Instrumentation)* module shown in Figure 1-20 (Holland, 1993). Developed as an outgrowth of a Cooperative Research and Development Agreement between Cybermotion and NCCOSC, the *SPI* uses a scanning configuration of microwave and passive infrared sensors to achieve the same 360-degree coverage at significantly reduced complexity and cost (DeCorte, 1994). A number of technical challenges associated with real-world operation have been uncovered and addressed during this rapid-prototyping test and development phase (Everett, et al., 1994; Gage, et al., 1995). Formal installation at an actual end-user site is scheduled to occur in the form of *Early User Experimentation* beginning in January 1997.

Figure 1-20. The *MDARS Interior* robot equipped with the Cybermotion *SPI Module* on patrol in the Camp Elliott warehouse in San Diego, CA (courtesy Naval Command Control and Ocean Surveillance Center).

1.2.9 Surrogate Teleoperated Vehicle (1990-1993)

The *Surrogate Teleoperated Vehicle (STV)*, a scaled-down follow-on version of the TOV concept, was developed under contract to NCCOSC by Robotic Systems Technology (RST), Inc., Westminster, MD, for the UGV/JPO in Huntsville, AL. The STV was intended to serve as a prototype system supporting the near-term development and evaluation of operational concepts for future unmanned ground vehicles, hence the terminology "Surrogate." A total of 14 vehicles was delivered to allow large numbers of military personnel to gain valuable hands-on robotics experience that could appropriately influence subsequent acquisition strategies. Figure 1-21 shows the STV fording a stream during the initial *Concept of Employment Exercise* at Fort Hunter Ligget, CA, in March 1992 (Metz et al., 1992).

From a technical perspective, the STV can be decomposed into four major inter-related subsystems: 1) the *Remote Platform*, 2) the *Mobility/RSTA Module*, 3) the *Operator Control Unit*, and 4) the *Communication System*.

The *Remote Platform* is built around a modified Polaris Industries *Big Boss* six-wheel-drive all-terrain vehicle measuring 117.5 inches long and 50.5 inches wide (Myers, 1992). The principle power source is a water-cooled three-cylinder 25-horsepower diesel engine built by Fuji Heavy Industries, capable of propelling the vehicle at speeds up to 35 miles per hour. The output shaft of the diesel drives a modified Polaris *variable belt transmission* that in turn is coupled to a gearbox providing neutral, reverse, low-forward, and high-forward speed ranges (RST, 1993). An auxiliary 3-horsepower electric golf-cart motor is also coupled to the gearbox input shaft (via an electric clutch) to provide for extremely quiet movement during surveillance operations at limited speeds up to 4 miles per hour. The gearbox output shaft powers the tandem rear axles through an exposed chain-drive arrangement. Two 12-volt sealed lead-acid batteries supply all required DC power, recharged by a 24-volt 60-amp engine-driven alternator.

Figure 1-21. Shown here crossing a stream at Fort Hunter Ligget, the *Surrogate Teleoperated Vehicle* is capable of traversing through water up to 2 feet deep (courtesy Unmanned Ground Vehicle Joint Program Office).

In similar fashion to its TOV predecessor, the STV *Mobility/RSTA Module* consists of a number of *reconnaissance, surveillance, and target acquisition* sensors mounted on a pan-and-tilt mechanism situated atop an extending scissors-lift mast (Figure 1-22). In a stowed configuration, the mast is only 24 inches high, but can raise the sensor pod when desired to a full height of 15 feet above ground level. Adjustable pneumatic springs in the rear of the vehicle allow for stiffening of the suspension when the surveillance mast is elevated, thus reducing sway and jitter during RSTA operations (Metz, et al., 1992). The mobility and RSTA sensors include:

- A stereo pair of 460-line day driving cameras.
- An image-intensified camera-pair for nighttime driving.
- A day targeting camera equipped with a 14-to-1 zoom lens.
- An image-intensified night targeting camera with a 10-to-1 zoom lens.
- An *IRIS-T* FLIR (forward-looking infrared).
- Either an *LTM-86* laser ranger/designator, or an *ESL-100* eye-safe laser ranger.

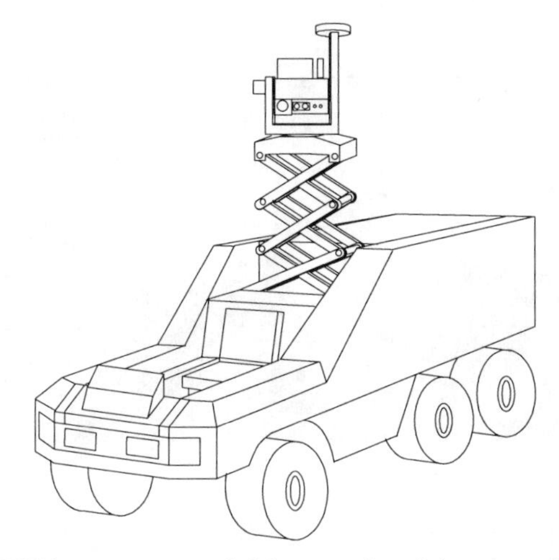

Figure 1-22. The STV incorporates a scaled-down version of the scissors lift developed for the TOV to raise the *Surveillance Module* 15 feet above ground level (courtesy Robotic Systems Technology).

The STV *Communications System* allows the vehicle to be controlled from the man-portable *Operator Control Unit* (Figure 1-23) using either a deployed fiber-optic tether or a back-up RF link (RST, 1993). The 10-kilometer inside-wound fiber-optic spool is packaged in a 3.5 cubic foot cargo-box area behind the engine compartment, with a hinged lid for easy access (Myers, 1992). A low-tension payout scheme feeds the 2.5-millimeter cable out the back as the vehicle moves forward. The RF back-up communications system consists of (RST, 1993):

- A 9600-baud full-duplex (dual-frequency) Repco *SLQ-96 Radio Modem* for command and status data.
- A Repco *Utility Data System (UDS)* FM transmitter for audio to the vehicle.
- A Dell-Star Technologies *900-Series* video transmitter for video and audio from the vehicle to the *Operator Control Unit.*

The maximum effective operating range under level-terrain conditions in the RF mode is approximately 2 kilometers.

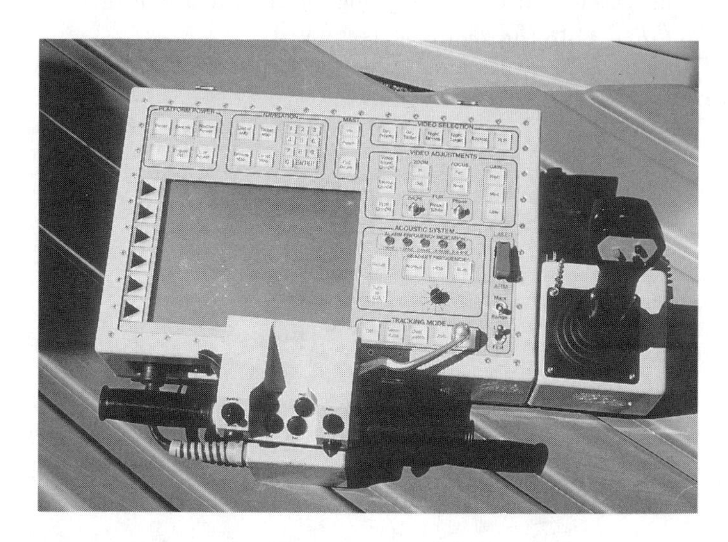

Figure 1-23. The *STV Operator Control Unit* uses a motorcycle-type steering device for vehicle mobility control, with a two-degree-of-freedom joystick for camera pan-and-tilt (courtesy Robotic Systems Technology).

1.2.10 ROBART III (1992-)

ROBART III (Figure 1-24) is intended to be an advanced demonstration platform for non-lethal response measures incorporating the reflexive teleoperated control concepts developed on *ROBART II.* I began work on this experimental system in my garage in July 1992 but was forced to suspend my efforts in December of that same year following a minor dirt-bike accident that put my right arm in a cast for about six months. That little inconvenience put me so far behind schedule on the preparation of this manuscript that further development of ROBART III was placed on hold for essentially the next two years. Recent government interest in dual-use technology reinvestment in a category known as *Operations Other Than War/Law Enforcement* have prompted renewed interest in completing the initial demonstration platform as soon as this book is finished.

Head-mounted sensors include two Polaroid sonar transducers, a Banner near-infrared proximity sensor, an AM Sensors microwave motion detector, and a video surveillance camera. The output of the CCD camera is broadcast to the operator over an analog RF link and simultaneously fed to an onboard video motion detector that provides azimuthal data allowing the head pan-axis controller to automatically track a moving target. Azimuthal and elevation information from the motion detector will be similarly fed to the pan-and-tilt controller for the six-barrel pneumatically fired dart gun for purposes of automated weapon positioning (Figure 1-25). Additional Polaroid sensors and near-infrared proximity detectors are strategically located to provide full collision avoidance coverage in support of the advanced teleoperation features desired.

Figure 1-24. Only the upper portion of *ROBART III* was completed before work was temporarily suspended in December 1992.

The non-lethal-response weapon chosen for incorporation into the system consists of a pneumatically powered dart gun capable of firing a variety of $^3/_{16}$-inch diameter projectiles. The simulated tranquilizer darts shown in the foreground of Figure 1-26 were developed to demonstrate a potential response application involving remote firing of temporarily incapacitating rounds by law enforcement personnel. The demonstration darts consist of a sharpened 20-gauge spring-steel wires approximately 3 inches long and terminated with $^3/_{16}$-inch plastic balls. A rotating-barrel arrangement was incorporated to allow for multiple firings (six) with minimal mechanical complexity. (The spinning-barrel mechanism also imparts a rather sobering psychological message during system initialization.)

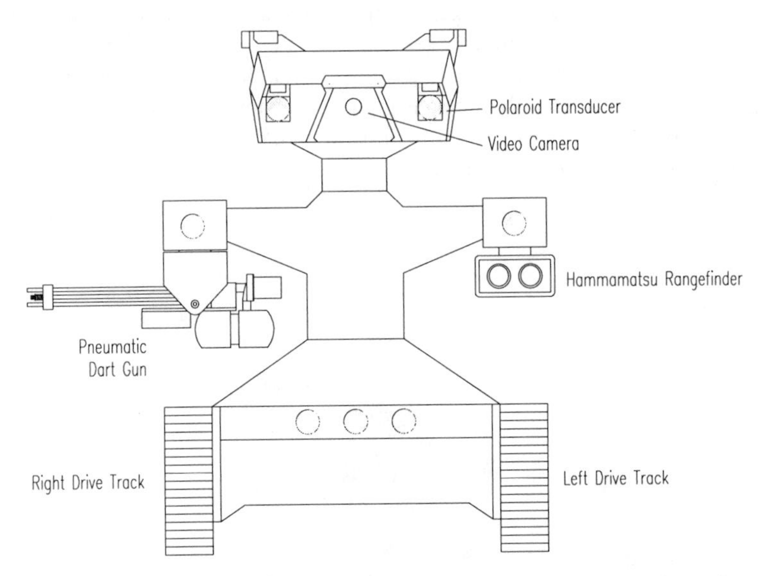

Figure 1-25. Intended to demonstrate the utility of an intelligent teleoperated security response vehicle, *ROBART III* is equipped with a laser-sighted six-barrel tranquilizer dart gun and video tracking.

The darts are expelled at high velocity from their 12-inch barrels by a release of compressed air from a pressurized accumulator at the rear of the gun assembly. To minimize air loss, the solenoid-operated valve linking the gun accumulator to the active barrel is opened under computer control for precisely the amount of time required to expel the projectile. The gun accumulator is monitored by a Micro Switch *242PC150G* electronic pressure transducer, and maintained at a constant pressure of 120 psi by a second solenoid valve connected to a 150-psi air source (see again Figure 1-26). All six darts can thus be fired in rapid succession (approximately 1.5 seconds) under highly repeatable launch conditions to ensure accurate performance. A visible-red laser sight is provided to facilitate manual operation under joystick control using video relayed from the head-mounted camera.

The left and right drive tracks are fashioned from 2.5-inch rubber timing belts turned inside out, driven by a pair of 12-volt electric wheelchair motors identical to those used on ROBART II. System power is supplied by a 80-amphour 12-volt gel-cell battery which provides for several hours of continuous operation between charges. A three-axis Precision Navigation *TCM Electronic Compass Module* (see Chapter 12) provides magnetic heading, temperature, and vehicle attitude (pitch and roll) information to the remote operator. Full-duplex data communication with the PC-based host control station is accomplished via a 9600-baud Telesystems spread-spectrum RF link.

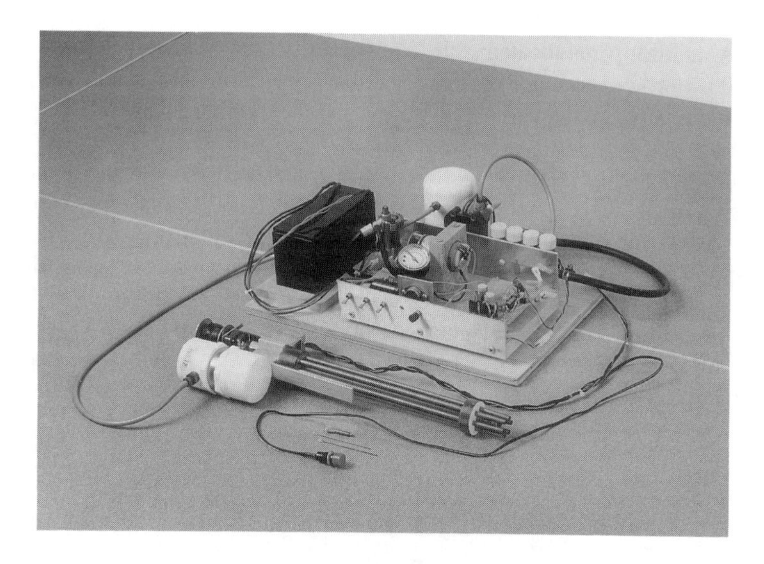

Figure 1-26. The gun accumulator is recharged after each firing from a 150-psi air tank supplied by a 12-volt Campbell Hausfeld automobile air compressor.

1.2.11 MDARS Exterior (1994-)

The *MDARS Exterior* program extends the robotic security and inventory control concepts of *MDARS Interior* into the realm of semi-structured (i.e., improved roads, defined fence lines, and standardized storage layouts) outdoor environments such as storage yards, dock facilities, and airfields. Intruder detection, assessment, and response, product inventories for theft prevention purposes, and lock/barrier checks are some of the physical security and inventory tasks currently performed by government personnel that will be replicated by the exterior robots. Inventory control will consist of verifying the contents of closed structures (i.e., warehouses, bunkers, igloos) without the need for opening. As is the case for the *Interior* program, the user's desire for minimum human involvement dictates that the exterior system operate in a supervised autonomous mode.

To perform the functions described above, it is envisioned that a basic exterior system will consist of the following:

- Two to eight exterior platforms patrolling the same or different areas on a site.

- RF-transponder tag-interrogation equipment on each of the remote platforms.
- A monitor panel located at the site's security command and control station.
- Position-location and communication subsystems for data, voice, and audio between the various platforms and the monitor panel.

The MDARS-E development effort began in early 1994 with the award of a three-year Broad Agency Announcement contract to Robotic Systems Technology (Myers, 1994) for the development of two brassboard platforms (Figure 1-27), with support from NCCOSC in the form of enhancements to the host architecture to accommodate exterior control (Heath-Pastore & Everett, 1994). The Phase-I effort will culminate with a technical feasibility demonstration at a Government site towards the end of 1996. The follow-up phase will provide enhancements such as intruder detection on the move and a non-lethal response capability.

Figure 1-27. The diesel-powered hydrostatic-drive prototype *MDARS Exterior* vehicle being demonstrated under pendant control in January 1995 (courtesy Robotic Systems Technology).

The MDARS Exterior platform currently weighs approximately 1700 pounds and measures 84 inches long by 35 inches high by 50 inches wide, with an 8-inch ground clearance. The four-wheel hydrostatic-drive configuration is powered by an 18-horsepower three-cylinder diesel engine with a 24-volt alternator and integral power steering pump. An Ackerman-steered design was chosen over a skid-steer arrangement for improved dead-reckoning capability. The water-cooled Kubota engine is directly coupled to a 50-cc/rev Rexroth hydrostatic pump that drives four White Industries rotary hydraulic wheel actuators with integral 200-line phase-quadrature encoders. The Rotac hydraulic steering actuator is independently supplied by the integrated power steering pump. The vehicle was carefully designed with an extremely low center of gravity (14.5 inches above ground level) for maximum stability on uneven terrain.

The MDARS-E vehicle is required to operate over unimproved roads and fairly rough terrain at speeds up to 9 miles per hour, automatically avoiding obstacles greater than 6 inches, breaches wider than 8 inches, and grades steeper than 10 percent. The collision avoidance strategy therefore incorporates a two-tier layered approach, wherein long-range (i.e., 0-100 feet) low-resolution sensors provide broad first-alert *obstacle-detection* coverage, and shorter-range (i.e., 0-30 feet typical) higher-resolution sensors are invoked for more precise *obstacle avoidance* maneuvering. Candidate systems currently being investigated include:

- Stereo vision (Burt, et al., 1992; 1993).
- Laser ranging (see Chapters 5 and 6).
- Millimeter-wave radar (see Chapter 6).
- Ultrasonic ranging (Hammond, 1994).

1.3 References

Aviles, W.A., Everett, H.R., Hughes, T.W., Koyamatsu, A.H., Laird, R.T., Martin, S.W., McArthur, S.P., Umeda, A.Y., "Issues in Mobile Robotics: The Unmanned Ground Vehicle Program TeleOperated Vehicle (TOV)," SPIE Vol. 1388, Mobile Robots V, Boston, MA, pp. 587-597, 8-9 November, 1990.

Burt, P.J., Anadan, P., Hanna, K., van der Wal, G., "A Front End Vision Processor for Unmanned Vehicles," Advanced Image Processing Group, David Sarnoff Research Center, Princeton, NJ, April, 1992.

Burt, P.J., Anadan, P., Hanna, K., van der Wal, G., Bassman, R., "A Front End Vision Processor for Vehicle Navigation," International Conference on Intelligent Systems, February, 1993.

Catling, I., *Advanced Technology for Road Transport: IVHS and ATT*, Artech House, Boston, MA, 1994.

DeCorte, C., "Robots train for Security Surveillance," *Access Control*, pp. 37-38, June, 1994.

Everett, H.R., "A Computer Controlled Sentry Robot," *Robotics Age*, March/April, 1982a.

Everett, H.R., "A Microprocessor Controlled Autonomous Sentry Robot", Masters Thesis, Naval Postgraduate School, Monterey, CA, October 1982b.

Everett, H.R., Gilbreath, G.A., Tran, T., Nieusma, J.M., "Modeling the Environment of a Mobile Security Robot," Technical Document 1835, Naval Command Control and Ocean Surveillance Center, San Diego, CA, June, 1990.

Everett, H.R., Gilbreath, G.A., Heath, T.A., Laird, R.T., "Coordinated Control of Multiple Security Robots," SPIE Vol. 2058, Mobile Robots VIII, Cambridge, MA, September, 1993.

Everett, H.R., Gage, D.W., Gilbreath, G.A., Laird, R.T., Smurlo, R.P., "Real-World Issues in Warehouse Navigation," SPIE Vol. 2352, Mobile Robots IX, Boston, MA, November, 1994.

Gage, D.W., Everett, H.R., Laird, R.T., Heath-Pastore, T.A., "Navigating Multiple Robots in Semi-Structured Environments," ANS 6th Topical Meeting on Robotics and Remote Systems, American Nuclear Society, Monterey, CA, February, 1995.

Gilbreath, G.A., Everett, H.R., "Path Planning and Collision Avoidance for an Indoor Security Robot," SPIE Mobile Robots III, Cambridge, MA, pp. 19-27, November, 1988.

Hammond, W., "Vehicular Use of Ultrasonic Systems," Technical Report, Cybermotion, Inc., Salem, VA, May, 1994.

Heath-Pastore, T.A., Everett, H.R., "Coordinated Control of Interior and Exterior Autonomous Platforms," ISRAM '94, Fifth International Symposium on Robotics and Manufacturing, Maui, HI, August, 1994.

Holland, J.M., "An Army of Robots Roams the Night," International Robot and Vision Automation Show and Conference, Detroit, MI, pp. 17.1-17.12, April, 1993.

Laird, R.T., Everett, H.R., "Reflexive Teleoperated Control," Association for Unmanned Vehicle Systems, Dayton, OH, July, 1990.

Laird, R.T., Everett, H.R., Gilbreath, G.A.., "A Host Architecture for Multiple Robot Control," ANS Fifth Topical Meeting on Robotics and Remote Handling, Knoxville, TN, April, 1993.

Martin, S.W., Hutchinson, R.C., "Low-Cost Design Alternatives for Head-Mounted Displays", Proceedings, SPIE 1083, Three Dimensional Visualization and Display Technologies, 1989.

Metz, C.D., Everett, H.R., Myers, S., "Recent Developments in Tactical Unmanned Ground Vehicles," Association for Unmanned Vehicle Systems, Huntsville, AL, June, 1992.

Myers, S.D., "Update on the Surrogate Teleoperated Vehicle (STV)," Association for Unmanned Vehicle Systems, Hunstville, AL, 1992

Myers, S.D., "Design of an Autonomous Exterior Security Robot," NASA Conference Publication 3251, Vol. 1, Conference on Intelligent Robotics in Field, Factory, Service, and Space, Houston, TX, pp. 82-87, March, 1994.

RST, "Surrogate Teleoperated Vehicle (STV) Technical Manual," Robotic Systems Technology, Westminster, MD, Contract No. N66001-91-C-60007, CDRL Item B001, Final Issue, 13 September, 1993.

Smurlo, R.P, Laird, R.T., "A Modular Robotic Architecture," SPIE Vol. 1388, Mobile Robots V, Boston, MA, 8-9 November, 1990.

Smurlo, R.P., Everett, H.R., "Intelligent Sensor Fusion for a Mobile Security Robot," *Sensors*, pp. 18-28, June, 1993.

2
Dead Reckoning

Dead reckoning (derived from "deduced reckoning" of sailing days) is a simple mathematical procedure for determining the present location of a vessel by advancing some previous position through known course and velocity information over a given length of time (Dunlap & Shufeldt, 1972). The concept was mechanically applied to automobile navigation as early as 1910, when the *Jones Live Map* was advertised as a means of replacing paper maps and eliminating the stress associated with route finding (Catling, 1994). This rather primitive but pioneering system counted wheel rotations to derive longitudinal displacement and employed a frictionally driven steering wheel encoder for calculating heading, and was thus subject to cumulative errors that precluded its ultimate success. The vast majority of land-based mobile robotic systems in use today rely on very similar dead-reckoning schemes to form the backbone of their navigational strategy, but like their nautical counterparts, periodically null out accumulated errors with recurring "fixes" from assorted navigational aids.

The most simplistic implementation of *dead reckoning* is sometimes termed *odometry*, the terminology implying vehicle displacement along the path of travel is directly derived from some onboard "odometer" as in the case of the *Jones Live Map*. A common means of *odometry* instrumentation involves optical encoders directly coupled to the motor armatures or wheel axles. In exterior applications, magnetic proximity sensors are sometimes used to sense a small permanent magnet (or magnets) attached to the vehicle driveshaft, as is typically done in the automotive industry to supply velocity feedback to cruise control equipment. Alternatively, inductive proximity sensors have been employed to detect cogs on a wheel or individual sections of a steel track when no rotational shafts are conveniently exposed, which is often the case when retrofitting conventional off-road equipment.

Heading information can be: 1) indirectly derived from an onboard steering angle sensor, 2) supplied by a magnetic compass or gyro, or 3) calculated from differential odometry as will be discussed below. Incremental displacement along the path is broken up into X and Y components, either as a function of elapsed time or distance traveled. For straight-line motion (i.e., no turns), periodic updates to vehicle-position coordinates are given by:

$$x_{n+1} = x_n + D\sin\theta$$
$$y_{n+1} = y_n + D\cos\theta$$

where:

D = vehicle displacement along path
θ = vehicle heading.

Klarer (1988) presents an excellent and detailed description of the appropriate algorithms for various types of steering configurations, some of which will be discussed in further detail later in this chapter.

2.1 Odometry Sensors

Since most (but not all!) mobile robots rely on some variation of wheeled locomotion, a basic understanding of sensors that accurately quantify angular position and velocity is an important prerequisite to further discussions of odometry. There are a number of different types of rotational displacement and velocity sensors in use today:

- Brush Encoders.
- Potentiometers.
- Synchros.
- Resolvers.
- Optical Encoders.
- Magnetic Encoders.
- Inductive Encoders.
- Capacitive Encoders.

A multitude of issues must be considered in choosing the appropriate device for a particular application. Aviolio (1993) points out that over 17 million variations of rotary encoders are offered by one company alone. We shall examine in more detail the three most common types as applied to odometry sensing: 1) *potentiometers*, 2) *resolvers*, and 3) *optical encoders*.

2.1.1 Potentiometers

Potentiometers, or *pots* for short, are often used as low-cost rotational displacement sensors in low-speed medium-accuracy applications not involving continuous rotation. (For example, both ROBART I and ROBART II used precision potentiometers to sense head pan position.) The principle of operation is that of a variable-resistance voltage divider (Figure 2-1), where the center tap is a mechanically coupled wiper that moves across the resistance element in conjunction with shaft rotation. A variety of relationships (tapers) defining resistance as a function of wiper displacement are employed in the fabrication of potentiometers (i.e., audio, logarithmic, sinusoidal, linear), with linear taper being the most common scheme in position sensing applications.

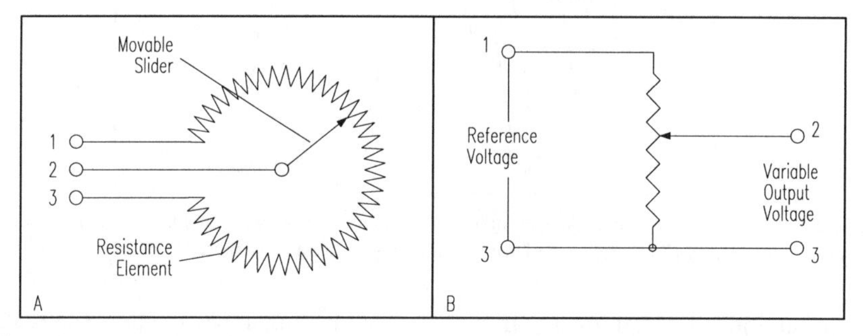

Figure 2-1. For a linear-taper pot, the output voltage V_o is directly related to the ratio of actual to full scale displacement.

The principle advantages of potentiometric sensors are very low cost and ease of interface. A regulated DC voltage is applied across the full resistance R as shown. Output voltage is given by the equation:

$$V_o = V_{ref} \frac{r}{R}$$

where:

V_o = output voltage from wiper
V_{ref} = reference voltage across pot
r = wiper-to-ground resistance
R = total potentiometer resistance.

For linear-taper devices, the quotient r/R is precisely equal to the ratio of actual to full-scale wiper displacement, assuming negligible loading effects. Since output voltage is also a linear function of the reference voltage V_{ref}, care must be taken to use a well-regulated noise-free supply.

Wire-wound pots can exhibit a piecewise quantification in performance as illustrated in Figure 2-2, since resistance is not continuously varied but instead incrementally stepped as a result of the coil design (Fraden, 1993). In addition, the wiper will temporarily "short" together adjacent windings in passing, changing the effective total resistance. The best precision potentiometers therefore employ a continuous resistive film fabricated from carbon, conductive plastic, or a ceramic-metal mix known as *cermet*. While a good wire-wound pot can provide an average resolution of about 0.1 percent of full scale, the high-quality resistive-film devices are generally limited only by manufacturing tolerances governing the uniformity of the resistance element (Fraden, 1993).

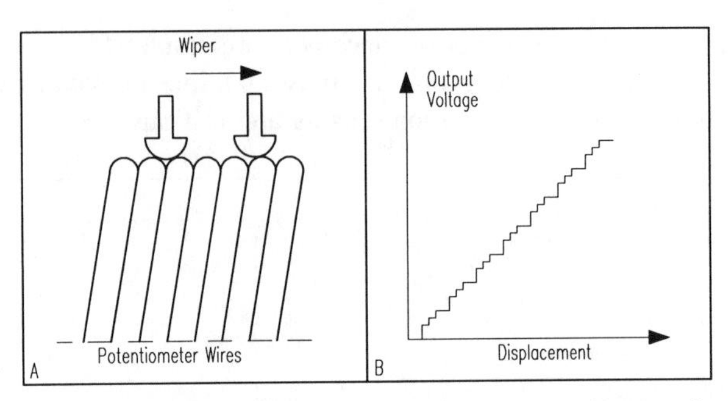

Figure 2-2. The sliding wiper (A) can alternately contact one or two wires at a time, introducing an output uncertainty (B) in the case of wire-wound potentiometers (adapted from Fraden, 1993).

In addition to significant frictional loading imparted to the shaft, the fundamental disadvantage of potentiometers is their relatively poor reliability due to dirt build-up and inevitable wiper wear, resulting in noisy and erratic operation. Other errors can be introduced by slack and/or elasticity in the belt drive if not directly coupled to the shaft, and electrical noise introduced into the analog output line. For these reasons, the use of potentiometers as rotation sensing devices has fallen off in recent years in favor of the more versatile incremental optical encoders, to be discussed in Section 2.1.3.

2.1.2 Synchros and Resolvers

Synchros are rotating electromechanical devices used to transmit angular information electrically from one place to another with great precision (Schwartz & Grafstein, 1971). In essence, the *synchro* forms a variable-coupling transformer consisting of an AC-excited rotor winding (primary) and two or more stator windings (secondaries) symmetrically oriented around the rotor. The effective magnetic coupling between the rotor winding and the surrounding stator windings varies as a function of rotor orientation. Maximum coupling occurs when the fields are parallel, while minimal coupling results when the rotor field is orthogonally aligned with respect to a particular stator winding. As a consequence, the stator outputs form a set of AC signals whose respective magnitudes uniquely define the rotor angle at any given point in time. A wide variety of *synchro* types exist:

- Transmitters.
- Differentials.
- Receivers.
- Control Transformers.
- Resolvers.
- Differential Resolvers.
- Linear Transformers.
- Transolvers.

Deirmengian (1990a) provides a comprehensive treatment of the theory of operation for the components listed above, followed by a detailed examination of the various design parameters and error sources that must be taken into account (1990b).

The most widely known *synchro* configuration is probably the three-phase *transmitter/receiver* pair commonly used for remote shaft-angle indication. The slave *synchro receiver* is electrically identical to the master *transmitter* and connected so that the stator windings for both devices are in parallel as shown in Figure 2-3. The rotor windings on both the transmitter and the remote-indicating receiver are excited by an AC current (400 Hz to several KHz) typically supplied through slip rings. When the receiver and transmitter rotors are in identical alignment with their respective stator windings, the individual stator outputs will be equal for the two devices, and consequently there will be no current flow.

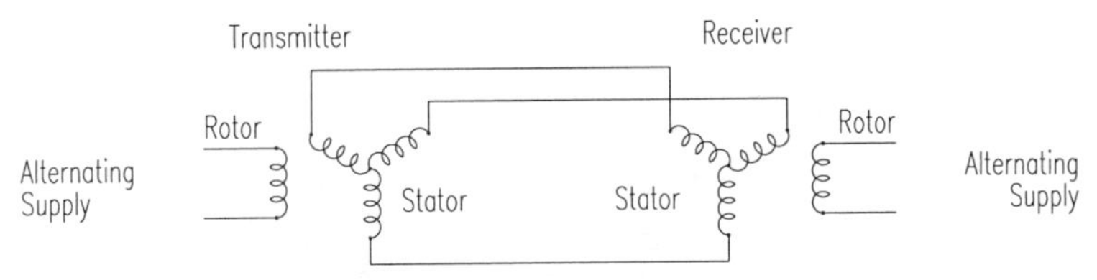

Figure 2-3. Schematic diagram of a typical remote-indicating synchro configuration (adapted from Schwartz & Graftstein, 1971).

If the transmitter rotor shaft is turned by some external force, the equilibrium conditions are upset, and the resulting voltage differences generate current flows in both sets of stator windings. These current flows induce an identical torque in both rotors, but since the transmitter rotor is constrained, the torque on the receiver rotor acts to restore alignment and thus equilibrium (Diermengian, 1990a). The observed effect is the receiver output shaft will precisely track any rotational displacement seen by the remotely located transmitter input shaft. More than one receiver can be driven by a common transmitter. For example, Navy ships are equipped with a number of remote heading indicators (directional gyro repeaters) located in the pilot house, on the port and starboard bridge wings, and up on the signal bridge, while the gyro itself is mounted deep inside the vessel to minimize effects of ships motion (i.e., pitch and roll).

The *resolver* is a special configuration of the *synchro* that develops voltages proportional to the sine and cosine of rotor angle, and thus is often used to break down a vector quantity into its associated components. A typical example is seen in the aircraft industry where resolvers are used to perform coordinate transforms between aircraft- and ground-based reference frames. Only two stator coils are involved, oriented 90 degrees apart as shown in Figure 2-4 (Tiwari, 1993).

The individual stator outputs as a function of input excitation and rotor position θ are given by the following equations (ILC, 1982):

$$V_x = K_x \sin\theta \sin(\omega t + a_x)$$
$$V_y = K_y \cos\theta \sin(\omega t + a_y)$$

where:

θ = the resolver input shaft angle
$\omega = 2\pi f$, where f is the excitation frequency
K_x and K_y are ideally equal transfer-function constants
a_x and a_y are ideally zero time-phase shifts between rotor and stator.

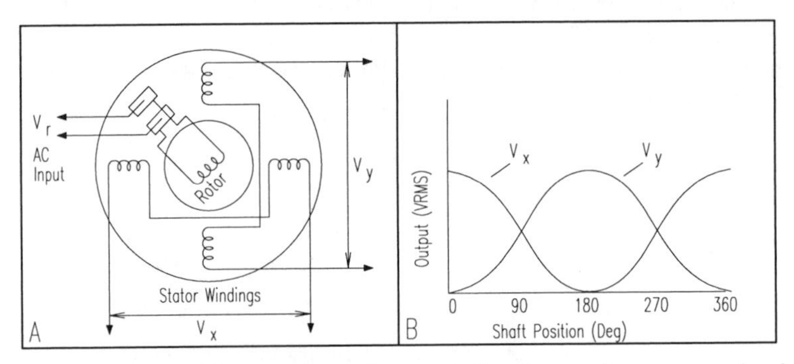

Figure 2-4. The outputs of the two orthogonal stator windings in a *resolver* are proportional to the sine and cosine of the applied rotor excitation (adapted from Tiwari, 1993).

The definitive mathematical relationship inherent in the *resolver* output signals means the transmitter can be used stand-alone (i.e., without a slave receiver) as an input transducer in a digital control system. A *resolver-to-digital converter* (RDC) is employed in place of the receiver to transform the output signals into an appropriate format for computer interface. This conversion is typically done in one of three ways: 1) phase-shift approach, 2) amplitude-ratio approach, or 3) multiplex approach. Grandner and Lanton (1986) present an excellent overview and comparison of these three techniques, of which the amplitude-ratio approach seems to be gaining the most popularity. The *ISN4* hybrid phase tracking RDC from Analog Devices provides a special velocity output in addition to absolute position information (Nickson, 1985).

In summary, resolvers offer a very rugged and reliable means for quantifying absolute angular position that is accurate, moderately cheap, and fairly small in terms of physical size. The advent of custom large-scale integrated (LSI) circuits has reduced the cost of associated electronics, making resolvers competitive with other alternatives (Grandner & Lanton, 1986). Brushless versions employ a special cylindrical transformer instead of slip rings to couple AC excitation to the rotor as shown in Figure 2-5 (Nickson, 1985). These configurations have essentially no wear and therefore zero maintenance, but at the expense of additional power consumption and increased length.

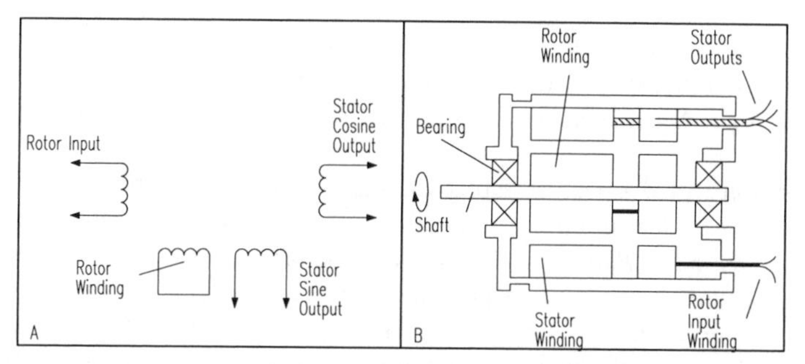

Figure 2-5. Brushless resolvers employ a rotating transformer instead of slip rings to couple excitation energy to the rotor and essentially require no maintenance (adapted from Nickson, 1985).

2.1.3 Optical Encoders

The first optical encoders were developed in the mid-1940s by the Baldwin Piano Company for use as "tone wheels" that allowed electric organs to mimic other musical instruments (Agent, 1991). Today's contemporary devices basically embody a miniaturized version of the *opposed-mode proximity sensor* (see Chapter 3). A focused beam of light aimed at a matched photodetector is periodically interrupted by a coded opaque/transparent pattern on a rotating intermediate disk attached to the shaft of interest. The rotating disk may take the form of chrome on glass, etched metal, or photoplast such as Mylar (Henkel, 1987). Relative to the more complex alternating-current resolvers, the straightforward encoding scheme and inherently digital output of the optical encoder results in a low-cost reliable package with good noise immunity.

There are two basic types of optical encoders: *incremental* and *absolute*. The *incremental* version measures rotational velocity and can infer relative position, while *absolute* models directly measure angular position and infer velocity. If non-volatile position information is not a consideration, *incremental encoders* generally are easier to interface and provide equivalent resolution at a much lower cost than *absolute optical encoders*.

Incremental Optical Encoders

The simplest type of *incremental encoder* is a single-channel *tachometer encoder,* which is basically an instrumented mechanical light chopper that produces a certain number of sine or square wave pulses for each shaft revolution. The greater the number of pulses, the higher the resolution (and subsequently the cost) of the unit. These relatively inexpensive devices are well suited as velocity feedback sensors in medium- to high-speed control systems, but run into noise and stability problems at extremely slow velocities due to quantization errors (Nickson, 1985). The tradeoff here is resolution versus update rate: improved transient response requires a faster update rate, which for a given line count reduces the number of possible encoder pulses per sampling interval. A typical limitation for a 2-inch diameter incremental encoder disk is 2540 lines (Henkel, 1987).

In addition to low-speed instabilities, single-channel *tachometer encoders* are also incapable of determining the direction of rotation and thus cannot be used as position sensors. *Phase-quadrature incremental encoders* overcome these problems by adding a second channel and displacing the detectors so the resulting pulse trains are 90 degrees out of phase as shown in Figure 2-6. This technique allows the decoding electronics to determine which channel is leading the other and hence ascertain the direction of rotation, with the added benefit of increased resolution. Holle (1990) provides an in-depth discussion of output options (single-ended TTL or differential drivers) and various design issues (i.e., resolution, bandwidth, phasing, filtering) for consideration when interfacing phase-quadrature incremental encoders to digital control systems.

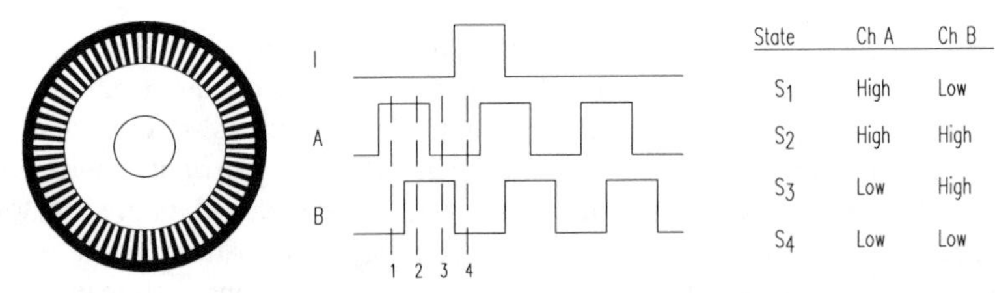

Figure 2-6. The observed phase relationship between Channel A and B pulse trains can be used to determine the direction of rotation with a phase-quadrature encoder, while unique output states S_1 - S_4 allow for up to a four-fold increase in resolution. The single slot in the outer track generates one index (I) pulse per disk rotation.

The incremental nature of the phase-quadrature output signals dictates that any resolution of angular position can only be relative to some specific reference, as opposed to absolute. Establishing such a reference can be accomplished in a number of ways. For applications involving continuous 360-degree rotation, most encoders incorporate as a third channel a special *index output* that goes high once for each complete revolution of the shaft (see Figure 2-6 above). Intermediate shaft positions are then specified by the number of encoder *up counts* or *down counts* from this known index position. One disadvantage of this approach is all relative position information is lost in the event of a power interruption.

In the case of limited rotation, such as the back-and-forth motion of a pan or tilt axis, electrical limit switches and/or mechanical stops can be used to establish a *home reference* position. To improve repeatability this homing action is sometimes broken up into two steps. The axis is rotated at reduced speed in the appropriate direction until the stop mechanism is encountered, whereupon rotation is reversed for a short predefined interval. The shaft is then rotated slowly back into the stop at a specified low velocity from this designated start point, thus eliminating any variations in inertial loading that could influence the final homing position. This two-step approach can usually be observed in the power-on initialization of stepper-motor positioners for dot-matrix printer heads.

Alternatively, the absolute indexing function can be based on some external referencing action that is decoupled from the immediate servo-control loop. A good illustration of this situation involves an incremental encoder used to keep track of platform steering angle. For example, when the Cybermotion *K2A Navmaster* robot is first powered up, the absolute steering angle is unknown and must be initialized through a "referencing" action with the docking beacon, a nearby wall, or some other identifiable set of landmarks of known orientation (see Chapters 15 and 16). The up/down count output from the decoder electronics is then used to modify the vehicle heading register in a relative fashion.

A growing number of very inexpensive off-the-shelf components have contributed to making the phase-quadrature incremental encoder the rotational sensor of choice within the robotics research and development community. Figure 2-7 shows an incremental optical encoder and PID motor-controller chip made by Hewlett Packard, along with a National Semiconductor H-bridge amplifier that collectively form the basis of a complete digital control system for a total package price of well under $100. Several manufacturers now offer small DC gearmotors with incremental encoders already attached to the armature shafts. Within the US automated guided vehicle (AGV) industry, however, resolvers are still generally preferred over optical encoders for their perceived superiority under harsh operating conditions, but the European AGV community seems to clearly favor the encoder (Manolis, 1993).

Figure 2-7. Shown here are the major components for a complete digital control system: (from left to right) a Hewlett Packard *HEDS-5500* incremental optical encoder, a Hewlett Packard *HCTL-1100* PID controller chip, and a National Semiconductor *LMD18200* H-bridge power amplifier (courtesy Naval Command Control and Ocean Surveillance Center).

Absolute Optical Encoders

Absolute encoders are typically used for slower rotational applications that require positional information when potential loss of reference from power interruption cannot be tolerated. Discrete detector elements in a photovoltaic array are individually aligned in break-beam fashion with concentric encoder tracks as shown in Figure 2-8, creating in effect a non-contact implementation of the earlier commutating brush encoder. The assignment of a dedicated track for each bit of resolution results in larger size disks (relative to incremental designs), with a corresponding decrease in shock and vibration tolerance. A general rule of thumb is that each additional encoder track doubles the resolution but quadruples the cost (Agent, 1991).

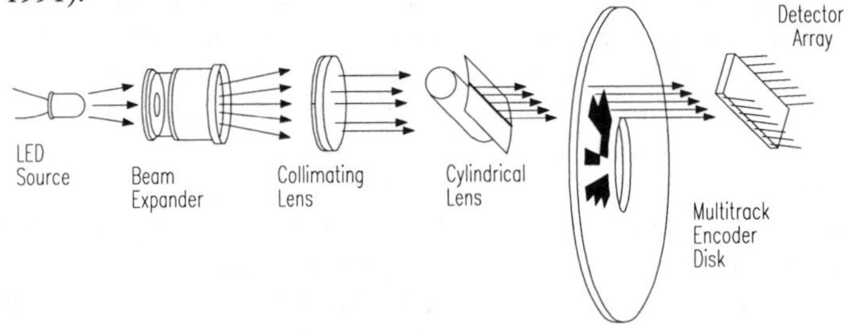

Figure 2-8. A line source of light passing through a coded pattern of opaque and transparent segments on the rotating encoder disk results in a parallel output that uniquely specifies the absolute angular position of the shaft (adapted from Agent, 1991).

Instead of the serial bit streams of incremental designs, absolute optical encoders provide a parallel word output with a unique code pattern for each quantized shaft position. The most common coding schemes are Gray code, natural binary, and binary-coded decimal (Avolio, 1993). The Gray code (for inventor Frank Gray of Bell Labs) is characterized by the fact that only one bit changes at a time, a decided advantage in eliminating asynchronous ambiguities caused by electronic and mechanical component tolerances. Binary code, on the other hand, routinely involves multiple-bit changes when incrementing or decrementing the count by one. For example, when going from position 255 to position 0 in Figure 2-9B, eight bits toggle from 1s to 0s. Since there is no guarantee all threshold detectors monitoring the detector elements tracking each bit will toggle at the same precise instant, considerable ambiguity can exist during state transition with a coding scheme of this form. Some type of handshake line signaling *valid data available* would be required if more than one bit were allowed to change between consecutive encoder positions.

Absolute encoders are best suited for slow and/or infrequent rotations such as steering angle encoding, as opposed to measuring high-speed continuous (i.e., drivewheel) rotations as would be required for calculating displacement along the path of travel. Although not quite as robust as resolvers for high-temperature, high-shock applications, operation at temperatures in excess of 125 degrees C is possible, and medium-resolution (1000 counts per revolution) metal or Mylar disk designs can compete favorably with resolvers in terms of shock resistance (Manolis, 1993).

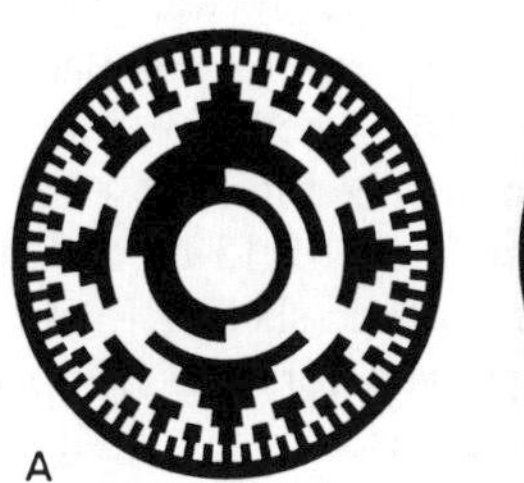

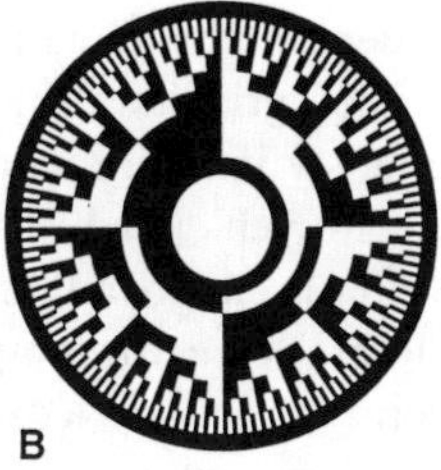

Figure 2-9. Rotating an 8-bit absolute Gray code disk (A) counterclockwise by one position increment will cause only one bit to change, whereas the same rotation of a binary-coded disk (B) will cause all bits to change in the particular case (255 to 0) illustrated by the reference line at 12 o'clock.

More complex interface issues due to the large number of leads associated with the parallel nature of the output pose a potential disadvantage. A 13-bit absolute encoder using complimentary output signals for noise immunity would require a 28-conductor cable (13 signal pairs plus power and ground), versus only six for a resolver or incremental encoder (Avolio, 1993).

2.2 Doppler and Inertial Navigation

The rotational displacement sensors discussed above derive navigational parameters directly from wheel rotation, and are thus subject to problems arising from slippage, tread wear, and/or improper tire inflation. In certain applications, Doppler and inertial navigation techniques are sometimes employed to reduce the effects of such error sources.

2.2.1 Doppler Navigation

Doppler navigation systems are routinely employed in maritime and aeronautical applications to yield velocity measurements with respect to the earth itself, thus eliminating dead-reckoning errors introduced by unknown ocean or air currents. The principle of operation is based on the Doppler shift in frequency observed when radiated energy reflects off a surface that is moving with respect to the emitter, as will be discussed in detail in Chapter 8. Maritime systems employ acoustical energy reflected from the ocean floor, while airborne systems sense microwave RF energy bounced off the surface of the earth. Both configurations typically involve an array of four transducers spaced 90 degrees apart in azimuth and inclined downward at a common angle with respect to the horizontal plane (Dunlap & Shufeldt, 1972).

Due to cost constraints and the reduced likelihood of transverse drift, most robotic implementations employ but a single forward-looking transducer to measure ground speed in the direction of travel. Similar configurations are sometimes used in the agricultural industry, where tire slippage in soft freshly plowed dirt can seriously interfere with the need to release seed or fertilizer at a rate commensurate with vehicle advance. The M113-based Ground Surveillance Robot (Harmon, 1986) employed an off-the-shelf RF system of this type manufactured by John Deere to compensate for track slippage (Figure 2-10). Milner (1990) reports a very low-cost ultrasonic unit designed to be worn by runners and skiers (marketed by Nike, Inc. as the *Nike Monitor*) that could measure a mile to within 10 feet.

Figure 2-10. A commercially available John Deere agricultural ground-speed sensor was employed on the Ground Surveillance Robot to improve dead-reckoning accuracy (courtesy Naval Command Control and Ocean Surveillance Center).

The microwave (or ultrasonic) sensor is aimed downward at a prescribed angle (typically 45 degrees) to sense ground movement as shown in Figure 2-11. Actual ground speed V_A is derived from the measured velocity V_D in accordance with the following equation (Schultz, 1993):

$$V_A = \frac{V_D}{\cos\alpha} = \frac{c\,F_D}{2\,F_o\,\cos\alpha}$$

where:

V_A = actual ground velocity along path
V_D = measured Doppler velocity
α = angle of declination
c = speed of light
F_D = observed Doppler shift frequency
F_O = transmitted frequency.

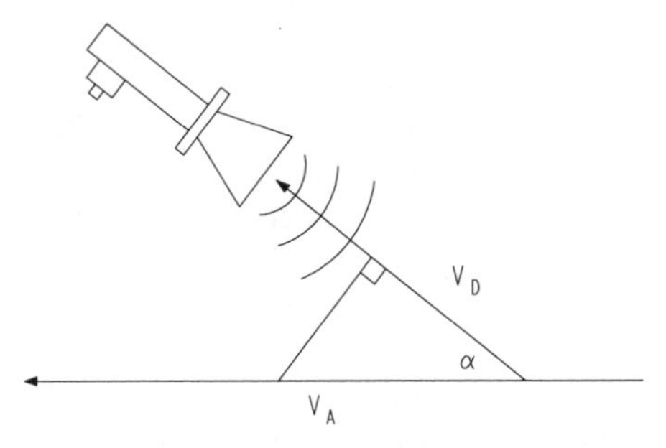

Figure 2-11. A Doppler ground speed sensor inclined at an angle α as shown measures the velocity component V_D of true ground speed V_A (adapted from Schultz, 1993).

Errors in determining true ground speed arise due to side-lobe interference, vertical velocity components introduced by vehicle reaction to road surface anomalies, and uncertainties in the actual angle of incidence due to the finite width of the beam. Since the Doppler frequency is proportional to the cosine of the angle, the far part of the beam returns a higher frequency than the near part, with a continuous distribution of frequencies in between (Milner, 1990). Signal processing techniques (i.e., square-root-of-frequency filters, centroid extractors, phase-lock loops) are necessary to extract a representative frequency from the spectrum.

Byrne, et al. (1992) point out another interesting scenario for potentially erroneous operation, involving a stationary vehicle parked over a stream of water. The Doppler ground-speed sensor in this case would misinterpret the relative motion between the stopped vehicle and the running water as vehicle travel.

2.2.2 Inertial Navigation

An alternative approach to augmenting the dead-reckoning solution beyond simple odometry is *inertial navigation*, initially developed for deployment on aircraft. The technology was quickly adapted for use on missiles and in outer space, and found its way to maritime usage when the nuclear submarines *Nautilus* and *Skate* were suitably equipped in support of their transpolar voyages in 1958 (Dunlap & Shufeldt, 1972). The principle of operation involves continuous sensing of minute accelerations in each of the three directional axes, and integrating over time to derive velocity and position. A gyroscopically stabilized sensor platform is used to maintain consistent orientation of the three accelerometers throughout this process.

Although fairly simple in basic concept, the specifics of implementation are rather demanding from the standpoint of minimizing the various error sources that adversely affect the stability of the gyros used to ensure correct attitude. The resulting high manufacturing and maintenance costs have effectively precluded any practical application of this technology in the automated guided vehicle industry (Turpin, 1986). For example, a "high-quality" *inertial navigation system* (INS) such as would be found in a commercial airliner will have a typical drift of about 1 nautical mile per hour of operation, and cost between \$50K and \$70K (Byrne, et al., 1992). High-end INS packages used in ground applications have shown performance of better than 0.1 percent of distance traveled but cost in the neighborhood of \$100K to \$200K, while lower performance versions (i.e., 1 percent of distance traveled) run between \$20K to \$50K (Dahlin & Krantz, 1988).

Experimental results by the Universite Montpellier in France (Vaganay, et al., 1993), Barsham and Durrant-Whyte (1993), Mononen, et al. (1994), and the University of Michigan (Borenstein, 1994a) indicate the inertial navigation approach is not realistically advantageous for the above reasons. As a consequence, the use of INS hardware in robotic applications has to date been generally limited to scenarios that aren't readily addressable by more practical alternatives. An example of such a situation is presented by Sammarco (1990; 1994), who reports preliminary results in the case of an INS used to control an autonomous vehicle in a mining application. The development of increasingly low-cost fiber-optic gyros and solid-state accelerometers, however, promises to open up new opportunities in the not too distant future.

The various gyro and accelerometer components that make up an *inertial navigation system* will be treated in some detail later in Chapter 13.

2.3 Typical Mobility Configurations

A number of design issues impact the selection of an appropriate drive and steering configuration for a mobile robotic vehicle:

- *Maneuverability* — The ability to translate and/or change direction of motion must be consistent with the constraints of the surrounding environment.
- *Controllability* — The hardware and software necessary to control the mobility scheme must be practical and not overly complex.
- *Traction* — Sufficient traction should be provided to minimize slippage under varying conditions in accordance with anticipated operational scenarios.

- *Climbing ability* — Indoor schemes must allow for traversal of minor discontinuities or gaps in the floor surface; exterior requirements are dictated by the range of terrain features associated with the application.
- *Stability* — The mobility base must provide sufficient stability for the payload with an adequate safety margin under expected conditions of acceleration, tilt, and roll.
- *Efficiency* — Power consumption must be in keeping with available energy reserves and duty cycle.
- *Maintenance* — The system should be easy to maintain with an acceptable mean time between failure, and not prone to excessive tire or component wear.
- *Environmental impact* — The drive and steering functions should not damage the floor or ground surface.
- *Navigational considerations* — Dead-reckoning accuracy should be in keeping with the needs of the application.

This next section will discuss only those sensor considerations in support of the last of these categories.

2.3.1 Differential Steering

A very common indoor propulsion system uses two individually controlled drive wheels on either side of the base, with casters in front and rear for stability. This configuration allows the robot to spin in place about the vertical axis for maneuvering in congested areas. Drivemotor velocities are monitored by optical encoders attached to the armature shafts (Figure 2-12). High-resolution phase-quadrature encoders produce hundreds of counts per turn of the motor armatures, which usually translates to thousands of counts per wheel revolution. Very precise displacement and velocity information is thus available for use in dead reckoning calculations, but the results are influenced by a number of systemic as well as external sources of error that must be taken into account.

Robot displacement D along the path of travel is given by the equation:

$$D = \frac{D_l + D_r}{2}$$

where:

D = displacement of platform
D_l = displacement of left wheel
D_r = displacement of right wheel.

Similarly, the platform velocity V is given by the equation:

$$V = \frac{V_l + V_r}{2}$$

where:

V = velocity of platform
V_l = velocity of left wheel
V_r = velocity of right wheel.

Figure 2-12. Early style incremental optical encoders attached to the left and right drive motor armatures provide differential odometry information for the drive controller on ROBART II.

Referring to Figure 2-13, arc D_l represents a portion of the circumference of a circle of radius $d + b$:

$$C_l = 2\pi(d + b)$$

where:

C_l = circumference of circle traced by left wheel
d = distance between left and right drive wheels
b = inner turn radius.

In addition, the relationship:

$$\frac{D_l}{C_l} = \frac{\theta}{2\pi} \quad \text{yields:} \quad C_l = \frac{2\pi D_l}{\theta}.$$

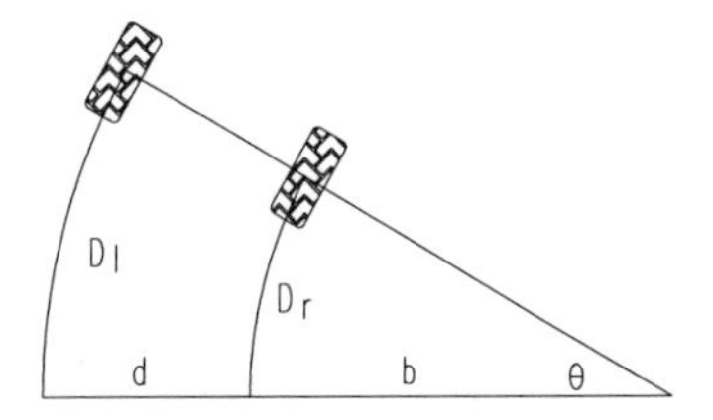

Figure 2-13. Arcs D_l and D_r are traced by the left and right wheels for change in robot heading θ.

Combining the above equations and solving for θ:

$$\theta = \frac{D_l}{d+b}.$$

Similarly, the shorter arc D_r represents a portion of the circumference of a circle of radius b:

$$C_r = 2\pi b$$

where:

C_r = circumference of circle traced by right wheel.

And the relationship: $\frac{D_r}{C_r} = \frac{\theta}{2\pi}$ yields: $C_r = \frac{2\pi D_r}{\theta}$.

Combining equations and solving for b:

$$b = \frac{D_r}{\theta}$$

Substituting this expression for b into the previous expression for θ:

$$\theta = \frac{D_l}{d + \frac{D_r}{\theta}} = \frac{D_l - D_r}{d}.$$

Note this expression for the change in vehicle orientation θ is a function of the displacements of the left and right drive wheels and is completely independent of the path taken. The variable d in the denominator, however, represents a significant source of error, due to uncertainties associated with the effective point of contact of the tires as illustrated in Figure 2-14. The assumption that wheel separation distance is simply the center-to-center distance d as shown is inappropriate. Non-planar irregularities in the floor surface can combine with variations in both tread wear and compliance to shift the effective point of contact in rather unpredictable fashion, with a very detrimental impact on vehicle heading.

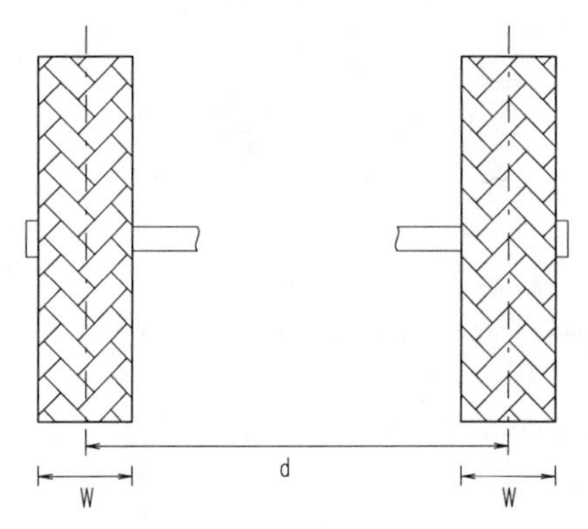

Figure 2-14. Uncertainties in the effective point of contact between tire and floor introduce an ambiguity $2W$ into wheel separation distance d.

Referring now to Figure 2-15, wheel displacement D_l is given by the equation:

$$D_l = \phi R_{el}$$

where:

ϕ = wheel rotation (radians)
R_{el} = effective left wheel radius.

Expressing in terms of encoder counts, this yields:

$$D_l = \frac{2\pi N_l}{C_t} R_{el}$$

where:

N_l = number of counts left encoder
C_t = encoder counts per wheel revolution.

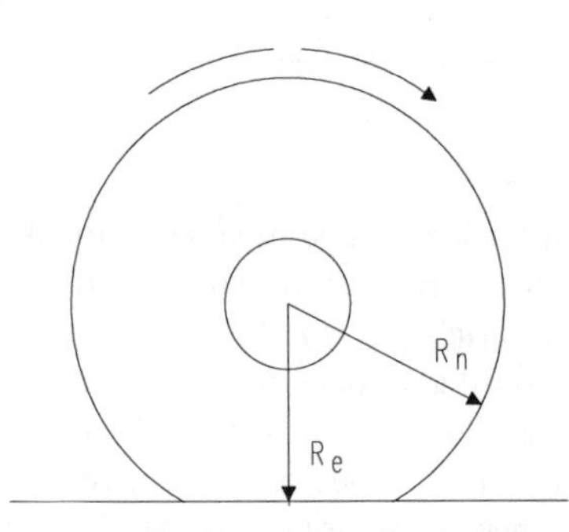

Figure 2-15. Due to tire compliance, effective wheel radius R_e is less than nominal wheel radius R_n.

Similarly, for the right drive wheel:

$$D_r = \frac{2\pi N_r}{C_t} R_{er}$$

where:

N_r = number of counts right shaft encoder
C_t = encoder counts per wheel revolution
R_{er} = effective right wheel radius.

The drive controller will attempt to make the robot travel a straight line by ensuring N_r and N_l are the same. Note, however, that effective wheel radius is a function of the compliance of the tire and the weight of the robot and must be determined empirically. In actuality, R_{el} may not be equal to R_{er}, as was the case when several tires were tested on ROBART II in an attempt to obtain a matched set. For some tires, the compliance (and hence effective radius) was found to vary as a function of wheel rotation ϕ.

Ignoring this situation momentarily for the sake of simplicity, let us next assume a non-compliant wheel of radius R traversing a step discontinuity of height h as shown in Figure 2-16 below. In climbing over the step, the wheel effectively rotates around the point C in the diagram until the axle is directly overhead C at point O' (Borenstein, 1994). The wheel encoder meanwhile measures an effective rotation ϕ corresponding to movement of the axle along path O-O', for a perceived distance D_m. As Borenstein points out, however, the actual horizontal distance traveled is only D_h, creating a linear error of magnitude $D_m - D_h$. In the case of bump traversal, a similar error will be incurred in rolling off the other side at point C' as the wheel drops an identical distance h in returning to floor level. This displacement differential between left and right drive wheels results in an instantaneous heading change (towards the side traversing the bump) equal to:

$$\Delta\theta = 2\frac{D_m - D_h}{d}$$

where:

D_m = measured distance traveled
D_h = actual horizontal distance traveled
d = wheel separation distance as before.

A similar effect is observed when traversing a crack in the floor surface, with the vertical drop h' determined by the relationship of wheel diameter to the width w of the crack.

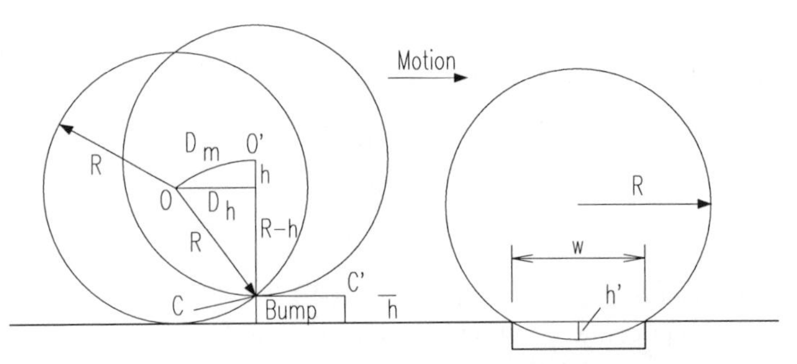

Figure 2-16. Traversal of a small bump creates a differential in the horizontal distance traveled versus the perceived distance measured by the encoder, resulting in a significant angular offset (adapted with changes from Borenstein, 1994a, © IEEE). A similar effect is experienced when rolling over a crack of width w.

Bumps and cracks in the floor can cause non-systemic errors that adversely affect dead reckoning performance. Another common error source is the inevitable slippage that occurs between tire and floor surfaces due to grease or oil build-up, fluid spills, excessive acceleration or deceleration, or even actual impact to the vehicle itself. This problem is especially noticeable in an exterior implementation of differential drive known as *skid steering*, routinely implemented in track form on bulldozers and armored vehicles. Such *skid-steer* configurations intentionally rely on track or wheel slippage for normal operation (Figure 2-17) and as a consequence provide rather poor dead-reckoning information. For this reason, *skid steering* is generally employed only in teleoperated as opposed to autonomous robotic applications, where the ability to surmount significant floor discontinuities is more desirable than accurate dead-reckoning information. An example is seen in the track drives popular with remote-controlled robots intended for explosive ordnance disposal.

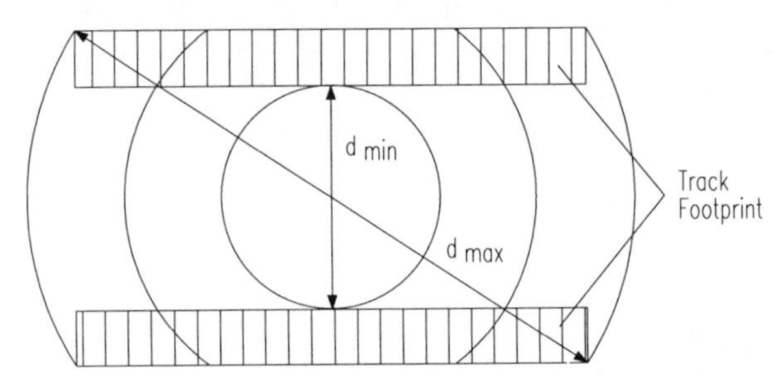

Figure 2-17. The effective point of contact for a skid-steer vehicle is roughly constrained on either side by a rectangular zone of ambiguity corresponding to the track footprint. As is implied by the concentric circles, considerable slippage must occur in order for the vehicle to turn.

2.3.2 Ackerman Steering

Used almost exclusively in the automotive industry, *Ackerman steering* (Jones & Flynn, 1993) is designed to ensure the inside front wheel is rotated to a slightly sharper angle than the outside wheel when turning, thereby eliminating geometrically induced tire slippage. As seen in Figure 2-18, the extended axes for the two front wheels intersect in a common point that lies on the extended axis of the rear axle. The locus of points traced along the ground by the center of each tire is thus a set of concentric arcs about this centerpoint of rotation P_1, and (ignoring for the moment any centrifugal accelerations) all instantaneous velocity vectors will subsequently be tangential to these arcs. Such a steering geometry is said to satisfy the *Ackerman equation* (Byrne, et al., 1992):

$$\cot\theta_i - \cot\theta_o = \frac{d}{l}$$

where:

θ_i = relative steering angle of inner wheel
θ_o = relative steering angle of outer wheel
l = longitudinal wheel separation
d = lateral wheel separation.

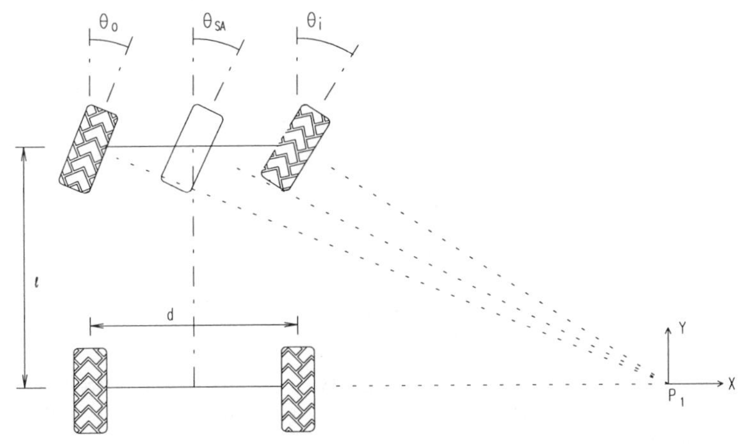

Figure 2-18. In an Ackerman-steered vehicle, the extended axes for all wheels intersect in a common point (adapted from Byrne, et al., 1992).

For sake of convenience, the vehicle steering angle θ_{SA} can be thought of as the angle (relative to vehicle heading) associated with an imaginary center wheel as shown in the figure above. θ_{SA} can be expressed in terms of either the inside or outside steering angles (θ_i or θ_o) as follows (Byrne, et al., 1992):

$$\cot\theta_{SA} = \frac{d}{2l} + \cot\theta_i \quad \text{or alternatively:} \quad \cot\theta_{SA} = \cot\theta_o - \frac{d}{2l}.$$

Ackerman steering provides a fairly accurate dead-reckoning solution while supporting the traction and ground clearance needs of all-terrain operation and is generally the method of choice for outdoor autonomous vehicles. Associated drive implementations typically employ a gasoline or diesel engine coupled to a manual or automatic transmission, with power applied to four wheels through a transfer case, differential, and a series of universal joints. A representative example is seen in the HMMWV-based prototype of the USMC Teleoperated Vehicle (TOV) Program (Aviles et al., 1990). From a military perspective, the use of existing-inventory equipment of this type simplifies some of the logistics problems associated with vehicle maintenance. In addition, reliability of the drive components is high due to the inherited stability of a proven power train. (Significant interface problems can be encountered, however, in retrofitting off-the-shelf vehicles intended for human drivers to accommodate remote or computer control.)

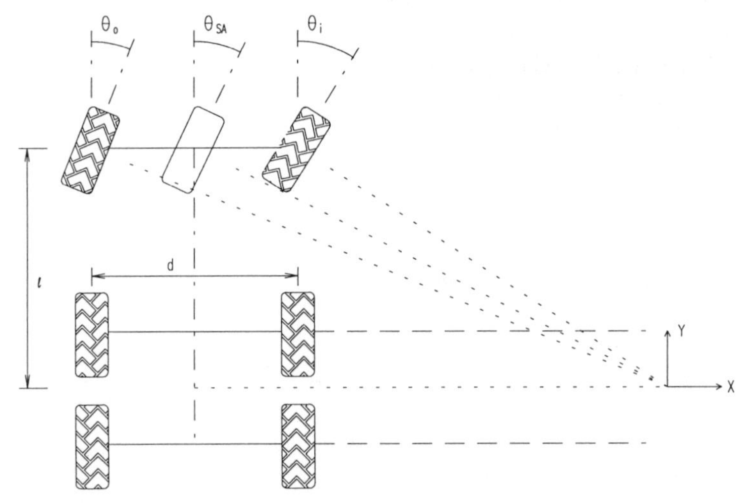

Figure 2-19. The six-wheel drive configuration employed on the Surrogate Teleoperated Vehicle suffers from excessive wheel slippage during turns as all extended axes do not intersect in a common point.

The Surrogate Teleoperated Vehicle (STV) developed by Robotic Systems Technology (Metz, et al., 1992; Myers, 1992) is loosely considered a six-wheel Ackerman-steered vehicle with twin rear axles, but the geometry (Figure 2-19) does not satisfy the Ackerman equation. The extra rear axle introduces some significant wheel slippage (and tire wear) during turns, even further aggravated by the fact that no differential action was incorporated in the chain-drive design. These detrimental effects on dead-reckoning accuracy were not all that significant from an operational standpoint since the vehicle was directly controlled by a remote human driver, but were a major factor in the decision not to use the STV in an MDARS Exterior role.

2.3.3 Synchro Drive

An innovative configuration known as *synchro drive* features three or more wheels (Figure 2-20) mechanically coupled in such a way that all rotate in the same direction at the same speed, and similarly pivot in unison about their respective steering axes when executing a turn. This drive and steering "synchronization" results in improved dead-reckoning accuracy through reduced slippage, since all wheels generate equal and parallel force vectors at all times.

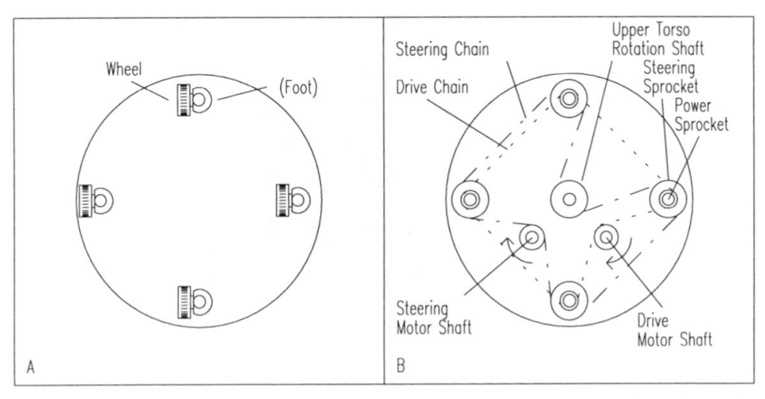

Figure 2-20. Bottom (A) and top (B) views of a four-wheel synchro-drive configuration (adapted from Holland, 1983).

The required mechanical synchronization can be accomplished in a number of ways, the most common being chain, belt, or gear drive. Carnegie Mellon University has implemented an electronically synchronized version on one of their *Rover* series robots (Figure 2-21), with dedicated drive motors for each of the three wheels. Chain- and belt-drive configurations experience some degradation in steering accuracy and alignment due to uneven distribution of slack, which varies as a function of loading and direction of rotation. In addition, whenever chains (or timing belts) are tightened to reduce such slack, the individual wheels must be realigned. These problems are eliminated with a completely enclosed gear-drive approach. An enclosed gear train also significantly reduces noise as well as particulate generation, the latter being very important in clean-room applications.

An example of a three-wheeled belt-drive implementation is seen in the Denning *MRV-2* and *Sentry* robots introduced by Denning Mobile Robots, Woburn, MA (Kadonoff, 1986). Referring to Figure 2-22, drive torque is transferred down through the three steering columns to polyurethane-filled rubber tires. The drivemotor output shaft is mechanically coupled to each of the steering-column power shafts by a heavy-duty timing belt to ensure synchronous operation. A second timing belt transfers the rotational output of the steering motor to the three steering columns, allowing them to synchronously pivot throughout a full 360-degree range (Everett, 1988).

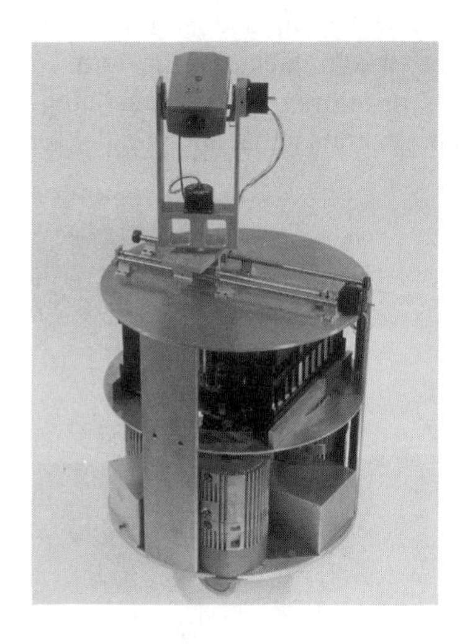

Figure 2-21. The CMU robot *Pluto* employs three electronically synchronized drive wheels (courtesy Carnegie Mellon University).

The three-point configuration ensures good stability and traction, while the actively driven large-diameter wheels provide more than adequate obstacle climbing capability for indoor scenarios. Disadvantages of this particular implementation include dead-reckoning errors introduced by compliance in the drive belts as well as by reactionary frictional forces exerted by the floor surface when turning in place.

To overcome these problems, the Cybermotion *K2A Navmaster* robot employs an enclosed gear-drive configuration with the wheels offset from the steering axis as shown in Figure 2-23. When a foot pivots during a turn, the attached wheel rotates in the appropriate direction to minimize floor and tire wear, power consumption, and slippage. Note that for correct compensation, the miter gear on the wheel axis must be on the opposite side of the power shaft gear from the wheel as illustrated. The governing equation for minimal slippage is (Holland, 1983):

$$\frac{A}{B} = \frac{r'}{r}$$

where:

A = number of teeth on the power shaft gear
B = number of teeth on the wheel axle gear
r' = wheel offset from steering pivot axis
r = wheel radius.

Figure 2-22. The Denning *MRV-2* mobility base incorporates a three-point *synchro-drive* configuration with each wheel located directly below the pivot axis of the associated steering column (courtesy Georgia Institute of Technology).

One drawback of this approach is seen in the decreased lateral stability that results when one wheel is turned in under the vehicle. Cybermotion's improved *K3A* design solves this problem (with an even smaller wheelbase) by incorporating a dual-wheel arrangement on each foot as shown in Figure 2-24 (Fisher, et al., 1994). The two wheels turn in opposite directions in differential fashion as the foot pivots during a turn, but good stability is maintained in the foregoing example by the outward swing of the additional wheel. In addition, the decreased lateral projection of the foot assembly significantly decreases the likelihood of a wheel climbing up the side of a projecting wall surface such as a column or corner, a situation that has on occasion caused the *Navmaster* to flip over.

The dead-reckoning calculations for *synchro drive* are almost trivial: vehicle heading is simply derived from the steering angle encoder, while displacement in the direction of travel is given as follows:

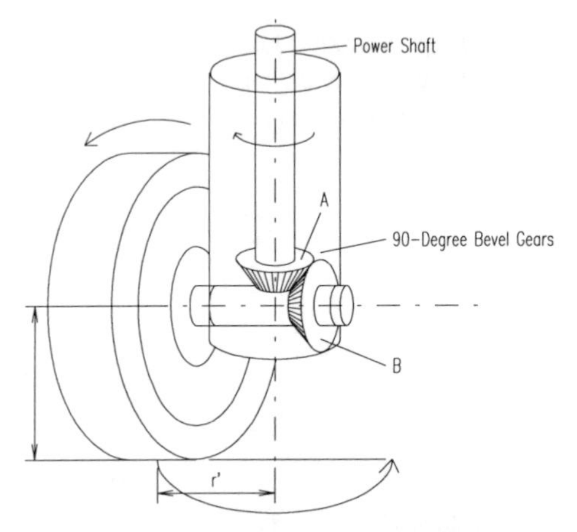

Figure 2-23. Slip compensation during a turn is accomplished through use of an offset foot assembly on the three-wheeled *K2A Navmaster* robot (adapted from Holland, 1983).

$$D = \frac{2\pi N}{C_e} R_e$$

where:

D = vehicle displacement along path
N = measured counts of drive motor shaft encoder
C_e = encoder counts per complete wheel revolution
R_e = effective wheel radius.

Figure 2-24. The new *K3A* dual-wheel foot assembly (bottom right) is shown in comparison to the original *K2A* assembly at the top right (courtesy Cybermotion, Inc.).

2.3.4 Tricycle Drive

Tricycle-drive configurations (Figure 2-25) employing a single driven front wheel and two passive rear wheels (or vice versa) are fairly common in AGV applications due to their inherent simplicity. For odometry instrumentation in the form of a steering angle encoder, the dead-reckoning solution is equivalent to that of an Ackerman-steered vehicle, where the drive wheel replaces the imaginary center wheel discussed in Section 2.3.2. Alternatively, if rear-axle differential odometry is used to determine heading, the solution is identical to the differential-drive configuration discussed in Section 2.3.1.

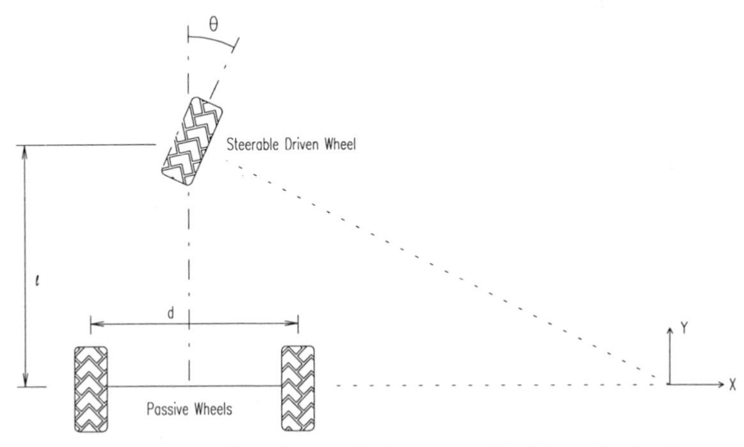

Figure 2-25. Tricycle-drive configurations employing a steerable driven wheel and two passive trailing wheels can derive heading information directly from a steering angle encoder or indirectly from differential odometry.

One problem associated with the tricycle-drive configuration is the vehicle's center of gravity tends to move away from the driven wheel when traversing up an incline, causing a loss of traction. As in the case of Ackerman-steered designs, some surface damage and induced heading errors are possible when actuating the steering while the platform is not moving.

2.3.5 Omni-Directional Drive

The dead-reckoning solution for most multiple-degree-of-freedom configurations is done in similar fashion to that for differential drive, with position and velocity data derived from the motor (or wheel) shaft- encoders. For the three-wheel example illustrated in Figure 2-26B, the equations of motion relating individual motor speeds to velocity components V_x and V_y in the reference frame of the vehicle are given by (Holland, 1983):

$$V_1 = \omega_1 r = V_x + \omega_p R$$
$$V_2 = \omega_2 r = -0.5 V_x + 0.867 V_y + \omega_p R$$
$$V_3 = \omega_3 r = -0.5 V_x - 0.867 V_y + \omega_p R$$

where:

V_1 = tangential velocity of wheel number 1
V_2 = tangential velocity of wheel number 2
V_3 = tangential velocity of wheel number 3
ω_1 = rotational speed of motor number 1
ω_2 = rotational speed of motor number 2
ω_3 = rotational speed of motor number 3
ω_p = rate of base rotation about pivot axis
r = effective wheel radius
R = effective wheel offset from pivot axis.

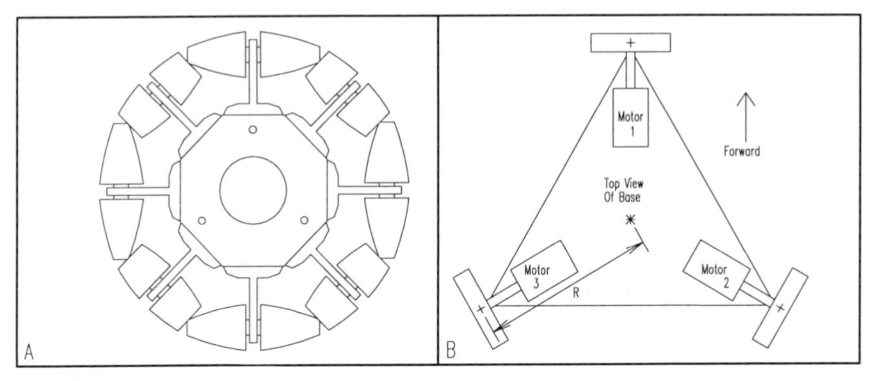

Figure 2-26. (A) Schematic of the wheel assembly used by the Veterans Administration (La, et al., 1981, © IEEE) on an omni-directional wheelchair. (B) Top view of base showing relative orientation of components in the three-wheel configuration (adapted from Holland, 1983).

The geometric relationships between wheel rotation and vehicle displacement are of course platform specific, as should be obvious from the alternative four-wheel design shown in Figure 2-27 below. Multiple-degree-of-freedom configurations display exceptional maneuverability in tight quarters in comparison to conventional 2-DOF mobility systems but have been found to be difficult to control due to their overconstrained nature (Reister, 1991; Kilough & Pin, 1992; Borenstein, 1994b). Resulting problems include increased wheel slippage, excessive tire wear, and inefficiency in operation, which can sometimes offset the not-always-required gain in maneuverability. A careful examination of all the tradeoffs involved should be made before committing to any particular drive configuration.

Figure 2-27. An example MDOF drive configuration developed for the Department of Energy nuclear waste inspection program (courtesy Martin Marietta, Denver).

2.4 Internal Position Error Correction

Partly motivated by the degraded navigational capabilities of MDOF vehicles, researchers at the University of Michigan have devised a clever way of significantly reducing dead-reckoning errors by one to two orders of magnitude without any external reference (Borenstein, 1994a). The *internal position error correction* scheme has been demonstrated on the *CLAPPER* (Compliant Linkage Autonomous Platform with Position Error Recovery), a 4-DOF robotic testbed consisting of two TRC *LabMate* vehicles joined together with a compliant linkage (Figure 2-28). The compliant linkage accommodates momentary controller errors without transferring any mutual force reactions between the *LabMates*, thereby eliminating excessive wheel slippage reported for alternative MDOF designs (Reister, 1991; Kilough & Pin, 1992).

More importantly, the linkage is instrumented as illustrated in Figure 2-29 to provide real-time feedback on the relative position and orientation of the two TRC platforms. An absolute encoder at each end measures the rotation of each

LabMate (with respect to the linkage) with a resolution of 0.3 degrees, while a linear encoder is used to measure the separation distance to within ±5 millimeters. A single supervisory computer reads the encoder pulses from all four drive wheels and computes each *LabMate's* dead-reckoned position and heading in conventional fashion. By examining these perceived solutions in conjunction with the known relative orientations of the two platforms, the *CLAPPER* system can detect and significantly reduce heading errors for both mobility bases.

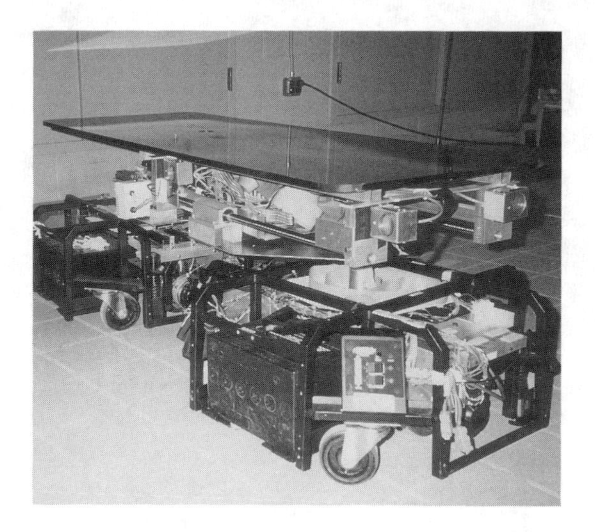

Figure 2-28. The *CLAPPER* is a dual-differential-drive multiple-degree-of-freedom vehicle consisting of two TRC *LabMates* coupled together with a compliant linkage (courtesy University of Michigan).

The principle of operation is centered on the concept of *error growth rate* presented by Borenstein (1994a), who makes a distinction between "fast-growing" and "slow-growing" dead-reckoning errors. For example, when a differentially steered robot traverses a floor discontinuity as discussed in Section 2.3.1, it will immediately experience an appreciable orientation error (i.e., a *fast-growing* error). The associated lateral displacement error, however, is initially very small (i.e., a *slow-growing* error) but grows in an unbounded fashion as a consequence of the orientation error. The internal error correction algorithm performs relative position measurements with sufficient update rate to allow each *LabMate* to detect *fast-growing* errors in orientation, while relying on the fact that the lateral position errors accrued by both platforms during the sampling interval were very small.

The compliant linkage in essence forms a pseudo-stable heading reference in world coordinates, its own orientation being dictated solely by the relative translations of its end points, which in turn are affected only by the lateral displacements of the two *LabMate* bases. Since the lateral displacements are *slow growing*, the linkage rotates only a very small amount between encoder samples. The *fast-growing* azimuthal disturbances of the bases, on the other hand, are not coupled through the rotational joints to the linkage, thus allowing the rotary encoders to detect and quantify the instantaneous orientation errors of the bases, even when both are in motion. Borenstein (1994a) provides a more complete description of this innovative concept and reports experimental results indicating improved dead-reckoning performance up to a factor of 100.

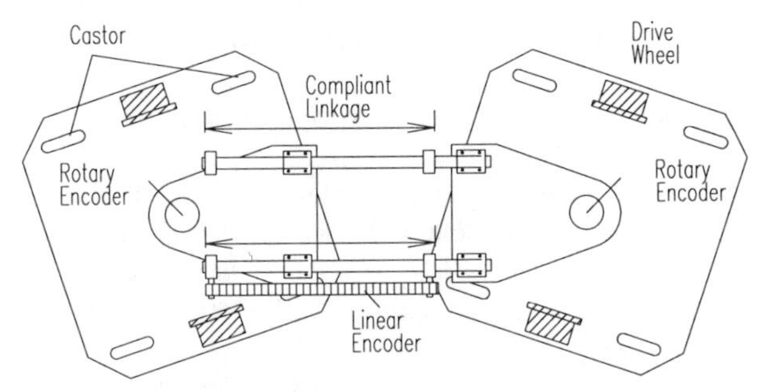

Figure 2-29. The compliant linkage is instrumented with two absolute rotary encoders and a linear encoder to measure the relative orientations and separation distance between the *LabMate* bases (adapted from Borenstein, 1994a, © IEEE).

2.5 References

Agent, A., "The Advantages of Absolute Encoders for Motion Control," *Sensors*, pp. 19-24, April, 1991.

Aviles, W.A., Everett, H.R., Hughes, T.W., Koyamatsu, A.H., Laird, R.T., Martin, S.W., McArthur, S.P., Umeda, A.Y., "Issues in Mobile Robotics: The Unmanned Ground Vehicle Program TeleOperated Vehicle (TOV)," SPIE Vol. 1388, Mobile Robots V, Boston, MA, pp. 587-597, 8-9 November, 1990.

Avolio, G., "Principles of Rotary Optical Encoders," *Sensors*, pp. 10-18, April, 1993.

Barshan, B, Durrant-Whyte, H.F., "An Inertial Navigation System for a Mobile Robot," Proceedings of the 1st IAV, Southampton, England, pp. 54-59, April 18-21, 1993.

Borenstein, J., "The CLAPPER: A Dual-drive Mobile Robot With Internal Correction of Dead Reckoning Errors," IEEE International Conference on Robotics and Automation, San Diego, CA, pp. 3085-3090, May, 1994a.

Borenstein, J., "Internal Correction of Dead-reckoning Errors with the Smart Encoder Trailer," International Conference on Intelligent Robots and Systems, Munchen, Germany, September, 1994b.

Byrne, R.H., Klarer, P.R., Pletta, J.B., "Techniques for Autonomous Navigation," Sandia Report SAND92-0457, Sandia National Laboratories, Albuquerque, NM, March, 1992.

Catling, I., *Advanced Technology for Road Transport: IVHS and ATT*, Artech House, Boston, MA, 1994.

Dahlin, T., Krantz, D., "Low-Cost, Medium Accuracy Land Navigation System," *Sensors*, pp. 26-34, February, 1988.

Deirmengian, C., "Synchros and Resolvers: Part I," *Sensors*, pp. 31-38, April, 1990a.

Deirmengian, C., "Synchros and Resolvers: Part II," *Sensors*, pp. 48-55, May, 1990b.

Dunlap, G.D., Shufeldt, H.H., *Dutton's Navigation and Piloting*, Naval Institute Press, pp. 557-579, 1972.

Everett, H.R., "A Computer Controlled Autonomous Sentry Robot," Masters Thesis, Naval Postgraduate School, Monterey, CA, October, 1982.

Everett, H.R., "Security and Sentry Robots", *International Encyclopedia of Robotics Applications and Automation*, R.C. Dorf, ed., John Wiley, pp. 1462-1476, March, 1988.

Fisher, D., Holland, J.M., Kennedy, K.F., "K3A Marks Third Generation Synchro-Drive," American Nuclear Society Winter Meeting, Proceedings of Robotics and Remote Systems, New Orleans, LA, June, 1994.

Fraden, J., *AIP Handbook of Modern Sensors*, ed., Radebaugh, R., American Institute of Physics, New York, 1993.

Grandner, W., Lanton, S., "Development of LSI Circuits for Position Encoders," *Sensors*, pp. 28, 32, April, 1986.

Harmon, S.Y., "USMC Ground Surveillance Robot (GSR): Lessons Learned," Mobile Robots, SPIE Vol. 727, Cambridge, MA, pp. 336-343, 1986.

Harrington, J.J., Klarer, P.R., "SIR-1: An Autonomous Mobile Sentry Robot," Sandia Report SAND87-1128, Sandia National Laboratories, Albuquerque, NM, May, 1987.

Henkel, S. vL., "Optical Encoders: A Review," *Sensors*, pp. 9-12, September, 1987.

Holland, J.M., *Basic Robotics Concepts*, Howard W. Sams, Macmillan, Inc., Indianapolis, IN, 1983.

Holle, S., "Incremental Encoder Basics," *Sensors*, pp. 22-30, April, 1990.

ILC Data Device Corporation, "Synchro Conversion Handbook", Bohemia, NY, April, 1982.

Jones, J.L., Flynn, A.M., *Mobile Robots: Inspiration to Implementation*, AK Peters, Ltd., Wellesley, MA, p. 141, 144, 1993.

Kadonoff, M.B., "Navigation Techniques for the Denning Sentry," MS86-757, RI/SME 2nd International Conference on Robotics Research, Scottsdale, AZ, August, 1986.

Kilough, S.M., Pin, F.G., "Design of an Omnidirectional Holonomic Wheeled Platform Prototype," IEEE Conference on Robotics and Automation, Nice, France, pp. 84-90, May, 1992.

Klarer, P.R., "Simple 2-D Navigation for Wheeled Vehicles," Sandia Report SAND88-0540, Sandia National Laboratories, Albuquerque, NM, April, 1988.

Klarer, P.R., Harrington, J.J., "Development of a Self-Navigating Mobile Interior Robot Application as a Security Guard/Sentry," Sandia Report SAND86-0653, Sandia National Laboratories, Albuquerque, NM, July, 1985.

La, W.H.T., Koogle, T.A., Jaffe, D.L., Leifer, L.J., "Microcomputer-Controlled Omnidirectional Mechanism for Wheelchairs," Proceedings, IEEE Frontiers of Engineering in Health Care, CH1621-2/81/0000-0326, 1981.

Manolis, S., Resolvers vs. Rotary Encoders For Motor Commutation and Position Feedback, *Sensors*, pp. 29-32, March, 1993.

Metz, C.D., Everett, H.R., Myers, S., "Recent Developments in Tactical Unmanned Ground Vehicles," Association for Unmanned Vehicle Systems, Huntsville, AL, June, 1992.

Milner, R., ""Measuring Speed and Distance with Doppler," *Sensors*, pp. 42-44, October, 1990.

Mononen, J., Nieminen, T., Puputti, J., "Teleoperation and Autonomous Guidance Systems for Off-Road Vehicles," International Off-Highway and Powerplant Congress and Exposition, Milwaukee, WI, Society of Automotive Engineers, ISSN 0148-7191, September, 1994.

Myers, S.D., "Update on the Surrogate Teleoperated Vehicle (STV)," Association for Unmanned Vehicle Systems, Huntsville, AL, 1992.

Nickson, P., "Solid-State Tachometry," *Sensors*, pp. 23-26, April, 1985.

Reister, D.B., "A New Wheel Control System for the Omnidirectional Hermies III Robot," IEEE Conference on Robotics and Automation, Sacramento, CA, pp. 2322-2327, April, 1991.

Sammarco, J.J., "Mining Machine Orientation Control Based on Inertial, Gravitational, and Magnetic Sensors," Report of Investigations 9326, US Bureau of Mines, Pittsburgh, PA, 1990.

Sammarco, J.J., "A Navigational System for Continuous Mining Machines," *Sensors*, pp. 11-17, January, 1994.

Schultz, W., "Traffic and Vehicle Control Using Microwave Sensors," *Sensors*, pp. 34-42, October, 1993.

Schwartz, O.B., Grafstein, P., *Pictorial Handbook of Technical Devices*, Chemical Publishing Co, Inc., New York, NY, pp. 272-275, 1971.

Tiwari, R., "Resolver-Based Encoders," *Sensors*, pp. 29-34, April, 1993.

Turpin, D.R., "Inertial Guidance: Is It a Viable Guidance System for AGVs?" 4th International Conference on AGVs (AGVS4), pp. 301-320, June, 1986.

Vaganay, J., Aldon, M.J., Fournier, A., "Mobile Robot Localization by Fusing Odometric and Inertial Measurements," Fifth Topical Meeting on Robotics and Remote Systems, Knoxville, TN, Vol. 1, pp. 503-510, April, 1993.

3

Tactile and Proximity Sensing

Tactile sensors are typically employed on automated guided vehicles (AGVs) and mobile robots to provide a last-resort indication of collisions with surrounding obstructions. As the name implies, the detection process involves *direct physical contact* between the sensor and the object of interest. *Proximity sensors*, on the other hand, are *non-contact devices* that provide advance warning on the presence of an object in close proximity to the sensing element.

3.1 Tactile Sensors

A number of different technologies are employed in various *tactile sensing* schemes (Harmon, 1983; Hall, 1984; Dario & DeRossi, 1985; Fielding, 1986; McAlpine, 1986; Pennywitt, 1986):

- Contact closure.
- Magnetic.
- Piezoelectric.
- Capacitive.
- Photoelectric.
- Magnetoresistive.
- Piezoresistive.
- Ultrasonic.

Furthermore, there are many different ways the above candidate sensing strategies can be physically configured. From a mobile robotics perspective, however, the actual embodiments can for the most part be broken down into three general areas: 1) *tactile feelers*, or *antennae*, 2) *tactile bumpers*, and 3) *distributed surface arrays*.

3.1.1 Tactile Feelers

My first experience with *tactile feelers* was in conjunction with the CRAWLER I robot introduced in Chapter 1. Tactile sensing was the only practical means available at the time (1966), due to existing limitations in both technology and my budget, for deriving any sort of collision avoidance feedback. The first implementation consisted of a short length of guitar string extended through the center of a small screw-eye; deflection of the wire due to physical contact caused contact closure with the surrounding screw-eye, completing a simple normally open circuit. Similar implementations are reported by Russell (1984), Schiebel, et al., (1986), and Brooks (1989).

An enhanced version of the CRAWLER sensor (Figure 3-1) involved looping the wire back on itself through a second screw-eye to form a circle, thus widening the protected area. This latter design was interfaced to a punched-card reader (see Chapter 1) to support programmable collision-recovery maneuvers. Small nylon spacers (not shown in the figure) were employed to limit the vertical motion of the wire and thus prevent false activation of the sensor due to vehicle vibration. Instantaneous sensor status was represented by four state variables.

Figure 3-1. Tactile sensors situated at the four corners of the CRAWLER robots (see Chapter 1) were fabricated from guitar strings looped through the center of a pair of small screw-eyes.

Like the CRAWLER robots, ROBART I also relied heavily on tactile sensing for collision detection feedback (see also Chapter 10). The guitar-string feeler probe was mechanically upgraded slightly (Figure 3-2) to make use of an off-the-shelf automobile curb feeler. A cylindrical metal sleeve was fitted around the lower end of the feeler and electrically insulated from it by means of a short length of plastic tubing wedged into the lower half of the metal sleeve as shown in the figure. Any significant deflection of the feeler probe caused it to come into contact with the upper lip of the coaxial sleeve, completing the circuit. Additional tactile sensors used on ROBART I are discussed later in section 3.1.2.

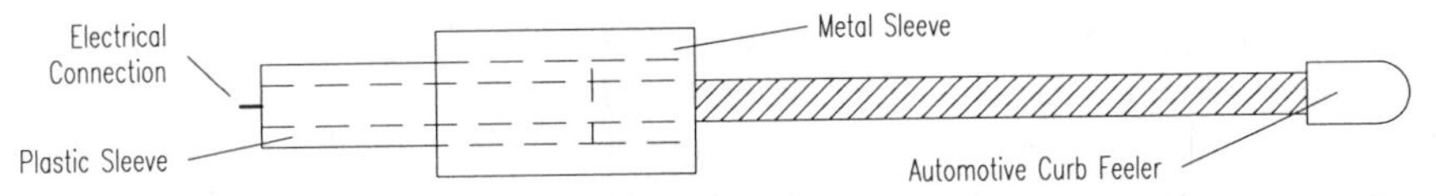

Figure 3-2. Tactile probes employed on *ROBART I* were fabricated from automobile curb feelers.

Patterned after the sensory antennae of certain insects, *active feelers* are an interesting variant of the *tactile feeler* category that incorporate some type of mechanical search strategy for increased utility and expanded coverage. Whereas the passive feelers described above rely on the relative motion between the robotic platform and the sensed object, active feelers are independently swept through a range of motion by their own dedicated actuation schemes. Kaneko (1994) describes such a system that uses a small rotary actuator to manipulate a *flexible feeler* (or *antenna*) in the horizontal plane (Figure 3-3). By careful analysis of the relationship between sensed torque and rotational displacement of the actuator after initial contact is made, the associated moment arm can be calculated. The length of this moment arm corresponds to the actual point of contact along the feeler. A similar *active-antenna* system reported by Ferrel (1994) is used on the six-legged robot *Attila* developed at MIT's Artificial Intelligence Lab in Cambridge, MA.

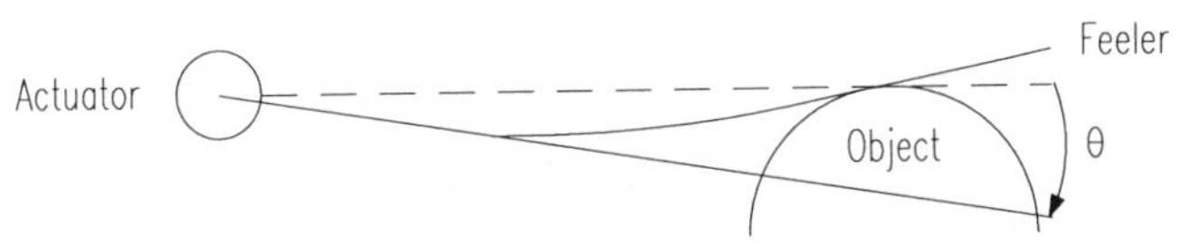

Figure 3-3. In this *active antenna* configuration, the point of actual contact along the flexible feeler can be determined by measuring the amount of rotation θ after initial contact, and the corresponding induced torque (adapted from Kaneko, 1994, © IEEE).

3.1.2 Tactile Bumpers

One of the simplest *tactile bumper* designs, common in AGV applications, consists of a flexible metal plate secured at one end and bowed out to form a protective bumper in front of the vehicle. The other end of the plate is constrained to a single degree of freedom in its motion such that any inward pressure on the plate causes the free edge to shift laterally away from the constrained edge, activating in the process some form of contact closure device (Gat, et al., 1993). This concept is similar in many respects to the previously discussed wire-loop sensors employed on the CRAWLER but with increased vertical coverage (i.e., some bumpers of this type are 18 inches high).

Rowan (1988) describes an alternative instrumentation method wherein a small retroreflective target is placed on the back of the flexible metal plate, directly in the beam of a forward-looking photoelectric sensor mounted on the front of the vehicle. If the bumper is deflected by physical contact with an obstruction, the retroreflector is shifted laterally out of the detector's footprint of illumination, breaking the beam. Another variation on this theme involves using an inflatable bladder in the void between the bumper and the front of the vehicle. Any subsequent impact causes an increase in air pressure within the bladder, actuating a diaphragm switch assembly that halts forward motion.

The entire front panel of ROBART I was mechanically floated on a spring suspension to form a contact plate for purposes of tactile feedback (Figure 3-4). In addition, all leading structural edges were protected by hinged sections of aluminum angle that would actuate recessed microswitches in the event of obstacle contact. Flexible nylon extensions protruding from either side of the base provided coverage for the rear wheels. Note also the vertically oriented tactile feelers described in the previous section.

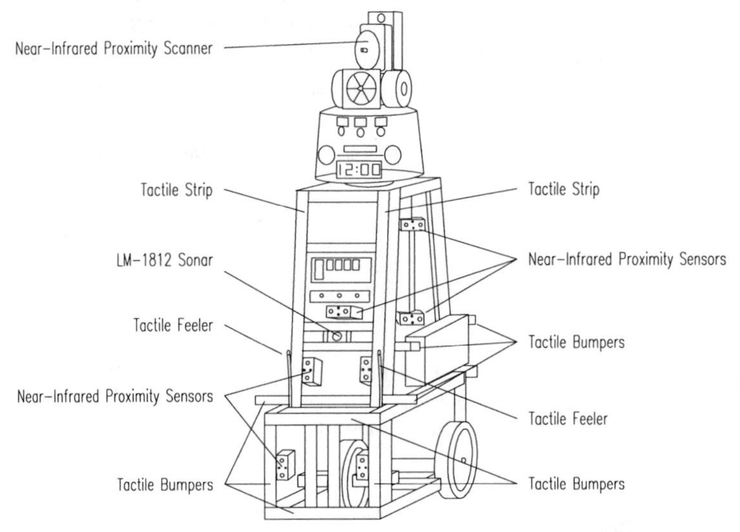

Figure 3-4. Constructed before the introduction of the Polaroid ranging module, *ROBART I* was generously equipped with tactile bumpers and feeler probes for collision detection feedback.

Relying more heavily on *sonar* and *proximity detectors* for collision avoidance protection, ROBART II employs only two tactile sensors in the form of circumferential bumpers situated around the periphery of its mobility base. Each bumper assembly consists of a free-floating plastic strip encased in a fixed housing, spring loaded to be normally in the extended position. A series of microswitches is arranged behind these housings such that individual switch elements are engaged by any displacement of the strip. When a bumper comes in contact with another surface, the floating strip is locally depressed and activates the appropriate microswitch to provide geometric resolution of the point of impact. This haptic *situation awareness* facilitates intelligent recovery by the collision avoidance software, while the housing configuration doubles as a protective bumper for the surface of the robot base.

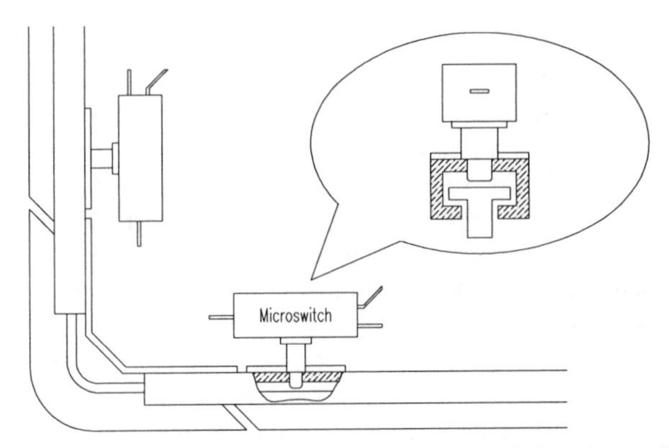

Figure 3-5. Spring-loaded tactile bumpers on ROBART II are designed to activate a series of microswitches when locally depressed, providing geometric resolution of the point of impact.

The most significant component of this continuous-bumper design is the corner piece (Figure 3-5), designed with an angled cut at both ends to mate with the floating strips in the linear encasings. When a corner comes in contact with another surface, it will press against a floating strip and thus activate the microswitch nearest the corner. The angled construction also permits lateral motion of the strips within their encasings when responding to oblique impacts.

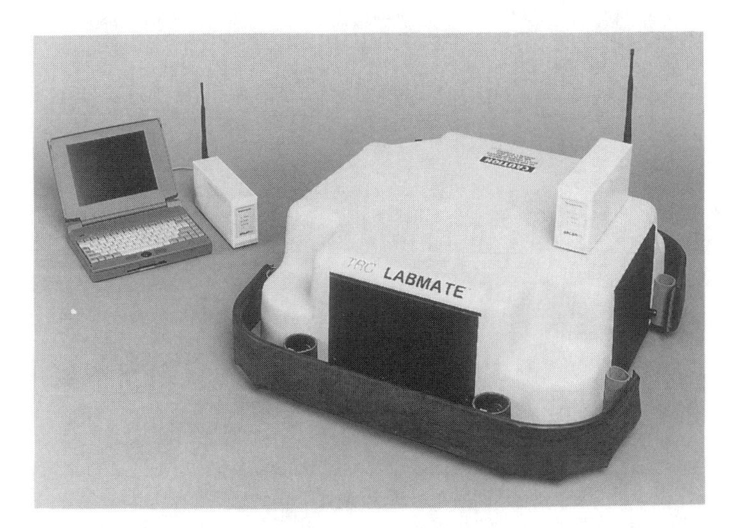

Figure 3-6. The tactile bumper employed on the *LabMate* uses a conductive foam material separated from a flexible metal backplate by an insulating mesh (courtesy Transitions Research Corp.).

Transitions Research Corporation (TRC), Danbury, CN, provides both front and rear tactile bumpers on their *LabMate* base (Figure 3-6). The sensing methodology is contact closure, but the implementation is a little more elegant than the myriad of electromechanical microswitches employed on ROBART II. Each bumper is a multi-layered assembly supported by a flexible metal backing plate attached to either side of the base as shown. A layer of conductive-foam material is placed in front of this metal backplate, electrically and mechanically isolated by an intermediate plastic mesh and covered on the outside by a protective rubber sheath. Any significant impact to the outer skin of this "sandwich" causes a deformation of the conductive foam, pushing it momentarily through the holes in the insulating mesh to make electrical contact with the metal backplate. One disadvantage to this scheme is the inherent lack of positional resolution in the strictly binary nature (i.e., contact/no-contact) of the resulting output.

Rather than instrument the entire bumper surface itself, the Cybermotion *K2A-Navmaster* design shown in Figure 3-7 below simply incorporates adjustable strain gauges in the cantilevered supporting arms to sense any impact. A minor problem with this approach is occasional false activation due to inertial loads created by vertical acceleration of the bumper assembly when traversing small cracks or bumps in the floor. When properly adjusted for actual site conditions, however, the concept works very well with minimal problems. The use of separate left and right sensors allows for some limited degree of geometric resolution of the point of impact.

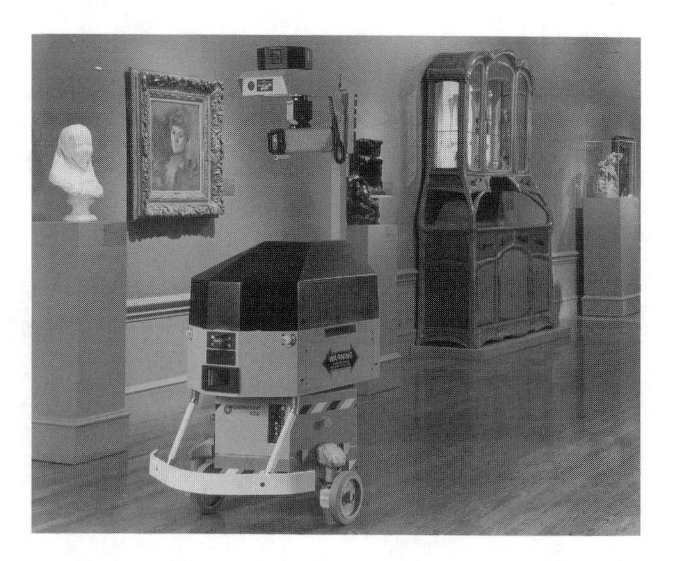

Figure 3-7. The tactile bumper on the Cybermotion *K2A Navmaster* robot is configured to activate adjustable-threshold strain sensors in the left and right supporting arms (courtesy Cybermotion, Inc.)

3.1.3 Distributed Surface Arrays

There is somewhat of a growing trend (on the research side of the house anyway) to move towards embedded tactile arrays that provide two-dimensional profiling of the contacting object. In addition to geometric resolution of the point of impact, many of these strategies also provide some quantification of the contact force magnitude. Early applications involved fairly small rectangular array structures geared towards component identification and/or orientation sensing for industrial robotic scenarios and are aptly summarized in surveys presented by Harmon (1983), Dario, et al., (1985), Pennywitt (1986), Nicholls and Lee (1989), and Grahn (1992).

More recently there has been emerging interest in the development of a continuous skin-like sensor array that could be incorporated directly into the entire outer covering of a manipulator arm or even a mobile robotic vehicle. Grahn (1992) describes a tactile array produced by Bonneville Scientific, Salt Lake City, UT, that uses rows of ultrasonic transmitters and receivers to measure the thickness of an overlying rubber pad. Each element of the sensor array transmits an ultrasonic pulse that reflects off the outer surface of the rubber and returns to the sensor, thereby providing a means of precisely measuring the round-trip path length. Contact with an external object causes compression of the rubber and subsequently reduces the measured time of flight (See also chapter 5). The current 256-element sensor array (Model 200-16 X 16A) is a rigid planar structure consisting of a ceramic substrate, the PVDF transducer material, and an elastomeric pad covering.

Merritt Systems, Inc., (MSI) Merritt Island, FL, is developing a continuous flexible array of tactile and temperature sensors under a Small Business Innovative Research program managed by the Naval Command Control and Ocean Surveillance Center, San Diego, CA. The goal is to produce a conformal skin-like material containing distributed arrays of temperature and tactile sensors that can be cut into any desired shape for attachment to robotic manipulator arms or to the structural housings of mobile robotic vehicles. The company has already developed a methodology for mounting miniature ultrasonic and near-infrared proximity sensors on a flexible base material incorporating an embedded matrix of power and communications buses (MSI, undated). Up to 1022 *SmartSensor* modules may be configured into a single *SensorSkin* (Wing, 1995). The skin can be custom wrapped around the robot in a single piece (PM, 1995).

3.2 Proximity Sensors

Proximity sensors, used to determine the presence (as opposed to actual range) of nearby objects, were developed to extend the sensing range beyond that afforded by direct-contact tactile or haptic sensors. Recent advances in electronic technology have significantly improved performance and reliability, thereby increasing the number of possible applications. As a result, many industrial installations that historically have used mechanical limit switches can now choose from a variety of alternative non-contact devices for their close (between a fraction of an inch and a few inches) sensing needs. Such *proximity sensors* are classified into several types in accordance with the specific properties used to initiate a switching action:

- Magnetic.
- Inductive.
- Ultrasonic.
- Microwave.
- Optical.
- Capacitive.

The reliability characteristics displayed by these sensors make them well suited for operation in harsh or otherwise adverse environments, while providing high-speed response and long service lives. Instruments can be designed to withstand significant shock and vibration, with some capable of handling forces over 30,000 Gs and pressures of nearly 20,000 psi (Hall, 1984). Burreson (1989) and Peale (1992) discuss advantages and tradeoffs associated with proximity sensor selection for applications in challenging and severe environments. In addition, proximity devices are valuable when detecting objects moving at high speed, when physical contact may cause damage, or when differentiation between metallic and non-metallic items is required. Ball (1986), Johnson (1987), and Wojcik (1994) provide general overviews of various alternative proximity sensor types with suggested guidelines for selection.

3.2.1 Magnetic Proximity Sensors

Magnetic proximity sensors include *reed switches*, *Hall-effect devices*, and *magnetoresistive sensors*.

Magnetic Reed Switches

The simplest form of magnetic proximity sensor is the *magnetic reed switch*, schematically illustrated in Figure 3-8. A pair of low-reluctance ferromagnetic reeds are cantilevered from opposite ends of a hermetically sealed tube, arranged such that their tips overlap slightly without touching. The extreme ends of the reeds assume opposite magnetic polarities when exposed to an external magnetic flux, and the subsequent attractive force across the gap pulls the flexible reed elements together to make electrical contact (Hamlin, 1988).

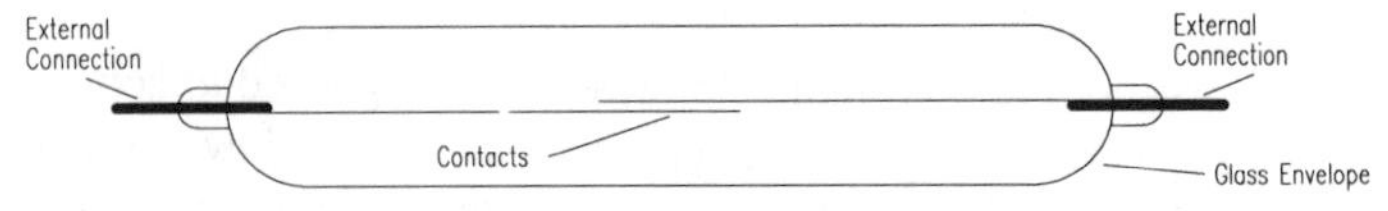

Figure 3-8. The hermetically sealed *magnetic reed switch*, shown here with normally open contacts, is filled with inert gas and impervious to dust and corrosion.

Some problems can be encountered with this type of sensor due to contact bounce, structural vibration, and pitting of the mating surfaces in the case of inductive or capacitive loads (Burreson, 1989).

Figure 3-9. Photo of rear-access doors on ROBART II, showing location (top outer corners) of the magnetic door-closure switch (left) and its associated permanent magnet (right).

Available in both *normally open* and *normally closed* configurations, these inexpensive and robust devices are commonly employed as door- and window-closure sensors in security applications. A magnetic reed switch of this type was installed on ROBART II to monitor the status of the rear access doors as shown in Figure 3-9.

Hall Effect Sensors

The *Hall effect*, as it has come to be known, was discovered by E.H. Hall in 1879. Hall noted a very small voltage was generated in the transverse direction across a conductor carrying a current in the presence of an external magnetic field (Figure 3-10), in accordance with the following equation (White, 1988):

$$V_h = \frac{R_h IB}{t}$$

where:

V_h = Hall voltage

R_h = material-dependent Hall coefficient
I = current in amps
B = magnetic flux density (perpendicular to I) in Gauss
t = element thickness in centimeters.

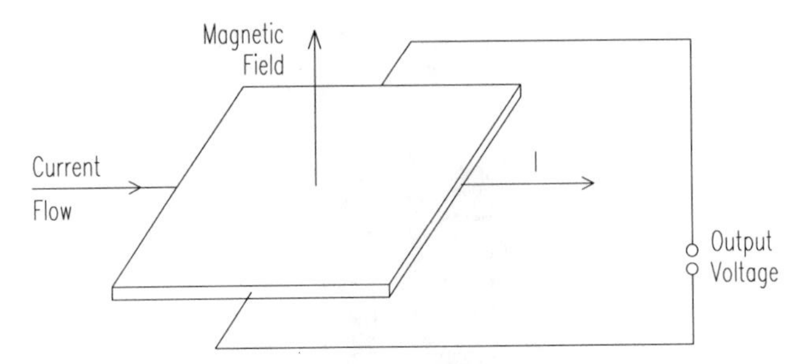

Figure 3-10. In 1879, E.H. Hall discovered a small transverse voltage was generated across a current-carrying conductor in the presence of a static magnetic field, a phenomenon now known as the *Hall effect* (adapted from Lenz, 1990).

It was not until the advent of semiconductor technology (heralded by the invention of the transistor in 1948) that this important observation could be put to any practical use. Even so, early silicon implementations were plagued by a number of shortcomings that slowed popular acceptance, including high cost, temperature instabilities, and otherwise poor reliability (McDermott, 1969). Subsequent advances in integrated circuit technology (i.e., monolithic designs, new materials, and internal temperature compensation) have significantly improved both stability and sensitivity. With a 100-milliamp current flow through *indium arsenide (InAs)*, for example, an output voltage of 60 millivolts can be generated with a flux density (B) of 10 kiloGauss (Hines, 1992). Large-volume applications in the automotive industry (such as distributor timing in electronic ignition systems) helped push the technology into the forefront in the late 1970s (White, 1988). Potential robotic utilization includes position and speed sensing, motor commutation (Manolis, 1993), guidepath following (Chapter 11), and magnetic compasses (Chapter 12).

The linear relationship of output voltage to transverse magnetic field intensity is an important feature contributing to the popularity of the modern *Hall-effect sensor*. To improve stability, *linear Hall-effect sensors* are generally packaged with an integral voltage regulator and output amplifier as depicted in the block diagram of Figure 3-11. The output voltage V_o fluctuates above and below a zero-field equilibrium position (usually half the power supply voltage V_{cc}), with the magnitude and direction of the offset determined by the field strength and polarity, respectively (White, 1988). (Note also that any deviation in *field direction* away from the perpendicular will also affect the magnitude of the voltage swing.) Frequency responses over 100 kiloHertz are easily achieved (Wood, 1986).

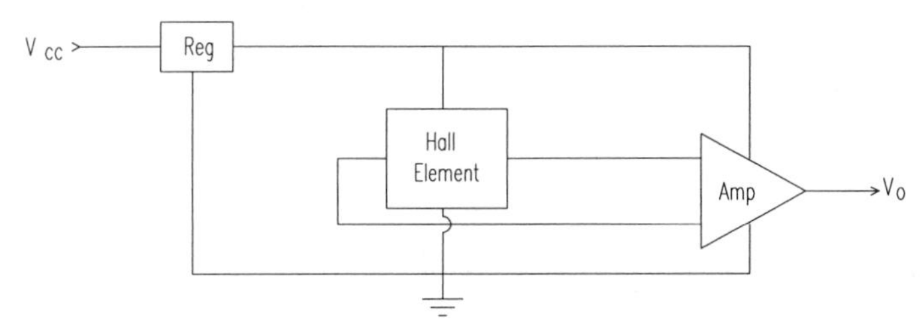

Figure 3-11. The *linear Hall-effect sensor* incorporates an integral voltage regulator and a stable DC output amplifier in conjunction with the *Hall-effect element* shown previously in Figure 3-10 above (adapted from White, 1988).

The addition of a *Schmitt-trigger* threshold detector and an appropriate output driver transforms the linear Hall-effect sensor into a digital *Hall-effect switch.* Most commercially available devices employ transistor drivers that provide an open-circuit output in the absence of a magnetic field (Wood, 1986). The detector trip point is set to some nominal value above the zero-field equilibrium voltage, and when this threshold is exceeded the output driver toggles to the *on* state (*source* or *sink*, depending on whether PNP or NPN transistor drivers are employed). A major significance of this design approach is the resulting insensitivity of the *Hall-effect switch* to reverse magnetic polarity. While the mere approach of the *south pole* of a permanent magnet will activate the device, even direct contact by the *north pole* will have no effect on switching action, as the amplified output voltage actually falls further away from the *Schmitt-trigger* setpoint. Switching response times are very rapid, typically in the 400-nanosecond range (Wood, 1986).

Magnetoresistive Sensors

For *anisotropic* materials, the value of a given property depends on the direction of measurement, in contrast to *isotropic* materials, which exhibit the same values for measured properties in all directions. *Anisotropy* may be related to the shape of a material, its crystalline structure, or internal strains (Graf, 1974). For example, the direction of magnetization in a ferromagnetic crystal will be oriented along a certain crystallographic axis known as the *easy axis*, referring to the "easy" or preferred direction of magnetization (Barrett, et al., 1973).

Changing this direction of magnetization (relative to the direction of current flow) in a conductive material through application of some external magnetic field H_y will result in a change in *resistivity* ρ of the material, a phenomenon known as the *magnetoresistive effect.* By way of illustration, rotating the magnetization state of thin-film anisotropic *permalloy* through 90 degrees causes a maximum change in resistivity of 2 to 3 percent (Dibburn & Petersen, 1986). At low temperatures, certain materials (such as bismuth) may be influenced by a factor as large as 10^6 (Fraden, 1994). The relationship of resistivity to the angle θ between the direction of magnetization and direction of current flow is given by (Dibburn & Petersen, 1986):

$$\rho = \rho_o + \Delta\rho_{max} \cos^2 \theta$$

where:

ρ = resistivity (resistance per unit volume)
ρ_o = isotropic resistivity
$\Delta\rho_{max}$ = maximum possible change in resistivity (resulting from 90-degree rotation)
θ = angle between magnetization and direction of current flow.

In the presence of a transverse field H_y (Figure 3-12A), a permalloy strip with an original direction of magnetization M_o will exhibit the behavior shown in Figure 3-12B. As the applied field H_y is increased, the change in resistivity increases as shown until a point of saturation is reached when the angle of rotation θ becomes equal to 90 degrees, after which no further increase is possible (Petersen, 1989). The symmetry of the plot (Figure 3-12B) with respect to the vertical axis implies the resistivity value is uninfluenced by the two possible directions of original magnetization (i.e., $+M_o$, $-M_o$) or the sign of the transverse field (i.e., $+H_y$, $-H_y$).

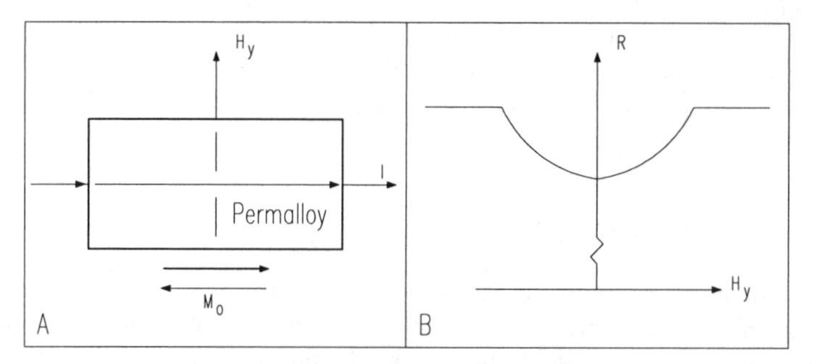

Figure 3-12. The *permalloy* strip with original direction of magnetization *Mo* as shown in (A) will exhibit a response (B) that is independent of the direction of the external transverse field H_y (adapted from Petersen, 1989).

If the demagnetizing and anisotropic fields tending to align the magnetization in the direction of current flow are represented by H_o, then:

$$\sin\theta = \frac{H_y}{H_o}$$

and so for $H_y < H_o$:

$$\rho = \rho_o + \Delta\rho_{\max}\left[1 - \frac{H_y^2}{H_o^2}\right]$$

while $\rho = \rho_o$ for saturation conditions where $H_y > H_o$ (Dibburn & Petersen, 1986).

The most immediate problem with this relationship (aside from a non-unique solution) is its nonlinearity. Kwiatkowski and Tumanski (1986) review a variety of ways for biasing the magnetoresistive device to achieve linear operation over a finite range (H_y much smaller than H_o). The most common method of biasing is the "barber-pole" configuration, where gold (Dibburn & Petersen, 1986) or aluminum (Petersen, 1989) stripes are affixed to the top of each permalloy strip at a 45-degree angle. The much higher conductivity of gold (or aluminum) relative to the permalloy results in a rotation of the current-flow direction by 45 degrees as illustrated in Figure 3-13A. The net effect on the transfer function is illustrated in Figure 3-13B, which shows ρ increases linearly with H_y, for small values of H_y relative to H_o (Dibburn & Petersen, 1986). The complementary barber-pole configuration, wherein the stripes are oriented -45 degrees to the strip axis, results in a linear decrease in ρ with an increasing H_y. In either case, measuring the change in resistivity Δρ provides a reliable and accurate means for detecting very small variations in the applied magnetic field along a specific axis.

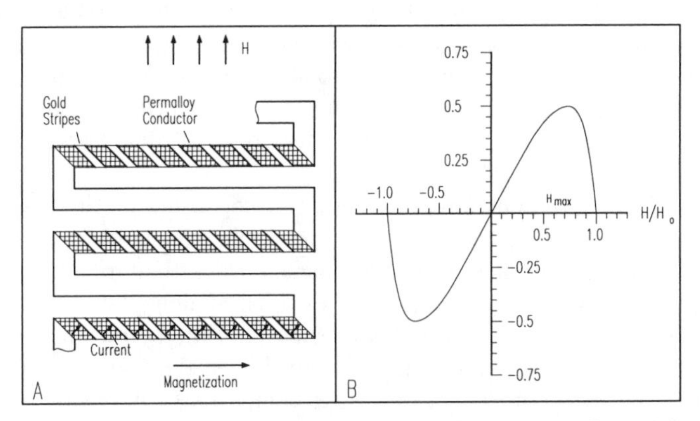

Figure 3-13. A barber-pole biasing scheme rotates the direction of current flow 45 degrees (A) to achieve a linear relationship (B) between resistivity ρ and sensed magnetic field H_y (Dibburn & Petersen, 1986).

A typical *anisotropic magnetoresistive (AMR)* sensor is constructed of four sets of *permalloy* strips, each laid down on a silicon substrate in a raster pattern and connected as shown in Figure 3-14 to form a Wheatstone bridge. Diagonally opposed elements in the bridge are biased such that for one pair ρ increases with *Hy*, while for the other pair ρ decreases. This complementary bridge configuration serves to largely eliminate effects of common-mode isotropic variations such as temperature, while effectively doubling device sensitivity (Dibburn & Petersen, 1986).

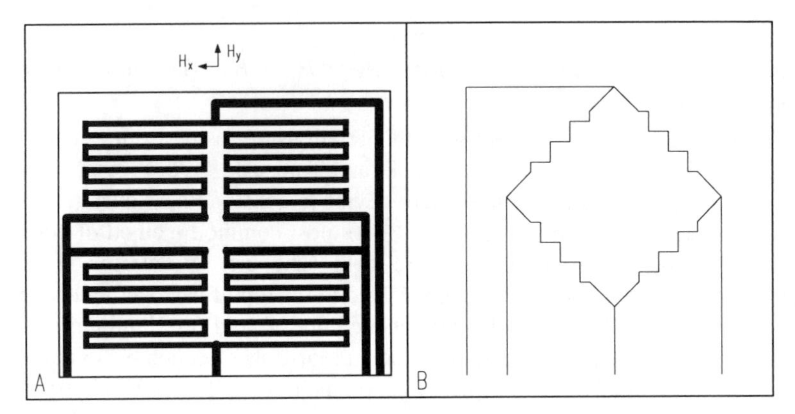

Figure 3-14. Four raster patterns of permalloy strips are connected to form a Wheatstone bridge in a typical *anisotropic magnetoresistive (AMR)* sensor (adapted from Fraden, 1994).

A second problem associated with magnetoresistive sensors is due to the bistable nature of the internal magnetization direction. A preferred magnetization is imparted along the axis of the *permalloy* strips through application of a strong magnetic field during the manufacturing process. The combination of this anisotropic structure and the geometric configuration (strip aspect ratio) means the magnetization direction will always tend to align with the longitudinal axis of the strips, even in the absence of any external magnetic field (Philips, undated). This axial alignment, however, could exist in two possible directions, 180 degrees apart.

Exposure to a strong external magnetic field opposing the internal magnetic field can cause the magnetization to "flip," reversing the internal magnetization of the strips and radically altering sensor characteristics (Figure 3-15). For most applications an auxiliary field H_x must be established along the length of the strip to ensure stability in the preferred internal magnetization direction so the sensor doesn't "flip" (Petersen, 1989). This "flipping" anomaly, however, can be put to good use in the design of a magnetoresistive compass, as will be discussed later in Chapter 12.

One way to provide this auxiliary magnetic field is through use of small permanent magnets or bias coils. The amount of bias is optimized to provide the desired sensitivity and linearity (see again Figure 3-13) but maintained sufficiently below the saturation point on the curve so as to preclude clipping (Lao, 1994). Figure 3-16 shows an example AMR device developed by Space Electronics, Inc., San Diego, CA, that incorporates an integral solenoidal bias coil in a 14-pin ceramic DIP package (SEI, 1994a; 1994b). The magnetoresistive element in the *MicroMag MMS101* is situated in a gap between two *permalloy* thin-film flux concentrators that magnify the sense-axis field component by a factor of 20, while simultaneously shielding and reducing orthogonal components by an order of magnitude (SEI, 1994b; Lao, 1994). Selected specifications for the device are listed in Table 3-1.

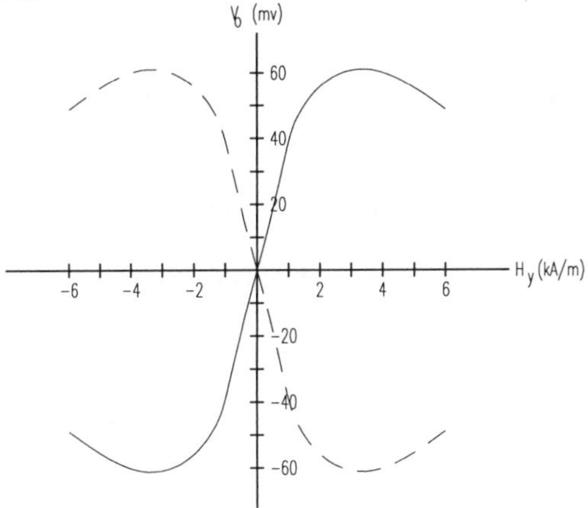

Figure 3-15. The transfer function for a "normal" magnetoresistive sensor (solid line) with magnetization oriented in the +X direction, and for a "flipped" sensor (dashed lines) oriented in the -X direction (adapted from Philips, undated).

In 1988 a French physicist by the name of Dr. Albert Fert at the University of Paris succeeded in efforts to amplify the magnetoresistive effect through fabrication of multiple thin layers of magnetoresistive materials (Baibich, et al., 1992). Such *giant magnetoresistance (GMR)* devices, as they are now called, exhibit a much larger magnetoresistive effect than do conventional *AMR* sensors (Henkel, 1994), resulting in output signals three to 20 times higher (Brown, 1994). More importantly, *GMR* devices are linear over most of their operating range, do not exhibit the characteristic "flipping" behavior of *AMR* sensors, and thus do not require a fixed-field biasing arrangement (Brown, 1994).

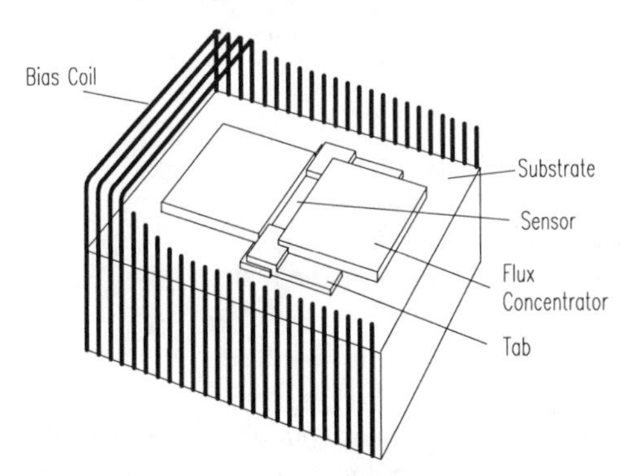

Figure 3-16. The Space Electronics, Inc. *MicroMag MMS101* monolithic AMR sensor employs integrated flux concentrators and bias coil in a 14-pin ceramic dual-inline package (SEI, 1994b).

Table 3-1. Selected specifications for *MicroMag MMS101* AMR sensor.

Parameter	Value	Units
Measurement range	10^{-5} - 0.65	Gauss
Maximum sensitivity	50	milliohms/ohm-Gauss
Bandwidth	DC - 50	MHz
Power	5	volts DC
	<100	milliwatts

The raw signal size is two orders of magnitude greater than that associated with *Hall-effect sensors* employed in similar applications (i.e., proximity, position, speed, orientation sensing, compassing), and requires less applied field for full output (NVE, undated). Brown (1994) cites three significant advantages of *GMR sensors* relative to comparably priced *Hall-effect devices* in position sensing applications:

- Increased standoff (gap) between sensor and magnet or geartooth.
- Improved high-temperature tolerance (> 200 degrees C).
- Superior temperature stability.

GMR sensors are fabricated as multiple thin-film layers of magnetic material sandwiched between alternating layers of non-magnetic conductive interlayers. The principle of operation is based on the fact that conduction electrons can have two opposite spin states, and their spin polarization (spin-state preference) in a magnetic material is determined by the direction in which the material is magnetized (Brown, 1994). The electrical *conductivity* of a material is directly proportional to the mean free path of its conduction electrons, in accordance with the following equation:

$$\sigma = \frac{nq^2 l}{mv}$$

where:

σ = material conductivity
n = number of conduction electrons
q = electron charge
l = mean free path of an electron in the material
m = mass of an electron
v = average electron velocity in the material.

GMR sensors basically change their *conductivity* by altering the mean free path of conducting electrons in the sandwich as a function of the applied magnetic field. To achieve this effect, the alternating magnetic layers in a GMR sandwich are magnetized (during fabrication) in antiparallel alignment, and consequently their conduction electrons are spin-polarized in opposite directions. (The intermediate non-magnetic interlayers serve to separate and decouple the two magnetic films.) Conduction electrons attempting to cross the boundary between layers with opposite spin polarizations have a high probability of being scattered at the interface, resulting in a relatively short mean free path and hence low conductivity (Brown, 1994). The presence of an external magnetic field tends to rotate the antiparallel magnetization axes of the alternating layers in opposite directions towards a common orthogonal axis. Conduction electrons attempting to traverse the multi-layer junctions under these conditions subsequently encounter fewer instances of scattering, with a corresponding increase in their mean free paths.

A representative example of a commercially available *GMR* device is seen in the *NVS5B50 GMR Bridge Sensor* offered by Nonvolatile Electronics (NVE) of Eden Prairie, MN. The NVE sensor consists of four 4.7K GMR "resistors" arranged in a Wheatstone bridge configuration as shown in Figure 3-17A (Henkel, 1994). Two of these resistors are shielded from the effects of external fields by a thick magnetic material, while the other two are situated in the gap between two flux concentrators as shown in Figure 3-17B. The full-scale resistance change of the two active "sense" resistors yields a bridge output of five percent of supply voltage, compared to less than one percent for similar AMR designs (NVE, undated). The *NVS5B50* provides a linear output over the range of 0 to ±35 Gauss and is available in an 8-pin surface-mount package.

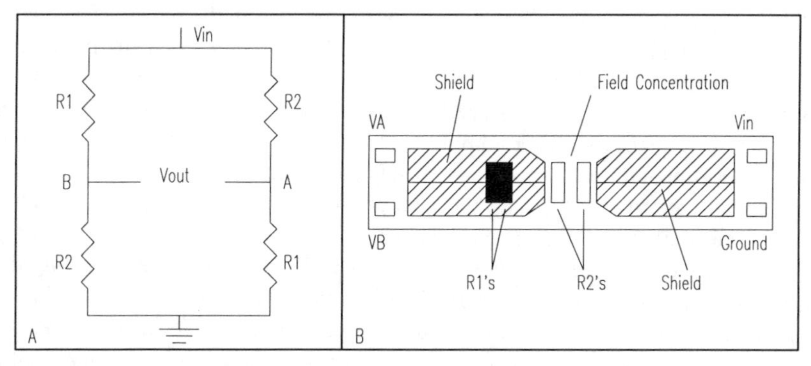

Figure 3-17. Equivalent circuit (A) and schematic drawing (B) for the Nonvolatile Electronics, Inc., *NVS5B50* GMR bridge sensor incorporating integral flux concentrators in an 8-pin surface-mount IC (adapted from Daughton, et al., 1994).

One of the most common robotic applications of AMR and GMR sensors is seen in the dead-reckoning wheel encoder application illustrated in Figure 3-18. Other uses include electronic compassing (Petersen, 1989), angle or position measurement, current sensing, and general magnetic field measurement (Henkel, 1994).

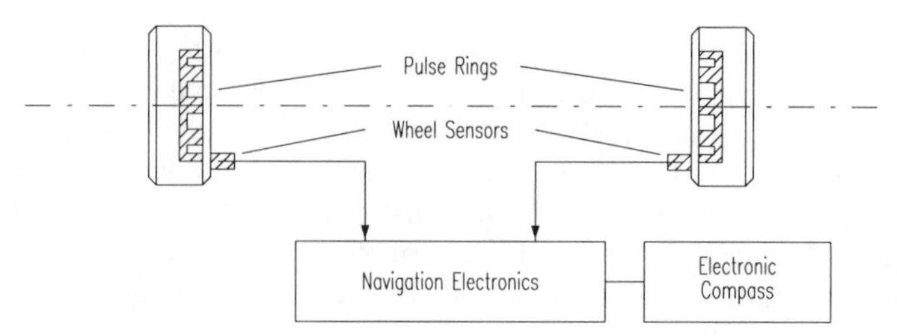

Figure 3-18. A proposed vehicle navigation system employing AMR sensors for electronic compassing (see Chapter 12) and differential wheel odometry (adapted from Petersen, 1989).

3.2.2 Inductive Proximity Sensors

Inductive proximity switches are today the most commonly employed industrial sensors (Moldoveanu, 1993) for detection of ferrous and non-ferrous metal objects (i.e., steel, brass, aluminum, copper) over short distances. Cylindrical configurations as small as 4 millimeters in diameter have been available for over a decade (Smith, 1985). Because of the inherent ability to sense through non-metallic materials, these sensors can be coated, potted, or otherwise sealed, permitting operation in contaminated work areas, or even submerged in fluids. Frequency responses up to 10 KHz can typically be achieved (Carr, 1987).

Inductive proximity sensors generate an oscillatory RF field (i.e., 100 KHz to 1 MHz) around a coil of wire typically wound around a ferrite core. When a metallic object enters the defined field projecting from the sensor face, eddy currents are induced in the target surface. These eddy currents produce a secondary magnetic field that interacts with field of the probe, thereby loading the probe oscillator. The effective impedance of the probe coil changes, resulting in an oscillator frequency shift (or amplitude change) that is converted into an output signal proportional to the sensed gap between probe and target.

A block diagram of a typical inductive proximity sensor is depicted in Figure 3-19A. The oscillator is comprised of an active device (i.e., a transistor or IC) and the sensor probe coil itself. An equivalent circuit (Figure 3-19B) representing this configuration is presented by Carr (1987), wherein the probe coil is modeled as an inductor L_p with a series resistor R_p, and the connecting cable between the coil and the active element shown as a capacitance C. In the case of a typical Collpitts oscillator, the probe-cable combination is part of a resonant frequency tank circuit.

As a conductive target enters the field, the effects of the resistive component R_p dominate, and resistive losses of the tank circuit increase, loading (i.e., damping) the oscillator (Carr, 1987). As the gap becomes smaller, the amplitude of the oscillator output continues to decrease, until a point is reached where oscillation can no longer be sustained. This effect gives rise to the special nomenclature of an *eddy-current-killed oscillator (ECKO)* for this type of configuration. Sensing gaps smaller than this minimum threshold (typically from 0.005 to 0.020 inch) are not quantified in terms of an oscillator amplitude that correlates with range, and thus constitute a dead-band region for which no analog output is available.

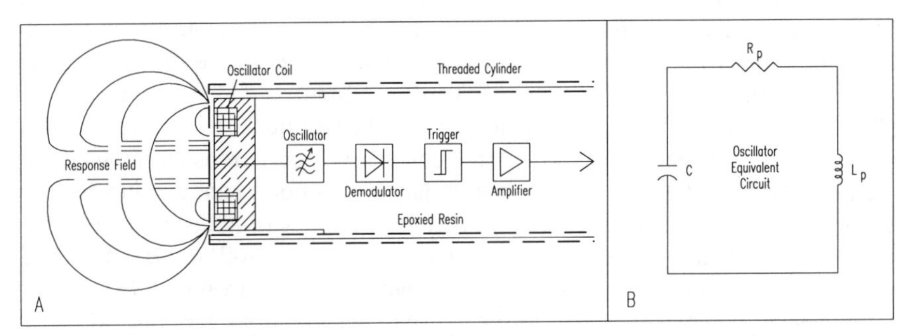

Figure 3-19. (A) Block diagram of a typical *ECKO-type* inductive proximity sensor (adapted from Smith, 1985), and (B) equivalent oscillator circuit (adapted from Carr, 1987).

Monitoring the oscillator output amplitude with an internal threshold detector (Figure 3-19A) creates an *inductive proximity switch* with a digital *on/off* output (Figure 3-20). As the metal target approaches the sensor face, the oscillator output voltage falls off as shown, eventually dropping below a preset *trigger level*, whereupon the threshold comparator toggles from an *off* state to an *on* state. Increasing the gap distance causes the voltage to again rise, and the output switches *off* as the *release level* is exceeded. The intentional small difference between the *trigger level* and the *release level*, termed *hysteresis*, prevents output instabilities near the detection threshold. Typical hysteresis values (in terms of gap distance) range from three to 20 percent of the maximum effective range (Damuck & Perrotti, 1993).

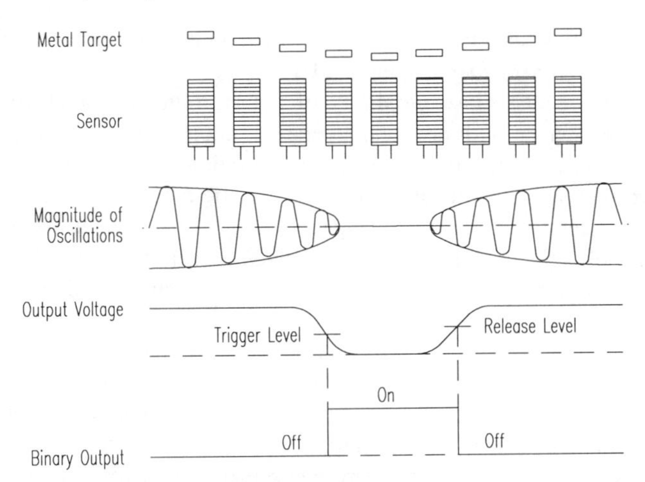

Figure 3-20. A small difference between the trigger and release levels (*hysteresis*) eliminates output instability as the target moves in and out of range (adapted from Moldoveanu, 1993).

Effective sensing range is approximately equal to the diameter of the sensing coil (Koenigsburg, 1982) and is influenced by target material, size, and shape. The industry standard target (for which the nominal sensing distance is specified) is a 1-millimeter-thick square of mild steel the same size as the diameter of the sensor, or three times the nominal sensing distance, whichever is greater (Flueckiger, 1992). For ferrous metals, increased target thickness has a negligible effect (Damuck & Perrotti, 1993). More conductive non-ferrous target materials such as copper and aluminum result in reduced detection range as illustrated in Table 3-2 below. For such non-ferrous metals, greater sensing distances (roughly equivalent to that of steel) can be achieved with thin-foil targets having a thickness less than their internal field attenuation distance (Smith, 1985). This phenomenon is known as the *foil effect* and results from the full RF field penetration setting up additional surface eddy currents on the reverse side of the target (Damuck & Perrotti, 1993).

Table 3-2. Nominal sensing ranges for material other than mild steel must be adjusted using the above attenuation factors (Smith, 1985).

Material	Attenuation Factor
Cast Iron	1.10
Mild Steel	1.00
Stainless Steel	0.70 - 0.90
Brass	0.45
Aluminum	0.40
Copper	0.35

There are two basic types of inductive proximity sensors: 1) *shielded* (Figure 3-21A), and 2) *unshielded* (Figure 3-21B). If an *unshielded* device is mounted in a metal surface, the close proximity of the surrounding metal will effectively saturate the sensor and preclude operation altogether (Swanson, 1985). To overcome this problem, the *shielded* configuration incorporates a coaxial metal ring surrounding the core, thus focusing the field to the front and effectively precluding lateral detection (Flueckiger, 1992). There is an associated penalty in maximum effective range, however, as *shielded* sensors can only detect out to about half the distance of an *unshielded* device of equivalent diameter (Swanson, 1985).

Mutual interference between inductive proximity sensors operating at the same frequency can result if the units are installed with a lateral spacing of less than twice the sensor diameter. This interference typically manifests itself in the form of an unstable pulsing of the output signal, or reduced effective range, and is most likely to occur in the situation where one sensor is undamped and the other is in the hysteresis range (Smith, 1985). Half the recommended $2d$ lateral spacing is generally sufficient for elimination of mutual interaction in the case of shielded sensors (Gatzios & Ben-Ari, 1986). When mounting in an opposed facing configuration, these minimal separation distances should be doubled.

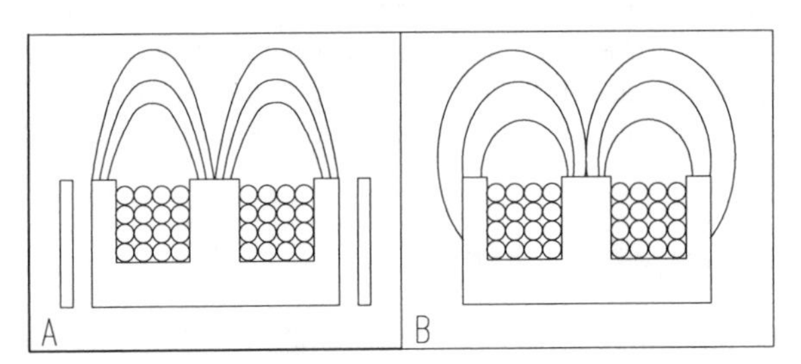

Figure 3-21. *Shielded* inductive sensors (A) can be embedded in metal without affecting performance, while the *unshielded* variety (B) must be mounted on non-metallic surfaces only (Flueckiger, 1992).

While most standard inductive proximity sensors operate on the *ECKO* principle and detect any and all metallic materials, some applications may call for differentiation between various types of metals. The Micro Switch Division of Honeywell Corporation offers an RF inductive *ECKO* sensor series that detects only ferrous (primary component iron) metals, with a 10 to 15 percent decrease in nominal ranges compared to an all-metals sensor (Dixon, 1990). Because of their selective targeting and extremely limited sensing ranges, inductive sensors in general have minimal use in mobile robotic systems for purposes of external object detection, except in application-specific instances.

Figure 3-22. This robotic shot-blasting device employs inductive proximity sensors to keep the closed-cycle end-effector in sealed contact with the ship's hull (courtesy Barnes and Reineke).

One such example involves a large industrial manipulator developed by Barnes and Reineke, Chicago, IL, that cleans the exterior hulls of ships in drydock with steel-shot abrasive (Figure 3-22). Three analog inductive sensors are used to sense the presence of the steel hull surface over a range of 0 to 1.75 inches, controlling a servomechanism that keeps the manipulator under preloaded contact as it traverses the hull removing rust and marine growth (Henkel, 1985).

3.2.3 Capacitive Proximity Sensors

The *capacitive proximity sensor* is very similar to the previously discussed *inductive proximity sensor*, except the capacitive type can reliably detect dielectric materials in addition to metals. Effective for short-range detection out to a few inches, such sensors react to the variation in electrical capacitance between a probe (or plate) and its surrounding environment. As an object draws near, the changing geometry and/or dielectric characteristics within the sensing region cause the capacitance to increase. This change in capacitance can be sensed in a number of different ways: 1) an increase in current flow through the probe (Hall, 1984), 2) initiation of oscillation in an RC circuit (McMahon, 1987), or 3) a decrease in the frequency of an ongoing oscillation (Vranish, et al., 1991). Typical industrial applications include level sensing for various materials (i.e., liquids, pellets, and powders) and product detection, particularly through non-metallic packaging.

An interesting application specifically intended for robotic collision avoidance is seen in the *Capaciflector* developed by the Robotics Branch at NASA Goddard Space Flight Center, Greenbelt, MD. The NASA objective was to produce a conformal proximity-sensing skin for use on robotic manipulator arms in both industrial and space applications, capable of sensing a human presence up to 12 inches away. Normally this type of range requirement would necessitate mounting the capacitive sensor plate with a stand-off displacement of about an inch from the grounded robot arm as illustrated in Figure 3-23A, creating unacceptable bulk and mechanical interference (Vranish, et al., 1991). The NASA design, based on an instrumentation technique for controlling stray capacitance (Webster, 1988), eliminates this offset requirement by introducing an intermediate *reflector* surface between the arm structure and the sensor plate as shown in Figure 3-23B.

In the conventional case (no reflector) illustrated in Figure 3-23A, the smaller the stand-off distance, the greater the capacitive coupling between the sensor plate and the robotic arm, with a corresponding decrease in the strength of the field projected away from the sensor in the direction of the object. The addition of an intermediate active reflector (driven in phase with the sensor plate) causes the sensor field lines to be reflected away from the robot structure, thereby significantly increasing the range of possible interaction with surrounding objects. The equivalent effect (in terms of increased detection range) of a large stand-off is achieved, but without adding unnecessary bulk to the robot's mechanical structure, since the effective offset is approximately equal to the reflective shield thickness of 0.06 inches (Vranish, et al., 1991). A single-element feasibility prototype attached to a *PUMA* industrial manipulator was demonstrated to routinely detect a human or aluminum structural element at distances out to 12 inches, and even smaller objects such as a graphite pencil lead at ranges of around 5 inches (Vranish, et al., 1991).

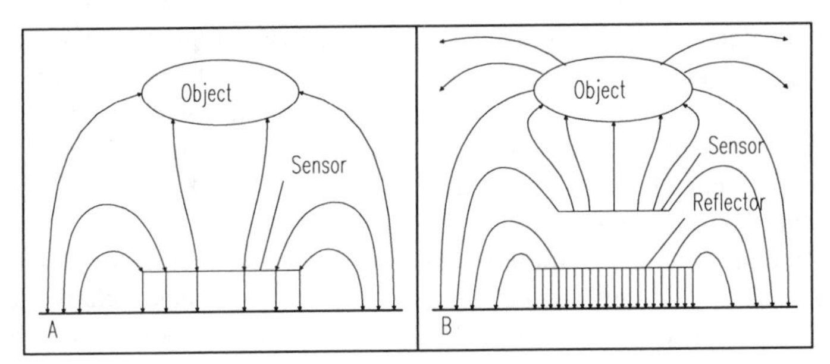

Figure 3-23. Electric field lines without a reflector are shown in (A), while the use of a reflector in (B) allows for greater detection range (adapted from Vranish, et al., 1991).

3.2.4 Ultrasonic Proximity Sensors

All of the preceding proximity sensors relied on target presence to directly change some electrical characteristic or property (i.e., inductance, capacitance) associated with the sense circuitry itself. The ultrasonic proximity sensor is an example of a *reflective* sensor that responds to changes in the amount of emitted energy returned to a detector after interaction with the target of interest. Typical systems consist of two transducers (one to transmit and one to receive the returned energy), although the relatively slow speed of sound makes it possible to operate in the transceiver mode with a common transducer. The transmitter emits a longitudinal wave in the ultrasonic region of the acoustical spectrum (typically 20-200 KHz), above the normal limits of human hearing. The receiver response is a function of the amplitude of the returned energy, as opposed to elapsed time before detection of an echo.

Ultrasonic proximity sensors are useful over distances out to several feet for detecting most objects, liquid and solid. If an object enters the acoustical field, energy is reflected back to the receiver. As is the case with any reflective sensor, maximum detection range is dependent not only on emitted power levels, but also on target cross-sectional area, reflectivity, and directivity. Once the received signal amplitude reaches a preset threshold, the sensor output changes state, indicating detection. Due in part to the advent of low-cost microcontrollers, such devices have for most situations been replaced by more versatile ultrasonic ranging systems (Chapter 5) that provide a quantitative indicator of distance to the detected object.

3.2.5 Microwave Proximity Sensors

Microwave proximity sensors operate at distances of 5 to 150 feet or more (Williams, 1989) and are very similar to the ultrasonic units discussed above, except that electromagnetic energy in the microwave region of the RF energy spectrum is emitted. The FCC has allocated 10.50 to 10.55 GHz and 24.075 to 24.175 GHz for microwave field-disturbance sensors of this type (Schultz, 1993). When the presence of a suitable target reflects sufficient energy from the transmitting antenna back to a separate receiving antenna (Figure 3-24), the output changes state to indicate an object is present within the field of view.

An alternative configuration employing a single transmit/receive antenna monitors the Doppler shift induced by a moving target to detect relative motion as opposed to presence. Such a setup is classified for our purposes as a motion sensor and treated in Chapter 17.

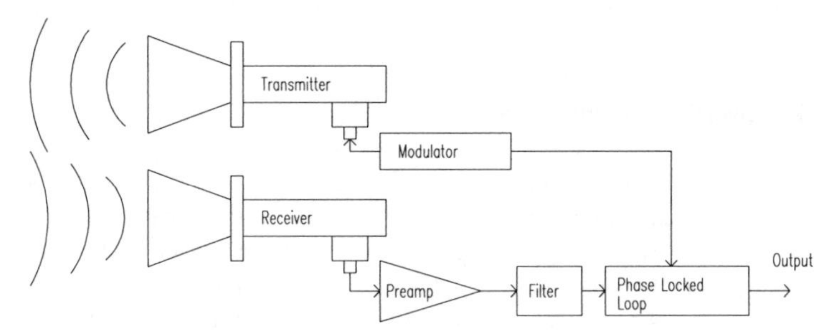

Figure 3-24. The microwave presence sensor, unlike the motion detector, requires a separate transmitter and receiver (adapted from Williams, 1989).

AM Sensors Microwave Proximity Sensors

AM Sensors, Inc., Salem, MA, offers a variety of proximity, direction of motion, displacement, level, and velocity sensors which cover numerous industrial applications. Their products include the *MSM10500* series of FMCW microwave sensors that provide non-contact position detection of metallic and non-metallic moving objects. The *MSM10500* sensor provides continuous distance information, range-gated position indication, and direction of motion. The *MSM10502* is preset to sense objects moving either toward or away from the sensor, and indicates distance as it passes through three range gates that can be adjusted to any fraction of the 50-foot maximum detection range. The microwave portion of the unit uses a *Gunn-diode* transmitter, two microwave *mixer-diode* receivers, and a *varactor diode* to vary the transmitted frequency. The output of the oscillator is focused by a horn antenna into a beam, and any object moving through this beam is detected.

The signal conditioning circuitry contains the power supply, amplifiers, comparator, and a microcontroller to drive the oscillator and convert the detected outputs into useful control signals. The amount of averaging applied to each reading is adjustable so the user may choose between maximum noise immunity and minimum output response time. The regulated power supply allows the module to operate with a wide range of input voltages, such as in automotive systems, and provide high electrical noise rejection. When the target is inside a given range, the corresponding output will turn on and remain on as long as the target is within this range, specified in normal environments to be accurate within 6 inches. This accuracy can be degraded if there are multiple targets moving in the range or if the target has low reflectivity. The point where a range gate will turn on for a given target is typically repeatable within 1 inch.

Table 3-3. Performance specifications of AM Sensors MSM10500 and MSM10502 microwave proximity sensors.

Parameter	MSM10500	MSM10502	Units
Range	50	50	feet
Resolution	6	6	inches
Size	6.5 by 6.5 by 4.25	4.25 by 4.25 by 3.5	inches
Weight	1	1	pound
Power	10 to 16	10 to 28	volts
	150	50	milliamps
Range gates	Adjustable	Preset at 3, 5, 10	feet
Frequency	10.525	10.525	GHz

3.2.6 Optical Proximity Sensors

Optical (photoelectric) sensors commonly employed in industrial applications can be broken down into three basic groups: (1) *opposed*, (2) *retroreflective*, and (3) *diffuse*. (The first two of these categories are not really "proximity" sensors in the strictest sense of the terminology.) Effective ranges vary from a few inches out to several feet. Common robotic applications include floor sensing, navigational referencing, and collision avoidance. Modulated near-infrared energy is typically employed to reduce the effects of ambient lighting, thus achieving the required signal-to-noise ratio for reliable operation. Visible-red wavelengths are sometimes used to assist in installation alignment and system diagnostics.

Actual performance depends on several factors. Effective range is a function of the physical characteristics (i.e., size, shape, reflectivity, and material) of the object to be detected, its speed and direction of motion, the design of the sensor, and the quality and quantity of energy it radiates or receives. Repeatability in detection is based on the size of the target object, changes in ambient conditions, variations in reflectivity or other material characteristics of the target, and the stability of the electronic circuitry itself. Unique operational characteristics of each particular type can often be exploited to optimize performance in accordance with the needs of the application.

Opposed Mode

Commonly called the "electric eye" at the time, the first of these categories was introduced into a variety of applications back in the early 1950s, to include parts counters, automatic door openers, annunciators, and security systems. Separate transmitting and receiving elements are physically located on either side of the region of interest; the transmitter emits a beam of light, often supplied in more recent configurations by an LED, that is focused onto a photosensitive receiver (Figure 3-25). Any object passing between the emitter and receiver breaks the beam, disrupting the circuit. Effective ranges of hundreds of feet or more are routinely possible and often employed in security applications.

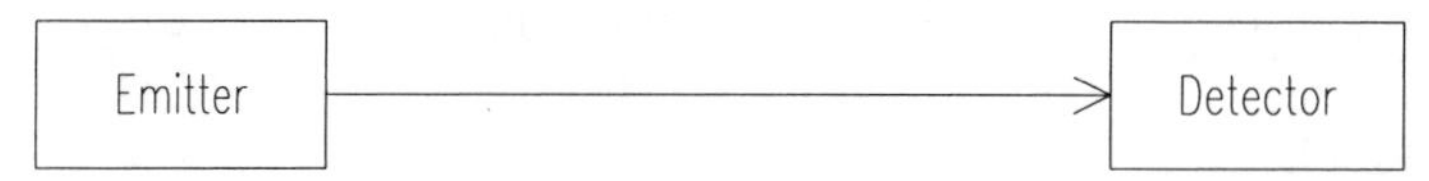

Figure 3-25. The *opposed-mode sensor* configuration relies on target passage between the emitter and detector to interrupt the beam.

Other than a few specialized cases of internal sensing (such as certain types of optical encoders) *opposed-mode sensors* have little applicability to mobile robotic systems due to their geometric configuration (i.e., opposed-pair transmitter and receiver elements).

Retroreflective Mode

Retroreflective sensors evolved from the *opposed* variety through the use of a mirror to reflect the emitted energy back to a detector located directly alongside the transmitter. *Corner-cube retroreflectors* (Figure 3-26) eventually replaced the mirrors to cut down on critical alignment needs. Corner-cube prisms have three mutually perpendicular reflective surfaces and a hypotenuse face; light entering through the hypotenuse face is reflected by each of the surfaces and returned back through the face to its source (Banner, 1993b).

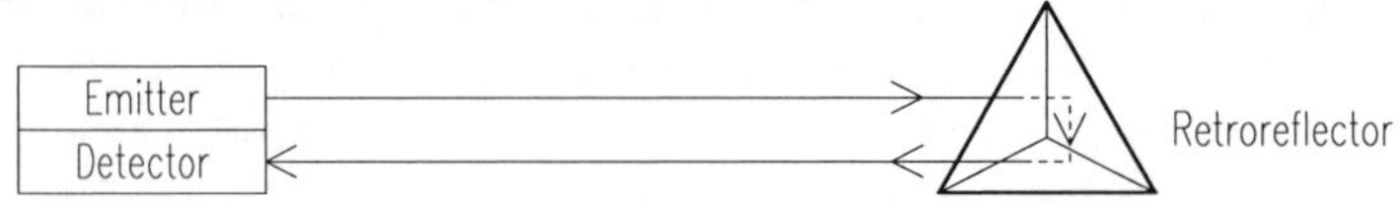

Figure 3-26. *Corner-cube retroreflectors* are employed to increase effective range and simplify alignment (adapted from Banner, 1993b).

In most factory automation scenarios, the object of interest is detected when it breaks the beam, although some applications call for placing the retroreflector on the item itself. A good retroreflective target will return about 3,000 times as much energy to the sensor as would be reflected from a sheet of white typing paper (Banner, 1993b).

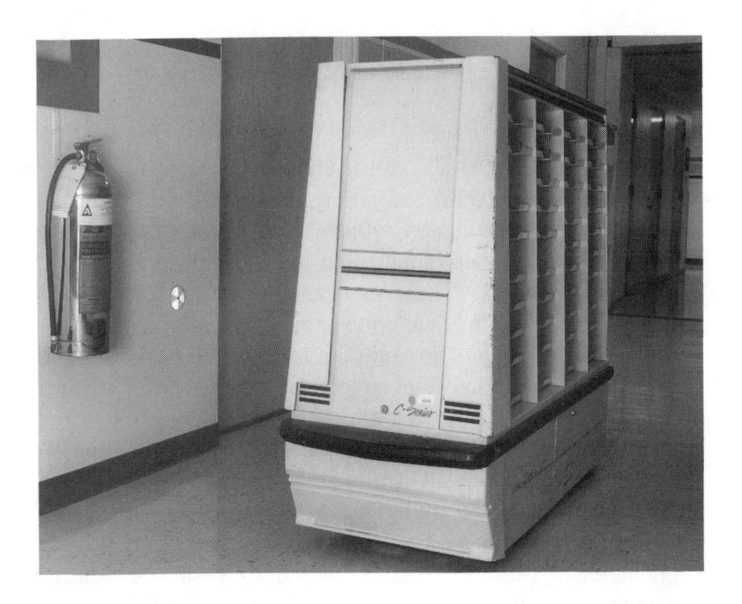

Figure 3-27. Earlier versions of this automated mail delivery cart relied on wall-mounted retroreflectors (left of doorway) to identify desired stops along the route; the current system senses longitudinal markers on the floor adjacent to the guidepath (courtesy Naval Command Control and Ocean Surveillance Center).

Collocation of the emitter and detector simplified installation in industrial assembly-line scenarios and opened up several applications for mobile systems as well. One common example is seen in the employment of fixed-location retroreflective reference markers for automated guided vehicles. Figure 3-27 shows a typical implementation in conjunction with a Bell and Howell mail delivery system in operation at NCCOSC. The circular retroreflective target mounted on the wall adjacent to the office doorway was used to mark a position along the route of travel where the platform is supposed to stop. (The present system actually senses a longitudinal marker on the floor adjacent to the guidepath, as will be discussed in Chapter 11). An onboard annunciator then alerts the secretarial staff to deposit outgoing mail in a collection bin and collect any mail intended for delivery at that particular station.

Diffuse Mode

Optical proximity sensors in the *diffuse* category operate in similar fashion to *retroreflective* types, except that energy is returned from the surface of the object of interest, instead of from a *cooperative reflector* (Figure 3-28). This feature facilitates random object detection in unstructured environments.

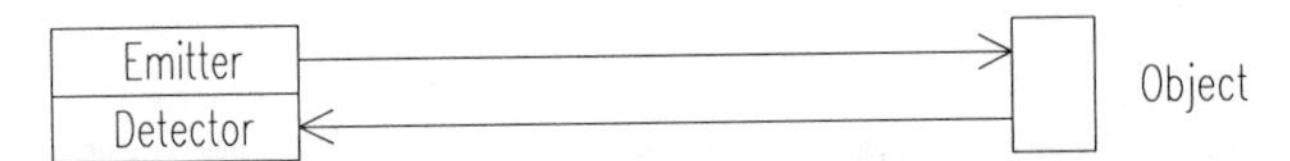

Figure 3-28. *Diffuse-mode proximity sensors* rely on energy reflected directly from the target surface.

Three Banner *Multi-Beam CX1-6* medium-range near-infrared proximity detectors (Banner, 1993a; 1993b) are arranged on ROBART II in a forward-looking horizontal array for collision avoidance purposes. Two additional units (*Mini-Beam* model *SM312D*) are mounted on the left and right sides of the front panel of the lower base unit. These modulated-beam sensors have adjustable maximum ranges, set for this application to about 30 inches for the *CX1-6* and 15 inches for the *SM312D*. The proximity sensors provide extended protection capability in the direction of travel and collectively can discern if an obstruction is directly ahead, to the right, or to the left of centerline.

There are several advantages of this type of sensor over ultrasonic ranging for close-proximity object detection. There is no appreciable time lag since optical energy propagates at the speed of light, whereas up to a full second can be required to update a sequentially fired ultrasonic array of only 12 sensors. In addition, optical energy can be easily focused to eliminate adjacent sensor interaction, thereby allowing multiple sensors to be fired simultaneously. Finally, the shorter wavelengths involved greatly reduce problems due to specular reflection, resulting in more effective detection of off-normal surfaces (see chapters 8 and 9). The disadvantage, of course, is that no direct range measurement is provided, and variations in target reflectivity can sometimes create erratic results. One method for addressing this limitation is discussed in the next section.

Convergent Mode

Diffuse proximity sensors can employ a special geometry in the configuration of the transmitter with respect to the receiver to ensure more precise positioning information. The optical axis of the transmitting LED is angled with respect to that of the detector, so the two intersect only over a narrowly defined region as illustrated in Figure 3-29. It is only at this specified distance from the device that a target can be in position to reflect energy back to the detector. Consequently, most targets beyond this range are not detected. This feature decouples the proximity sensor from dependence on the reflectivity of the target surface, and is useful where targets are not well displaced from background objects.

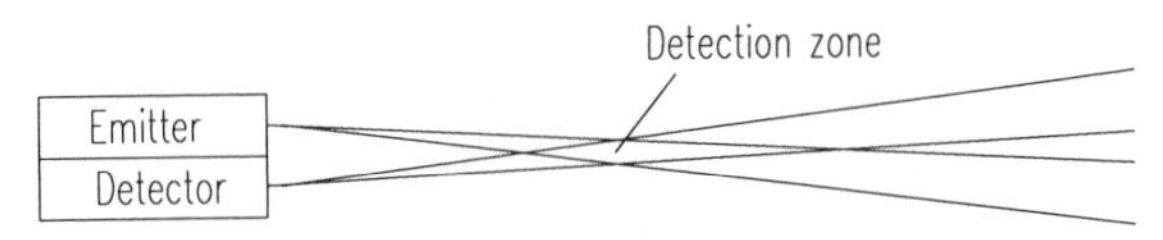

Figure 3-29. *Diffuse proximity sensors* configured in the *convergent mode* can be used to ascertain approximate distance to an object.

Convergent-mode sensors were used on ROBART II to detect discontinuities in the floor, such as a descending stairway, where significant variations in floor surface reflectivities precluded the use of diffuse-mode sensors. A Banner *SM512DB* near-infrared sensor is installed on the front and a model *SE612* on each corner of the base. The sensors are positioned to detect the normal presence of the floor, to preclude attempts to traverse unsuitable discontinuities that could entrap or even damage the unit. Any undesirable conditions detected by these sensors cause the drive motors to be immediately disabled, and the controlling processor is alerted to which corner sensor detected the problem.

In the early 1980s, Honeywell Visitronics, Englewood, CO, developed an interesting non-contact proximity gauge that employed the convergent sensing technique to determine relative distance as well as the presence or absence of an object. The *HVS-300 Three Zone Distance Sensor* (no longer available) would indicate whether a target surface was close to the sensor, at an intermediate distance, far from the sensor, or out of range. Conventional diffuse proximity detectors based on return signal intensity display high repeatability only when target surface reflectivity is maintained constant. The *HVS-300* was capable of higher range accuracy under varying conditions of reflectivity and ambient lighting through use of the triangulation ranging scheme.

The *HVS-300* proximity sensor consisted of a pair of 820-nanometer near-infrared LED sources, a dual-element silicon photodetector, directional optics, and control logic circuitry. The LEDs emitted coded light signals at differing angles through one side of a directional lens as shown in Figure 3-30. If an outgoing beam struck an object, a portion of the reflected energy was returned through the other side of the lens and focused onto the detector assembly.

The detector employed two photodiode elements placed side by side, separated by a narrow gap. Depending on the range to the reflective surface, a returning reflection would either fall on one photodetector (indicating the reflecting surface was close to the sensor), or the other (indicating the surface was far from the sensor), or equally on both (meaning the object was on the boundary between these two regions). With two separate transmissions projected onto the scene at different angles of incidence, two such boundaries were created. The first distinguished between the *near* and *intermediate* regions, while the second distinguished between the *intermediate* and *far* regions. Because both transmissions used the same detector, the sources were uniquely coded for positive identification by the control electronics.

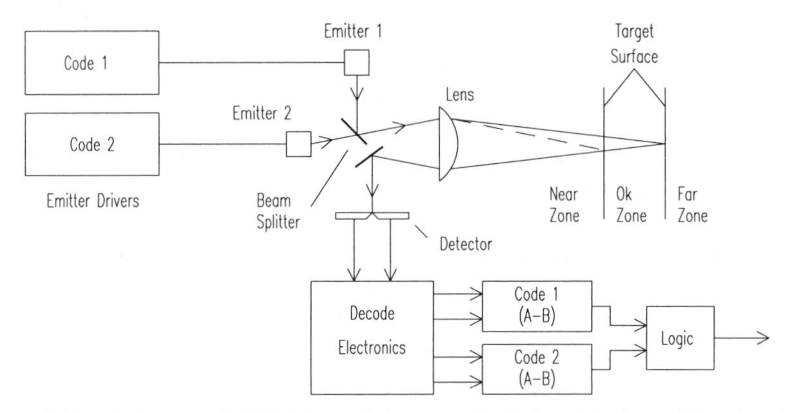

Figure 3-30. The Honeywell *HVS-300* proximity sensor incorporated dual near-infrared emitters to determine if an object was in the adjustable OK zone (courtesy Honeywell Visitronics, Inc.).

Generally insensitive to changes in surface texture or color and unaffected by ambient lighting conditions, the *HVS-300* seemed well suited to the collision avoidance needs of an indoor mobile robot, where the limited speed of advance was in keeping with the sensor's maximum range of 30 inches. In addition, the four discrete range bins would give a relative feel for the distance to a threatening object, allowing for more intelligent evasive maneuvering. Unfortunately, the higher unit cost relative to ultrasonic ranging systems and conventional diffuse-mode proximity detectors was hard to justify, and subsequent advances in position sensitive detectors (PSDs) made competing triangulation ranging systems even more attractive (see Chapter 4).

3.3 References

Baibich, M., Broto, J., Fert, A., Van Dau, F.N., Petroff, F., Eitenne, P., Creuzet, G., Friederich, A., Chazelas, J., "Giant Magnetoresistance of (001) Fe/(001) Cr Magnetic Super Lattices," Physical Review Letters, Vol. 61, No. 21, p. 2472, November, 1992.

Ball, D., "Sensor Selection Guide," *Sensors*, pp. 50-53, April, 1986.

Banner, *Photoelectric Controls*, Product Catalog, Banner Engineering Corp., Minneapolis, MN, 1993a.

Banner, *Handbook of Photoelectric Sensing*, Banner Engineering Corp., Minneapolis, MN, 1993b.

Barrett, C.R., Nix, W.D., Tetelman, A.S., *The Principles of Engineering Materials*, Prentice Hall, Englewood Cliffs, NJ, 1973.

Brooks, R.A., "A Robot that Walks: Emergent Behaviors from a Carefully Evolved Network," *Neural Computation*, Vol. 1, pp. 253-262, 1989.

Brown, J., "GMR Materials: Theory and Applications," *Sensors*, pp. 42-48, September, 1994.

Burreson, B., "Magnetic Proximity Switches in Severe Environments," *Sensors*, pp. 28-36, June, 1989.

Carr, W.W., "Eddy Current Proximity Sensors," *Sensors*, pp. 23- 25, November, 1987.

Damuck, N., Perrotti, J., "Getting the Most out of Your Inductive Proximity Switch," *Sensors*, pp. 25-27, August, 1993.

Dario, P., DeRossi, D., "Tactile Sensors and the Gripping Challenge," *IEEE Spectrum*, pp. 46-52, August, 1985.

Daughton, J., Brown, J., Chen, E., Beech, R., Pohm, A., Kude, W., "Magnetic Field Sensors Using GMR Multilayer," *IEEE Transactions on Magnetics*, Vol. 30, No. 6, pp. 4608-4610, November, 1994.

Dibburn, U., Petersen, A., "The Magnetoresistive Sensor - a Sensitive Device for Detecting Magnetic Field Variations," *Electronic Components and Applications*, Vol. 5, No. 3, June, 1983.

Dixon, D., "Ferrous Metals Only Inductive Proximity Sensors," *Sensors*, pp. 18-20, April, 1990.

Ferrel, C.L., "An Autonomous Mobile Robot, a Planetary Microrover," *Sensors*, pp. 37-47, February, 1994.

Fielding, P.J., "Evaluation of Robotic Tactile Sensing System," *Sensors*, pp. 35-46, April, 1986.

Fraden, J., *AIP Handbook of Modern Sensors*, ed., Radebaugh, R., American Institute of Physics, New York, 1993.

Flueckiger, N., "Inductive Proximity Sensors: Theory and Applications," *Sensors*, pp. 11-13, May, 1992.

Gat, E., Behar, A., Desai, R., Ivlev, R., Loch, J., Miller, D.P., "Behavior Control for Planetary Exploration," IEEE International Conference on Robotics and Automation, Atlanta, GA, Vol. 2, pp. 567-571, May, 1993.

Gatzios, N.E., Ben-Ari, H., "Proximity Control Primer," *Sensors*, pp. 47-49, April, 1986.

Graf, R.F., *Dictionary of Electronics*, Howard W. Sams, Indianapolis, IN, 1974.

Grahn, J.M., "Robotic Applications for Tactile Sensors," Sensors Expo, Chicago, IL, pp. 63-73, September, 1992.

Hall, D.J., "Robotic Sensing Devices," Report No. CMU-RI-TR-84-3, Carnegie-Mellon University, Pittsburgh, PA, March, 1984.

Hamlin, "The Versatile Magnetic Proximity Sensor," *Sensors*, pp. 16-22, May, 1988.

Harmon, L.D., "Tactile Sensing for Robots," Recent Advances in Robotics, Vol. 1, John Wiley and Sons, 1983.

Henkel, S. vL., "Hull Cleaning System for Trident-Class Submarines," *Robotics Age*, p. 11, November, 1985.

Henkel, S. vL., "GMR Materials Advance Magnetic Field Detection," *Sensors*, p. 8, June, 1994.

Hines, R., "Hall Effect Sensors in Paddlewheel Flowmeters," *Sensors*, pp. 32-33, January, 1992.

Johnson, R.F., "A Refresher in Position Sensing," *Sensors*, pp. 18-24, September, 1987.

Kaneko, M., "Active Antenna," IEEE International Conference on Robotics and Automation, San Diego, CA, pp. 2665-2671, May, 1994.

Koenigsburg, W.D., "Noncontact Distance Sensor Technology," GTE Laboratories, Inc., 40 Sylvan Rd., Waltham, MA, 02254, pp. 519-531, March, 1982.

Kwiatkowski, W., Tumanski, S., "The Permalloy Magnetoresistive Sensors - Properties and Applications," *Journal of Physics E: Scientific Instruments*, Vol. 19, pp. 502-515, 1986.

Lao, R., "A New Wrinkle in Magnetoresistive Sensors," *Sensors*, pp. 63-65, October, 1994.

Lenz, J.E., "A Review of Magnetic Sensors," *Proceedings of the IEEE*, Vol. 78, No. 6, June, 1990.

Manolis, S., Resolvers vs. Rotary Encoders For Motor Commutation and Position Feedback, *Sensors*, pp. 29-32, March, 1993.

McAlpine, G.A., "Tactile Sensing," *Sensors*, pp. 7-16, April, 1986.

McDermott, J., "The Hall Effect: Success at 90," *Electronic Design 21*, pp. 38-45, 11 October, 1969.

McMahon, V.C., "Solutions from Capacitive Proximity Switches," *Sensors*, pp. 31-33, May, 1987.

Moldoveanu, A., "Inductive Proximity Sensors: Fundamentals and Standards," *Sensors*, pp. 11-14, June, 1993.

MSI, "Proximity Sensing Products," Product Literature, Merritt Systems, Inc., Merritt Island, FL, undated.

Nicholls, H.R., Lee, M.H., "A Survey of Robotic Tactile Sensing Technology," *The International Journal of Robotics Research*, Vol. 8, No. 3, June, 1989.

NVE, "NVS5B50 GMR Bridge Sensor," Preliminary Product Literature, Nonvolatile Electronics, Inc., Eden Prairie, MN, undated.

Peale, S., "Speed/Motion Sensing in Challenging Environments," *Sensors*, pp. 45-46, January, 1992.

Pennywitt, K.E., "Robotic Tactile Sensing," *Byte*, pp. 177-200, January, 1986.

Petersen, A., "Magnetoresistive Sensors for Navigation," Proceedings, 7th International Conference on Automotive Electronics, London, England, pp. 87-92, October, 1989.

Philips, "The Magnetoresistive Sensor: A Sensitive Device for Detecting Magnetic-Field Variations," Technical Publication 268, Philips Components, undated.

PM, "Robots Get Sensitive," *Popular Mechanics*, p. 20, February, 1995.

Rowan, J., "The Decade of the Sensor in Materials Handling," *Sensors*, pp. 11-13, April, 1988.

Russell, R.A., "Closing the Sensor-Computer-Robot Control Loop," *Robotics Age*, pp. 15-20, April, 1984.

Schiebel, E.N., Busby, H.R., Waldren, K.J., "Design of a Mechanical Proximity Sensor," IEEE International Conference on Robotics and Automation, pp. 1941-1946, 1986.

SEI, "*Micromag* Application Note AN-1," Space Electronics, Inc., San Diego, CA, May, 1994a.

SEI, "High-Sensitivity Magnetoresistive Magnetometer," Product Literature, MMS101, Space Electronics, Inc., San Diego, CA, June, 1994b.

Smith, J.W., "Design and Application of Inductive Proximity Sensors," *Sensors*, pp. 9-14, November, 1985.

Swanson, R., "Proximity Switch Application Guide," *Sensors*, pp. 20-28, November, 1985.

Vranish, J.M., McConnel, R.L., Mahalingam, S., "Capaciflector Collision Avoidance Sensors for Robots," Product Description, NASA Goddard Space Flight Center, Greenbelt, MD, February, 1991.

Vranish, J.M., McConnell, R.L., "Driven Shielding Capacitive Proximity Sensor, *NASA Tech Briefs*, p. 16, March, 1993.

Webster, J.G., *Tactile Sensors for Robotics and Medicine*, John Wiley and Sons, New York, NY, 1988.

White, D., "The Hall Effect Sensor: Basic Principles of Operation and Application," *Sensors*, pp. 5-11, May, 1988.

Williams, H., "Proximity Sensing with Microwave Technology," *Sensors*, pp. 6-15, June, 1989.

Wingo, W., "Freckled 'Skin' Gives Keener Senses to Robots," *Design News*, p. 16, January 9, 1995.

Wojcik, S., "Noncontact Presence Sensors for Industrial Environments," *Sensors*, pp. 48-54, February, 1994.

Wood, T., "The Hall Effect Sensor," *Sensors*, pp. 27-36, March, 1986.

4
Triangulation Ranging

One of the first areas for concern in the evolution of a mobile robot design is the need to provide the system with sufficient situational awareness to support intelligent movement. The first step towards this end consists of the acquisition of appropriate information regarding ranges and bearings to nearby objects, and the subsequent interpretation of that data. Proximity sensors represent a first step in this direction, but by themselves fall short of the mark for a number of reasons previously discussed, not the least of which is the inability to quantify range.

Sensors that measure the actual distance to a target of interest with no direct physical contact can be referred to as non-contact ranging sensors. There are at least seven different types of ranging techniques employed in various implementations of such distance measuring devices (Everett, et al., 1992):

- Triangulation.
- Time of flight (pulsed).
- Phase-shift measurement (CW).
- Frequency modulation (CW).
- Interferometry.
- Swept focus.
- Return signal intensity.

Furthermore, there are a number of different variations on the theme for several of these techniques, as for example in the case of triangulation ranging:

- Stereo disparity.
- Single-point active triangulation.
- Structured light.
- Known target size.
- Optical flow.

Non-contact ranging sensors can be broadly classified as either *active* (radiating some form of energy into the field of regard) or *passive* (relying on energy emitted by the various objects in the scene under surveillance). The commonly used terms *radar* (radio direction and ranging), *sonar* (sound navigation and ranging), and *lidar* (light direction and ranging) refer to *active* methodologies that can be based on any of several of the above ranging techniques. For example, *radar* is usually implemented using time-of-flight, phase-shift measurement, or frequency modulation. *Sonar* typically is based on time-of-flight ranging, since the speed of sound is slow enough to be easily measured with fairly inexpensive electronics. *Lidar* generally refers to laser-based schemes using time-of-flight or phase-shift measurement.

For any such active (reflective) sensors, effective detection range is dependent not only on emitted power levels, but also the following target characteristics:

- *Cross-sectional area* — Determines how much of the emitted energy strikes the target.
- *Reflectivity* — Determines how much of the incident energy is reflected versus absorbed or passed through.
- *Directivity* — Determines how the reflected energy is redistributed (i.e., scattered versus focused).

Triangulation ranging is based upon an important premise of plane trigonometry that states given the length of a side and two angles of a triangle, it is possible to determine the length of the other sides and the remaining angle. The basic *Law of Sines* can be rearranged as shown below to represent the length of side B as a function of side A and the angles θ and ϕ:

$$B = A\frac{\sin\theta}{\sin\alpha} = A\frac{\sin\theta}{\sin(\theta+\phi)}.$$

In ranging applications, length B would be the desired distance to the object of interest at point P_3 (Figure 4-1) for known sensor separation baseline A.

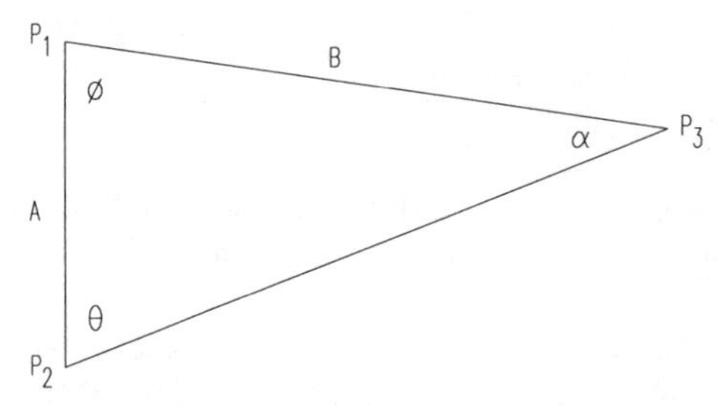

Figure 4-1. Triangulation ranging systems determine range B to target point P_3 by measuring angles ϕ and θ at points P_1 and P_2.

Triangulation ranging systems are classified as either *passive* (use only the ambient light of the scene) or *active* (use an energy source to illuminate the target). Passive stereoscopic ranging systems position directional detectors (video cameras, solid-state imaging arrays, or position sensitive detectors) at positions corresponding to locations P_1 and P_2 (Figure 4-2). Both imaging sensors are arranged to view the same object point, P_3 forming an imaginary triangle. The measurement of angles θ and ϕ in conjunction with the known orientation and lateral separation of the cameras allows the calculation of range to the object of interest.

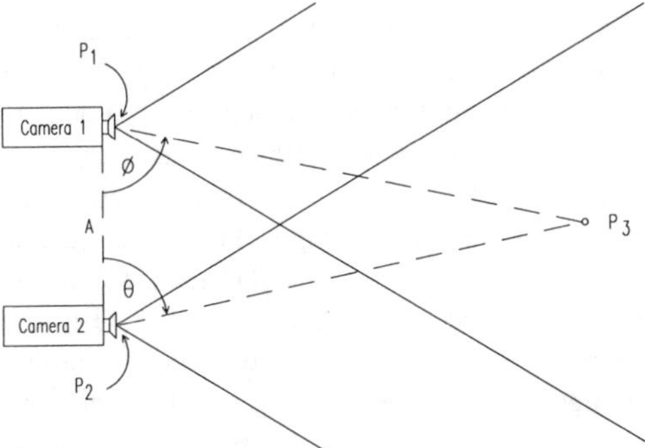

Figure 4-2. Passive stereoscopic ranging system configuration.

Active triangulation systems, on the other hand, position a controlled light source (such as a laser) at either point P_1 or P_2, directed at the observed point P_3. A directional imaging sensor is placed at the remaining triangle vertex and is also aimed at P_3. Illumination from the source will be reflected by the target, with a portion of the returned energy falling on the detector. The lateral position of the spot as seen by the detector provides a quantitative measure of the unknown angle ϕ, permitting range determination by the *Law of Sines*.

The performance characteristics of triangulation systems are to some extent dependent on whether the system is active or passive. Passive triangulation systems using conventional video cameras require special ambient lighting conditions that must be artificially provided if the environment is too dark. Furthermore, these systems suffer from a correspondence problem resulting from the difficulty in matching points viewed by one image sensor with those viewed by the other. On the other hand, active triangulation techniques employing only a single detector do not require special ambient lighting, nor do they suffer from the correspondence problem. Active systems, however, can encounter instances of no recorded strike because of specular reflectance or surface absorption of the light.

Limiting factors common to all triangulation sensors include reduced accuracy with increasing range, angular measurement errors, and a *missing parts* (also known as *shadowing*) problem. *Missing parts* refers to the scenario where particular portions of a scene can be observed by only one viewing location (P_1 or P_2). This situation arises because of the offset distance between P_1 and P_2, causing partial occlusion of the target (i.e., a point of interest is seen in one view but otherwise occluded or not present in the other). The design of triangulation systems must include a tradeoff analysis of the offset: as this baseline measurement increases, the range accuracy increases, but problems due to directional occlusion worsen.

4.1 Stereo Disparity

The first of the triangulation schemes to be discussed, *stereo disparity*, (also called *stereo vision*, *binocular vision*, and *stereopsis*) is a passive ranging technique modeled after the biological counterpart. When a three-dimensional object is viewed from two locations on a plane normal to the direction of vision, the image as observed from one position is shifted laterally when viewed from the other. This displacement of the image, known as *disparity*, is inversely proportional to the distance to the object. Humans subconsciously *verge* their eyes to bring objects of interest into rough registration (Burt, et al., 1992). Hold up a finger a few inches away from your face while focusing on a distant object and you can simultaneously observe two displaced images in the near field. In refocusing on the finger, your eyes actually turn inward slightly to where their respective optical axes converge at the finger instead of infinity.

Most robotic implementations use a pair of identical video cameras (or a single camera with the ability to move laterally) to generate the two disparity images required for stereoscopic ranging. The cameras are typically aimed straight ahead viewing approximately the same scene, but (in simplistic cases anyway) do not possess the capability to *verge* their center of vision on an observed point as can human eyes. This limitation makes placement of the cameras somewhat critical because stereo ranging can take place only in the region where the fields of view overlap. In practice, analysis is performed over a selected range of disparities along the *Z* axis on either side of a perpendicular plane of zero disparity called the *horopter* (Figure 4-3). The selected image region in conjunction with this disparity range defines a three-dimensional volume in front of the vehicle known as the *stereo observation window* (Burt, et al., 1993).

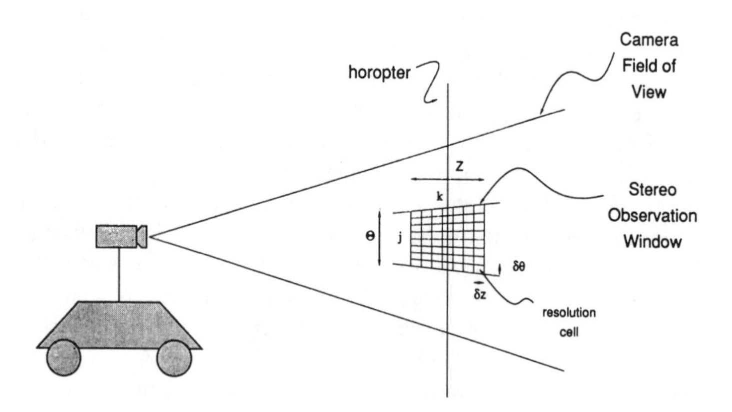

Figure 4-3. The *stereo observation window* is that volume of interest on either side of the plane of zero disparity known as the *horopter* (courtesy David Sarnoff Research Center).

More recently there has evolved a strong interest within the research community for dynamically reconfigurable camera orientation (Figure 4-4), often termed *active vision* in the literature (Aloimonos, et al., 1987; Swain & Stricker, 1991; Wavering, et al., 1993). The widespread acceptance of this terminology is perhaps somewhat unfortunate in view of potential confusion with stereoscopic systems employing an active illumination source (see Section 4.1.3). *Verging stereo*, another term in use, is perhaps a more appropriate choice. *Mechanical verging* is defined as the process of rotating one or both cameras about the vertical axis in order to achieve zero disparity at some selected point in the scene (Burt, et al., 1992).

Figure 4-4. This stereoscopic camera mount uses a pair of lead-screw actuators to provide reconfigurable baseline separation and vergence as required (courtesy Robotic Systems Technology, Inc.).

There are four basic steps involved in the stereo ranging process (Poggio, 1984):

- A point in the image of one camera must be identified (Figure 4-5, left).
- The same point must be located in the image of the other camera (Figure 4-5, right).
- The lateral positions of both points must be measured with respect to a common reference.
- Range Z is then calculated from the disparity in the lateral measurements.

On the surface this procedure appears rather straightforward, but difficulties arise in practice when attempting to locate the specified point in the second image

(Figure 4-5). The usual approach is to match "interest points" characterized by large intensity discontinuities (Conrad & Sampson, 1990). Matching is complicated in regions where the intensity and/or color are uniform (Jarvis, 1983b). Additional factors include the presence of shadows in only one image (due to occlusion) and the variation in image characteristics that can arise from viewing environmental lighting effects from different angles. The effort to match the two images of the point is called *correspondence*, and methods for minimizing this computationally expensive procedure are widely discussed in the literature (Nitzan, 1981; Jarvis, 1983a; Poggio, 1984; Loewenstein, 1984; Vuylsteke, et al., 1990; Wildes, 1991).

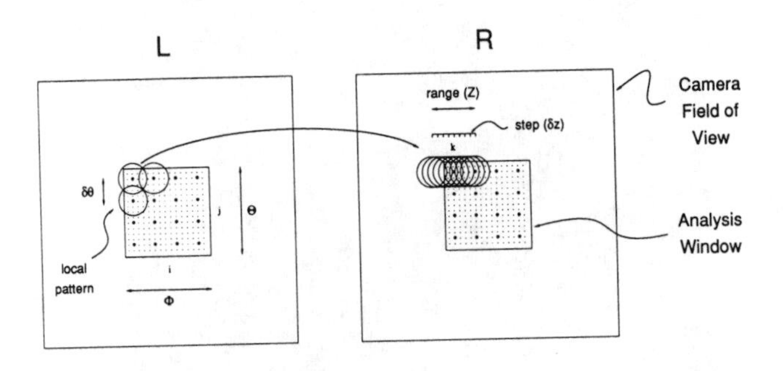

Figure 4-5. Range Z is derived from the measured disparity between interest points in the left and right camera images (courtesy David Sarnoff Research Center).

Probably the most basic simplification employed in addressing the otherwise overwhelming *correspondence* problem is seen in the *epipolar restriction* that reduces the two-dimensional search domain to a single dimension (Vuylsteke, et al., 1990). The *epipolar surface* is a plane defined by the point of interest P and the positions of the left and right camera lenses at L and R as shown in Figure 4-6. The intersection of this plane with the left image plane defines the *left epipolar line* as shown. As can be seen from the diagram, since the point of interest P lies in the *epipolar plane*, its imaged point P_l must lie somewhere along the *left epipolar line*. The same logic dictates the imaged point P_r must lie along a similar *right epipolar line* within the right image plane. By carefully aligning the camera image planes such that the *epipolar lines* coincide with identical scan lines in their respective video images, the correspondence search in the second image is constrained to the same horizontal scan line containing the point of interest in the first image. This effect can also be achieved with non-aligned cameras by careful calibration and rectification (resampling), as is done in real time by JPL's stereo vision system (see below) using a Datacube *Miniwarper* module.

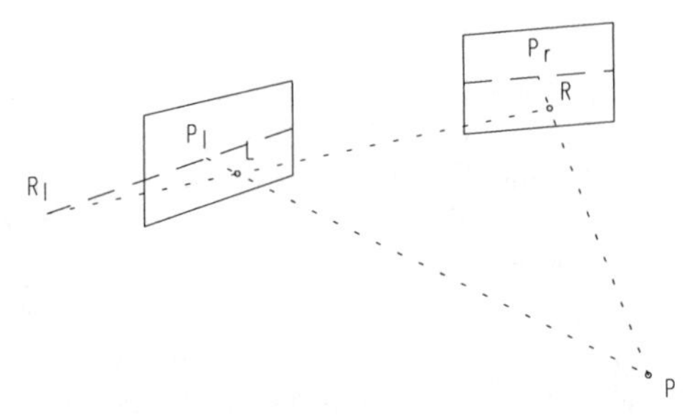

Figure 4-6. The *epipolar surface* is a plane defined by the lens centerpoints *L* and R and the object of interest at *P* (adapted from Vuylsteke, et al., 1990).

To reduce the image processing burden, most correspondence schemes monitor the overall scene at relatively low resolution and examine only selected areas in greater detail. A *foveal representation* analogous to the acuity distribution in human vision is generally employed as illustrated in Figure 4-1, allowing an extended field-of-view without loss of resolution or increased computational costs (Burt, et al., 1993). The high-resolution *fovea* must be shifted from frame to frame in order to examine different regions of interest individually. Depth acuity is greatest for small disparities near the horopter and falls off rapidly with increasing disparities (Burt, et al., 1992).

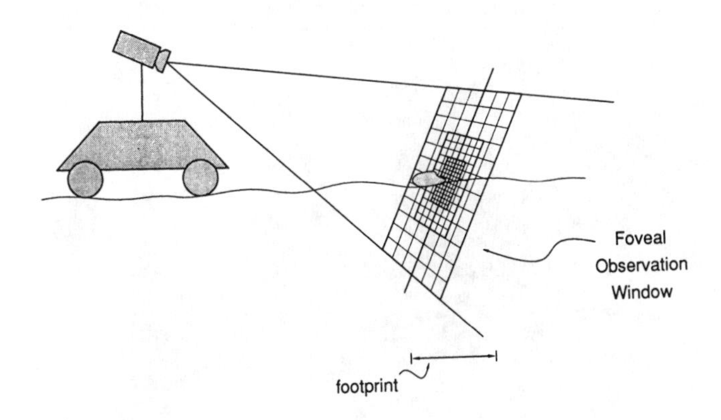

Figure 4-7. The *foveal* stereo representation provides high acuity near the center of the *observation window*, with decreasing resolution towards the periphery (courtesy David Sarnoff Research Center).

4.1.1 JPL Stereo Vision

The Jet Propulsion Laboratory (JPL), Pasadena, CA, is developing a passive stereo vision system for use on board the NASA Planetary Rover and for US

Army robotic land vehicle applications (Bedard, et al., 1991a, 1991b; Slack, 1989). In 1990, JPL developed a vision system that computed Laplacian image pyramids using Datacube hardware, followed by a method of stereo matching which applies a sum-of-squared-differences operator to 8-bit greyscale images. Originally, the sum-of-squared-differences operation was performed at the 64- by 60-pixel image level of the pyramid using a *68020* processor, producing range and confidence images in approximately two seconds. Subsequent implementations perform the correlation at much higher speeds (see below).

An alternate version of the algorithm augments the cross-correlation with a one-dimensional smooth-surface model, allowing interpolation over textureless image areas. Disparity estimates are performed independent of each scan line, requiring approximately six seconds per 64- by 60-pixel image-pair. This system has been implemented on the *Planetary Rover Navigation Testbed* vehicle (Figure 4-8) and performed reliably in off-road navigation tests. Both algorithms assume that the cameras are well aligned, confining the matching search to corresponding scan lines of the two images.

Figure 4-8. NASA Planetary Rover navigation testbed (courtesy Jet Propulsion Laboratory).

The US Army Tank Automotive Command is applying this technology for obstacle detection and reflexive obstacle avoidance within the context of computer-aided remote driving of a HMMWV. Disparity field estimation at 7.5 frames per second has been achieved from the 64- by 60-pixel level of Laplacian image pyramids, using a *Datacube MaxVideo-20* board and a 68040 host processor. Obstacle detection is performed at the rate of three frames per second with postprocessing, triangulation, and a very simple detection algorithm (Matthies, 1992a, 1992b).

This system was demonstrated at the US Army *Demo I* at Aberdeen, MD, in April/May of 1992. The vehicle successfully detected obstacles of about 50 centimeters on a side while driving at several kilometers per hour on gravel roads. Continued development of the system is taking place under the *Unmanned Ground Vehicle (UGV) Demo II* program sponsored by the Advanced Research Projects Agency (ARPA). Under this program, the need for precise camera alignment has been eliminated by performing real-time image resampling before computing image pyramids, greatly simplifying implementation. The system was used to detect obstacles on relatively smooth off-road terrain during the UGV *Demo B* in June, 1994 (Figure 4-9), using the 128- by 120-level of the Laplacian pyramid. Higher speeds, higher resolution, and rougher terrain are anticipated for *Demo C* in June 1995 and *Demo II* in June 1996.

Figure 4-9. The JPL stereo vision system was used to detect obstacles for the HMMWV-based *Surrogate Semiautonomous Vehicle (SSV)* at *Demo B* in June 1994 (courtesy Martin Marietta Denver Aerospace).

4.1.2 David Sarnoff Stereo Vision

Conventional application of stereoscopic ranging to mobile robot collision avoidance generally involves creating a dense range map over an appropriate field of view dictated in size by the vehicle dynamics (Chapter 1). Sufficient resolution must be provided to detect small hazards at distances typically 10 to 20 meters ahead of the vehicle. From a practical standpoint, this combination of high-resolution processing over a large field of regard is computationally intensive, resulting in low throughput rates and expensive hardware (Burt, et al., 1993).

The Advanced Image Processing Research Group at David Sarnoff Research Center, Princeton, NJ, is developing a specialized image processing device called the *vision front end (VFE).* This dedicated hardware performs image preprocessing functions faster and more efficiently than a general purpose computer, and thus opens the door for more practical solutions in keeping with the needs of a mobile robotic system. The *VFE* concept is characterized by four innovative features:

- Laplacian pyramid processing (Burt & Adelson, 1983; Anderson, et al., 1985).
- Electronic vergence (Burt, et al., 1992; 1993).
- Controlled horopter (Burt, et al., 1992; 1993).
- Image stabilization (Burt & Anandan, 1994; Hansen, et al., 1994).

The Sarnoff *VFE* approach emulates the *vergence* and *foveal organization* attributes of human vision through electronic warping and local disparity estimation within a pyramid data structure, thus providing appropriate resolution where required, but at lower cost. This approach is algorithmically accomplished as follows (Burt, et al., 1992):

- The right image is warped to bring it into alignment with the left image within a designated region of analysis (Figure 4-10).
- Residual stereo disparity between the partially aligned images is then estimated.
- Global displacement is used to refine the alignment.
- The global displacement and local disparity field are passed to the main vision computer.

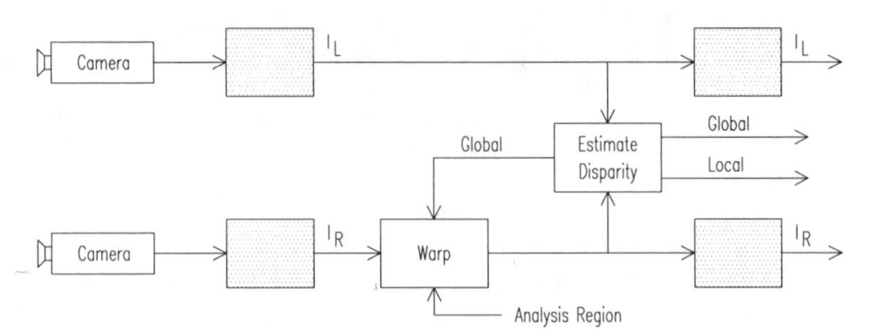

Figure 4-10. After warping the right image into alignment with the left, global alignment (electronic vergence) and local disparity estimations are passed to the main vision processor (adapted from Burt, et al., 1992).

Mechanical vergence of the stereo camera pair results in a horopter oriented perpendicular to the system optical axis as shown earlier in Figure 4-3. *Electronic vergence* is used by the *VFE* to dynamically warp images such that the horopter is tilted to provide optimal stereo acuity for a particular vision task (Burt, et al., 1992). This *controlled horopter* approach thus goes a step beyond the flexibility of interactive mechanical vergence, in that the orientation of the horopter can be varied in addition to its stand-off distance. For example, Figure 4-11 shows a situation where the horopter is made coincident with the ground plane. When *foveal stereo* is employed in conjunction with the tilted horopter technique, maximum sensitivity to small topographical features can be achieved due to the increased clustering of high-resolution stereo cells along the ground surface (Burt, et al., 1993). The ability to detect low-lying obstacles or potholes in the path of the vehicle is subsequently greatly enhanced.

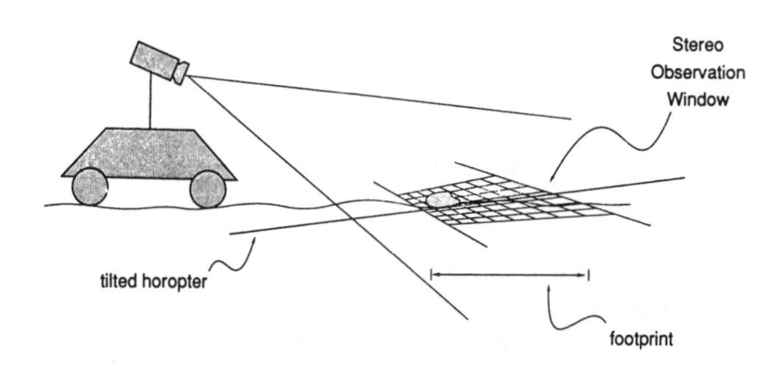

Figure 4-11. Through electronic warping of both images, the plane of zero disparity (horopter) can be made coincident with the ground surface for enhanced acuity in the region of most interest (courtesy David Sarnoff Research Center).

A fundamental challenge to employing any image-based system in a collision avoidance role is the inherent jitter introduced by vehicle pitch and roll. Effective image stabilization has proven to be a critical factor in successful implementation of vehicle-based stereo ranging capabilities (Burt & Anandan, 1994). Mechanically stabilized optics have been used with good results in aerial photography and for filming sporting events, but are very expensive and rather bulky, which limits their utility from a robotics perspective. Low-end electronic-stabilization techniques have recently been introduced into some consumer camcorders (Uomori, et al., 1990) but are generally limited to a small amount of translational compensation only. Using pyramid-based motion estimation and electronic warping, the Sarnoff *VFE* provides real-time removal of first-order deformations between sequential images, and can assemble aligned components over time to produce a *scene mosaic* (Hansen, et al., 1994).

4.2 Active Triangulation

Rangefinding by *active triangulation* is a variation on the *stereo disparity* method of distance measurement. In place of one camera is a laser (or LED) light source aimed at the surface of the object of interest. The remaining camera is offset from this source by a known distance A and configured to hold the illuminated spot within its field of view (Figure 4-12).

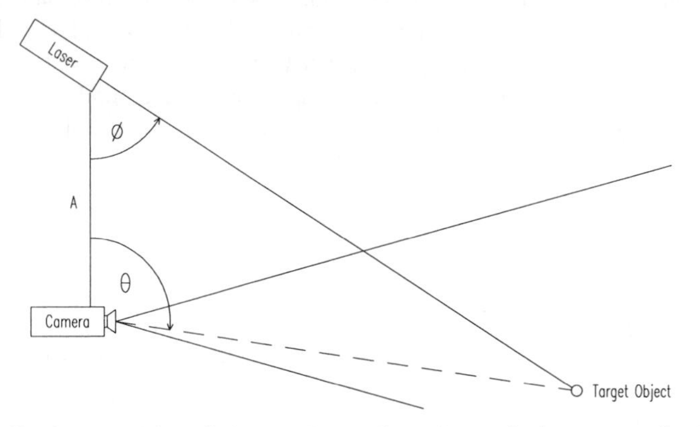

Figure 4-12. An *active triangulation-ranging configuration* employing a conventional CCD array as the detector.

For one- or two-dimensional array detectors such as vidicon or CCD cameras, the range can be determined from the known baseline distance A and the relative position of the laser-spot image on the image plane. For mechanically scanned single-element detectors such as photodiodes or phototransistors, the rotational angles of the detector and/or source are measured at the exact instant the detector observes the illuminated spot. The trigonometric relationships between these angles and the baseline separation are used (in theory) to compute the distance. To obtain three-dimensional information for a volumetric region of interest, laser triangulators can be scanned in both azimuth and elevation. In systems where the source and detector are self-contained components, the entire configuration can be moved mechanically as illustrated in Figure 4-13. In systems with movable optics, the mirrors and lenses are generally scanned in synchronization while the laser and detector remain stationary.

In practice, the actual baseline separation distance A as well as the angles θ and ϕ are difficult to measure with any precision, and therefore most designers simply calibrate the ranging system with test targets placed at known distances along the Z axis. Nguyen and Blackburn (1995) present a typical procedure illustrated in Figure 4-14 below. The line uP passing through the lens focal point O can be represented by:

$$y = \frac{u}{f} z$$

where:

y = height above Z axis
u = vertical projection of point P on the image plane
f = focal length of the lens,

while the laser path is similarly of the form: $y = mz + c$. Combining these equations and simplifying eventually yields the desired expression for range z along the camera optical axis (Nguyen & Blackburn, 1995):

$$z = \frac{N}{ud - k}$$

where N, d, and k are obtained from the calibration setup as follows:

$$N = (u_1 - u_2)z_1 z_2$$
$$d = z_2 - z_1$$
$$k = u_2 z_2 - u_1 z_1$$

Figure 4-13. A 5-milliwatt laser source used in conjunction with a CCD camera is mounted on a 2-DOF pan and tilt mechanism on the ModBot research prototype (courtesy Naval Command Control and Ocean Surveillance Center).

In other words, calibration targets are placed at distances z_1 and z_2 from the camera, and their associated offsets u_1 and u_2 (i.e., where the laser spot is observed striking the targets in the image plane) used to calculate *d*, *k*, and *N*, yielding a general expression for range *z* as a function of pixel offset *u*. Note this calibration approach does not require any information regarding the baseline separation distance *A* or lens focal length *f*.

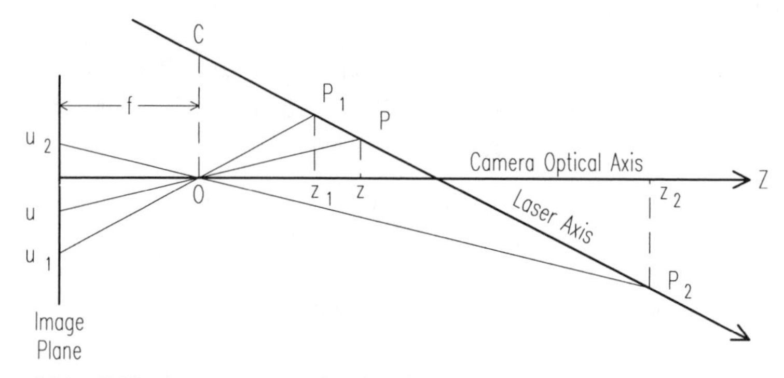

Figure 4-14. Calibration targets are placed at known distances z_1 and z_2 to derive a general expression for range z as a function of image displacement u (adapted from Nguyen & Blackburn, 1995).

Drawbacks to active triangulation include the *missing parts* situation, where points illuminated by the light source cannot be seen by the camera and vice versa (Jarvis, 1983b), as well as surface absorption or specular reflection of the irradiating energy (see Chapter 9). On the positive side, however, point-source illumination of the image effectively eliminates the correspondence problem encountered in stereo disparity rangefinders. There is also no dependence on scene contrast, and reduced influence from ambient lighting effects. (Background lighting is effectively a noise source that can limit range resolution.)

4.2.1 Hamamatsu Rangefinder Chip Set

The block diagram for a triangulation rangefinding chip set manufactured by Hamamatsu Corporation is shown in Figure 4-15. This 16-step rangefinder offers a maximum sample rate of 700 Hz and consists of three related components: a *position sensitive detector (PSD)*, a *rangefinder IC*, and an *LED light source*. Near-infrared energy is emitted by the LED source and reflected by the target back to the PSD, a continuous light-spot position detector (basically a light-sensitive diode combined with a distributed resistance). A small injected current flows from the center to both ends of the detector element with a distribution determined by the footprint of illumination; the ratio of the respective current flows can be used to determine the location of the spot centroid (Vuylsteke, et al., 1990). The sensitive receiver circuitry is capable of detecting pulsed light returns generating as little as 1 nanoamp of output current in the PSD.

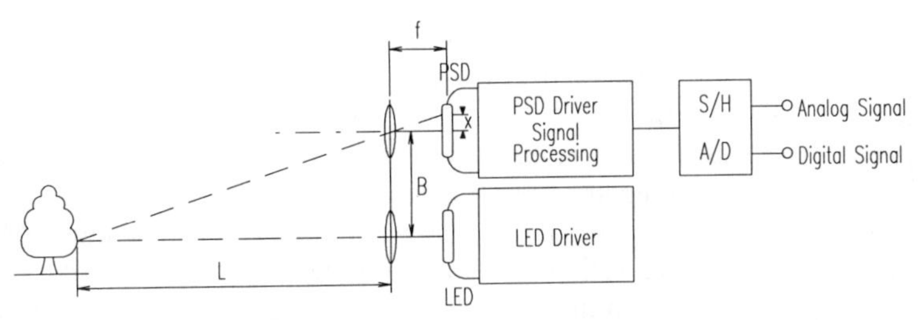

Figure 4-15. Block diagram of the Hamamatsu triangulation rangefinding chip set (courtesy Hamamatsu Corp.).

The Hamamatsu rangefinder chip operates from a 3-volt DC supply and provides both analog and digital signal outputs. The 0.24-to-0.46-volt analog output is produced by a sample-and-hold circuit, while the digital output is determined by an integral A/D converter with 4-bit resolution corresponding to 16 discrete range zones (Hamamatsu, 1990).

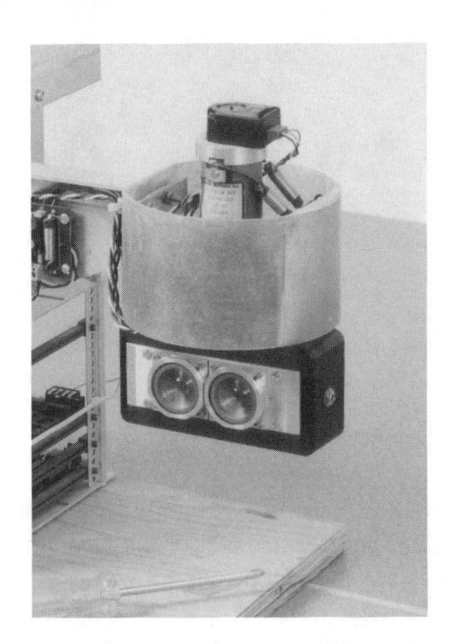

Figure 4-16. The active triangulation ranging system employed on ROBART III is based on the Hamamatsu chip set (courtesy Naval Command Control and Ocean Surveillance Center).

4.2.2 Draper Laboratory Rangefinder

A team of MIT students at the Charles Stark Draper Laboratory has recently designed and built a small (5 kilogram) autonomous microrover for exploration of the Martian surface (Malafeew & Kaliardos, 1994). In the process, the need for a compact, short-range, and inexpensive non-contact rangefinder emerged. The limited energy and computational resources aboard the rover dictated that potential candidates operate on a low power budget, with an output signal supporting simple range extraction. Simplicity in the electronics was also desired, since the rover will have to endure the harsh environments found in space. It was decided that an 180-degree azimuthal scan was necessary in the direction of forward travel, but that an elevation scan was not necessary. A five-percent range error was deemed acceptable in light of the inherent navigational errors associated with dead reckoning. From these requirements, an active triangulation rangefinder was developed using a near-infrared laser source and a one-dimensional *position-sensitive detector* (Figure 4-17).

Figure 4-17. This active triangulation ranging system uses a 500-milliwatt near-infrared laser diode as the active source (courtesy Charles Stark Draper Laboratory).

The initial prototype was constructed slightly larger than necessary to simplify mounting and machining, but the diameter of the receiving lens was intentionally kept small (15 millimeters) to demonstrate an ability to collect returned energy with sufficient signal-to-noise ratio. Due to developmental time constraints, the electronics employed on the prototype are typical of those routinely suggested for DC operation of a standard *position-sensitive detector* circuit, hence this rangefinder is very similar in concept to the previously described Hamamatsu system. Signal currents from the detector are read immediately before and during the firing of the active source, a common method for subtracting off ambient background noise. Due to the slow vehicle speed, there is no need for an extremely fast ranging system, and a 25-Hz sampling rate should suffice.

The large amount of electronic noise associated with the rest of the rover systems combined with the small-diameter receiving lens made detection of weak signals difficult, requiring a relatively high-power (>250 milliwatts) illumination source. The source also needed to be well collimated, since triangulation systems work best when the footprint of illumination is small. To meet these needs, a 920-nanometer laser diode with a beam divergence of under 15 milliradians was selected. The laser provides an optical power output of about 500 milliwatts for 1-millisecond intervals. This power level is not eye-safe, of course, but that is of little concern on Mars.

With a matched interference filter, the rangefinder is able to operate under direct sunlight conditions. Initial test results show a ranging accuracy that is about five percent at the maximum range of 3 meters. As with any triangulation system, this normalized accuracy improves as the range is decreased. Azimuthal scanning on the rover is currently accomplished by servoing the entire rangefinder unit through 180-degree sweeps.

4.2.3 Quantic Ranging System

A novel LED-based triangulation ranging system was developed for the Navy by Quantic Industries, Inc. under the Small Business Innovative Research (SBIR) Program (Moser & Everett, 1989). The prototype unit shown in Figure 4-18 was specifically designed for use on a mobile robotic platform, under the following general guidelines:

- Coverage of 100-degrees azimuth and 30-degrees elevation.
- No moving parts.
- 10-Hz update rate.
- Real-time range measurements out to 20 feet.
- Minimal power consumption.
- Small size and weight.

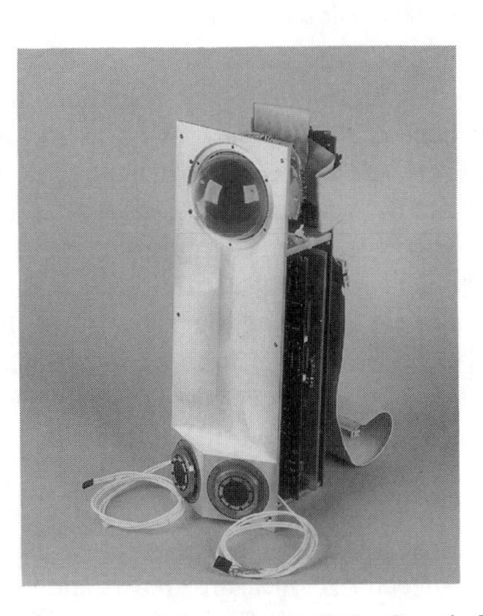

Figure 4-18. Prototype triangulation ranging sensor built by Quantic Industries, Inc. (courtesy Naval Command Control and Ocean Surveillance Center).

Active triangulation ranging is employed with about 5-degree spatial resolution over a nominal field of regard of 100 degrees in azimuth and 30 degrees in elevation. Under typical indoor conditions, fairly accurate target detection and range measurements are obtained to about 24 feet in the dark and about 15 feet under daylight conditions. No mechanical scanning is employed, and the protected envelope can be covered in 0.1 to 1 second, depending upon the required accuracy.

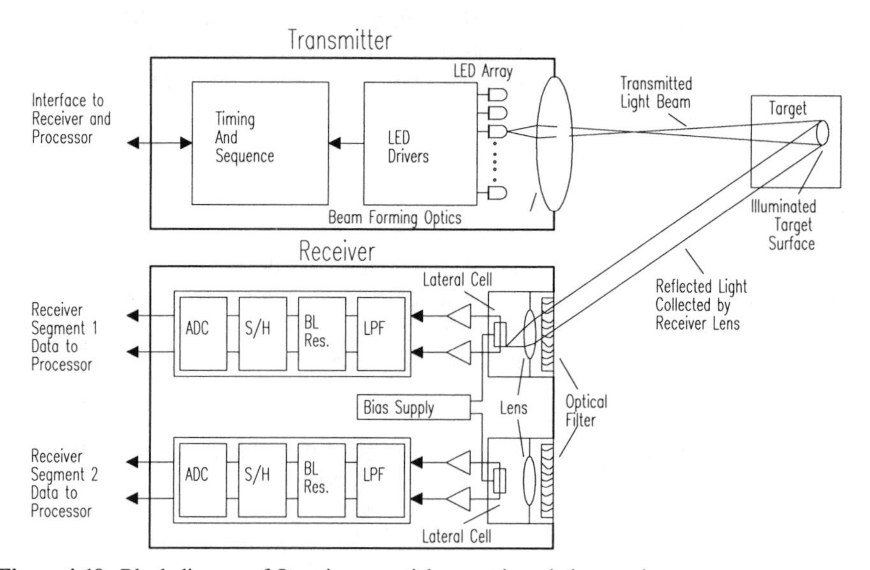

Figure 4-19. Block diagram of Quantic sequential-array triangulation ranging system.

The transmitter consists of 164 high-power, gallium-aluminum-arsenide LEDs mounted in an array behind a spherical lens so as to produce a corresponding number of narrow, evenly spaced beams that interrogate the volume of interest. The LEDs are sequentially activated at a particular repetition rate while a synchronous receiver detects reflected energy from targets within its field of view. The self-lensed LEDs yield relatively narrow beams, so most of their power is projected within the critical angle of the sphere lens for high power-transfer efficiency. Figure 4-20 shows the pattern of the beams and their positioning behind the lens for the desired 5-degree spatial sampling.

The optical receiver consists of two identical units, each covering a field of view of about 50 by 50 degrees. Both units contain a Fresnel lens, an optical bandpass filter, a position-sensitive detector, and the associated electronics to process and digitize the analog signals. The receiver uses a silicon lateral-effect position-sensing photodetector to measure the location (in the image plane) of transmitted light reflected (scattered) from a target surface. The transmitter and receiver are vertically separated by a 10-inch baseline.

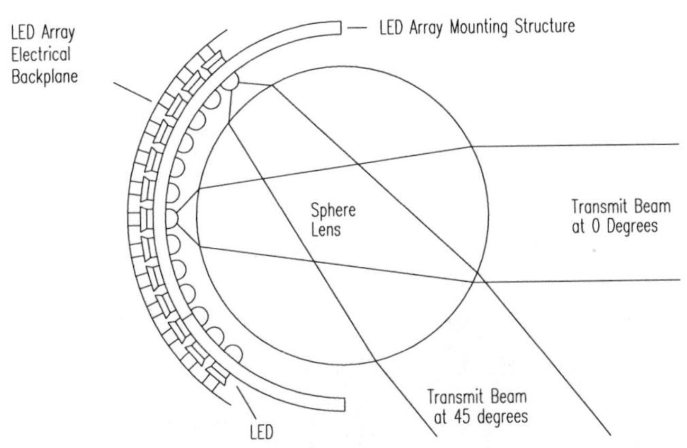

Figure 4-20. Arrangement of near-infrared LED array behind spherical lens in the Quantic ranging system.

The location of the centroid of reflected energy focused on the position-sensing detector is a function of the particular beam that is active and the range to the target being illuminated by that beam. The position signals from the detector (resulting from the sequential activation of LEDs in the transmitter) are collectively processed by a dedicated microcomputer to determine the ranges to targets throughout the sensor's detection envelope. Target azimuth and elevation are a function of the position of the LED (in the transmitter array) active at the time of detection. A look-up table derived from calibration data is used to perform the position-to-range conversions and to compensate for receiver non-uniformities.

4.3 Active Stereoscopic

Due to the computationally intensive complexities and associated resources required for establishing correspondence, passive stereoscopic methods were initially limited in practical embodiments to very simple scenes (Blais, et al., 1988). One way around these problems is to employ an active source in conjunction with a pair of stereo cameras. This active illumination greatly improves system performance when viewing scenes with limited contrast. Identification of the light spot becomes a trivial matter; a video frame representing a scene illuminated by the source is subtracted from a subsequent frame of the same image with the light source deactivated. Simple thresholding of the resultant difference image quickly isolates the region of active illumination. This process is performed in rapid sequence for both cameras, and the lateral displacement of the centroid of the spot is then determined.

Such an active stereoscopic vision system was initially employed on ROBART II for ranging purposes (Figure 4-21). A 6-volt incandescent source was pulsed at about a 10-Hz rate, projecting a sharply defined V-shaped pattern across the intersection of the camera plane with the target surface. The incandescent source was chosen over a laser-diode emitter because of simplicity, significantly lower cost (at the time), and the limited range requirements for an indoor system.

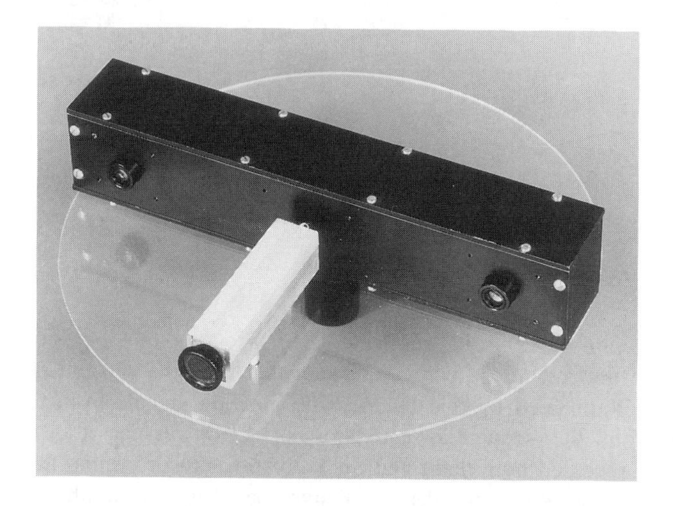

Figure 4-21. The active stereoscopic ranging system employed on ROBART II used a pair of linear CCD arrays in conjunction with an incandescent structured-light source (center).

The configuration did not represent a true three-dimensional capability in that each of the cameras consisted of a horizontal linear (as opposed to two-dimensional) CCD array. (Linear arrays were employed in order to keep the image processing requirements realistic with respect to the available 8-bit computational hardware on board.) The twin cameras provided no vertical resolution, but furnished range and bearing information on interest points detected in the horizontal plane coincident with their respective optical axes. This limitation was consistent with the two-dimensional simplified world model employed by the robot; objects were represented by their projection on the X-Y plane, and height information was not taken into account. The linear cameras were removed in 1987 and replaced with a single two-dimensional high-resolution CCD camera coupled to a line-oriented video digitizer.

4.3.1 HERMIES

Alignment between the source and cameras is not critical in active stereoscopic ranging systems; in fact, the source does not even have to be located on board the robot. For example, Kilough and Hamel (1989) describe two innovative configurations using external sources for use with the robot HERMIES IIB, built at Oak Ridge National Laboratory. A pair of wide-angle black-and-white CCD cameras are mounted on a pan-and-tilt mechanism atop the robot's head as shown in Figure 4-22. Analog video outputs from the cameras are digitized by a frame grabber into a pair of 512- by 384-pixel arrays, with offboard image processing performed by a *Hypercube* at a scaled-down resolution of 256 by 256. The initial application of the vision system was to provide control of a pair of robotic arms (from the Heathkit *HERO-1* robot) employed on HERMIES.

Figure 4-22. HERMIES IIB employed an active stereo ranging system with an external laser source that could be used to designate objects of interest in the video image (courtesy Oak Ridge National Laboratory).

To accomplish this task, a near-infrared LED is attached to the end of the HERO-1 arm near the manipulator and oriented so as to be visible within the field of view of the stereo camera pair. A sequence of images is then taken by each camera, with the LED first *on* and then *off*. The *off* representations are subtracted from the *on* representations, leaving a pair of difference images, each comprised of a single bright dot representing the location of the LED. The centroids of the dots are calculated to precisely determine their respective coordinates in the difference-image arrays. A range vector to the LED can then be easily calculated, based on the lateral separation of the dots as perceived by the two cameras. This technique establishes the actual location of the manipulator in the reference frame of the robot. Experimental results indicated a 2-inch accuracy with a 0.2-inch repeatability at a distance of approximately 2 feet (Kilough and Hamel, 1989).

A near-infrared solid-state laser mounted on a remote tripod was then used by the operator to designate a target of interest within the video image of one of the cameras. The same technique described above was repeated, only this time the imaging system toggled the laser power *on* and *off*. A subsequent differencing operation enabled calculation of a range vector to the target, also in the robot's reference frame. The difference in location of the gripper and the target object could then be used to effect both platform and arm motion. The imaging processes would alternate in near-real-time for the gripper and the target, enabling the HERMIES robot to drive over and grasp a randomly designated object under continuous closed-loop control.

4.3.2 Dual-Aperture 3-D Range Sensor

A novel implementation of active stereoscopic ranging employing only one camera is presented by Blais, et al. (1988; 1990), wherein a dual-aperture pin-hole mask is substituted for the diaphragm iris of a standard camera lens as shown in Figure 4-23 below. A Pulnix model *TM-540* CCD camera (512 by 492 pixels) is employed as the detector. The basic principle of operation for the *BIRIS* (i.e., bi-iris) system is described by Rioux and Blais (1986). Lens focus is adjusted such that a point located at position *A* is in focus at *A'* in the image plane of the detector; ignoring the mask for a moment, any ray traveling from point *A* through the lens will arrive at the image point *A'*. Under these conditions, a second point *B* at a further distance *z* from the lens will be imaged at *B'*.

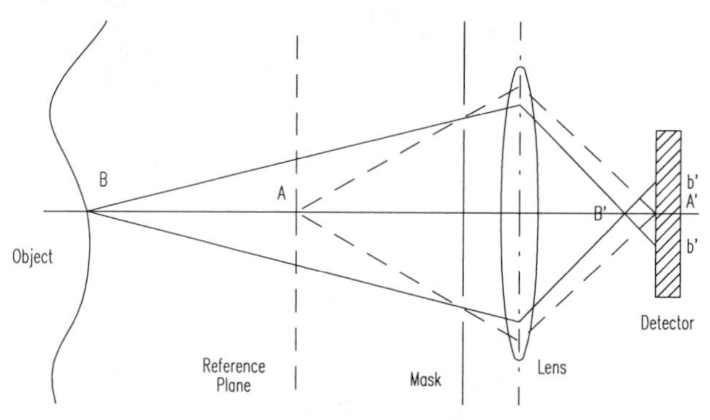

Figure 4-23. The dual-aperture pin-hole mask allows an active stereoscopic ranging capability to be implemented with a single CCD camera (adapted from Blais, et al., 1988).

With the mask inserted in front of the lens, two separate rays originating at point B will pass through the two apertures and be redirected by the lens to point *B'* in front of the detector. These two rays will continue on to strike the CCD array at points b_1 and b_2 as indicated in the figure. The lateral separation between points b_1 and b_2 is directly proportional to the range *Z* from the lens to point *B*. The *X* and *Y* displacement of the spots on the CCD array with respect to its center can be used to calculate the vector direction to the target with respect to the optical axis of the lens. The magnitude of this vector is of course the measured range Z.

Projecting a pattern of stripes (instead of a single spot of light) perpendicular to an imaginary line drawn between the two apertures in the mask enables acquisition of multiple range profiles from a single video frame (Rioux & Blais, 1986). Each projected stripe will be appear as a pair of lines on the detector. Similarly, the lateral separation between line pairs can be used to derive the range value Z. A 256- by 240-point range image can be acquired in under 4.3 seconds when a single stripe is projected; the same image will be acquired in approximately one second if four profiles are projected simultaneously (Blais, et al., 1988). Discontinuities in the imaged lines will be generated by objects illuminated by the structured pattern of light. This *structured-light* illumination technique will be described in more detail in the next section.

4.4 Structured Light

Ranging systems that employ *structured light* are a further refined case of active triangulation. A pattern of light (either a line, a series of spots, or a grid pattern) is projected onto the object surface while the camera observes the pattern from its offset vantage point. Range information manifests itself in the distortions visible in the projected pattern due to variations in the depth of the scene. The use of these special lighting effects tends to reduce the computational complexity and improve the reliability of three-dimensional object analysis (Jarvis, 1983b; Vuylsteke, et al., 1990). The technique is commonly used for rapid extraction of limited quantities of visual information of moving objects (Kent, 1985), and thus lends itself well to collision avoidance applications. Besl (1988) provides a good overview of *structured-light* illumination techniques, while Vuylsteke, et al. (1990) classify the various reported implementations according to the following characteristics:

- The number and type of sensors.
- The type of optics (i.e., spherical or cylindrical lens, mirrors, multiple apertures).
- The dimensionality of the illumination (i.e., point or line).
- Degrees of freedom associated with scanning mechanism (i.e., zero, one, or two).
- Whether or not the scan position is specified (i.e., the instantaneous scanning parameters are not needed if a redundant sensor arrangement is incorporated).

The most common *structured-light* configuration entails projecting a line of light onto a scene, originally introduced by P. Will and K. Pennington of IBM Research Division Headquarters, Yorktown Heights, NY (Schwartz, undated). Their system created a plane of light by passing a collimated incandescent source through a slit, thus projecting a line across the scene of interest. (More recent systems create the same effect by passing a laser beam through a cylindrical lens or by rapidly scanning the beam in one dimension.) Where the line intersects an object, the camera view will show displacements in the light stripe that are proportional to the depth of the scene. In the example depicted in Figure 4-24, the lower the reflected illumination appears in the video image, the closer the target object is to the laser source. The exact relationship between stripe displacement and range is dependent on the length of the baseline between the source and the detector. Like any triangulation system, when the baseline separation increases, the accuracy of the sensor increases, but the *missing parts* problem worsens.

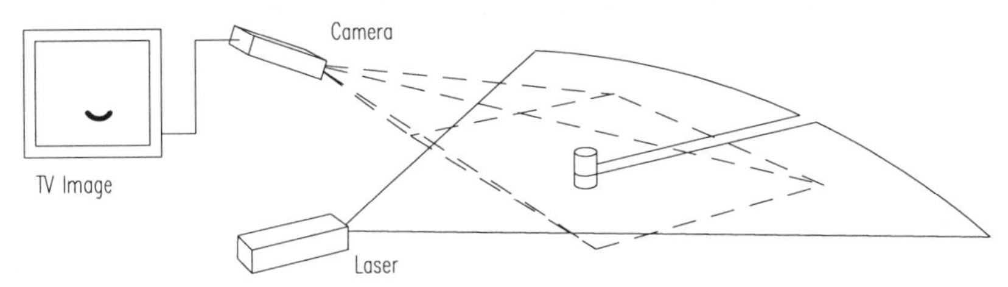

Figure 4-24. A common structured-light configuration used on robotic vehicles projects a horizontal line of illumination onto the scene of interest and detects any target reflections in the image of a downward-looking CCD array.

Three-dimensional range information for an entire scene can be obtained in relatively simple fashion through striped lighting techniques. By assembling a series of closely spaced two-dimensional contours, a three-dimensional description of a region within the camera's field of view can be constructed. The third dimension is typically provided by scanning the laser plane across the scene. Compared to single-point triangulation, striped lighting generally requires less time to digitize a surface, with fewer moving parts because of the need to mechanically scan only in one direction. The drawback to this concept is that range extraction is time consuming and difficult due to the necessity of storing and analyzing many frames.

An alternative structured-light approach for three-dimensional applications involves projecting a rectangular grid of high-contrast light points or lines onto a surface. Variations in depth cause the grid pattern to distort, providing a means for range extraction. The extent of the distortion is ascertained by comparing the displaced grid with the original projected patterns as follows (LeMoigue & Waxman, 1984):

- Identify the intersection points of the distorted grid image.
- Label these intersections according to the coordinate system established for the projected pattern.
- Compute the disparities between the intersection points and/or lines of the two grids.
- Convert the displacements to range information.

The comparison process requires correspondence between points on the image and the original pattern, which can be troublesome. By correlating the image grid points to the projected grid points, this problem can be somewhat alleviated. A critical design parameter is the thickness of the lines that make up the grid and the spacing between these lines. Excessively thin lines will break up in busy scenes, causing discontinuities that adversely affect the intersection points labeling process. Thicker lines will produce less observed grid distortion resulting in reduced range accuracy (LeMoigue & Waxman, 1984). The sensor's intended domain of operation will determine the density of points required for adequate scene interpretation and resolution.

4.4.1 TRC Strobed-Light Triangulation System

Transitions Research Corporation (TRC), Danbury, CN, has incorporated a structured light system to detect and measure the position of objects lying within or adjacent to the forward path of their *HelpMate* mobile platform (Evans, et al., 1990; King, 1990). The TRC system (Figure 4-25) is comprised of a CCD camera and two 700-nanometer near-infrared strobes. The strobes alternately fire with a low (3 Hz) duty cycle, resulting in a 300-millisecond update rate. A bandpass filter is employed at the camera end to enhance the received signal-to-noise ratio, thereby minimizing noise contributions from outside the near-infrared spectrum. By performing a pixel-by-pixel subtraction of a non-flashed image from a flashed image, that portion of the scene resulting from reflected energy is emphasized.

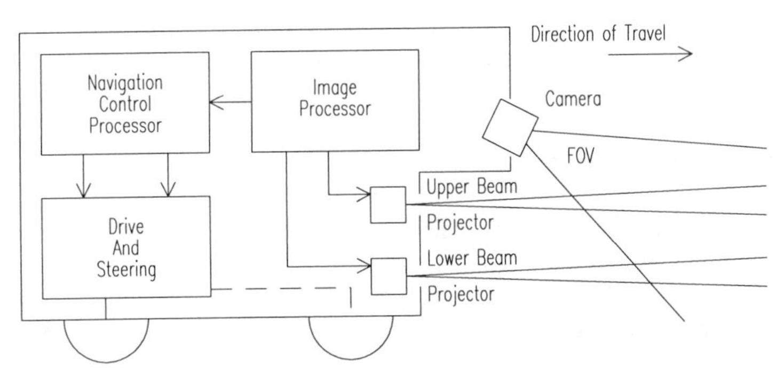

Figure 4-25. Block diagram of the TRC *Strobed Light Triangulation System* installed on the company's *HelpMate* mobile platform (courtesy Transitions Research Corp.).

The reflected light planes are viewed across the horizontal pixel lines of the camera. An object approaching the mobile platform first appears at the top of the field-of-view and then moves down the image plane as the distance closes. In this way, each pixel in the image plane corresponds to a predetermined range and bearing derived through simple triangulation. To ensure real-time computation,

TRC has implemented a thresholding algorithm that uses every sixth pixel in an image of 512 by 480 pixels. Effective range of the system is out to 6 feet with a resolution of 1 to 3 inches, and an angular resolution of 2 degrees. Power consumption (including the frame grabber, camera, AT computer, and strobes) is around 40 watts.

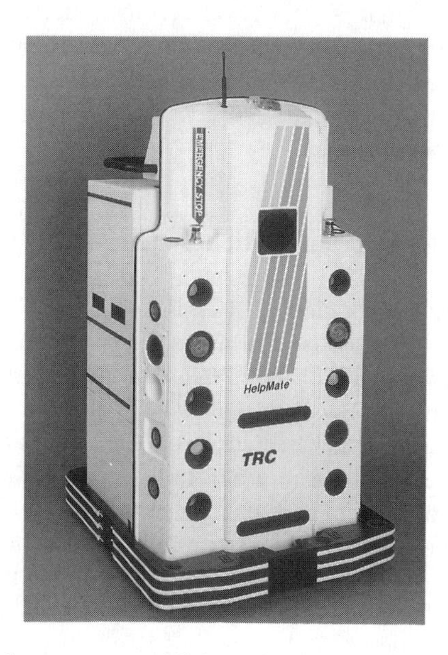

Figure 4-26. Slots for the two structured-light strobes are visible directly above and below the company's logo on the front panel of TRC *HelpMate* (courtesy Transitions Research Corp.).

4.5 Known Target Size

A stadimeter is a hand-held nautical instrument used for optically measuring the distance to objects of known heights, typically between 50 and 200 feet, covering ranges from 200 to 10,000 yards. The stadimeter measures the angle subtended by the object, and converts it into a range reading taken directly from a micrometer drum (Dunlap & Shufeldt, 1972).

The final variation on the triangulation ranging method to be discussed makes use of this same technique. Range is calculated through simple trigonometry; the known baseline, instead of being between two cameras (or a detector and a light source) on the robot, is now the target itself. The concept is illustrated in Figure 4-27. The only limiting constraint (besides knowing the size of the target) is the target must be normal to the optical axis of the sensor, which in the case of a passive system can be an ordinary CCD camera. The standard lens equation applies:

$$\frac{1}{r}+\frac{1}{s}=\frac{1}{f}$$

where:

r = distance from lens to object viewed
s = distance from lens to image plane
f = focal length of the lens.

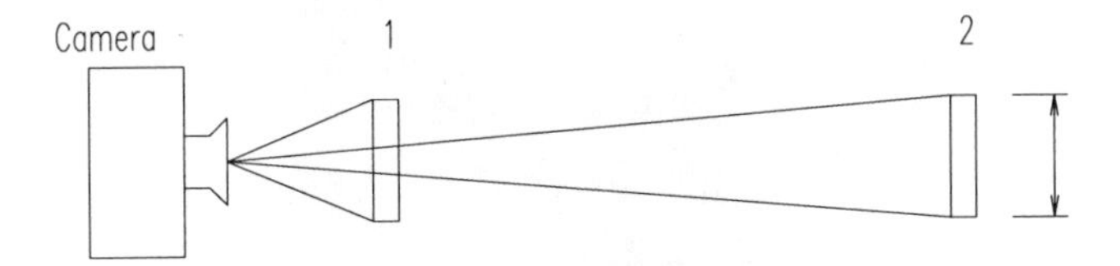

Figure 4-27. The angle subtended by an object of known size is observed to increase as the distance decreases in moving from position 2 to position 1, and can be used to derive the unknown range.

Now suppose the camera views an open doorway of known width A. If A is relatively small compared to the unknown distance r, then the range can be approximated by the formula (Nitzan, et al., 1986):

$$r=\frac{Af}{w}$$

where:

A = known width
w = perceived width in image plane.

If the view angle for the object of interest is wide (i.e., A is not small with respect to r), then local geometric features should be examined (Nitzan, et al., 1986).

4.5.1 NAMCO *Lasernet®* *Scanning Laser Sensor*

One implementation of this ranging concept employs a scanning laser source mechanically coupled to a photodiode detector. NAMCO Controls, Mentor, OH, developed the *Lasernet®* *Scanning Laser Sensor* (Figure 4-28) for automated guided vehicle (AGV) applications in industrial environments (see also Chapter 15). A retroreflective target of known width is placed in a strategically located position to serve as a navigational aid (Laskowski, 1988). As the rotating laser scans across the retroreflector, energy is returned to the collocated detector. The length of the arc of rotation during which the detector senses reflected energy is directly related to the distance to the target: the closer the target, the longer the perceived arc. Multiple targets can be processed simultaneously, and it is also possible to specifically identify objects through the use of uniquely identifiable codes.

Figure 4-28. The *Lasernet*® system detects retroreflective targets with a scanning near-infrared laser to provide bearing and range information used in the navigation of automated guided vehicles (courtesy NAMCO Controls).

A solid-state diode laser source, photodetector, mechanical scanner, beam-forming optics, and control electronics are housed in an enclosure measuring 5 by 6.5 by 3.4 inches for the standard range unit, and 5 by 9 by 3.4 inches for the long-range unit. The photodiode detector has an operational bandwidth of 1.0 MHz, tailored to receive inputs only from the 670-nanometer region of the spectrum. A servo-controlled rotating mirror horizontally pans the laser beam through a 90-degree field of view (45 degrees off either side of centerline) at a rate of 20 scans per second. A directional mirror routes the beam from the laser diode to the scanning mirror; a collecting lens focuses the return signal onto the photodetector.

The standard retroreflective test target used by the developer is a 4- by 4-inch square surface of corner-cube prisms with an overall 90-percent reflection coefficient. When the laser beam sweeps across a retroreflective target, a return signal of finite duration is sensed by the detector. Since the targets are all the same size, the return generated by a close target will be of longer duration than that from a distant one (Figure 4-29). In effect, the closer target appears larger.

The standard model of *Lasernet*® can process up to eight retroreflective targets simultaneously for range and/or angle information. Range is calculated from the equation (NAMCO, 1989):

$$d = \frac{W}{2\tan(\frac{vT_a}{2})}$$

where:

d = range to target
W = target width
v = scan velocity (7200 degrees/second)
T_a = duration of the returned pulse.

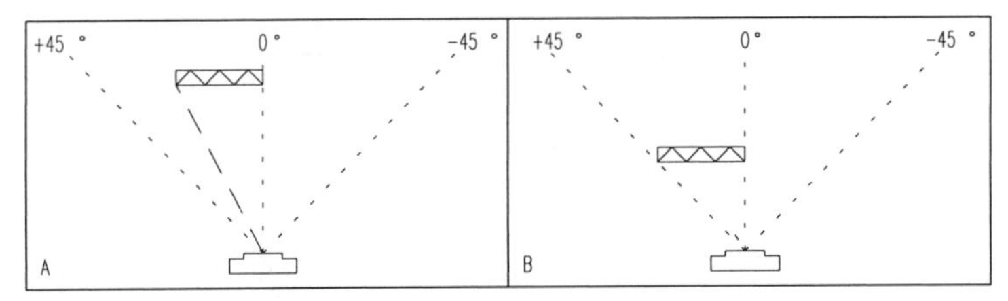

Figure 4-29. The NAMCO *Lasernet*® system determines both range (A) and bearing (B) for standard-size retroreflective targets (adapted from NAMCO, 1989).

Because the target width and angular scan velocity are known, the equation reduces to an inverse function of the pulse duration T_a. With 4-inch targets, the effective range of the sensor is from 1 to 20 feet (2 to 50 feet for the long-range model), and range resolution for either model is 9.6 inches (1.57 inches using digital output) at 20 feet down to 0.1 inch (0.017 inch using digital output) at 1 foot. *Lasernet*® produces an analog output ranging from 0 to 10 volts over the range 0 to 20 feet, and an inverse range function (representing T_a rather than d) digital output on an RS-232 serial port.

The above calculation assumes the target is positioned perpendicular to the angle of incidence of the laser source. If a planar target happens to be rotated or otherwise skewed away from the perpendicular, the resulting decrease in apparent cross-section will induce a range measurement error. Cylindrical targets are sometimes used to overcome this problem.

4.6 Optical Flow

The observed two-dimensional displacement of the brightness pattern in a video image known as optical flow represents a promising new method of obstacle avoidance. The perceived "flow" results from the relative motion between the moving camera and the viewed objects in the surrounding environment, as seen over a sequence of images. Each pixel has an associated instantaneous velocity vector representing the image motion at that point. For example, Figure 4-30 shows an optical flow field resulting from the translational motion of a camera mounted on a vehicle traveling on a planar surface. The optical-flow vectors from closer objects will have greater magnitudes than the vectors from distant objects.

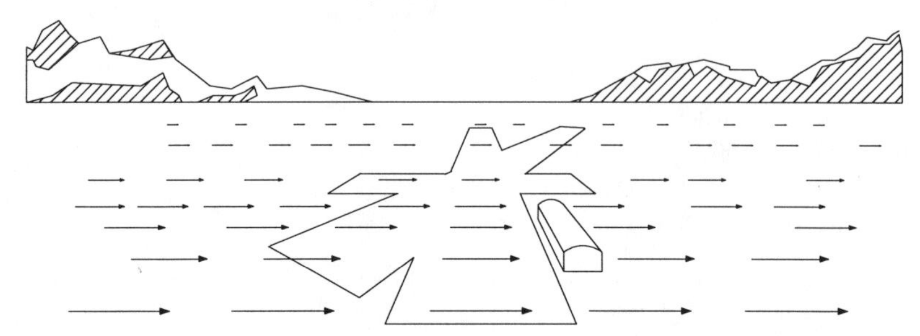

Figure 4-30. The optical flow field due to translation in a direction perpendicular to the camera optical axis will show decreased flow magnitudes with increasing range (reprinted with permission from Gibson, 1950, © Houghton Mifflin Company).

One of the main advantages of using optical flow is that the ratio of distance to speed (e.g., time-to-collision) can be easily obtained and used to generate avoidance maneuvers (Young et al., 1992; Heeger & Jepson, 1990a, 1990b). Disadvantages are seen in the required computational hardware: to achieve real-time results requires processing power on the order of a 50-MHz 80486-based system, which eats up energy at a pretty significant rate.

The optical flow often cannot be found by local computations on the image pixels due to a phenomenon known as the *aperture problem.* However, the component of the optical flow in the direction of the *local brightness gradient* (also known as the *normal flow*, since it is perpendicular to the brightness edge) can always be computed locally without a great deal of difficulty. The magnitude of the *normal flow vector* is:

$$M_n = \frac{E_t}{\sqrt{E_x^2 + E_y^2}}$$

where:

M_n = normal flow vector magnitude
E_t = time derivative of pixel brightness
E_x = spatial derivative along x axis
E_y = spatial derivative along y axis.

When the motion of the camera is known, distances to points in the scene can be computed directly from the normal flow, with most accurate results at points where both the brightness gradient and the normal flow are greatest (Nguyen, 1993).

When camera motion is not known, the camera motion and distances to points in the scene can be recovered from the optical flow, but only up to a scaling factor. That is, it is possible to find the ratios between the distances to different points in the image, but not their absolute distances. If the distance to one point can be pin-pointed by another method (such as active sonar), however, then the distances to all points will be known. The computations are easiest if the camera motion is purely translational or purely rotational (Horn, 1986). Iterative and approximation schemes for estimating camera motion and distances from visual motion are still being actively investigated (Fermuller, 1991; Duric, 1991).

4.6.1 NIST Passive Ranging and Collision Avoidance

The method of flow extraction employed by the National Institute for Standards and Technology (NIST) reduces the computational burden by assuming that the camera is moving in a known fashion in a stationary world (Herman & Hong, 1991). These assumptions lead to two conclusions:

- The optical-flow field in the image (i.e., the flow direction at every point) can be predicted.
- Once the optical flow has been extracted, the flow vectors can be easily converted to range values.

These conclusions are generally true for arbitrary camera motion, including pure translation, pure rotation, and a combination of translation and rotation. The assumption that the flow field can be predicted enables precalculation of the true flow-vector directions; to extract optical flow, only the magnitudes of the flow vectors need to be computed. Knowledge of the flow field also enables the use of local image operators (for extracting information) that can run in parallel at all points in the image, further minimizing computation time. Additional details on the algorithms are presented by Lau, et al., (1992) and Liu, et al. (1993).

4.6.2 David Sarnoff Passive Vision

Researchers at David Sarnoff Research Center have developed algorithms for recovering scene geometry (range, 3-D orientation, and shape) from passively acquired binocular and motion imagery. Distance measurements are derived from intensity derivatives of two or more images of the same scene. The approach combines a local-brightness-constancy constraint with a global-camera-motion constraint to relate local range values with a global camera model and local image intensity derivatives.

Beginning with initial estimates of the camera motion and local range, the range is refined using the camera motion model as a constraint, whereupon the model is refined using local range estimates as constraints. This estimation procedure is iterated several times until convergence. The entire procedure is performed within a (spatially) coarse-to-fine algorithmic framework. Demonstration of this technology has made use of a commercial CCD camera and frame grabber for image capture coupled with a workstation to perform the actual range recovery in non-real-time. By way of illustration, Figure 4-31 (left) shows one image from a stereo pair; the brighter regions in the recovered range map depicted in Figure 4-31 (right) represent those regions closer to the cameras. The range values are plausible almost everywhere except at the image border and in the vicinity of the focus of expansion (near the image center).

Figure 4-31. One image of a pair is shown at left; pixel intensity in the resulting 3-D range map (right) is inversely related to range (courtesy David Sarnoff Research Center).

Limitations of this approach are two-fold. First, the basic formulation assumes that camera motion is small between captured images and that the image intensity of the same point between images is constant (brightness constancy); violation of either of these constraints can lead to erroneous results. Second, current estimates for a real-time implementation in commercially available hardware suggest that power requirements will be approximately 60 watts. Additional technical details on this technology are presented by Hanna (1991) and Wildes (1990a, 1190b, & 1991).

4.7 References

Aloimonos, J., Weiss, I., Bandyopadhyay, A., "Active Vision," First International Conference on Computer Vision, pp. 35-54, 1987.

Anderson, C.H., Burt, P.J., van der Wal, G. S., "Change Detection and Tracking Using Pyramid Transform Techniques," SPIE Vol. 579, Intelligent Robots and Computer Vision, pp. 72-78, 1985.

Bedard, R.J., et al., "Navigation of Military and Space Unmanned Ground Vehicles in Unstructured Terrains," 3rd Military Robotic Vehicle Conference, Medicine Hat, Canada, September 9-12, 1991a.

Bedard, R.J., et al., "The 1991 NASA Planetary Rover Program," 42nd International Astronautical Federation, Montreal, Canada, October 6-9, 1991b.

Besl, P.J., "Range Imaging Sensors," GMR-6090, General Motors Research Laboratory, 1988.

Blais, F., Rioux, M., Domey, J., Beraldin, J.A., "A Very Compact Real Time 3-D Range Sensor for Mobile Robot Applications," SPIE Vol. 1007, Mobile Robots III, Cambridge, MA, November, 1988.

Blais, F., Rioux, M., Domey, J., Beralsin, J.A., "On the Implementation of the BIRIS Range Sensor for Applications in Robotic Control and Autonomous Navigation," Canadian Conference on Electrical and Computer Engineering, Attawa, Ontario, Canada, pp. 37.1.1-37.1.4, September, 1990.

Burt, P.J., Adelson, E.H., "The Laplacian Pyramid as a Compact Image Code," *IEEE Transactions on Communications*, Vol. COM-31, No. 4, pp. 532-540, April, 1983

Burt, P.J., Anadan, P., Hanna, K., van der Wal, G., "A Front End Vision Processor for Unmanned Vehicles," Advanced Image Processing Group, David Sarnoff Research Center, Princeton, NJ, April, 1992.

Burt, P.J., Anadan, P., Hanna, K., van der Wal, G., Bassman, R., "A Front End Vision Processor for Vehicle Navigation," International Conference on Intelligent Systems, pp. 653-662, February, 1993.

Burt, P.J., Anadan, P., "Image Stabilization by Registration to a Reference Mosaic," 1994 Image Understanding Workshop, Monterey, CA, pp. 425-434, November, 1994.

Conrad, D.J., Sampson, R.E., "3D Range Imaging Sensors," in *Traditional and Non-Traditional Robotic Sensors*, T.C. Henderson, ed., NATO ASI Series, Vol. F63, Springer-Verlag, pp. 35-47, 1990.

Dunlap, G.D., Shufeldt, H.H., *Dutton's Navigation and Piloting*, Naval Institute Press, p. 1013, 1972.

Duric, Z., Aloimonos Y., "Passive Navigation: An Active and Purposive Solution," Technical Report CAR-TR-560, Center for Automation Research, University of Maryland, College Park, MD, 1991.

Evans, J.M., King, S.J., Weiman, C.F.R., "Visual Navigation and Obstacle Avoidance Structured Light Systems," U.S. Patent No. 4,954,962, 4 September, 1990.

Everett, H.R., DeMuth, D.E., Stitz, E.H., "Survey of Collision Avoidance and Ranging Sensors for Mobile Robots," Technical Report 1194, Naval Command Control and Ocean Surveillance Center, San Diego, CA, December, 1992.

Fermuller, C., Aloimonos, Y., "Estimating 3-D Motion from Image Gradients," Technical Report CAR-TR-564, Center for Automation Research, University of Maryland, College Park, MD, 1991.

Gibson, J., *The Perception of the Visual World*, Houghton Mifflin, Boston, MA, 1950.

Hanna, K.J., "Direct Multi-Resolution Estimation of Ego-Motion and Structure from Motion," Proceedings of IEEE Workshop on Visual Motion, pp. 156-162, 1991.

Hansen, M., Anandan, P., Dana, K., van der Wal, G, Burt, P., Real-Time Scene Stabilization and Mosaic Construction," 1994 Image Understanding Workshop, Monterey, CA, pp. 457-465, November, 1994.

Hamamatsu, "16 Step Range-Finder IC H2476- 01," Product Literature , Hamamatsu Corporation, January, 1990.

Heeger, D.J., Jepson, A., "Method and Apparatus for Image Processing to Obtain Three Dimensional Motion and Depth," U.S. Patent 4,980,762, MIT, Cambridge, MA, 25 December, 1990a

Heeger, D.J., Jepson, A., "Subspace Methods for Recovering Rigid Motion I: Algorithm and Implementation," Technical Report RBCV-TR- 90-35, University of Toronto, Ontario, Canada, November, 1990b.

Herman, M., Hong, T., "Visual Navigation using Optical Flow," Proceedings NATO Defense Research Group Seminar on Robotics in the Battlefield, Paris, France, March, 1991.

Horn, B. K. P., *Robot Vision*, The MIT Press, Cambridge, MA, 1986.

Jarvis, R.A., "A Perspective on Range Finding Techniques for Computer Vision," IEEE Transactions on Pattern Analysis and Machine Intelligence, Vol. PAMI-1, No. 2, pp. 122-139, March, 1983a.

Jarvis, R.A., "A Laser Time-of-Flight Range Scanner for Robotic Vision," IEEE Transactions on Pattern Analysis and Machine Intelligence, Vol. PAMI-5, No. 5, pp. 505-512, September, 1983b.

Kent, E.W., et al., Real-time Cooperative Interaction Between Structured Light and Reflectance Ranging for Robot Guidance," Robotica, Vol. 3, pp. 7-11, January-March, 1985.

Kilough, S.M., Hamel, W.R., "Sensor Capabilities for the HERMIES Experimental Robot," American Nuclear Society, Third Topical Meeting on Robotics and Remote Systems, Charleston, SC, CONF-890304, Section 4-1, pp. 1-7, March, 1989.

King, S.J., Weiman, C.F.R., "HelpMate Autonomous Mobile Robot Navigation System," SPIE Vol. 1388, Mobile Robots V, Boston, MA, pp. 190-198, November, 1990.

Laskowski, E.L., "Range Finder Wherein Distance Between Target and Source is Determined by Measuring Scan Time Across a Retroreflective Target," U.S. Patent No. 4,788,441, 29 November, 1988.

Lau, H., Hong, T., Herman, M., "Optimal Estimation of Optical Flow, Time-to-Contact and Depth," NISTIR 4919, National Institute of Standards and Technology, Gaithersburg, MD, September, 1992.

LeMoigue, J., Waxman, A.M., "Projected Light Grids for Short Range Navigation of Autonomous Robots," Proceedings, 7th IEEE Conference on Pattern Recognition, Montreal, Canada, pp. 203-206, 30 July - 2 August, 1984.

Liu, H., Hong, T., Herman, M., Chellappa, R., "A Reliable Optical Flow Algorithm Using 3-D Hermite Polynomials," NISTIR 5333, National Institute of Standards and Technology, Gaithersburg, MD, December, 1993.

Loewenstein, D., "Computer Vision and Ranging Systems for a Ping Pong Playing Robot," *Robotics Age*, pp. 21-25, August, 1984.

Malafeew, E., Kaliardos, W., "The MITy Micro-Rover: Sensing, Control, and Operation," AIAA/NASA Conference on Intelligent Robots in Field, Factory, Service, and Space, Houston, TX, pp. 696-704, March, 1994.

Matthies, L.H., "Stereo Vision for Planetary Rovers: Stochastic Modeling to Near-Real-time Implementation," International Journal of Computer Vision, Vol. 8, No. 1, July, 1992a.

Matthies, L.H., "Passive Stereo Range Imaging for Semi-Autonomous Land Navigation," Journal of Robotic Systems, Vol. 9, No. 6, September, 1992b.

Moser, J., Everett, H.R., "Wide-Angle Active Optical Triangulation Ranging System," SPIE Vol. 1195, Mobile Robots IV, Philadelphia, PA, November, 1989.

NAMCO, "LNFL03-A 5M/4-90," *Lasernet* Product Bulletin, NAMCO Controls, Mentor, OH, November, 1989.

Nitzan, D., "Assessment of Robotic Sensors," Proceedings of 1st International Conference on Robotic Vision and Sensory Controls, pp. 1-11, 1-3 April, 1981.

Nitzan, D. et al., "3-D Vision for Robot Applications." NATO Workshop: Knowledge Engineering for Robotic Applications, Maratea, Italy, 12-16 May, 1986.

Nguyen, H. G., "Summary of Auto-Landing Problem Analysis and Proposal," NRaD Memorandum 943/11-93, Naval Command Control and Ocean Surveillance Center, February, 1993.

Nguyen, H.G., Blackburn, M.R., "A Simple Method for Range Finding via Laser Triangulation," NCCOSC Technical Document, TD-2734, Naval Command Control and Ocean Surveillance Center, San Diego, CA, January, 1995.

Poggio, T., "Vision by Man and Machine," *Scientific America*, Vol. 250, No. 4, pp. 106-116, April, 1984.

Rioux, M., Blais, F., "Compact Three-Dimensional Camera for Robotic Applications," Journal of Optical Society of America, Vol. 3, pp. 1518-1521, September, 1986.

Schwartz, J.T., "Structured Light Sensors for 3-D Robot Vision," Technical Report No. 65, Courant Institute of Mathematical Sciences, New York University, undated.

Slack, M., "Generating Symbolic Maps from Grid Based Height Maps," JPL-D-6948, December 7, 1989.

Swain, M.J., Stricker, M., eds., *Promising Directions in Active Vision*, Report from the National Science Foundation Active Vision Workshop, University of Chicago, IL, 1991.

Uomori, K., Morimura, A., Ishii, A., Sakaguchi, H., Kitamura, Y., "Automatic Image Stabilizing System by Full Digital Signal Processing," *IEEE Transactions on Consumer Electronics*, Vol. 36, No. 3, pp. 510-519, 1990.

Vuylsteke, P., Price, C.B., Oosterlinck, A., "Image Sensors for Real-Time 3D Acquisition, Part 1," *Traditional and Non-Traditional Robotic Sensors*, T.C. Henderson, ed., NATO ASI Series, Vol. F63, Springer-Verlag, pp. 187-210, 1990.

Wavering, A.J., Fiala, J.C., Roberts, K.J., Lumia, R., "TRICLOPS: A High-Powered Trinocular Active Vision System," IEEE International Conference on Robotics and Automation, pp. 410-417, 1993.

Wildes, R.P., "Qualitative 3D Shape from Stereo," SPIE Intelligent Robots and Computer Vision Conference, pp. 453-463, 1990a.

Wildes, R.P., "Three-Dimensional Surface Curvature from Binocular Stereo Disparity," *Optical Society of America Technical Digest*, Vol. 25, p. 58, 1990b.

Wildes, R.P., "Direct Recovery of Three-Dimensional Scene Geometry from Binocular Stereo Disparity," *IEEE Transactions on Pattern Analysis and Machine Intelligence*, Vol. 13, No. 8, pp. 761-774, August, 1991.

Young, G.S., Hong, T.H., Herman, M., Yang, J.C.S., "Obstacle Avoidance for a Vehicle Using Optical Flow," Technology Description, NIST, Gaithersburg, MD, July, 1992.

5

Time of Flight

Time-of-flight (TOF) ranging systems measure the round-trip time required for a pulse of emitted energy to travel to a reflecting object, then echo back to a receiver. Ultrasonic, RF, or optical energy sources are typically employed; the relevant parameters involved in range calculation, therefore, are the speed of sound in air (roughly 1 foot per millisecond), and the speed of light (1 foot per nanosecond). Using elementary physics, distance is determined by multiplying the velocity of the energy wave by the time required to travel the round-trip distance:

$$d = v\,t$$

where:

d = round-trip distance
v = speed of propagation
t = elapsed time.

The measured time is representative of traveling twice the separation distance (i.e., out and back) and must therefore be reduced by half to result in actual range to the target.

The advantages of *TOF* systems arise from the direct nature of their straight-line active sensing. The returned signal follows essentially the same path back to a receiver located coaxially with or in close proximity to the transmitter. In fact, it is possible in some cases for the transmitting and receiving transducers to be the same device. The absolute range to an observed point is directly available as output with no complicated analysis required, and the technique is not based on any assumptions concerning the planer properties or orientation of the target surface. The *missing parts* problem seen in triangulation does not arise because minimal or no offset distance between transducers is needed. Furthermore, *TOF* sensors maintain range accuracy in a linear fashion as long as reliable echo detection is sustained, while triangulation schemes suffer diminishing accuracy as distance to the target increases.

Potential error sources for *TOF* systems include the following:

- Variations in the speed of propagation, particularly in the case of acoustical systems.
- Uncertainties in determining the exact time of arrival of the reflected pulse (Figueroa & Lamancusa, 1992).
- Inaccuracies in the timing circuitry used to measure the round-trip time of flight.
- Interaction of the incident wave with the target surface.

Each of these areas will be briefly addressed below, and discussed later in more detail along with other factors influencing performance in Chapters 8 and 9.

Propagation Speed — For mobile robotic applications, changes in the propagation speed of electromagnetic energy are for the most part inconsequential and can basically be ignored, with the exception of satellite-based position-location systems as presented in Chapter 14. This is not the case, however, for acoustically based systems, where the speed of sound is markedly influenced by temperature changes, and to a lesser extent by humidity. (The speed of sound is actually proportional to the square root of temperature in degrees Rankine; an ambient temperature shift of just 30 degrees can cause a 1-foot error at a measured distance of 35 feet.)

Detection Uncertainties — So called *time-walk errors* are caused by the wide dynamic range in returned signal strength as a result of: 1) varying reflectivities of target surfaces, and, 2) signal attenuation to the fourth power of distance due to spherical divergence. These differences in returned signal intensity influence the rise time of the detected pulse, and in the case of fixed-threshold detection will cause the less reflective targets to appear further away (Lang, et al., 1989). For this reason, *constant fraction timing discriminators* are typically employed to establish the detector threshold at some specified fraction of the peak value of the received pulse (Vuylsteke, et al., 1990). See also Chapter 8.

Timing Considerations — The relatively slow speed of sound in air makes TOF ranging a strong contender for low-cost acoustically-based systems. Conversely, the propagation speed of electromagnetic energy can place severe requirements on associated control and measurement circuitry in optical or RF implementations. As a result, TOF sensors based on the speed of light require subnanosecond timing circuitry to measure distances with a resolution of about a foot (Koenigsburg, 1982). More specifically, a desired resolution of 1 millimeter requires a timing accuracy of 3 picoseconds (Vuylsteke, et al., 1990). This capability is somewhat expensive to realize and may not be cost effective for certain applications, particularly at close range where high accuracies are required.

Surface Interaction — When light, sound, or radio waves strike an object, any detected echo represents only a small portion of the original signal. The remaining energy reflects in scattered directions and can be absorbed by or pass through the target, depending on surface characteristics and the angle of incidence of the beam. Instances where no return signal is received at all can occur because of specular reflection at the object surface, especially in the ultrasonic region of the energy spectrum. If the transmission source approach angle meets or exceeds a certain critical value, the reflected energy will be deflected outside of the sensing envelope of the receiver. Scattered signals can reflect from secondary objects as well, returning to the detector at various times to generate false signals that can yield questionable or otherwise noisy data. To compensate, repetitive measurements are usually averaged to bring the signal-to-noise ratio within acceptable levels, but at the expense of additional time required to determine a single range value.

5.1 Ultrasonic TOF Systems

Ultrasonic TOF ranging is today the most common technique employed on indoor mobile robotic systems, primarily due to the ready availability of low-cost systems and their ease of interface. Over the past decade, much research has been conducted investigating applicability in such areas as world modeling and collision avoidance (Chapter 10), position estimation (Chapter 15), and motion detection (Chapter 17). Several researchers have more recently begun to assess the effectiveness of ultrasonic sensors in exterior settings (Pletta, et al., 1992; Langer & Thorpe, 1992; Pin & Watanabe, 1993; Hammond, 1994). In the automotive industry, BMW now incorporates four piezoceramic transducers (sealed in a membrane for environmental protection) on both front and rear bumpers in its Park Distance Control system (Siuru, 1994).

Four of the most popular commercially available ultrasonic ranging systems will be reviewed in the following sections.

5.1.1 National Semiconductor's *LM1812 Ultrasonic Transceiver*

The *LM1812,* discontinued in 1990, was a general purpose ultrasonic transceiver IC originally designed to support fish- and depth-finding products in the recreational electronics industry (Frederiksen & Howard, 1974). The 18-pin chip contained a pulse-modulated class-C transmitter, a high-gain receiver, a pulse modulation detector, and noise rejection circuitry as shown in Figure 5-1 (National, 1991). Maximum range was 100 feet in water and 20 feet in air, at typical operating frequencies of 20 to 350 KHz.

The chip's specifications (National, 1989) listed the following features:

- *Monostatic* (single transducer) or *bistatic* (dual transducer) operation.
- Transducers interchangeable without realignment.
- No external transistors required.

- Impulse noise rejection.
- No heat sinking required.
- 12 watts peak transmit power.
- Power consumption of 50 milliamps at 18 volts DC.

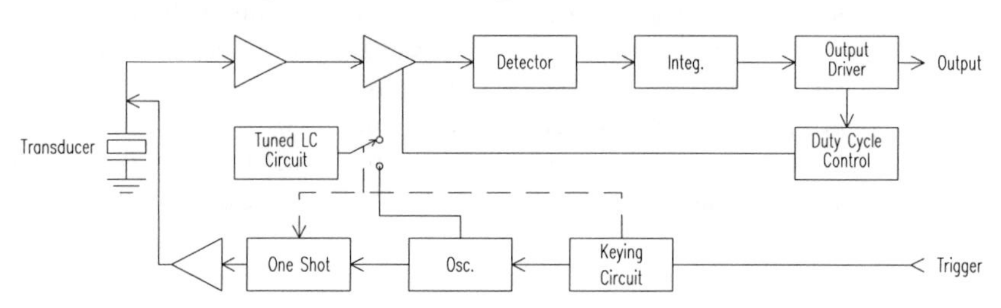

Figure 5-1. Block diagram of *LM1812* monolithic sonar transceiver (courtesy National Semiconductor Corporation).

Two different types of ultrasonic transducers, *electrostatic* and *piezoceramic* (also known as *piezoelectric*), were commonly used with the *LM1812* (Everett, 1982; Pletta, et al., 1992). *Electrostatic* transducers transmit an outgoing signal and act as an electrostatic microphone in order to receive the reflected waveform (National, 1991). *Piezoceramic* transducers are electrically similar to quartz crystals and resonant at only two frequencies: the *resonant* and *antiresonant* frequencies (Pletta, et al., 1992). Transmission is most efficient at the resonant frequency while optimum receiving sensitivity occurs at the antiresonant frequency (National, 1989). In bistatic systems, the resonant frequency of the transmitting transducer is matched to the antiresonant frequency of the receiver for optimal performance.

The majority of practical applications, however, use a monostatic configuration for which the maximum echo sensitivity occurs at a frequency close to resonance. The ultrasonic ranging system on ROBART I, for example, was based on the *LM1812* in conjunction with a single 40-KHz piezoceramic transducer (see Chapter 10). Pletta, et al. (1992) elected to use three Massa piezoceramic transducers operating at 26 KHz in an *LM1812*-based collision-avoidance sonar for Sandia's *Telemanaged Mobile Security System*. Effective range to favorable targets (rough surfaced or normal to the beam) was approximately 12 meters.

The receiver gain could be varied over time by attenuating the signal between pin 3 (first-stage amplifier output) and pin 2 (second-stage amplifier input) using external circuitry as shown in Figure 5-2. The 12-volt trigger pulse that keyed the transmitter simultaneously charged C_1 to a preset voltage determined by R_8, thereby turning off the FET to block the transducer ring-down signal (National, 1989). C_1 then slowly discharged through R_1, decreasing the gate voltage and allowing the FET to conduct. The resulting attenuation of the received signal thus decreased as the voltage on C_1 fell, effectively increasing overall receiver gain as a function of elapsed time. This feature served to both block the unwanted ring-down effect as well as keep the amplifier gain proportionally matched to the decay in returned-echo intensity resulting from the inverse square law.

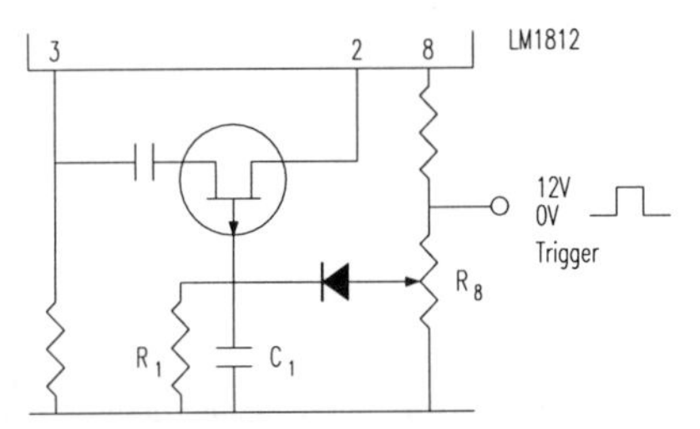

Figure 5-2. An optional time-variable FET attenuator could be connected between pins 2 and 3 of the *LM1812* to implement a ramped-gain response (adapted from National, 1989).

5.1.2 Massa Products *Ultrasonic Ranging Module Subsystems*

Massa Products Corporation, Hingham, MA, offers a full line of ultrasonic ranging subsystems with maximum detection ranges from 2 to 30 feet (Massa, undated). The *E-201B series* sonar operates in the bistatic mode with separate transmit and receive transducers, either side by side for echo ranging or as an opposed pair for unambiguous distance measurement between two uniquely defined points. This latter configuration is sometimes used in ultrasonic position location systems (see Chapter 15) and provides twice the effective operating range with respect to that advertised for conventional echo ranging. The *E-220B series* (Figure 5-3) is designed for monostatic (single-transducer) operation but is otherwise functionally identical to the *E-201B*. Either version can be externally triggered on command, or internally triggered by a free-running oscillator at a repetition rate determined by an external resistor.

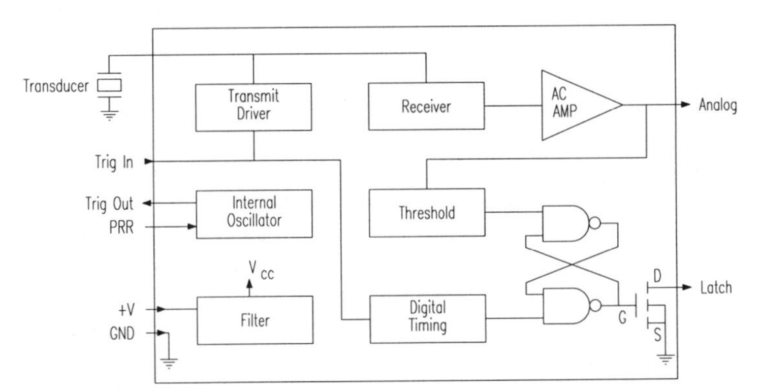

Figure 5-3. The single-transducer Massa *E-220B-series* ultrasonic ranging module can be internally or externally triggered and offers both analog and digital outputs (courtesy Massa Products Corp.).

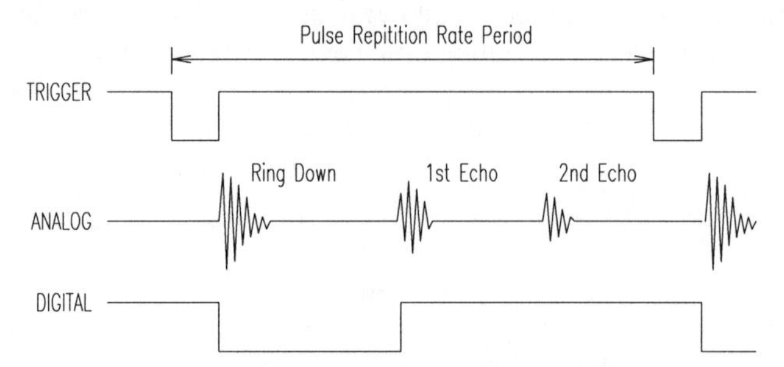

Figure 5-4. Timing diagram for the *E-220B series* ranging module showing analog and digital output signals in relationship to the trigger input (courtesy Massa Products Corp.).

Selected specifications for the four operating frequencies available in the *E-220B series* are listed in Table 5-1 below. A removable focusing horn is provided for the 26- and 40-KHz models that decreases the effective beamwidth (when installed) from 35 to 15 degrees. The horn must be in place to achieve the maximum listed range.

Table 5-1. Selected specifications for the monostatic *E-220B Ultrasonic Ranging Module Subsystems.* The *E-201 series* is a bistatic configuration with very similar specifications.

Parameter	220B/215	220B/150	220B/40	220B/26	Units
Range	4 - 24	8 - 60	24 - 240	24 - 360	inches
Beam width	10	10	35 (15)	35 (15)	degrees
Frequency	215	150	40	26	KHz
Max rep rate	150	100	25	20	Hz
Resolution	0.03	0.04	0.3	0.4	inches
Power	8 - 15	8 - 15	8 - 15	8 - 15	volts DC
Weight	4 - 8	4 - 8	4 - 8	4 - 8	ounces

5.1.3 Polaroid *Ultrasonic Ranging Modules*

The Polaroid ranging module is an active TOF device developed for automatic camera focusing and determines the range to target by measuring elapsed time between transmission of an ultrasonic waveform and the detected echo (Biber, et al., 1980). Probably the single most significant sensor development from the standpoint of its catalytic influence on the robotics research community, this system is the most widely found in the literature (Koenigsburg, 1982; Moravec & Elfes, 1985; Everett, 1985; Kim, 1986; Arkin, 1989; Borenstein & Koren, 1990). Representative of the general characteristics of a number of such ranging devices, the Polaroid unit soared in popularity as a direct consequence of its extremely low cost (Polaroid offers both the transducer and ranging module circuit board for less than $50), made possible by high-volume usage in its original application as a camera autofocus sensor.

The most basic configuration consists of two fundamental components: 1) the ultrasonic transducer and 2) the ranging module electronics. A choice of transducer types is now available. In the original instrument-grade electrostatic version (Figure 5-5), a very thin metalized diaphragm mounted on a machined backplate forms a capacitive transducer (Polaroid, 1981). A smaller diameter electrostatic transducer (*7000-Series*) has also been made available, developed for the Polaroid *Spectra* camera (Polaroid, 1987). A ruggedized piezoelectric (*9000-Series*) *environmental transducer* introduced for applications that may be exposed to rain, heat, cold, salt spray, and vibration is able to meet or exceed guidelines set forth in SAE J1455 January 1988 specification for heavy-duty trucks.

Figure 5-5. From left to right: 1) the original instrument grade electrostatic transducer, 2) *9000-Series* environmental transducer, and 3) *7000 Series* electrostatic transducer (courtesy Polaroid Corp.).

The original Polaroid ranging module (607089) functioned by transmitting a *chirp* of four discrete frequencies in the neighborhood of 50 KHz (see also Chapter 8). The *SN28827* module was later developed with a reduced parts count, lower power consumption, and simplified computer interface requirements. This second-generation board transmits only a single frequency at 49.1 KHz. A third-generation board (*6500 series*) introduced in 1990 provided yet a further reduction in interface circuitry, with the ability to detect and report multiple echoes (Polaroid, 1990). An *Ultrasonic Ranging Developer's Kit* based on the Intel *80C196* microprocessor is now available (Figure 5-6) that allows software control of transmit frequency, pulse width, blanking time, amplifier gain, and achieved range measurements from 1 inch to 50 feet (Polaroid, 1993).

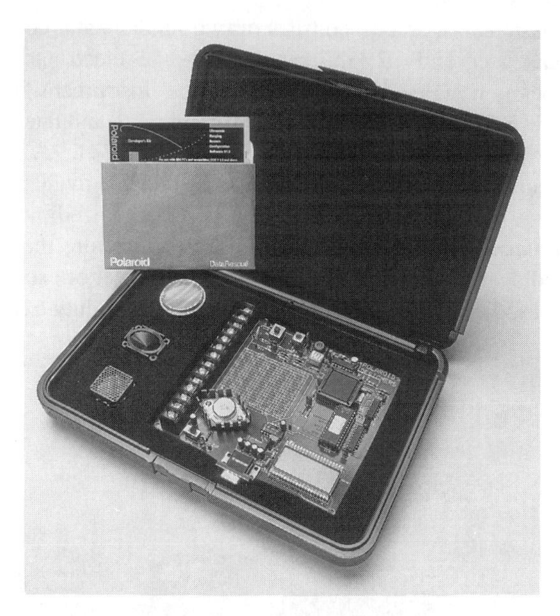

Figure 5-6. The Polaroid *Ultrasonic Ranging Developer's Kit* offers programmable pulse, frequency, and gain parameters, with the ability to detect multiple echoes (courtesy Polaroid Corp.).

The range of the Polaroid system runs from about 1 foot out to 35 feet, with a half-power (-3dB) beam dispersion angle of approximately 12 degrees for the original instrument-grade electrostatic transducer. A typical operating cycle is as follows.

- The control circuitry fires the transducer and waits for an indication that transmission has begun.
- The receiver is blanked for a short period of time to prevent false detection due to residual transmit signal ringing in the transducer.
- The received signals are amplified with increased gain over time to compensate for the decrease in sound intensity with distance.
- Returning echoes that exceed a fixed-threshold value are recorded and the associated distances calculated from elapsed time.

In the *single-echo* mode of operation for the *6500-series* module, the *blank* (BLNK) and *blank-inhibit* (BINH) lines are held low as the *initiate* (INIT) line goes high to trigger the outgoing pulse train. The *internal blanking* (BLANKING) signal automatically goes high for 2.38 milliseconds to prevent transducer ringing from being misinterpreted as a returned echo. Once a valid return is received, the echo (ECHO) output will latch high until reset by a high-to-low transition on INIT. For *multiple-echo* processing, the *blank* (BLNK) input must be toggled high for at least 0.44 milliseconds after detection of the first return signal to reset the *echo* output for the next return as shown in Figure 5-7 (Polaroid, 1990).

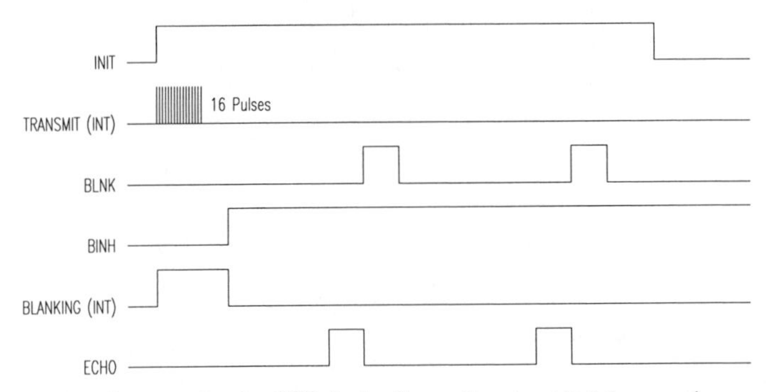

Figure 5-7. Timing diagrams for the 6500-*Series Sonar Ranging Module* executing a multiple-echo-mode cycle with blanking input (courtesy Polaroid Corp.).

The ultrasonic ranging capability of ROBART II is based entirely on the Polaroid system (three *SN28827* ranging modules each multiplexed to 12 electrostatic transducers). For obstacle avoidance purposes, a fixed array of 11 transducers is installed on the front of the body trunk to provide distance information to objects in the path of the robot as shown in Figure 5-8. A ring of 24 additional ranging sensors (15 degrees apart) is mounted just below the robot's head and used to gather range information for position estimation. A final ranging unit is located on the rotating head assembly, allowing for distance measurements to be made in various directions. Reliability of the Polaroid components has been exceptional, with no failures or degraded performance of any type in over eight years of extended operation.

Table 5-2. Selected specifications for the various Polaroid ultrasonic ranging modules.

Parameter	607089	SN28827	6500	Units
Maximum range	35	35	35	feet
Minimum range	10.5	6	6	inches
Number of pulses	56	16	16	
Blanking time	1.6	2.38	2.38	milliseconds
Resolution	1	2	1	percent
Gain steps	16	12	12	
Multiple echo	no	yes	yes	
Programmable frequency	no	no	yes	
Power	4.7 - 6.8	4.7 - 6.8	4.7 - 6.8	volts
	200	100	100	milliamps

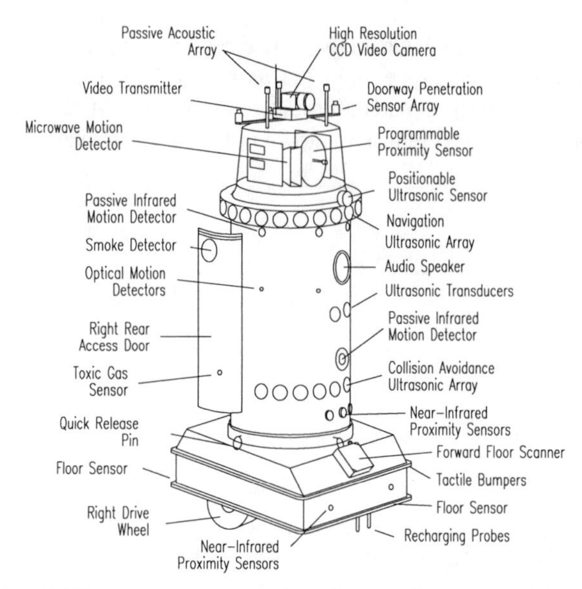

Figure 5-8. ROBART II, an autonomous security robot, employs a total of 132 external sensors for navigation and intruder detection, including 36 Polaroid electrostatic transducers.

5.1.4 Cybermotion *CA-2 Collision Avoidance System*

The *CA-2 Collision Avoidance System* is a dual-channel ultrasonic ranging module developed by Cybermotion, Inc., Salem, VA, for use on indoor vehicles operating at speeds up to 10 miles per hour. The *CA-2* achieves a maximum detection range of 8 feet at a 10-Hz update rate, with programmable resolution (0.084 inch standard) over the span of interest (Cybermotion, 1991). Two broad-beam (70-degree) ceramic transducers are employed for maximum protection in the direction of vehicle travel. Four operating modes are provided:

- OFF — The system is powered up but no transducers are fired.
- LEFT — The left transducer only is fired.
- RIGHT — The right transducer only is fired.
- BOTH — The left and right transducers are alternately fired.

Hammond (1993) reports that most man-made noise sources have energy peaks below 50 KHz, and much of this industrial noise spectrum is avoided by choosing an operating frequency of 75 KHz. In addition, the *CA-2* employs a number of specialized techniques for improving the generally poor signal-to-noise ratio experienced by wide-beam transducers in order to achieve higher immunity to sources of ultrasonic interference (i.e., rotating machinery, leaking or rushing air, fluorescent lighting, other ultrasonic equipment). Referring now to Figure 5-9, the received echo signal generated by the ultrasonic transducer is passed through a narrow-band gain-controlled amplifier before presentation to an envelope detector and two additional stages of baseband filtering. The output signal is then digitized and stored in memory, whereupon five different filtering algorithms are invoked to eliminate transducer ring-down, white noise, impulse noise, residual echoes from previous ranging operations, and interference from other robots (Hammond, 1993).

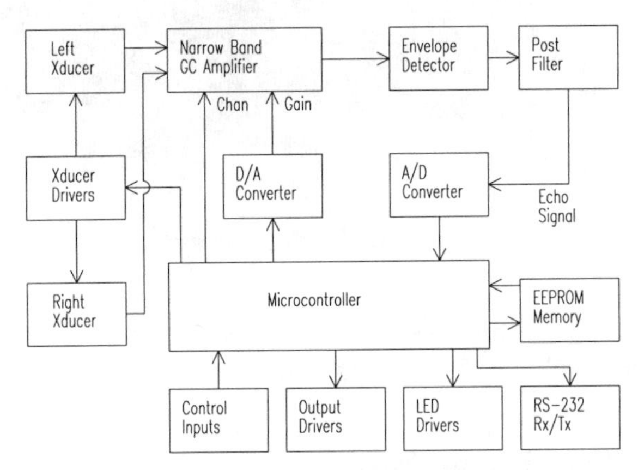

Figure 5-9. Block diagram of the Cybermotion *CA-2* Collision Avoidance System (adapted from Hammond, 1993).

The resulting digital signature is then compared to a complex threshold generated from a programmable baseline and several dynamically calculated components, with distance computed for the first point in time where signal amplitude exceeds the threshold value. This *range to first echo* is compared to three preset variables downloaded from system EEPROM on initialization (or subsequently reset by an external command):

- SLOW — Range threshold for causing the vehicle to decrease speed.
- STOP — Range threshold below which the vehicle should stop.
- HORN — Range threshold for enabling a warning enunciator.

If the measured range is less than any of the threshold values listed above for any two of five consecutive readings, the appropriate output signal is generated. The measured range must then exceed the prespecified threshold for five consecutive pings to cancel the indicated condition. Red LED status lights are associated with both the SLOW and STOP outputs for convenience.

The *CA-2* (Figure 5-10) is offered by Cybermotion as a stand-alone unit measuring 7.25 wide, 5.75 inches deep, and 1 inch high, with both parallel and serial interfaces. System sensitivity is programmable, down to as small as a 1-

inch-square surface at a distance of 5 feet. Power consumption is 150 milliamps at 12 volts DC.

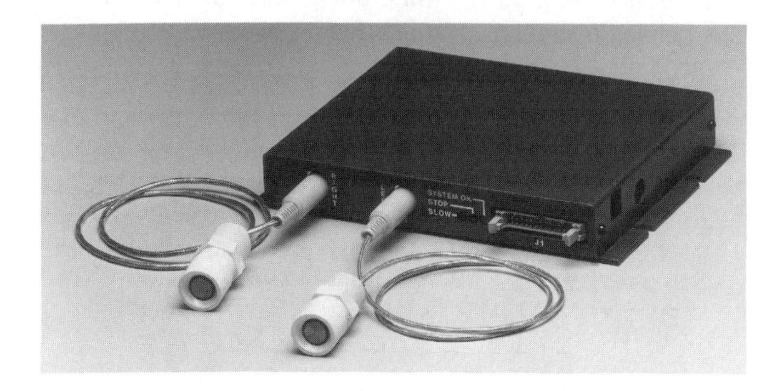

Figure 5-10. The *CA-2 Collision Avoidance System* is available as a stand-alone non-contact ranging subsystem (courtesy Cybermotion, Inc.).

5.2 Laser-Based TOF Systems

Laser-based TOF ranging systems, also known as *laser radar* or *lidar*, first appeared in work performed at the Jet Propulsion Laboratory, Pasadena, CA, in the 1970s (Lewis & Johnson, 1977). Laser energy is emitted in a rapid sequence of short bursts aimed directly at the object being ranged. The time required for a given pulse to reflect off the object and return is measured and used to calculate distance to the target based on the speed of light. Accuracies for early sensors of this type could approach a few centimeters over the range of 1 to 5 meters (NASA, 1977; Depkovich & Wolfe, 1984).

5.2.1 Schwartz Electro-Optics Laser Rangefinders

Schwartz Electro-Optics, Inc. (SEO), Orlando, FL, produces a number of laser TOF rangefinding systems employing an innovative *time-to-amplitude-conversion* scheme to overcome the subnanosecond timing requirements necessitated by the speed of light. As the laser fires, a precision film capacitor begins discharging from a known setpoint at a constant rate, with the amount of discharge being proportional to the round-trip time-of-flight (Gustavson & Davis, 1992). An analog-to-digital conversion is performed on the sampled capacitor voltage at the precise instant a return signal is detected, whereupon the resulting digital representation is converted to range and time-walk corrected using a look-up table.

SEO *LRF-X* Series Rangefinders

The *LRF-X* series rangefinder shown in Figure 5-11 features a compact size, high-speed processing, and an ability to acquire range information from most surfaces (i.e., minimum 10-percent Lambertian reflectivity) out to a maximum of 100 meters. The basic system uses a pulsed InGaAs laser diode in conjunction with an avalanche photodiode detector and is available with both analog and digital (RS-232) outputs. The following general specifications detail the sensor's performance (SEO, 1991a).

Figure 5-11. The *LRF-200* series rangefinder (courtesy Schwartz Electro Optics, Inc.).

Table 5-3. Selected specifications for the *LRF-200* laser rangefinder.

Parameter	Value	Units
Maximum range	100	meters
Minimum range	1	meter
Accuracy	±0.3	meter
Range jitter	±12	centimeters
Wavelength	902	nanometers
Diameter	8.9	centimeters
Length	17.75	centimeters
Weight	1	kilogram
Power	8 to 24	volts DC
	5	watts

The *High Accuracy Altitude Measurement System (HAAMS)* is an enhanced variation of the basic *LRF* concept intended as a *lidar altimeter* for aircraft. The *HAAMS* system operates at a 3-KHz update rate with a maximum range of 200 meters and is available in the same 8.9-centimeter-diameter cylindrical package as the *LRF-200*. An inclinometer was added to automatically compensate for aircraft angle of bank. In addition, peak-detection feedback was incorporated to reduce *time-walk errors* for an increased accuracy of 3 to 4 inches.

SEO Hover Obstacle Proximity Sensor System

The *Hover Obstacle Proximity Sensor System (HOPSS)* was developed for the US Army (SEO, 1991c) as an onboard pilot alert to the presence of surrounding obstructions. Located on the bottom of the fuselage directly below the main-rotor driveshaft (Figure 5-12), the *HOPSS* system provides continuous distance and azimuth measurements in the horizontal plane of a helicopter.

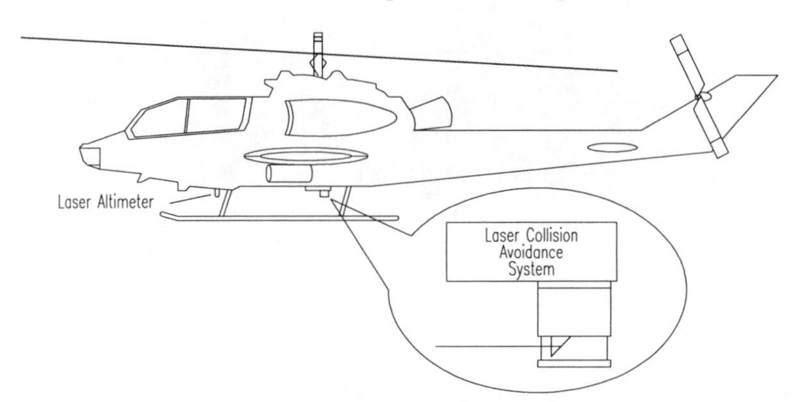

Figure 5-12. Placement of the *Hover Optical Proximity Sensor System* on a US Army helicopter (courtesy Schwartz Electro-Optics, Inc.).

A high-pulse-repetition-rate GaAs laser-diode emitter shares a common aperture with a sensitive avalanche photodiode detector. The transmit and return beams are reflected from a motor-driven prism rotating at 300 rpm as depicted in Figure 5-13. Range measurements are taken at 1.5-milliradian intervals and correlated with the azimuth angle using an optical encoder. The detection range for a 3/8-inch cable is greater than 75 feet, while larger targets can be reliably sensed out to 250 feet or more. Detected obstacles are displayed in a format similar to a radar plan-position indicator, and visual and audible warnings are provided in the event the measured range within prespecified warning zones falls below an established threshold. To achieve broader three-dimensional sensor coverage, a concept employing two counter-rotating wedge-prisms is under investigation (SEO, 1991d).

Figure 5-13. Distance measurements are correlated with the azimuth angle of a rotating prism to yield range and bearing information for detected obstacles (courtesy Schwartz Electro-Optics, Inc.).

Table 5-4. Selected specifications for the *Hover Obstacle Avoidance Proximity Sensor System.*

Parameter	Value	Units
Wavelength	904	nanometers
Output energy	50	nanojoules
Pulse width	7	nanoseconds
Minimum range	5	feet
Maximum range	250	feet
Accuracy	±0.5	feet
Scan angle	360	degrees
Scan rate	5	Hz
Samples per scan	2048	
Diameter	7	inches
Length	11.75	inches
Weight (sensor)	< 10	pounds
(display)	< 10	pounds
Power	18 to 36	volts DC
	< 2	amps

SEO *TreeSense*

TreeSense was developed by SEO for automating the selective application of pesticides to orange trees, where the goal was to enable individual spray nozzles only when a tree was detected within their associated field of coverage. The sensing subsystem consists of a horizontally oriented HAAMS unit mounted on the back of an agricultural vehicle, suitably equipped with a rotating mirror arrangement that scans the beam in a vertical plane orthogonal to the direction of travel. The scan rate is controllable up to 40 revolutions per second (35 typical). The ranging subsystem is gated on and off twice during each revolution to illuminate two 90-degree fan-shaped sectors to a maximum range of 25 feet on either side of the vehicle as shown in Figure 5-14. (The existing hardware is theoretically capable of ranging to 100 feet using a PIN photodiode and can be extended further through an upgrade option that incorporates an avalanche photodiode detector.)

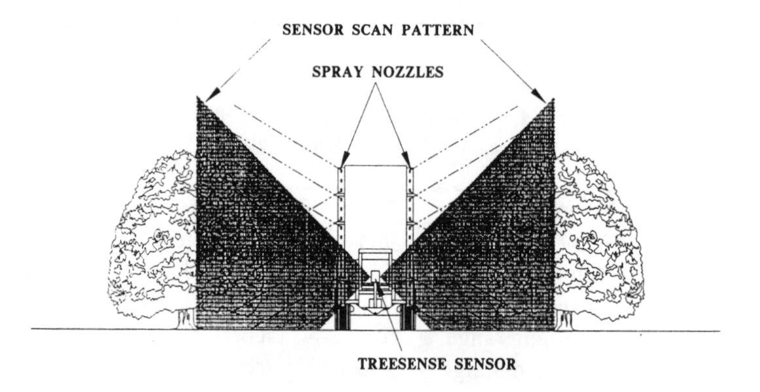

Figure 5-14. The *TreeSense* system illuminates two fan-shaped sectors (±45 degrees with respect to horizontal) on either side of the path to determine the location of trees for precision application of pesticides (courtesy Schwartz Electro-Optics, Inc.).

The *TreeSense* system is hard-wired to a valve manifold to enable/disable a vertical array of nozzles for the spraying of insecticides, but analog as well as digital (RS-232) output can easily be made available for other applications. (A *TreeSense* unit was purchased by Robotic Systems Technology, Inc. for evaluation as a possible collision avoidance sensor on the MDARS Exterior robot.) The system is housed in a rugged fiberglass enclosure (Figure 5-15) with a total weight of only 5 pounds. Power requirements are 12 watts at 12 volts DC.

Table 5-5. Selected specifications for the SEO *TreeSense* system.

Parameter	Value	Units
Maximum range	100	feet
Accuracy	3-4	inches
Wavelength	902	nanometers
Pulse repetition rate	18	KHz
Length	9	inches
Width	9	inches
Height	4.5	inches
Weight	5	pounds
Power	12	volts DC

Figure 5-15. The *TreeSense* system is enclosed in a fiberglass housing with two rectangular windows on either side for the left and right fan-shaped beams (courtesy Schwartz Electro-Optics, Inc.).

SEO *AutoSense*

The *AutoSense I* system was developed by SEO under a Department of Transportation Small Business Innovative Research (SBIR) effort as a replacement for buried inductive loops for traffic signal control. (Inductive loops don't always sense motorcyclists and some of the smaller cars with fiberglass or plastic body panels, and replacement or maintenance can be expensive as well as disruptive to traffic flow.) The system is configured to look down at about a 30-degree angle on moving vehicles in a traffic lane as illustrated in Figure 5-16. The ability to accurately measure vehicle height profiles as well as velocities opens up new possibilities for classifying vehicles as part of the *intelligent vehicle highway systems (IVHS)* concept (Olson, et al., 1994).

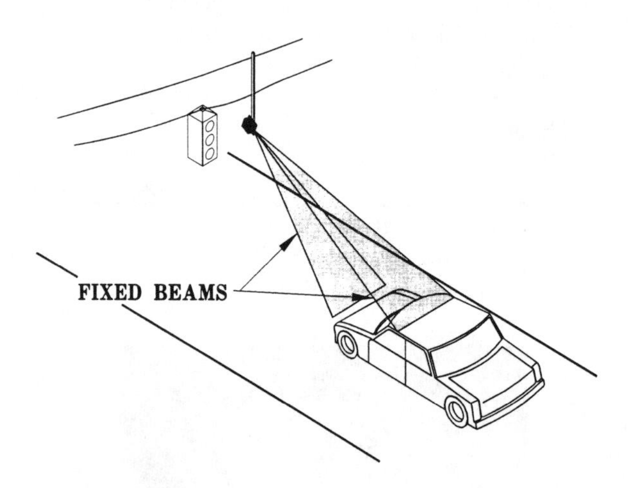

Figure 5-16. Two fan-shaped beams look down on moving vehicles for improved target detection (courtesy Schwartz Electro-Optics, Inc.).

AutoSense I uses a PIN photodiode detector and a pulsed (8 nanosecond) InGaAs near-infrared laser-diode source with peak power of 50 watts. The laser output is directed by a beam splitter into a pair of cylindrical lenses to generate two fan-shaped beams 10 degrees apart in elevation for improved target detection. (The original prototype projected only a single spot of light but ran into problems due to target absorption and specular reflection.) As an added benefit, the use of two separate beams makes it possible to calculate the speed of moving vehicles to an accuracy of 1 mile per hour. In addition, a two-dimensional image (i.e., length and width) is formed of each vehicle as it passes through the sensor's field of view, providing accurate data for numerous vehicle classification applications.

An improved second-generation unit (*AutoSense II*) uses an avalanche photodiode detector instead of the PIN photodiode for greater sensitivity, and a multifaceted rotating mirror with alternating pitches on adjacent facets to create the two beams. Each beam is scanned across the traffic lane 720 times per second, with 15 range measurements made per scan. This azimuthal scanning action allows for generation of a precise three-dimensional profile to better facilitate vehicle classification in automated toll booth applications. An abbreviated system block diagram is depicted in Figure 5-17.

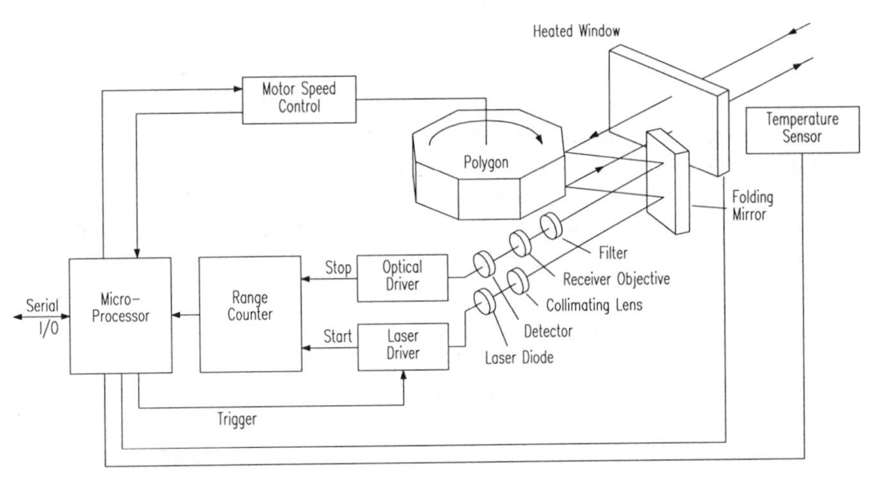

Figure 5-17. Simplified block diagram of the *AutoSense II* time-of-flight 3-D ranging system (courtesy Schwartz Electro-Optics, Inc.).

Intensity information from the reflected signal is used to correct the *time-walk* error in threshold detection resulting from varying target reflectivities, for an improved range accuracy of ±3 inches over a 5- to 30-foot field of regard. The scan resolution is 1 degree, and vehicle velocity can be calculated with an accuracy of 2 mph at speeds up to 60 mph. High-speed RS-232 and RS-422 outputs are provided. A third-generation *AutoSense III* is now under development for an application in Canada that requires three-dimensional vehicle profile generation at speeds up to 100 miles per hour.

Table 5-6. Selected specifications for the SEO *AutoSense II* system.

Parameter	Value	Units
Maximum range	30	feet
Accuracy	±3	inches
Wavelength	904	nanometers
Pulse repetition rate	15	KHz
Scan rate	29.29	rps
Length	9	inches
Width	4.5	inches
Height	9	inches
Weight	5	pounds
Power	12	volts DC
	1	amp

5.2.2 RIEGL Laser Measurement Systems

RIEGL Laser Measurement Systems, Horn, Austria, offers a number of commercial products (i.e., laser binoculars, surveying systems, "speed guns," level sensors, profile measurement systems, and tracking laser scanners) employing short-pulse TOF laser ranging. Typical applications include lidar altimeters, vehicle speed measurement for law enforcement, collision avoidance for cranes and vehicles, and level sensing in silos. All RIEGL products are distributed in the United States by RIEGEL USA, Orlando, FL.

LD90-3 Laser Rangefinder

The RIEGL *LD90-3 series* laser rangefinder employs a near-infrared laser diode source and a photodiode detector to perform TOF ranging out to 500 meters with diffuse surfaces, and to over 1000 meters in the case of cooperative targets. Round-trip propagation time is precisely measured by a quartz-stabilized clock and converted to measured distance by an internal microprocessor using one of two available algorithms. The *clutter suppression* algorithm incorporates a combination of range measurement averaging and noise rejection techniques to filter out backscatter from airborne particulates, and is therefore useful when operating under conditions of poor visibility (Riegel, 1994). The *standard measurement* algorithm, on the other hand, provides rapid range measurements without regard for noise suppression, and can subsequently deliver a higher update rate under more favorable environmental conditions. Worst case range measurement accuracy is ±5 centimeters, with typical values of around ±2 centimeters. The pulsed near-infrared laser is Class-1 eye-safe under all operating conditions.

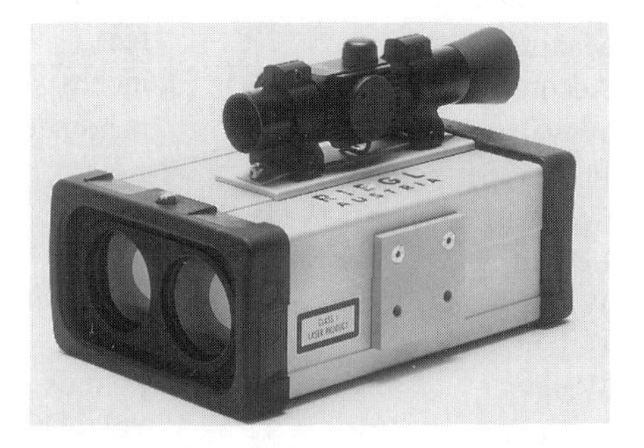

Figure 5-18. The Class 1 (eye-safe) *LD90-3 series* TOF laser rangefinder is a self-contained unit available in several versions with maximum ranges of 150 to 500 meters under average atmospheric conditions (courtesy RIEGL USA).

A nominal beam divergence of 2 milliradians for the LD90-3100 unit (see Table 5-7 below) produces a 20-centimeter footprint of illumination at 100 meters (Riegl, 1994). The complete system is housed in a small light-weight metal enclosure weighing only 1.5 kilograms and draws 10 watts at 11 to 18 volts DC. The standard output format is serial RS-232 at programmable data rates up to 19.2 kilobits per second, but RS-422 as well as analog options (0 to 10 volts DC and 4 to 20 milliamps current-loop) are available upon request.

Table 5-7. Typical specifications for two popular models of the *LD90-3 series* rangefinders.

Parameter		LD90-3100	LD90-3300	Units
Maximum range	(diffuse)	150	400	meters
	(cooperative)	>1000	>1000	meters
Minimum range		1	3-5	meters
Accuracy	(distance)	±2	±5	centimeters
	(velocity)	±0.3	±0.5	meters/sec
Beam divergence		2	2.8	milliradians
Power		11-18	11-18	volts DC
		10	10	watts
Size		22 x 13 x 7.6	22 x 13 x 7.6	centimeters
Weight		1.5	1.5	kilograms

Scanning Laser Rangefinders

The *LRS90-3 Laser Radar Scanner* is an adaptation of the basic LD90-3 electronics, fiber-optically coupled to a remote *scanner unit* as shown in Figure 5-19. The scanner package contains no internal electronics and is thus very robust under demanding operating conditions typical of industrial or robotic scenarios. The motorized scanning head pans the beam back and forth in the horizontal plane at a 10-Hz rate, resulting in 20 data-gathering sweeps per second. Beam divergence is 5 milliradians, with the option of expanding in the vertical direction if desired up to 2 degrees.

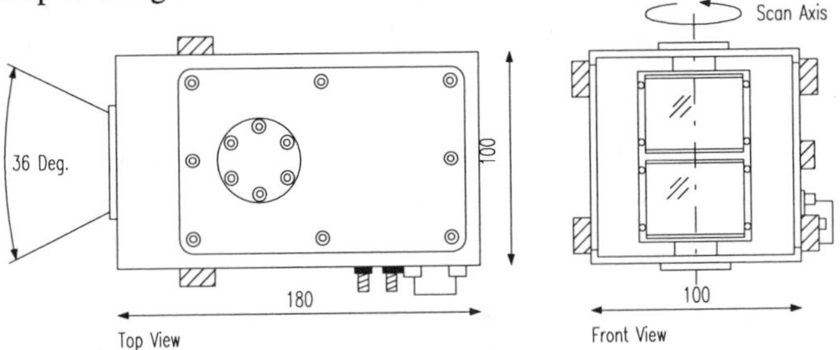

Figure 5-19. The *LRS90-3 Laser Radar Scanner* consists of an *electronics unit* (not shown) connected via a duplex fiber-optic cable to the remote *scanner unit* depicted above (courtesy RIEGL USA).

Figure 5-20 shows a representative plot of actual range data output taken along a curved section of roadway with the scanner fixed in a stationary position.

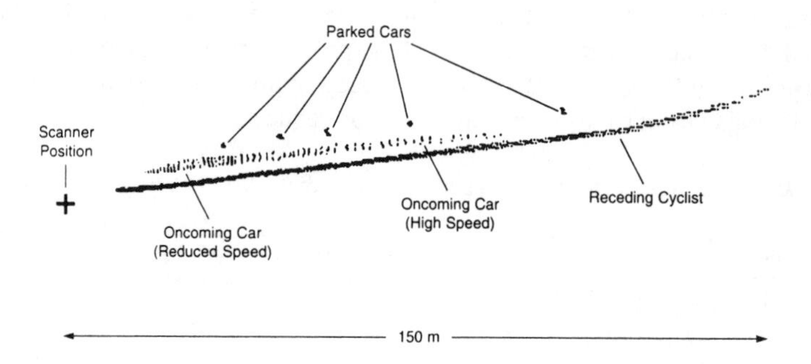

Figure 5-20. Plot of measured range values for a fixed sensor position at X, showing the path of an oncoming vehicle and a receding bicyclist along a curved roadway section (courtesy RIEGL USA).

The *LSS390 Laser Scanning System* is very similar to the *LRS90-3* but scans a more narrow field of view (±10 degrees) with a faster update rate (2000 Hz) and a more tightly focused beam. Range accuracy is typically ±10 centimeters, ±20 centimeters worst case. The *LSS390* unit is available with an RS-422 digital output (19.2 kilobits standard, 150 kilobits optional) or a 20-bit parallel TTL interface. Selected specifications for the *LRS90-3* and *LSS390* scanners are presented in Table 5-8.

Table 5-8. Typical specifications for the *LRS90-3 Laser Radar Scanner* and the *LSS390 Laser Scanner System* (courtesy RIEGL USA).

Parameter	LRS90-3	LSS390	Units
Maximum range	80	60	meters
Minimum range	2	1	meters
Accuracy	±3	±10	centimeters
Beam divergence	5	3.5	milliradians
Sample rate	1000	2000	Hz
Scan range	±18	±10	degrees
Scan rate	10	10	scans/second
Output (digital)	RS-232, -422	parallel, RS-422	
Power	11-15	9-16	volts DC
	880	880	milliamps
Size (electronics)	22 x 13 x 7.6	22 x 13 x 7.6	centimeters
(scanner)	18 x 10 x 10	6 x 9 x 12	centimeters
Weight (electronics)	1.5	1.3	kilograms
(scanner)	1.6	0.9	kilograms

5.2.3 Odetics *Fast Frame Rate 3-D Laser Imaging System*

Odetics, Inc., Anaheim, CA, has designed and partially fabricated a fast-frame-rate, pulsed TOF laser imager for use in high-speed autonomous land vehicle navigation and other machine vision applications. Three-dimensional range images out to 300 feet are captured by a pulsed laser rangefinder capable of acquiring one million range pixels per second. A GaAlAs diode laser produces a peak power output of approximately 50 watts, but the extremely narrow pulse width (12 nanoseconds) permits eye-safe operation within the maximum permissible exposure limits pursuant to the ANSI standard for the safe use of lasers.

A 60-degree azimuth and 30-degree elevation field of view is provided by a high-speed polygonal scanner mechanism as illustrated in the block diagram of Figure 5-21. Line scanning is programmable in elevation for random line access and faster scan functions, with a minimum capability of 12 frames/second of 256 pixels by 128 lines.

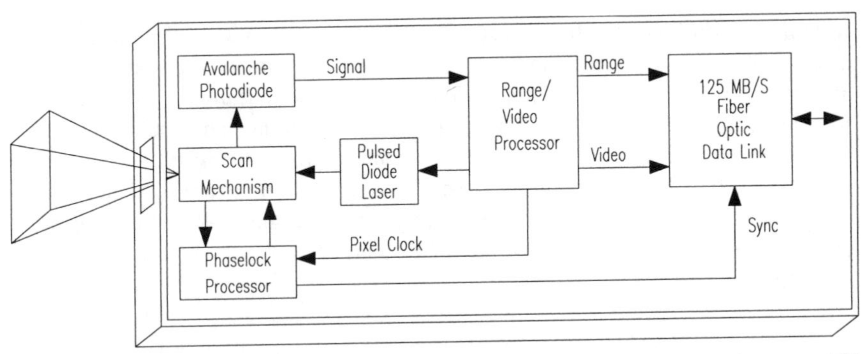

Figure 5-21. The *Fast Frame Rate 3-D Laser Imaging System* provides 3-D range data out to 300 feet for a 256-pixel by 128-line format at a 12-Hz frame rate (courtesy Odetics, Inc.).

Simultaneous range and reflectance images are processed and stored in a *VME Bus* frame buffer for direct pixel access by user image processors. Range data is processed by a pipelined picosecond *emitter coupled logic (ECL)* time interpolator. Range resolution is 0.5 inches (78 picoseconds) with a single-pulse *noise-equivalent range* of less than 1 foot at the *minimum-discernible-signal* range of 306 feet. Multiple-pulse averaging can reduce this noise scatter as required. The self-contained imaging system will be packaged in a compact (less than 1 cubic foot) enclosure suitable for vehicle mounting, with a full-duplex high-speed user interface provided by a 125 megabit/second fiber-optic data link. Selected specifications are listed in Table 5-9.

Note: Odetics also offers a previously developed phase-shift-measurement laser ranging system discussed in the next chapter.

Table 5-9. Selected specifications for the Fast Frame Rate 3-D Laser Imaging System (courtesy Odetics, Inc.).

Parameter	Value	Units
Maximum range	306	feet
Minimum range	2	feet
Range resolution	0.5	inches
Noise equivalent range	<1	foot
Frame rate	12	Hz
Format	256 pixels x 128 lines	
Field of view (azimuth)	60	degrees
(elevation)	30	degrees
Wavelength	820	nanometers
Output power	<50	watts
Pulsewidth	12 (nominal)	nanoseconds

5.2.4 RVSI *Long Optical Ranging and Detection System*

Robotic Vision Systems, Inc., Haupauge, NY, has conceptually designed a laser-based TOF ranging system capable of acquiring three-dimensional image data for an entire scene without scanning. The *Long Optical Ranging and Detection System (LORDS)* is a patented concept incorporating an optical encoding technique with ordinary vidicon or solid-state camera(s), resulting in precise distance measurement to multiple targets in a scene illuminated by a single laser pulse. The design configuration is relatively simple (Figure 5-22) and comparable in size and weight to traditional TOF and phase-shift measurement laser rangefinders.

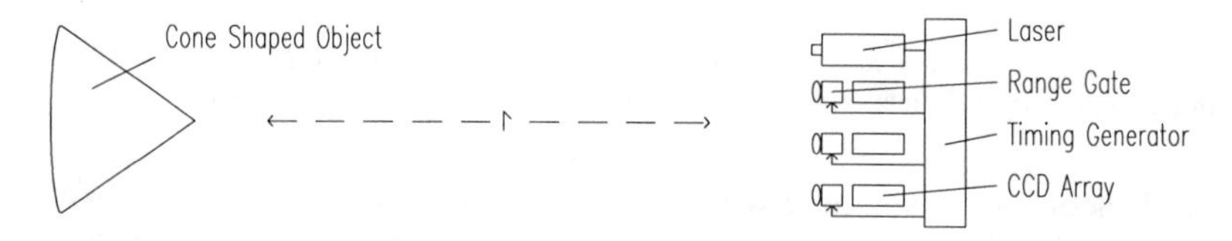

Figure 5-22. Simplified block diagram of a three-camera configuration of the *LORDS* 3-D laser TOF rangefinding system (courtesy Robotic Vision Systems, Inc.).

Major components include a single laser-energy source, one or more imaging cameras, each with an electronically implemented shuttering mechanism, and the associated control and processing electronics. In a typical configuration, the laser will emit a 25-millijoule pulse lasting 1 nanosecond, for an effective transmission of 25 megawatts. The anticipated operational wavelength will lie between 532 and 830 nanometers, due to the ready availability within this range of the required laser source and imaging arrays.

The cameras will be two-dimensional CCD arrays spaced closely together, side by side, with parallel optical axes resulting in nearly identical, multiple views of the illuminated surface. Lenses for these cameras will be of the standard photographic varieties between 12 and 135 millimeters. The shuttering function will be performed by microchannel plate image intensifiers (MCPs), 18 or 25 millimeters in size, which will be gated in a binary encoding sequence, effectively turning the CCDs on and off during the detection phase. Control of the system will be handled by a single-board processor based on the Motorola *MC-68040*.

LORDS obtains three-dimensional image information in real time by employing a novel time-of-flight technique requiring only a single laser pulse to collect all the information for an entire scene. The emitted pulse journeys a finite distance over time; hence, light traveling for 2 milliseconds will illuminate a scene a greater distance away than light traveling only 1 millisecond.

The entire sensing range is divided into discrete distance increments, each representing a distinct range plane. This is accomplished by simultaneously gating the MCPs of the observation cameras according to their own unique on-off encoding pattern over the duration of the detection phase. This binary gating alternately blocks and passes any returning reflection of the laser emission off objects within the field-of-view. When the gating cycles of each camera are aligned and compared, there exists a uniquely coded correspondence which can be used to calculate the range to any pixel in the scene.

For instance, in a system configured with only one camera, the gating MCP would be cycled on for half the detection duration, then off the remainder of the time. Figure 5-23 shows that any object detected by this camera must be positioned within the first half of the sensor's overall range (half the distance the laser light could travel in the allotted detection time). However, significant distance ambiguity exists because the exact time of reflected-energy detection could have occurred at any point within this relatively long interval.

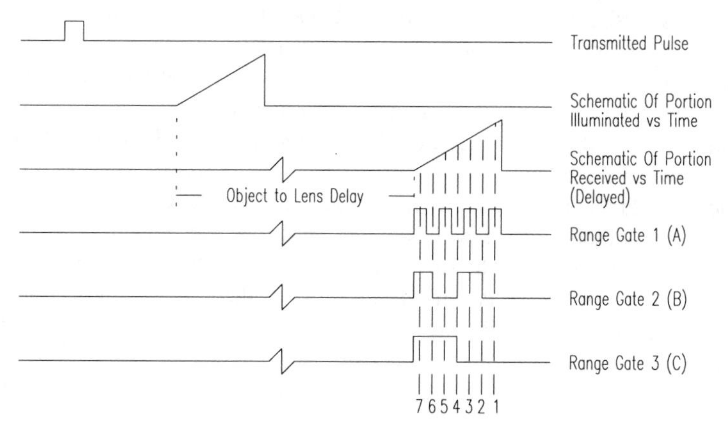

Figure 5-23. Range ambiguity is reduced by increasing the number of binary range gates (courtesy Robotic Vision Systems, Inc.).

This ambiguity can be reduced by a factor of two through the use of a second camera with its associated gating cycled at twice the rate of the first. This scheme would create two complete *on-off* sequences, one taking place while the first camera is on and the other while the first camera is off. Simple binary logic can be used to combine the camera outputs and further resolve the range (Figure 5-24). If the first camera did not detect an object but the second did, then by examining the instance when the first camera is off and the second is on, the range to the object can be associated with a relatively specific time frame. Incorporating a third camera at again twice the gating frequency (i.e., two cycles for every one of camera 2, and four cycles for every one of camera 1) provides even more resolution. For each additional CCD array incorporated into the system, the number of distance divisions is effectively doubled.

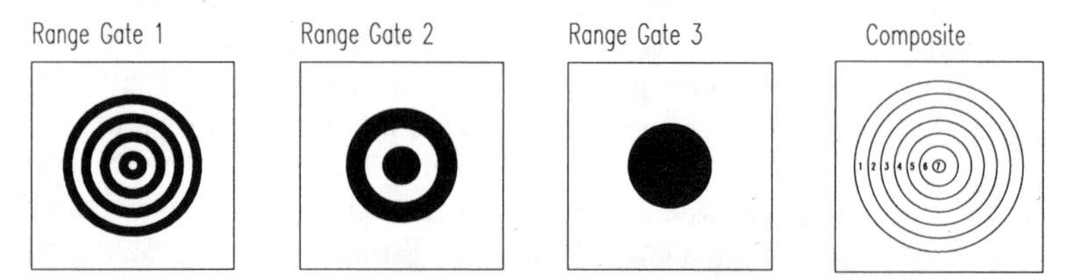

Figure 5-24. Binary coded images from range gates 1-3 are combined to generate the composite range map on the far right (courtesy Robotic Vision Systems, Inc.).

Alternatively, the same encoding effect can be achieved using a single camera when little or no relative motion exists between the sensor and the target area. In this scenario, the laser is pulsed multiple times, and the gating frequency for the single camera is sequentially changed at each new transmission. This creates the same detection intervals as before but with an increase in the time required for data acquisition.

LORDS is designed to operate over distances between 1 meter and several kilometers. An important characteristic is the projected ability to range over selective segments of an observed scene to improve resolution, in that the depth of field over which a given number of range increments is spread can be variable. The entire range of interest is initially observed, resulting in the maximum distance between increments (coarse resolution). An object detected at this stage is thus localized to a specific, abbreviated region of the total distance.

The sensor is then electronically reconfigured to cycle only over this region, which significantly shortens the distance between increments, thereby increasing resolution. A known delay is introduced between the time of transmission and initiation of the detection/gating process. The laser light thus travels to the region of interest without concern for objects positioned in the foreground. This feature can be especially helpful in eliminating backscatter from fog or smoke in outdoor applications.

5.3 References

Arkin, R.C., "Motor-Schema-Based Mobile Robot Navigation," *International Journal of Robotics Research*, Vol. 8., No. 4, pp. 92-112, August, 1989.

Biber, C., Ellin, S., Shenk, E., "The Polaroid Ultrasonic Ranging System," Audio Engineering Society, 67th Convention, New York, NY, October-November, 1980.

Borenstein, J., Koren, Y., "Real-Time Obstacle Avoidance for Fast Mobile Robots in Cluttered Environments," IEEE International Conference on Robotics and Automation, Vol. CH2876-1, Cincinnati, OH, pp. 572-577, May, 1990.

Cybermotion, "Ultrasonic Collision Avoidance System," Cybermotion Product Literature, Salem, VA, 1991.

Depkovich, T., W. Wolfe, "Definition of Requirements and Components for a Robotic Locating System," Final Report MCR-83-669, Martin Marietta Denver Aerospace, Denver, CO, February, 1984.

Everett, H.R., "A Microprocessor Controlled Autonomous Sentry Robot," Masters Thesis, Naval Postgraduate School, Monterey, CA, October, 1982.

Everett, H.R., "A Multi-Element Ultrasonic Ranging Array," *Robotics Age*, pp. 13-20, July, 1985.

Figueroa, J.F., Lamancusa, J.S., "A Method for Accurate Detection of Time of Arrival: Analysis and Design of an Ultrasonic Ranging System," Journal of the Acoustical Society of America, Vol. 91, No. 1, pp. 486-494, January, 1992.

Frederiksen, T.M., Howard, W.M., "A Single-Chip Monolithic Sonar System," *IEEE Journal of Solid State Circuits*, Vol. SC-9, No. 6, December, 1974.

Gustavson, R.L., Davis, T.E., "Diode-Laser Radar for Low-Cost Weapon Guidance," SPIE Vol. 1633, Laser Radar VII, Los Angeles, CA, pp. 21-32, January, 1992.

Hammond, W., "Smart Collision Avoidance Sonar Surpasses Conventional Systems," *Industrial Vehicle Technology '93: Annual Review of Industrial Vehicle Design and Engineering*, UK and International Press, pp. 64-66, 1993.

Hammond, W., "Vehicular Use of Ultrasonic Systems," Technical Report, Cybermotion, Inc., Salem, VA, May, 1994.

Kim, E.J., "Design of a Phased Sonar Array for a Mobile Robot," Bachelor's Thesis, MIT, Cambridge, MA, May, 1986.

Koenigsburg, W.D., "Noncontact Distance Sensor Technology," GTE Laboratories, Inc., 40 Sylvan Rd., Waltham, MA, pp. 519-531, March, 1982.

Lang, S., Korba, L., Wong, A., "Characterizing and Modeling a Sonar Ring," SPIE Mobile Robots IV, Philadelphia, PA, pp. 291-304, 1989.

Langer, D., Thorpe, C., "Sonar Based Outdoor Vehicle Navigation and Collision Avoidance," International Conference on Intelligent Robots and Systems, IROS '92, Raleigh, NC, July , 1992.

Lewis, R.A., Johnson, A.R., "A Scanning Laser Rangefinder for a Robotic Vehicle," 5th International Joint Conference on Artificial Intelligence, pp. 762-768, 1977.

Massa, "E-201B & E-220B Ultrasonic Ranging Module Subsystems Product Selection Guide," Product Literature 891201-10M, Massa Products Corporation, Hingham, MA, undated.

Moravec, H. P., Elfes, A., "High Resolution Maps from Wide Angle Sonar," IEEE International Conference on Robotics and Automation, St. Louis, MO, pp. 116-121, March, 1985.

NASA, "Fast, Accurate Rangefinder, *NASA Tech Brief*, NPO-13460, Winter, 1977.

National, "LM1812 Ultrasonic Transceiver," *Special Purpose Linear Devices Databook*, National Semiconductor Corp., Santa Clara, CA, Section 5, pp. 103-110, 1989.

National, "Electrostatic Transducers Provide Wide Range Ultrasonic Measurement," *Linear Applications Handbook*, National Semiconductor Corp., Santa Clara, CA, pp. 1172-1173, 1991.

Olson, R.A., Gustavson, R.L., Wangler, R.J., McConnell, R.E., "Active Infrared Overhead Vehicle Sensor," IEEE Transactions on Vehicular Technology, Vol. 43, No. 1, pp. 79-85, February, 1994.

Pin, F. G., Watanabe, Y., "Using Fuzzy Behaviors for the Outdoor Navigation of a Car with Low-Resolution Sensors," IEEE International Conference on Robotics and Automation, Atlanta, GA, pp. 548-553, 1993.

Pletta, J.B., Amai, W.A., Klarer, P., Frank, D., Carlson, J., Byrne, R., "The Remote Security Station (RSS) Final Report," Sandia Report SAND92-1947 for DOE under Contract DE-AC04-76DP00789, Sandia National Laboratories, Albuquerque, NM, October, 1992.

Polaroid, "Polaroid Ultrasonic Ranging System User's Manual," Publication No. P1834B, Polaroid Corporation, Cambridge, MA, December, 1981.

Polaroid, "Technical Specifications for Polaroid Electrostatic Transducer," 7000-Series Product Specification ITP-64, Polaroid Corporation, Cambridge, MA, June, 1987.

Polaroid, "6500-Series Sonar Ranging Module," Product Specifications PID 615077, Polaroid Corporation, Cambridge, MA, 11 October, 1990.

Polaroid, "Polaroid Ultrasonic Ranging Developer's Kit," Publication No. PXW6431 6/93, Polaroid Corporation, Cambridge, MA, June, 1993.

RIEGL, "Laser Distance, Level, and Speed Sensor LD90-3," Product Data Sheet 3/94, RIEGL Laser Measurement Systems, RIEGL USA, Orlando, FL, March, 1994.

SEO, "LRF-X Laser Rangefinder Series,", Product Literature, Schwartz Electro-Optics, Inc., Orlando, FL, October, 1991a.

SEO, "Scanning Laser Rangefinder,", Product Literature, Schwartz Electro-Optics, Inc., Orlando, FL, October, 1991b.

SEO, "Helicopter Optical Proximity Sensor System,", Product Literature, Schwartz Electro-Optics, Inc., Orlando, FL, October, 1991c.

SEO, Process Report for US Army Contract DAAJ02-91-C-0026, Schwartz Electro-Optics, Inc., Orlando, FL, December, 1991d.

Siuru, B., "The Smart Vehicles Are Here," *Popular Electronics*, Vol. 11, No. 1, pp. 41-45, January, 1994.

Vuylsteke, P., Price, C.B., Oosterlinck, A., "Image Sensors for Real-Time 3D Acquisition, Part 1," ," in *Traditional and Non-Traditional Robotic Sensors*, T.C. Henderson, ed., NATO ASI Series, Vol. F63, Springer-Verlag, pp. 187-210, 1990.

6
Phase-Shift Measurement and Frequency Modulation

6.1 Phase-Shift Measurement

The *phase-shift measurement* (or *phase-detection*) ranging technique involves continuous-wave (CW) transmission as opposed to the short-duration pulsed outputs used in the time-of-flight systems discussed in Chapter 5. (The transmission of short pulses may also be used if synchronized to a continuous-wave reference against which the phase of the returning signal is measured.) One advantage of continuous-wave systems over pulsed methods is the ability to measure the direction and velocity of a moving target, in addition to its range. In 1842, an Austrian by the name of Johann Doppler published a paper describing what has since become known as the Doppler effect: the frequency of an energy wave reflected from an object in motion is a function of the relative velocity between the object and the observer. This subject will be discussed in more detail in Chapter 8.

In practice, a beam of amplitude-modulated laser, RF, or acoustical energy is directed towards the target as illustrated in Figure 6-1. A small portion of this wave (potentially up to six orders of magnitude less in amplitude) is reflected by the object surface back to the detector (Chen, et al., 1993). Improved measurement accuracy and increased range can be achieved when cooperative targets are attached to the objects of interest to increase the power density of the reflected signal. The returned energy is compared to a simultaneously generated reference that has been split off from the original signal, and the relative phase shift between the two is measured to ascertain the round-trip distance the wave has traveled. As with time-of-flight rangefinders, the paths of the source and the reflected beam are essentially coaxial, preventing the *missing parts* problem.

For high-frequency RF- or laser-based systems, detection is usually preceded by heterodyning the reference and received signals with an intermediate frequency (the relative phase shift is preserved) to allow the phase detector to operate at a more convenient lower frequency (Vuylsteke, 1990). The phase shift expressed as a function of distance to the reflecting target surface is (Woodbury, et al., 1993):

$$\phi = \frac{4\pi d}{\lambda}$$

where:

ϕ = phase shift
d = distance to target
λ = modulation wavelength.

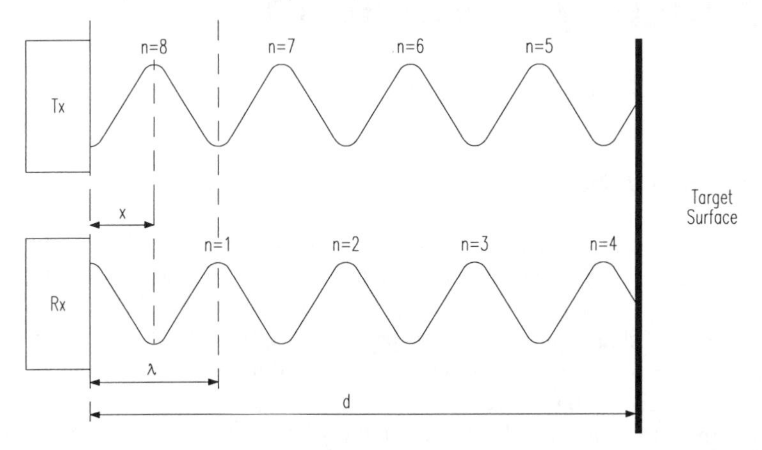

Figure 6-1. Relationship between outgoing and reflected waveforms, where x is the distance corresponding to the differential phase ϕ (adapted from Woodbury, et al., 1993).

The desired distance to target d as a function of the measured phase shift ϕ is therefore given by:

$$d = \frac{\phi\,\lambda}{4\pi} = \frac{\phi\,c}{4\pi f}$$

where:

c = speed of light.
f = modulation frequency.

The phase shift between outgoing and reflected sine waves can be measured by multiplying the two signals together in an electronic mixer, then averaging the product over many modulation cycles (Woodbury, et al., 1993). This integrating process can be relatively time consuming, making it difficult to achieve extremely rapid update rates. The result can be expressed mathematically as follows (Woodbury, et al., 1993):

$$\lim_{T \to \infty} \frac{1}{T} \int_0^T \sin\left(\frac{2\pi c}{\lambda} t + \frac{4\pi d}{\lambda}\right) \sin\left(\frac{2\pi c}{\lambda}\right) dt$$

which reduces to:

$$A \cos\left(\frac{4\pi d}{\lambda}\right)$$

where:

t = time
T = averaging interval
A = amplitude factor from gain of integrating amplifier.

From the earlier expression for ϕ, it can be seen that the quantity actually measured is in fact the *cosine* of the phase shift and not the phase shift itself (Woodbury, et al., 1993). This situation introduces a so-called *ambiguity interval* for scenarios where the round-trip distance exceeds the modulation wavelength λ (i.e., the phase measurement becomes ambiguous once ϕ exceeds 360 degrees). Conrad and Sampson (1990) define this ambiguity interval as the maximum range that allows the phase difference to go through one complete cycle of 360 degrees:

$$R_a = \frac{c}{2f}$$

where:

R_a = ambiguity range interval.

Referring again to Figure 6-1, it can be seen that the total round-trip distance $2d$ is equal to some integer number of wavelengths $n\lambda$ plus the fractional wavelength distance x associated with the phase shift. Since the cosine relationship is not single-valued for all of ϕ, there will be more than one distance d corresponding to any given phase-shift measurement (Woodbury, et al., 1993):

$$\cos\phi = \cos\left(\frac{4\pi d}{\lambda}\right) = \cos\left(\frac{2\pi(x + n\lambda)}{\lambda}\right)$$

where:

$d = (x + n\lambda)/2$ = true distance to target
x = distance corresponding to differential phase ϕ
n = number of complete modulation cycles.

Careful re-examination of Figure 6-1, in fact, shows that the cosine function is not single-valued even within a solitary wavelength interval of 360 degrees.

Accordingly, if only the cosine of the phase angle is measured, the ambiguity interval must be further reduced to half the modulation wavelength, or 180 degrees (Scott, 1990). In addition, the slope of the curve is such that the *rate of change* of the non-linear cosine function is not constant over the range of $0 \leq \phi \leq 180$ degrees, and is in fact zero at either extreme. The achievable accuracy of the phase-shift measurement technique thus varies as a function of target distance, from best-case performance for a phase angle of 90 degrees to worst case at 0 and 180 degrees. For this reason, the useable measurement range is typically even further limited to 90 percent of the 180-degree ambiguity interval (Chen, et al., 1993).

A common solution to this problem involves taking a second measurement of the same scene but with a 90-degree phase shift introduced into the reference waveform, the net effect being the sine of the phase angle is then measured instead of the cosine. This additional information (i.e., both sine and cosine measurements) can be used to expand the phase angle ambiguity interval to the full 360 degree limit previously discussed (Scott, 1990). Furthermore, an overall improvement in system accuracy is achieved, as for every region where the cosine measurement is insensitive (i.e., zero slope), the complementary sine measurement will be at peak sensitivity (Woodbury, et al., 1993).

Nevertheless, the unavoidable potential for erroneous information as a result of the ambiguity interval is a detracting factor in the case of phase-detection schemes. Some applications simply avoid such problems by arranging the optical path in such a fashion as to ensure the maximum possible range is always less than the ambiguity interval (Figure 6-2). Alternatively, successive measurements of the same target using two different modulation frequencies can be performed, resulting in two equations with two unknowns, allowing both x and n (in the previous equation) to be uniquely determined. Kerr (1988) describes such an implementation using modulation frequencies of 6 and 32 MHz.

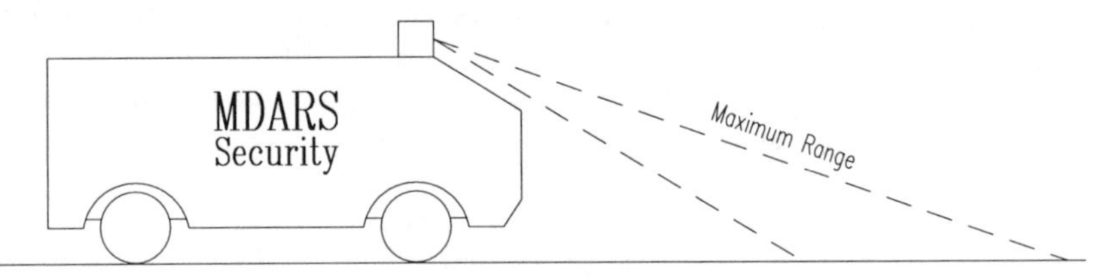

Figure 6-2. By limiting the maximum distance measured to be less than the range *ambiguity interval* R_a, erroneous distance measurements can be avoided.

For square-wave modulation at the relatively low frequencies typical of ultrasonic systems (20-200 KHz), the phase difference between incoming and outgoing waveforms can be measured with the simple linear circuit shown in Figure 6-3 (Figueroa & Barbieri, 1991a). The output of the *exclusive-or* gate goes high whenever its inputs are at opposite logic levels, generating a voltage across capacitor C_1 that is proportional to the phase shift. For example, when the two signals are in phase (i.e., $\phi = 0$), the gate output stays low and V is zero; maximum output voltage occurs when ϕ reaches 180 degrees. While easy to implement, this simplistic approach is limited to very low frequencies and may require frequent calibration to compensate for drifts and offsets due to component aging or changes in ambient conditions (Figueroa & Lamancusa, 1992).

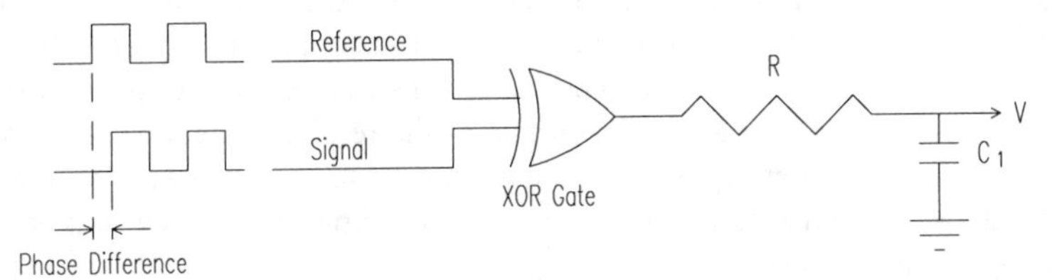

Figure 6-3. At low frequencies typical of ultrasonic systems, a simple phase-detection circuit based on an *exclusive-or* gate will generate an analog output voltage proportional to the phase difference seen by the inputs (adapted from Figueroa & Barbieri, 1991a).

Figueroa and Barbieri (1991a; 1991b) report an interesting method for extending the ambiguity interval in ultrasonic phase-detection systems through frequency division of the received and reference signals. Since the span of meaningful comparison is limited (best case) to one wavelength, λ, it stands to reason that decreasing the frequency of the phase detector inputs by some common factor will increase λ by a similar amount. The concept is illustrated in Figure 6-4 below. Due to the very short wavelength of ultrasonic energy (i.e., about 0.25 inches for the Polaroid system at 49.1 KHz), the total effective range is still only 4 inches after dividing the detector inputs by a factor of 16. Due to this inherent range limitation, ultrasonic phase-detection ranging systems are not extensively applied in mobile robotic applications, although Figueroa and Lamancusa (1992) describe a hybrid approach used to improve the accuracy of TOF ranging for three-dimensional position location (see Chapter 15).

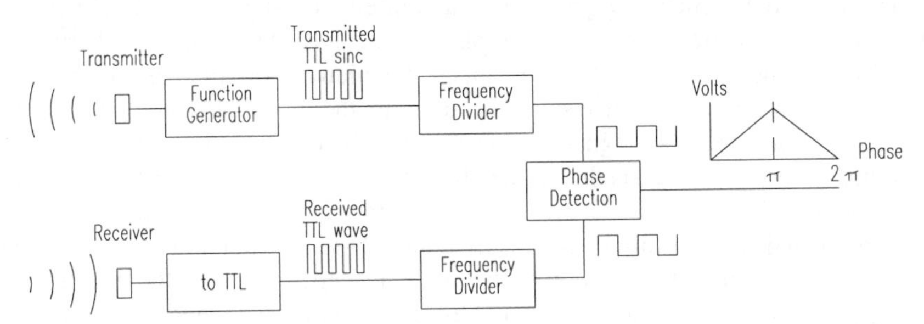

Figure 6-4. Dividing the input frequencies to the phase comparator by some common integer value will extend the ambiguity interval by the same factor, at the expense of resolution (adapted from Figueroa & Barbieri, 1991a).

Laser-based continuous-wave ranging originated out of work performed at the Stanford Research Institute in the 1970s (Nitzan, et al., 1977). Range accuracies approach those achievable by pulsed laser TOF methods. Only a slight advantage is gained over pulsed TOF rangefinding, however, since the time-measurement problem is replaced by the need for fairly sophisticated phase-measurement electronics (Depkovich & Wolfe, 1984). In addition, problems with the phase-shift measurement approach are routinely encountered in situations where the outgoing energy is simultaneously reflected from two target surfaces at different distances from the sensor, as for example when scanning past a prominent vertical edge (Hebert & Krotkov, 1991). The system electronics are set up to compare the phase of a single incoming wave with that of the reference signal and are not able to cope with two superimposed reflected waveforms. Adams (1993) describes a technique for recognizing the occurrence of this situation in order to discount the resulting erroneous data.

6.1.1 ERIM 3-D Vision Systems

The *Adaptive Suspension Vehicle (ASV)* developed at Ohio State University (Patterson, et al., 1984) and the *Autonomous Land Vehicle (ALV)* developed by Martin Marietta Denver Aerospace were the premier mobile robot projects sponsored by the Defense Advanced Research Projects Agency (DARPA) in the 1980s under the *Strategic Computing Program.* In support of these efforts, the Environmental Research Institute of Michigan (ERIM) was tasked to develop an advanced three-dimensional vision system to meet the close-in navigation and collision avoidance needs of a mobile platform. The initial design, known as the *Adaptive Suspension Vehicle Sensor* (Figure 6-5), operates on the principle of optical radar and determines range to a point through phase-shift measurement using a CW laser source (Beyer, et al., 1987).

The ranging sequence begins with the transmission of an amplitude-modulated laser beam that illuminates an object and is partially reflected back to the detector, generating a representative signal that is amplified and filtered to extract the 16-MHz modulation frequency. The amplitude of the signal is picked off at this point to produce a reflectance video image for viewing or for two-dimensional image processing. A reference signal is output by the modulation oscillator; both the detector and reference signals are then sent to the comparator electronics. The resulting phase difference is determined by a time-measurement technique, where the leading edge of the reference signal initiates a counting sequence that is terminated when the leading edge of the returned signal enters the counter. The resulting count value is a function of the phase difference between the two signals and is converted to an 8-bit digital word representing the range to the scene.

Three-dimensional images are produced by the *ASV* sensor through the use of scanning optics. The mechanism consists of a nodding mirror and a rotating polygonal mirror with four reflective surfaces as shown in Figure 6-6. The polygonal mirror pans the transmitted laser beam in azimuth across the ground, creating a scan line at a set distance in the front of the vehicle. The scan line is deflected by the objects and surfaces in the observed region and forms a contour of the scene across the sensor's horizontal field of view. The third dimension is added by the nodding mirror which tilts the beam in discrete elevation increments. A complete image is created by scanning the laser in a left-to-right and bottom-to-top raster pattern.

Figure 6-5. The *Adaptive Suspension Vehicle Sensor* (courtesy Environmental Research Institute of Michigan).

The returning signals share the same path through the nodding mirror and rotating polygon (actually slightly offset) but are split through a separate glass optical chain to the detector. The scan rate of 180 lines per second is a function of the field of view and desired frame rate, determined by the vehicle's maximum forward velocity (10 feet/second in this case). The size, weight, and required velocities of the mirrors precluded the use of galvanometers in the system design; the rotating and nodding mirrors therefore are servo driven.

An 820-nanometer gallium arsenide (GaAs) laser diode with collimating and expansion optics is used to produce a 6-inch diameter laser footprint at 30 feet. The detector is a silicon avalanche photodiode, optically filtered to match the laser wavelength. The laser source, detector, scanning optics, and drive motors are housed in a single enclosure situated at a height of 8 feet, looking down upon the field of view. The scanning laser beam strikes the ground between 2 and 30 feet in front of the vehicle, with a 22-foot wide horizontal scan line at the maximum distance of 30 feet. (The major factor limiting the useful range of the system is the measurement ambiguity that occurs when the phase difference between the reference and returned energy exceeds 360 degrees.) The 2-Hz system update rate creates a new image of the scene for every 5 feet of forward motion at the vehicle's maximum speed of 10 feet/second.

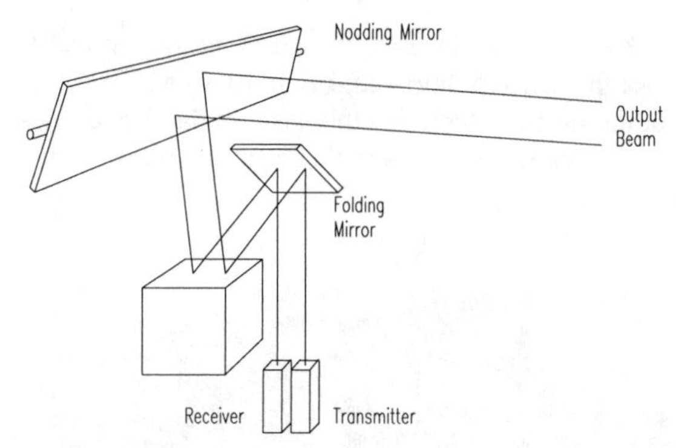

Figure 6-6. Scanning and nodding mirror arrangement in the ERIM laser rangefinder for the *Adaptive Suspension Vehicle* (courtesy Environmental Research Institute of Michigan).

Following the design and fabrication of the *ASV sensor*, ERIM undertook the task of developing a similar device known as the *ALV sensor* for DARPA's autonomous land vehicle. The two instruments were essentially the same in configuration and function but with modified performance specifications to meet the needs of the individual mobile platforms (Table 6-1).

Table 6-1. Selected specifications for the *Adaptive Suspension Vehicle* and *Autonomous Land Vehicle* scanning laser rangefinders.

Parameter	ASV	ALV	Units
Horizontal FOV	80	80	degrees
Vertical FOV	60	30	degrees
Beamwidth	1	0.5	degrees
Frame rate	2	2	Hz
Scan lines per frame	128	64	
Pixels per scan line	128	256	
Maximum range	32	64	feet
Vertical scan	10	20	degrees
Wavelength	820	820	nanometers
Power	24	24	volts
	450	450	watts
Size	14 by 26 by 22	14 by 29 by 22	inches
Weight	85	85	pounds

An advanced ranging device known as the *Multispectral ALV Sensor* was later developed for exterior applications addressing rugged cross-country traversal as opposed to the relatively uniform road surfaces seen in the initial tests of the autonomous land vehicle concept. The variations in terrain, surface cover, and vegetation encountered in off-road scenarios require an effective means to distinguish between earth, rocks, grass, trees, water, and other natural features.

The scanner mechanism for the multispectral sensor was essentially identical to the scanners developed for the earlier *ASV* and *ALV* sensors, the only significant difference being the substitution of a hexagonal rotating mirror instead of a square mirror for panning the beam in azimuth. This configuration caused the transmitted and returned signals to impinge on separate mirrored surfaces, resulting in reduced crosstalk and simplified sensor alignment (Figure 6-7). The nodding mirror for tilting the beam in elevation remained largely unchanged.

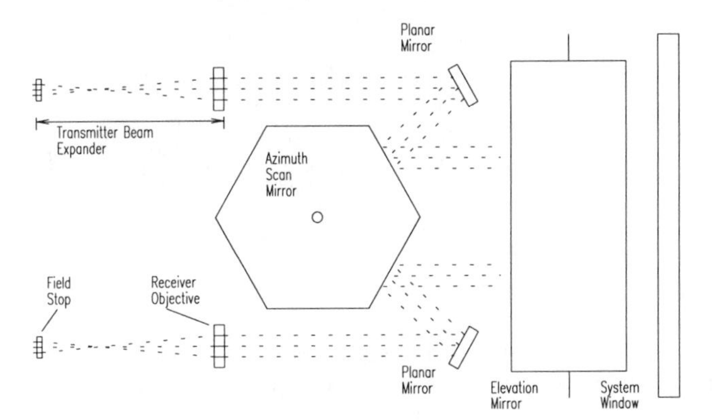

Figure 6-7. Hexagonal rotating mirror used in the multispectral scanner reduces crosstalk and simplifies mirror alignment (courtesy Environmental Research Institute of Michigan).

The mass of the scanning mechanism plus the plurality of lasers, optics, and detectors made the multispectral sensor large (12 by 3 by 2 feet) and heavy (600 pounds), increasing the complexity of the control and analysis required to produce results. The multiple frequency sources, corresponding detectors, detector cooling system, and scanner resulted in significant power consumption: 15 kilowatts!

6.1.2 Perceptron *LASAR*

Perceptron Corporation, Farmington Hills, MI, has developed and is currently in production of *LASAR*, the AM-modulated 3-D laser scanner shown in Figure 6-8. Intended for industrial machine vision applications, versions of this device have already been used in navigational guidance, bin-picking, hazardous inspection, and mining scenarios. The sensor employs a nodding mirror in conjunction with a rotating-polygon assembly to achieve a 45-degree symmetrical field of view. At full-frame (1024 x 1024) resolution, a single update takes 6.4 seconds, with increased frame rates possible at lower resolutions. The maximum operating range of the *LASAR* system is around 40 meters, with an advertised single-frame range accuracy of ±2 millimeters at a distance of 2 meters. Frame rates up to 10 Hz and operating ranges in excess of 100 meters have been demonstrated in specially configured versions of the device.

Figure 6-8. The *LASAR* 3-D scanner achieves a range-measurement accuracy of 2 millimeters over a 45- by 45-degree field of view at a stand-off distance of 2 meters (courtesy Perceptron Corp.).

6.1.3 Odetics Scanning Laser Imaging System

Odetics, Inc., Anaheim, CA, developed an adaptive and versatile scanning laser rangefinder in the early 1980s for use on *ODEX 1*, the six-legged walking robot shown in Figure 6-9 (Binger & Harris, 1987; Byrd & DeVries, 1990). The system determines distance by phase-shift measurement, constructing three-dimensional range pictures by panning and tilting the sensor across the field of view. The phase-shift measurement technique was selected over acoustic-ranging, stereo-vision, and structured-light alternatives because of the inherent accuracy and fast update rate.

The imaging system is broken down into the two major subelements depicted in Figure 6-10: the *scan unit* and the *electronics unit*. The *scan unit* houses the laser source, the photodetector, and the scanning mechanism. The laser source is a GaAlAs laser diode emitting at a wavelength of 820 nanometers, with power output adjustable under software control between 1 to 50 milliwatts. Detection of the returned energy is achieved through use of an avalanche photodiode whose output is routed to the phase-measuring electronics.

The second subelement, the *electronics unit*, contains the range calculating and video processor as well as a programmable frame buffer interface. The range and video processor is responsible for controlling the laser transmission, activation of the scanning mechanism, detection of the returning energy, and determination of range values. Distance is calculated through a proprietary phase-detection scheme, reported to be fast, fully digital, and self-calibrating with a high signal-to-noise ratio. The minimum observable range is 1.5 feet, while the maximum range without ambiguity due to phase shifts greater than 360 degrees is 30.74 feet.

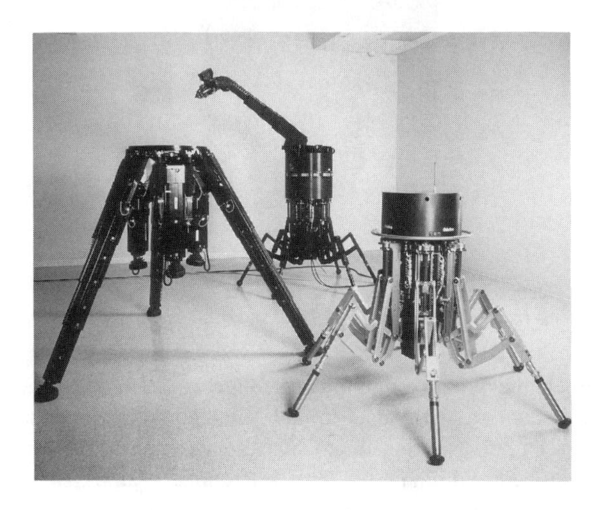

Figure 6-9 The *Scanning Laser Imaging System* was initially developed for use on the *Odex Series* of six-legged walking robots (courtesy Odetics, Inc.).

The scanning hardware consists of a rotating polygonal mirror that pans the laser beam across the scene and a planar mirror whose back-and-forth nodding motion tilts the beam for a realizable field of view of 60 degrees in azimuth and 60 degrees in elevation. The scanning sequence follows a raster-scan pattern and can illuminate and detect an array of 128 by 128 pixels at a frame rate of 1.2 Hz (Boltinghouse, et al., 1990).

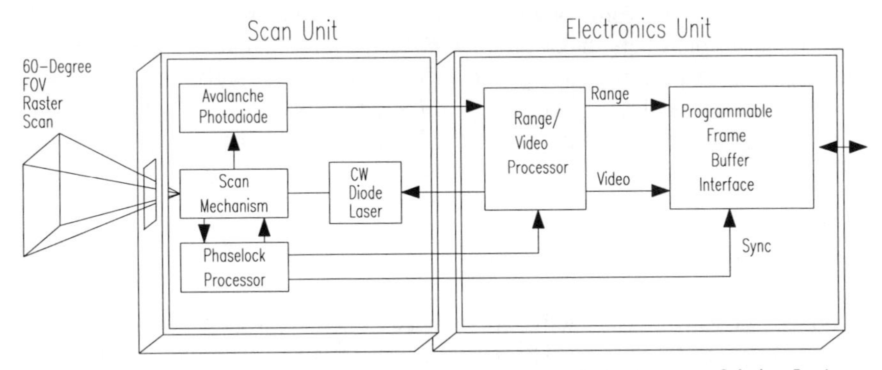

Figure 6-10. Block diagram of the Odetics scanning laser rangefinder (courtesy Odetics, Inc.).

Figure 6-11. The Odetics *Scanning Laser Imaging System* captures a 128- by 128-pixel image every 835 milliseconds (courtesy Odetics, Inc.).

For each pixel, the processor outputs a range value and a video reflectance value. The video data are equivalent to that obtained from a standard black-and-white television camera, except that interference due to ambient light and shadowing effects are eliminated. The reflectance value is compared to a prespecified threshold to eliminate pixels with insufficient return intensity to be properly processed, thereby eliminating potentially invalid range data; range values are set to maximum for all such pixels (Boltinghouse & Larsen, 1989). A three-by-three *neighborhood median filter* is used to further filter out noise from data qualification, specular reflection, and impulse response (Larson & Boltinghouse, 1988).

The output format is a 16-bit data word consisting of the range value in either 8 or 9 bits, and the video information in either 8 or 7 bits, respectively. The resulting range resolution for the system is 1.44 inches for the 8-bit format, and 0.72 inch with 9 bits. A buffer interface provides interim storage of the data and can execute single-word or whole-block direct-memory-access transfers to external host controllers under program control. Information can also be routed directly to a host without being held in the buffer. Currently, the interface is designed to support *VAX, VME-Bus, Multibus*, and IBM-*PC/AT* equipment. The scan and electronics unit together weigh 31 pounds and require 2 amps at 28 volts DC.

6.1.4 Sandia *Scannerless Range Imager*

Originally conceived as an active seeker head for smart weapons, the *Scannerless Range Imager* (Figure 6-12) developed at Sandia National Laboratories, Albuquerque, NM, computes three-dimensional range information without need for mechanical or solid-state scanning. A laser diode or LED array is used to illuminate an entire scene in similar fashion to Robotic Vision System's pulsed TOF *Long Optical Range and Detection System* described at the end of the previous chapter. The Sandia approach, however, employs an amplitude-modulated continuous-wave source in conjunction with a single CCD camera, and determines ranges to all pixel elements in essentially simultaneous fashion based on the perceived round-trip phase shift (Frazier, 1994). Phase-shift measurement is rather elegantly accomplished by converting the phase difference to a more easily quantified intensity representation through use of a *microchannel-plate image intensifier* as shown in the block diagram of Figure 6-13.

Figure 6-12. The Sandia *Scannerless Range Imager* employs an amplitude-modulated CW laser source in conjunction with a single CCD camera (courtesy Sandia National Laboratories).

Reflected energy from the illuminated scene is focused by the receiver optics upon a photocathode element that creates a stream of electrons modulated in accordance with the amplitude variations of the incoming light. The sinusoidal laser-modulation signal f_m is coupled to a thin conductive sheet (i.e., analogous to the *grid* of a vacuum tube), as shown in the above figure, to control the flow of electrons from the photocathode into the microchannel plate (Scott, 1990). The electron stream is amplified through secondary emissions as it passes through the microchannel plate, and converted back to optical energy upon striking the phosphor screen as illustrated. Since the gain of the image intensifier stage is in this fashion modulated at the same frequency as the outgoing optical energy, the magnitude of phosphor radiance is thus a function of the *cosine* of the range-dependent phase angle (i.e., due to constructive and destructive interference). A 210-frames/second Dalsa CCD camera is coupled to the phosphor screen by way of a coherent fiber-optic bundle to serve as an integrating 256-by-256 detector array (Weiss, 1994).

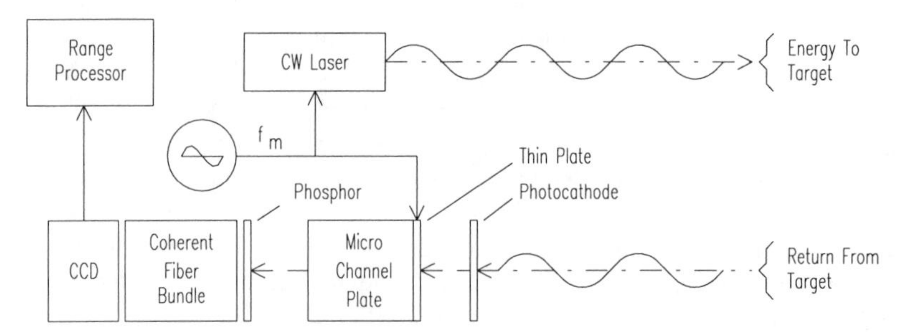

Figure 6-13. Range values are computed for all pixels in the CCD detector array based on the observed phase shift (adapted from Scott, 1990).

To expand the phase ambiguity interval and improve resolution, a second image is obtained with the image intensifier modulated 90 degrees out of phase with respect to the light source, effectively measuring the *sine* of the phase angle. These "sine" and "cosine" images are processed together with a baseline image taken under conditions of no receiver or transmitter modulation in order to eliminate non-range-related intensity variations (Scott, 1990). The current system update rate using a 68040-based PC running at 40 MHz is one frame per second, but will be expanded to 8 Hz in the very near future through incorporation of *TI-C40* digital signal processor (DSP) hardware.

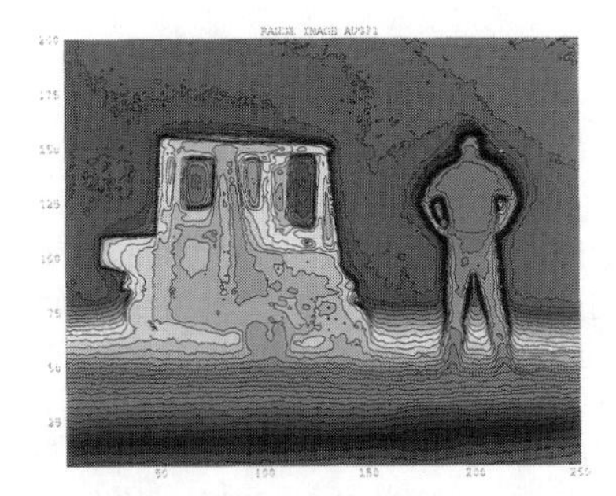

Figure 6-14. Resulting range image (left) and reflectance image (right) for a typical outdoor scene using an array of 660-nanometer (red) LEDs (courtesy Sandia National Laboratories).

Due to its structural simplicity, relatively low cost, and demonstrated potential for high-bandwidth, medium-resolution range data, the Sandia *Scannerless Range Imager* is being investigated for use on a number of robotic platforms, including the MDARS-Exterior system. One existing prototype of the sensor employs a 20-watt laser diode modulated at 5 MHz, resulting in a 90-foot ambiguity interval with a range resolution of 1 foot and a maximum range of 2,000 feet at night

(Weiss, 1994). Nighttime operation using eye-safe LED emitters has also been demonstrated out to 200 feet; representative range and reflectance images at a distance of approximately 60 feet are presented in Figure 6-14. Potential problems still being investigated include the significant power and cooling requirements for the laser source, and attainment of sufficient signal-to-noise ratio for reliable daytime operation.

6.1.5 ESP *Optical Ranging System*

The *Optical Ranging System (ORS-1)* is a low-cost near-infrared rangefinder (Figure 6-15) developed in 1989 by ESP Technologies, Inc., Lawrenceville, NJ, for use in autonomous robotic cart navigation in factories and similar environments. A 2-milliwatt 820-nanometer LED source, 100-percent modulated at 5 MHz, forms a collimated 1-inch diameter transmit beam that is unconditionally eye-safe. Reflected radiation is focused by a 4-inch diameter coaxial Fresnel lens onto a photodetector; the measured phase shift is proportional to the round-trip distance to the illuminated object. An adaptation of an earlier prototype developed by AT&T (Miller & Wagner, 1987), the *ORS-1* provides three outputs: range and angle of the target and an automatic gain control (AGC) signal. Range resolution at 20 feet is approximately 2.5 inches, while angular resolution is about 1 inch at a distance of 5 feet.

Figure 6-15. The *ORS-1 Optical Ranging System* determines range through phase-shift measurement using an eye-safe near-infrared LED source (courtesy ESP Technologies, Inc.).

The AGC output signal is inversely proportional to the received signal strength and provides information about a target's near-infrared reflectivity, warning against insufficient or excessive signal return (ESP, 1992). Useable range results are produced only when the corresponding gain signal is within a predetermined operating range. A rotating mirror mounted at 45 degrees to the optical axis provides 360-degree polar-coordinate coverage (Figure 6-16). It is driven at one to two revolutions per second by a motor fitted with an integral incremental encoder and an optical indexing sensor that signals the completion of each revolution. The system is capable of simultaneous operation as a wideband optical communication receiver (Miller & Wagner, 1987).

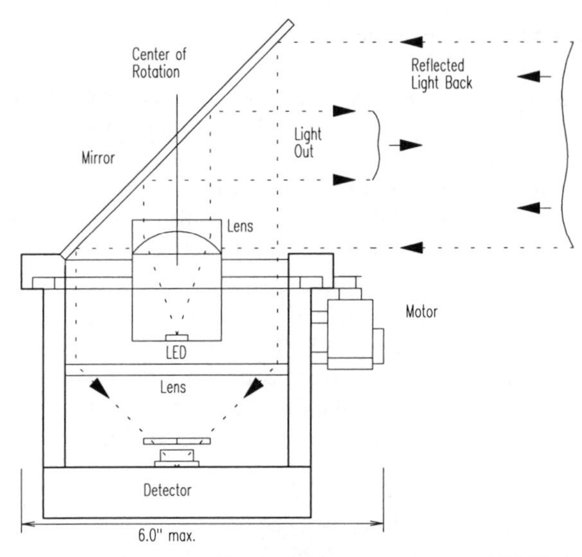

Figure 6-16. Schematic drawing of the *ORS-1* ranging system (courtesy ESP Technologies, Inc.).

Table 6-2. Selected specifications for the LED-based near-infrared *Optical Ranging System.*

Parameter	Value	Units
Maximum range	20	feet
Minimum range	2	feet
Accuracy	< 6	inches
AGC output	1-5	volts
Output power	2	milliwatts
Beamwidth	1	inch
Dimensions	6 x 6 x 12	inches
Weight		pounds
Power	12	volts DC
	2	amps

A representative ranger scan taken in a laboratory environment is shown in Figure 6-17. The ranger is mounted on the robot cart and located at the (0,0) position marked by a cross in the center of the plot. The data collection for this plot corresponds to a single mirror rotation taking approximately one second. Note the absence of any data points in regions (i.e., segment AB in the lower right corner), where the return signal is outside the AGC window; and therefore no data are accumulated. All of the objects in the room are found to correspond accurately to their positions as indicated by the range measurements (Miller & Wagner, 1987).

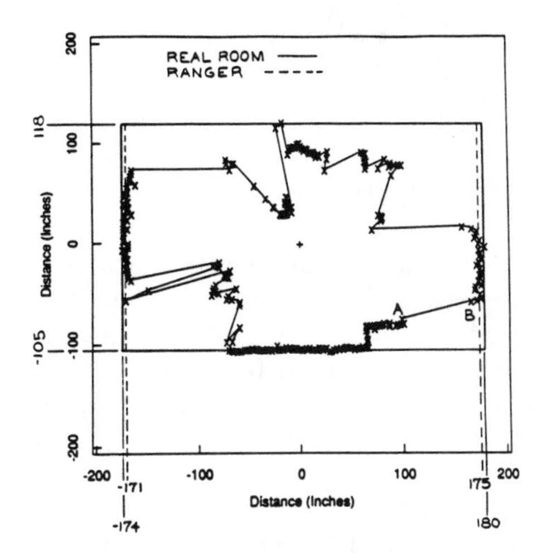

Figure 6-17. This actual scan of a laboratory environment resulted from a single mirror rotation taking approximately one second (courtesy ESP Technologies, Inc.).

6.1.6 Acuity Research *AccuRange 3000*

Acuity Research, Inc., Menlo Park, CA, has recently introduced an interesting product capable of acquiring unambiguous range data from 0 to 20 meters using a proprietary technique wherein the optical beam path is part of an oscillatory feedback loop. The *AccuRange 3000* (Figure 6-18) projects a collimated beam of near-infrared or visible laser light, amplitude modulated with a non-sinusoidal waveform at a 50-percent duty cycle (Clark, 1994). A 2.5-inch collection aperture surrounding the laser diode emitter on the front face of the cylindrical housing gathers any reflected energy returning from the target and subsequently disables the laser source. Once the sensed energy disappears as a consequence of this action, the laser is re-energized, whereupon the cycle repeats. The net effect of this innovative approach, which requires minimal circuitry, is a square-wave output with a period of oscillation proportional to the measured range.

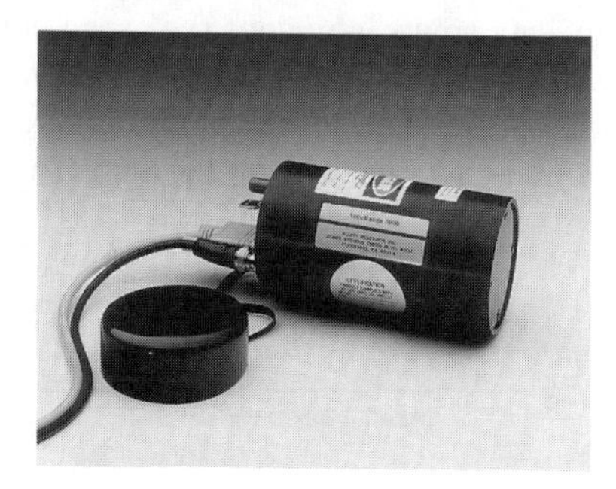

Figure 6-18. The *AccuRange 3000* distance measuring sensor provides a square-wave output that varies inversely in frequency as a function of range (courtesy Acuity Research, Inc.).

The frequency of the output signal varies from approximately 50 MHz at zero range to 4 MHz at 20 meters. Distance to target can be determined through use of a frequency-to-voltage converter, or by measuring the period with a hardware or software timer (Clark, 1994). Separate 0- to 10-volt analog outputs are provided for returned signal amplitude, ambient light, and temperature to facilitate dynamic calibration for optimal accuracy in demanding applications. The range output changes within 250 nanoseconds to reflect any change in target distance, and all outputs are updated within a worst-case time frame of only three microseconds. This rapid response rate (up to 312.5 KHz for all outputs with the optional SCSI interface) allows the beam to be manipulated at a 1000- to 2000-Hz rate with the mechanical-scanner option shown in Figure 6-19 below. A 45-degree balanced-mirror arrangement is rotated under servo control to deflect the coaxial outgoing and incoming beams for full 360-degree planar coverage.

Figure 6-19. A 360-degree beam-deflection capability is provided by an optional single axis rotating scanner (courtesy Acuity Research, Inc.).

Table 6-3. Selected specifications for the Acuity *AccuRange 3000* distance measurement sensor.

Parameter	Value	Units
Laser output	5	milliwatts
Beam divergence	0.5	milliradians
Wavelength	780/670	nanometers
Maximum range	20	meters
Minimum range	0	meters
Accuracy	2	millimeters
Sample rate	up to 312.5	KHz
Diameter	3	inches
Length	5.5	inches
Weight	18	ounces
Power	5 and 12	volts DC
	250 and 50	milliamps

6.1.7 TRC *Light Direction and Ranging System*

Transitions Research Corporation (TRC), Danbury, CT, offers a low-cost *lidar* system for detecting obstacles in the vicinity of a robot and/or estimating position from local landmarks or retroreflective targets (see Chapter 15), based on the previously discussed Acuity Research *AccuRange 3000* unit. TRC adds a two-degree-of-freedom scanning mechanism employing a gold front-surfaced mirror specially mounted on a vertical pan axis that rotates between 200 and 900 rpm (Figure 6-20). The tilt axis of the scanner is mechanically synchronized to nod one complete cycle (down 45 degrees and back to horizontal) per 10 horizontal scans, effectively creating a protective spiral of detection coverage around the robot (TRC, 1994). The tilt axis can be mechanically disabled if so desired for 360-degree azimuthal scanning at a fixed elevation angle.

Table 6-4. Selected specifications for the TRC *Light Direction and Ranging System.*

Parameter	Value	Units
Maximum range	12	meters
Modulation frequency	2	MHz
Accuracy (range)	25	millimeters
Resolution (range)	5	millimeters
(azimuth)	0.18	degrees
Sample rate	25	KHz
Size (scanner)	13 x 13 x 35	centimeters
(electronics)	30 x 26 x 5	
Weight	2	kilograms

A *68HC11* microprocessor automatically compensates for variations in ambient lighting and sensor temperature, and reports range, bearing, and elevation data via an Ethernet or RS-232 interface. Power requirements are 500 milliamps at 12 volts DC and 100 milliamps at 5 volts DC. Typical operating parameters are listed in Table 6-4.

Figure 6-20. The TRC *Light Direction and Ranging System* incorporates a two-axis scanner to provide full-volume coverage sweeping 360 degrees in azimuth and 45 degrees in elevation (courtesy Transitions Research Corp.).

6.2 Frequency Modulation

A closely related alternative to the amplitude-modulated phase-shift-measurement ranging scheme is frequency-modulated (FM) radar. This technique involves transmission of a continuous electromagnetic wave modulated by a periodic triangular signal that adjusts the carrier frequency above and below the mean frequency f_0 as shown in Figure 6-21. The transmitter emits a signal that varies in frequency as a linear function of time:

$$f(t) = f_o + at$$

where:

a = some constant
t = elapsed time.

This signal is reflected from a target and arrives at the receiver at time $t + T$.

$$T = \frac{2d}{c}$$

where:

T = round-trip propagation time
d = distance to target
c = speed of light.

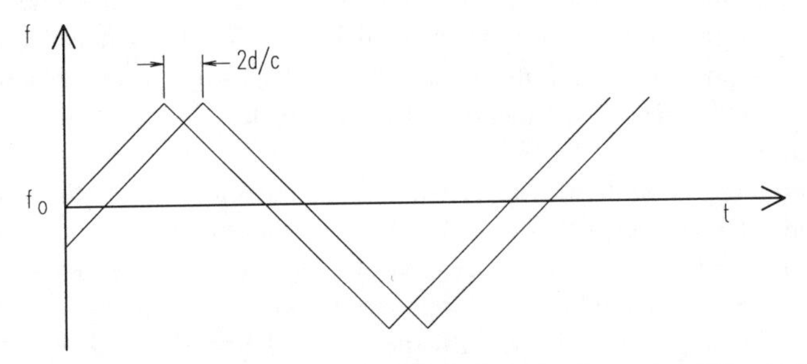

Figure 6-21. The received frequency curve is shifted along the time axis relative to the reference frequency.

The received signal is compared with a reference signal taken directly from the transmitter. The received frequency curve will be displaced along the time axis relative to the reference frequency curve by an amount equal to the time required for wave propagation to the target and back. (There might also be a vertical displacement of the received waveform along the frequency axis, due to the Doppler effect.) These two frequencies when combined in the mixer produce a beat frequency F_b:

$$F_b = f(t) - f(T + t) = aT$$

where:

a = constant.

This beat frequency is measured and used to calculate the distance to the object:

$$d = \frac{F_b c}{4F_r F_d}$$

where:

d = range to target
c = speed of light
F_b = beat frequency
F_r = repetition (modulation) frequency
F_d = total FM frequency deviation.

Distance measurement is therefore directly proportional to the difference or beat frequency, and is as accurate as the linearity of the frequency variation over the counting interval.

Advances in wavelength control of laser diodes now permit this radar ranging technique to be used with lasers. The frequency or wavelength of a laser diode can be shifted by varying its temperature. Consider an example where the wavelength of an 850-nanometer laser diode is shifted by 0.05 nanometers in four seconds: the corresponding frequency shift is 5.17 MHz per nanosecond. This laser beam, when reflected from a surface 1 meter away, would produce a beat frequency of 34.5 MHz. The linearity of the frequency shift controls the accuracy of the system.

The frequency-modulation approach has an advantage over the phase-shift-measurement technique in that a single distance measurement is not ambiguous. (Recall phase-shift systems must perform two or more measurements at different modulation frequencies to be unambiguous.) However, frequency modulation has several disadvantages associated with the required linearity and repeatability of the frequency ramp, as well as the coherence of the laser beam in optical systems. As a consequence, most commercially available FMCW ranging systems are radar based, while laser devices tend to favor TOF and phase-detection methods.

6.2.1 VRSS Automotive Collision Avoidance Radar

One of the first practical automotive collision avoidance radar systems was developed by Vehicle Radar Safety Systems (VRSS) of Mt. Clemens, MI. This modified Doppler radar unit is intended to alert drivers to potentially dangerous situations. A grill-mounted miniaturized microwave horn antenna sends out a narrow-beam signal that detects only those objects directly in the path of the vehicle, ignoring targets (such as road signs and parked cars) on either side. When the radar signal is reflected from a slower-moving or stationary target, it is detected by the antenna and passed to an under-the-hood electronic signal processor (VRSS, 1983).

The signal processor continuously computes the host vehicle speed and acceleration, distance to the target, relative velocity, and target acceleration. If these parameters collectively require the driver to take any corrective or precautionary action, a warning buzzer and signal light are activated on a special dashboard monitor. An *alert* signal lights up when an object or slower-moving vehicle is detected in the path of the host vehicle. If the target range continues to decrease, and the system determines that a collision is possible, a *warning* light and buzzer signal the driver to respond accordingly. If range continues to decrease with no reduction in relative velocity, then a *danger* light illuminates indicating the need for immediate action.

A filter in the signal processor provides for an optimum operating range for the system, based on the relative velocity between the vehicle and the perceived object. The response window corresponds to a calculated difference in speed of between 0.1 and 30 miles per hour (VRSS, 1983). If the speed differential exceeds 30 miles per hour, the filter circuit delays signals to the dashboard monitor. This helps to eliminate false signals and signals that might otherwise be caused by approaching vehicles when passing another vehicle on a two-lane highway.

The VRSS collision warning system has been tested in over a million miles of driving conditions in fog, rain, snow, and ice with good results. The present model was perfected in 1983 after 36 years of research, and approved by the FCC in 1985. Although aimed initially at the bus and trucking industries, the low-cost unit offers convincing proof that small, low-power radar systems offer a practical alternative to ultrasonic rangefinders for the collision avoidance needs of a mobile robot operating in outdoor scenarios.

Table 6-5. Selected specifications for the VRSS automotive radar.

Parameter	Value	Units
Effective range	1-300	feet
Accuracy	1.5	percent
Update rate	200	Hz
Operating frequency	24	GHz
RF power	10	milliwatts
Beamwidth (horizontal)	6	degrees
(vertical)	6	degrees
Size (antenna)	3 x 4	inches
(electronics unit)	4 x 5 x 2	inches
Weight (total)	4	pounds
Power	12	volts DC
	12	watts

6.2.2 VORAD Vehicle Detection and Driver Alert System

VORAD (Vehicle Onboard Radar) Safety Systems, Inc., San Diego, CA, has also developed a commercial millimeter-wave FMCW Doppler radar system designed for use on a motor vehicle (VORAD, undated). The *Vehicle Collision Warning System* employs a 5- by 5-inch antenna/transmitter/receiver package mounted on the front grill of a vehicle to monitor speed and distance to other traffic or obstacles on the road (Figure 6-22). The flat etched-array antenna radiates approximately 0.5 milliwatts of power at 24.725 GHz directly down the roadway in a narrow directional beam. A GUNN diode is used for the transmitter, while the receiver employs a balanced-mixer detector (Woll, 1993).

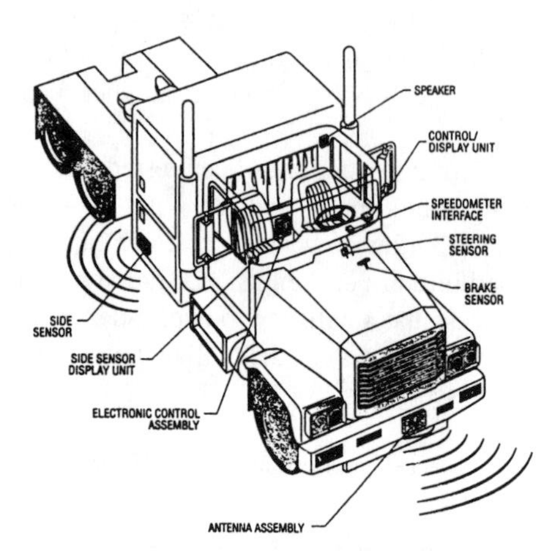

Figure 6-22. The forward-looking antenna/transmitter/receiver module is mounted on the front of the vehicle at a height between 50 and 125 centimeters, while an optional side antenna can be installed as shown for blind-spot protection (courtesy VORAD Safety Systems, Inc.).

The *Electronics Control Assembly* located in the passenger compartment or cab (see again Figure 6-22) can individually distinguish up to 20 moving or stationary objects (Siuru, 1994) out to a maximum range of 350 feet; the closest three targets within a prespecified warning distance are tracked at a 30-Hz rate. A Motorola *DSP 56001* and an Intel *87C196* microprocessor calculate range and range-rate information from the RF data and analyze the results in conjunction with vehicle-velocity, braking, and steering-angle information. The *Control Display Unit* alerts the operator if warranted of potentially hazardous driving situations with a series of caution lights and audible beeps.

As an optional feature, the *Vehicle Collision Warning System* offers blind-spot detection along the right-hand side of the vehicle out to 15 feet. The *Side Sensor* transmitter employs a *dielectric resonant oscillator* operating in pulsed Doppler mode at 10.525 GHz, using a flat etched-array antenna with a beamwidth of about 70 degrees (Woll, 1993). The system microprocessor in the *Electronics Control Assembly* analyzes the signal strength and frequency components from the *Side Sensor* subsystem in conjunction with vehicle speed and steering inputs, and activates audible and visual LED alerts if a dangerous condition is thought to exist.

A standard recording feature stores 20 minutes of the most recent historical data on a removable EEPROM memory card for post-accident reconstruction, including steering, braking, and idle time. VORAD Safety Systems also offers complete trip reporting for commercial vehicle operators, to include engine rpm, idle time, number and severity of warnings, over-rpm, over-speed conditions, etc.

Greyhound Bus Lines recently completed installation of the VORAD radar on all of its 2,400 buses (Bulkeley, 1993), and subsequently reported a 25-year low accident record (Greyhound, 1994). The entire system weighs just 6.75 pounds and operates from 12 or 24 volts DC with a nominal power consumption of 20 watts. An RS-232 digital output is available. Selected specifications are listed in Table 6-6 below.

Table 6-6. Selected specifications for the *Collision Warning System.*

Parameter	Value	Units
Effective range	1-350	feet
Accuracy	3	percent
Update rate	30	Hz
Host platform speed	0.5-120	miles per hour
Closing rate	0.25-100	miles per hour
Operating frequency	24.725	GHz
RF power	0.5	milliwatts
Beam width (horizontal)	4	degrees
(vertical)	5	degrees
Size (antenna)	6 x 8 x 1.5	inches
(electronics unit)	8 x 6 x 5	inches
Weight (total)	6.75	pounds
Power	12-24	volts DC
	20	watts
Mean-time-between-failure	17,000	hours

Along with their new joint-venture partner, Eaton Corporation, VORAD is now shipping "ruggedized" commercial systems to the heavy truck industry, with an extended temperature range of -40 to +85°C, 100 volts/meter EMI susceptibility, and full compliance to environmental standard SAE J1455 (VORAD, 1994).

6.2.3 Safety First Systems *Vehicular Obstacle Detection and Warning System*

Safety First Systems, Ltd., Plainview, NY, and General Microwave, Amityville, NY, have teamed to develop and market a 10.525-GHz microwave unit for use as an automotive blind-spot alert for drivers when backing up or changing lanes (Siuru, 1994). The narrow-band (100-KHz) modified-FMCW technique uses patent-pending phase discrimination augmentation for a 20-fold increase in achievable resolution. For example, a conventional FMCW system operating at 10.525 GHz with a 50-MHz bandwidth is limited to a best-case range resolution of approximately 10 feet, while the improved approach can resolve distance to within 0.6 feet out to 40 feet (SFS, undated). Even greater accuracy and maximum ranges (i.e., 160 feet) are possible with additional signal processing.

A prototype of the system delivered to Chrysler Corporation uses conformal bistatic microstrip antennae mounted on the rear side panels and rear bumper of a minivan and can detect both stationary and moving objects within the coverage patterns shown in Figure 6-23. Coarse range information about reflecting targets is represented in four discrete range bins with individual TTL output lines: 0 to 6 feet, 6 to 11 feet, 11 to 20 feet, and greater than 20 feet. Average radiated power is about 50 microwatts with a three-percent duty cycle, effectively eliminating adjacent-system interference. The system requires 1.5 amps from a single 9- to 18-volt DC supply.

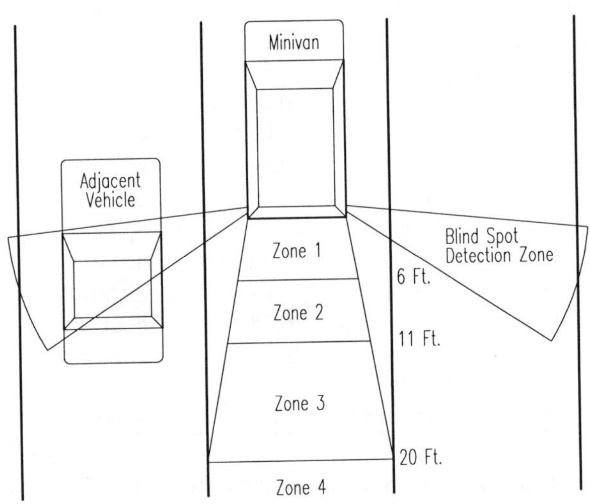

Figure 6-23. The *Vehicular Obstacle Detection and Warning System* employs a modified FMCW ranging technique for blind-spot detection when backing up or changing lanes (courtesy Safety First Systems, Ltd.).

6.2.4 Millitech Millimeter Wave Radar

Millitech Corporation, Deerfield, MA, has designed a number of millimeter-wave FMCW systems aimed at satisfying the short-distance non-contact ranging needs for robotic collision avoidance. These sensors operate at wavelengths of 3.2 millimeters (94 GHz) to 8.6 millimeters (35 GHz), and are superior to infrared devices under all weather conditions, since performance is not significantly degraded by environmental conditions such as fog, rain, dust, and blowing sand (see Chapter 9). Figure 6-24 shows a scanned imaging and data acquisition system in which four vertically stacked beams are mechanically scanned in azimuth to produce a 256-pixel frame of range data at a 5-Hz rate. Each individual pixel contains 512 range bins spaced 0.5 meters apart.

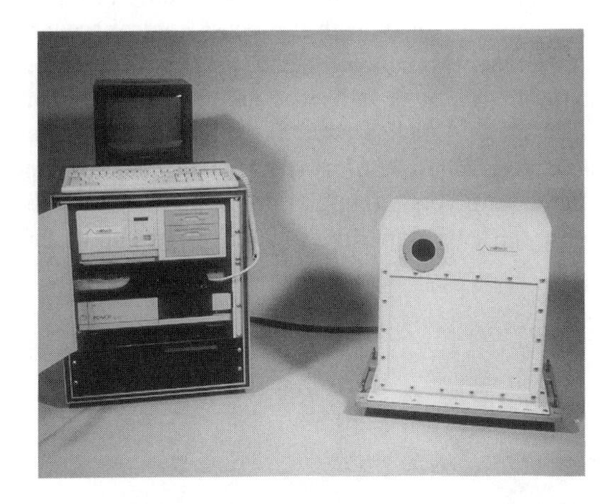

Figure 6-24. Four vertically stacked millimeter-wave beams are mechanically scanned in azimuth to produce a 4- by 64-pixel image over a 12- by 64-degree field of view (courtesy Millitech Corp.).

An innovative feature of Millitech's design is the use of closed-loop control of the oscillator to generate the basic transmitter waveform, yielding stable low-cost performance that will not degrade over time. Operation in the millimeter-wave region of the RF spectrum allows higher-resolution performance in smaller-package configurations than can be achieved with lower-frequency microwave counterparts. A two-degree field of view, for example, requires an aperture size of only 110 millimeters at 94 GHz, as compared to 300 millimeters at 35 GHz. Representative range data collected by the scanning sensor shown above is depicted in Figure 6-25 for two sets of target objects.

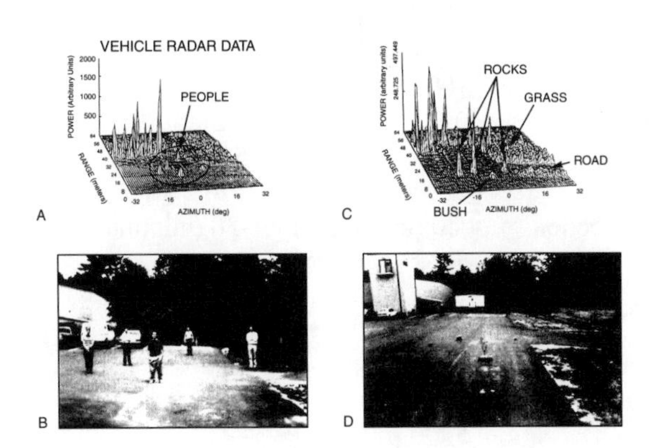

Figure 6-25. Range data acquired by the 256-pixel scanned sensor is shown for (A & B) human targets, and, (C & D) inanimate objects such as rocks and grass (courtesy Millitech Corp.).

Fixed-orientation single-beam versions of Millitech's FMCW radar (Figure 6-26) have also been produced for industrial process-control scenarios where severe dust, smoke and/or steam preclude the use of conventional laser-based or acoustical ranging systems. Such industrial applications generally require higher range resolution (i.e., 50 millimeters typical) over shorter operating distances (i.e., ≤30 meters) than needed for purposes of robotic collision avoidance (see Table 6-7). In addition, the extreme operating conditions associated with some industrial processes can expose the sensor aperture to temperatures up to 200°C.

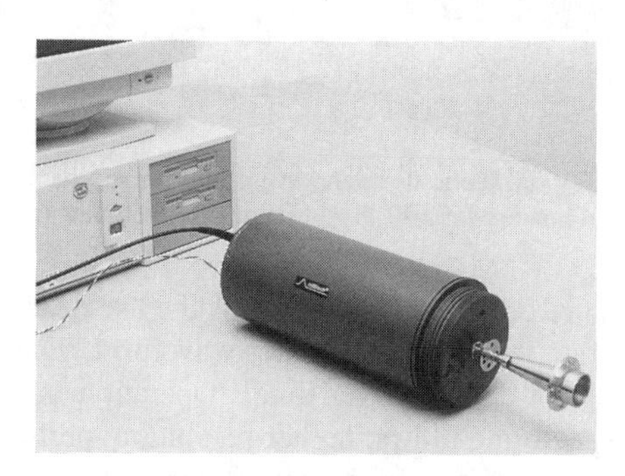

Figure 6-26. Typical industrial millimeter-wave FMCW range sensors are often exposed to ambient temperatures in excess of 200°C (courtesy Millitech Corp.).

Table 6-7. Selected specifications for Millitech millimeter-wave radar prototypes.

Parameter	256-Pixel Scanned Sensor	Fixed-Beam Industrial Sensor	Units
Maximum range:	100	30	meters
Minimum range	0.5	0.2	meters
Output power	10	5-10	milliwatts
Field of View	12 by 64	1 to 2	degrees
Radar cross-section	-40 (minimum)	-30 (minimum)	dBsm
Resolution: range	50	5	centimeters
azimuth	1	1-2	degrees
elevation	3	1-2	degrees
Center frequency	77	94	GHz
Sweep bandwidth	300	400	MHz
Frame rate	5	--	Hz
Data output	Digital	External A/D	
Power	24	±18 to ±28	volts DC
	3	0.5	amps

6.3 References

Adams, M.D., "Amplitude Modulated Optical Range Data Analysis in Mobile Robotics," IEEE International Conference on Robotics and Automation, Atlanta, GA, pp. 8-13, 1993.

Beyer, J., Jacobus, C., Pont, F., "Autonomous Vehicle Guidance Using Laser Range Imagery," SPIE Vol. 852, Mobile Robots II, p. 34-43, Nov., 1987.

Binger, N, Harris, S.J., "Applications of Laser Radar Technology," *Sensors*, pp. 42-44, April, 1987.

Boltinghouse, S., Larsen, T., "Navigation of Mobile Robotic Systems Employing a 3D Laser Imaging Radar," ANS Third Topical Meeting on Robotics and Remote Systems, Section 2-5, pp. 1-7, Charleston, SC, March, 1989.

Boltinghouse, S., Burke, J., Ho, D., "Implementation of a 3D Laser Imager Based Robot Navigation System with Location Identification," SPIE Vol. 1388, Mobile Robots V, Boston, MA, pp. 14-29, November, 1990.

Bulkeley, D., "The Quest for Collision-Free Travel," *Design News*, Oct. 4, 1993.

Byrd, J.S., DeVries, K.R., "A Six-Legged Telerobot for Nuclear Applications Development, *International Journal of Robotics Research*, Vol. 9, pp. 43-52, April, 1990.

Chen, Y.D., Ni, J., Wu, S.M., "Dynamic Calibration and Compensation of a 3D Lasar Radar Scanning System," IEEE International Conference on Robotics and Automation, Atlanta, GA, Vol. 3, pp. 652-664, May, 1993.

Clark, R.R., "A Laser Distance Measurement Sensor for Industry and Robotics," *Sensors*, pp. 43-50, June, 1994.

Conrad, D.J., Sampson, R.E., "3D Range Imaging Sensors," in *Traditional and Non-Traditional Robotic Sensors*, T.C. Henderson, ed., NATO ASI Series, Vol. F63, Springer-Verlag, pp. 35-47, 1990.

Depkovich, T., Wolfe, W., "Definition of Requirements and Components for a Robotic Locating System," Final Report No. MCR-83-669, Martin Marietta Aerospace, Denver, CO, February, 1984.

ESP, "ORS-1 Optical Ranging System," Product Literature, ESP Technologies, Inc., Lawrenceville, NJ 08648, 23 March, 1992.

Figueroa, F., Barbieri, E., "Increased Measurement Range Via Frequency Division in Ultrasonic Phase Detection Methods," *Acustica*, Vol. 73, pp. 47-49, 1991a.

Figueroa, J.F., Barbieri, E., "An Ultrasonic Ranging System for Structural Vibration Measurements," *IEEE Transactions on Instrumentation and Measurement*, Vol. 40, No. 4, pp. 764-769, August, 1991b.

Figueroa, J.F., Lamancusa, J.S., "A Method for Accurate Detection of Time of Arrival: Analysis and Design of an Ultrasonic Ranging System," *Journal of the Acoustical Society of America*, Vol. 91, No. 1, pp. 486-494, January, 1992.

Frazier, K., "Innovative Range Imager Sees How Targets Measure Up," *Sandia LabNews*, Vol. 46, No. 19, September 16, 1994.

Greyhound, "Accident Rate Keeps Falling," *Greyhound Connections*, Vol. 4, No. 2, March/April, 1994.

Hebert, M., Krotkov, E., "3-D Measurements from Imaging Laser Radars: How Good Are They?" International Conference on Intelligent Robots and Systems, pp. 359-364, 1991.

Kerr, J.R., "Real Time Imaging Rangefinder for Autonomous Land Vehicles," SPIE Vol. 1007, Mobile Robots III, , pp. 349-356, November, 1988.

Larson, T.R., Boltinghouse, S., "Robotic Navigation Within Complex Structures," SPIE Vol. 1007, Mobile Robots III, Cambridge, MA, pp. 339-348, November, 1988.

Miller, G.L., Wagner, E.R., "An Optical Rangefinder for Autonomous Robot Cart Navigation," Proceedings of the Advances in Intelligent Robotic Systems: SPIE Mobile Robots II, 1987.

Nitzan, D. et al. "The Measurement and Use of Registered Reflectance and Range Data in Scene Analysis," Proceedings of IEEE, vol. 65, no. 2, pp. 206-220, February, 1977.

Patterson, M.R., Reidy, J.J., Rudolph, R.C., "Guidance and Actuation Systems for an Adaptive-Suspension Vehicle," Final Technical Report, Battelle Columbus Division, OH, AD #A139111, 20 March, 1994.

Scott, M.W., "Range Imaging Laser Radar," US Patent 4,935,616, June 19, 1990.

SFS, "Vehicular Obstacle Detection and Warning System," Product Literature, Safety First Systems, Ltd., Plainview, NY, undated.

Siuru, B., "The Smart Vehicles Are Here," *Popular Electronics*, Vol. 11, No. 1, pp. 41-45, January, 1994.

TRC, "TRC Light Ranger," Product Literature 940823, Transitions Research Corp., Danbury, CN, August, 1994.

VORAD, "The VORAD Vehicle Detection and Driver Alert System," Product Literature, VORAD Safety Systems, Inc., San Diego, CA, undated.

VORAD, "EATON VORAD Collision Warning System," Product Literature VOSL-0100, VORAD Safety Systems, Inc., San Diego, CA, 1994.

VRSS, "Rashid Radar Safety Brake," Product Literature, Vehicle Radar Safety Systems, Inc., Mount Clemens, MI, 1983.

Vuylsteke, P., Price, C.B., Oosterlinck, A., "Image Sensors for Real-Time 3D Acquisition, Part 1," ," in *Traditional and Non-Traditional Robotic Sensors*, T.C. Henderson, ed., NATO ASI Series, Vol. F63, Springer-Verlag, pp. 187-210, 1990.

Weiss, S.A., Lasar-Radar Imaging Without Scanners," *Photonics Spectra*, pp. 28-29, April, 1994.

Woll, J.D., "A Review of the Eaton VORAD Vehicle Collision Warning System," Reprinted from International Truck and Bus Meeting and Exposition, Detroit, MI, SAE Technical Paper Series 933063, ISSN 0148-7191, pp. 1-4, November, 1993.

Woodbury, N., Brubacher, M., Woodbury, J.R., "Noninvasive Tank Gauging with Frequency-Modulated Laser Ranging," *Sensors*, pp. 27-31, Sep., 1993.

7
Other Ranging Techniques

This chapter discusses three final non-contact ranging techniques of potential interest: 1) *interferometry*, 2) *range from focus*, and 3) *return signal intensity*.

7.1 Interferometry

One of the most accurate and precise distance ranging methods known, *interferometry* has existed for many years in laboratory scenarios that afforded the necessary controlled or otherwise structured environment (Brown, 1985). Under such non-turbulent atmospheric conditions, laser interferometers can achieve fractional wavelength accuracy. Recent developments in optical technologies are making possible applications of this technique outside of laboratory scenarios.

The concept is based on the resulting interference patterns that occur when two energy waves caused to travel different paths are compared. If the length of one of the paths is changed, the two beams will interact in such a way that clearly visible constructive and destructive interference fringes are produced. (Fringes are patterns or disturbances in the combined waveform that alternate between maximum and minimum intensity.) Figure 7-1 shows a typical system consisting of a laser emitter, a series of beam splitters and directional mirrors, and a fringe counter. The output of a single coherent light source is split into a reference beam and an output beam. The reference beam is immediately directed into the fringe counter for future recombination with the reflected beam.

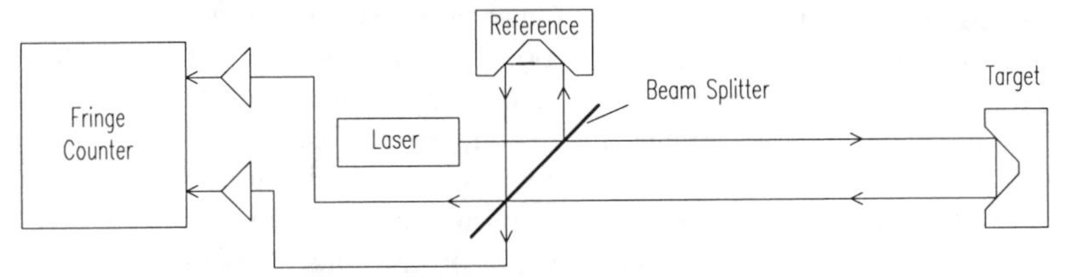

Figure 7-1. Block diagram of typical interferometer ranging system.

The second beam exits the instrument and travels through the air to a retroreflector located on the object of interest. (Retroreflectors must be employed to provide a reliable return signal for the interferometer.) The reflected energy is optically recombined with the reference beam in the fringe counter. Movement of the target through a distance equal to half the source wavelength results in the detection of one fringe (Beesley, 1971). By counting the number of fringes passing the detector it is possible to calculate with extreme accuracy the distance the retroreflector (and thus the object) has traveled along the line of the source beam.

Interferometers do not measure absolute range, but the relative distance an object has moved from its previous location; therefore, the distance from the sensor to the target is not directly known. By initializing the retroreflector to a specified reference point, however, it becomes possible to determine absolute distance to an object. All subsequent measurements will be relative to this reference point, provided the beam is never broken and the target momentarily lost.

In conventional laser interferometry, target displacement of 1 centimeter can result in the movement of approximately 10 million fringes past a detector capable of measuring changes on the order of one tenth of a fringe (Beesley, 1971). Potential accuracies over a distance of 10 meters can approach one part in 1,000,000, provided similar accuracy is available for the wavelength of the energy source. The maximum distance that can be measured by such instruments is therefore dependent on the coherent qualities of the source used. In theory, distances of hundreds of kilometers can be measured; however, this goal cannot be practically achieved using current technology (Beesley, 1971).

While extremely precise, limiting factors of interferometry include the relatively high cost, the need for a continuous line of sight between the source and retroreflector, and the limitation to relative only (as opposed to absolute) distance measurement. Air turbulence effectively reduces the practical range of such systems to around 10 meters (Beesley, 1971). The turbulence causes sufficiently large variations in the path lengths of the light beams so that no spatial coherence exists between the interfering beams; therefore, there are no fringes produced. Temperature changes and microphonic disturbances can cause fluctuations in components of the light source delivery system that alter the wavelength and intensity of the output (Beesley, 1971). The laser output must therefore be well stabilized to realize the full potential of interferometric measuring.

The use of interferometers in robotic applications was initially limited to measurement of single-axis linear motion. Recent developments have expanded their applicability to three-dimensional six-degree-of-freedom systems, known as tracking interferometers because the returning beam is also used by the system to track the lateral motion of retroreflective mirrors mounted on the object. Systems currently in existence are capable of precision tracking of robotic manipulators performing non-rectilinear motions in six degrees of freedom (Everett, 1985; Brown, 1985; Lau, et al., 1985).

7.1.1 CLS Coordinate Measuring System

Chesapeake Laser Systems began development in 1983 of the *CMS-1000*, a laser-based tracking interferometer system that can measure the location of a moving object to better than 10 microns over a volume of 3 by 3 by 3 meters (Brown, 1985; Cleveland, 1986; Brown, et al., 1987). The system employs a servo-controlled beam-steering mechanism to track a randomly moving retroreflective target with a 50-Hz update rate. After a brief calibration routine, three tracking interferometers are used to continuously measure the distance to a number of retroreflectors as shown in Figure 7-2 to calculate the X-Y-Z coordinates of the robotic end effector through trilateration (Everett, 1988).

Figure 7-2. The *CMS-1000* uses multiple HeNe laser beams to track retroreflective targets attached to the moving end effector of the *Intelligent Robotic Inspection System* built for the Navy by MTS Systems Corp. (courtesy Applied Research Lab, Penn State University).

The trilateration solution (i.e., superposition of three range arcs) is inherently more accurate than conventional triangulation using angle measurement alone, in which case position errors appear as:

$$\text{error} = r\,d\theta$$

where:

r = radial distance from tracker to retroreflector
$d\theta$ = angular error.

The position error for trilateration, on the other hand, shows up as:

$$\text{error} = r(1-\cos d\theta) = r\,d\theta^2$$

which is orders of magnitude smaller than $r\,d\theta$.

An improved system, the *CMS-2000*, combines laser interferometry with servo-controlled trackers to measure movement with submicron resolution at ranges up to 35 feet. The *CMS-2000* was initially designed for use by the US Air Force as part of the Strategic Defense Initiative to track (with a 100-Hz update rate) a retroreflector mounted on a hovering rocket. The data obtained by the *CMS-2000* is then used to check the vehicle's onboard control systems (CLS, 1991). This application was unsuccessful due to the excessive heat and dust present in the hanger during launch.

7.2 Range From Focus

The driving thrust behind most of the early work in *range from focus* has been the automatic-focus interest within the commercial video and still camera industry (Marnheim, 1980; Goldberg, 1982; Ctein, 1982; Stauffer & Wilwerding, 1982; Wolpert, 1987). Some more recent efforts have concentrated on the development of sensor subsystems intended specifically for robotic applications (Krotkov & Martin, 1986; Farsaie, et al., 1987). The principle of operation is based on the *Gauss thin lens law* (Conrad & Sampson, 1990):

$$\frac{1}{r}=\frac{1}{f}-\frac{1}{s}$$

where:

r = distance from lens to object viewed
s = distance from lens to image plane
f = focal length of the lens.

Rearranging in terms of the desired distance to object r yields:

$$r=\frac{s\,f}{s-f}.$$

Krotkov and Martin (1986) summarize the *range-from-focus* technique as follows:

- The system is initially calibrated with a point source to establish the relationship between focus motor shaft position and lens focal length f.
- The sharpness of focus is maximized by varying f until the detector plane coincides with the image plane.
- The resulting value of f is read.

- The measured value of f and the constant s are substituted into the thin lens equation, which is then solved for r.

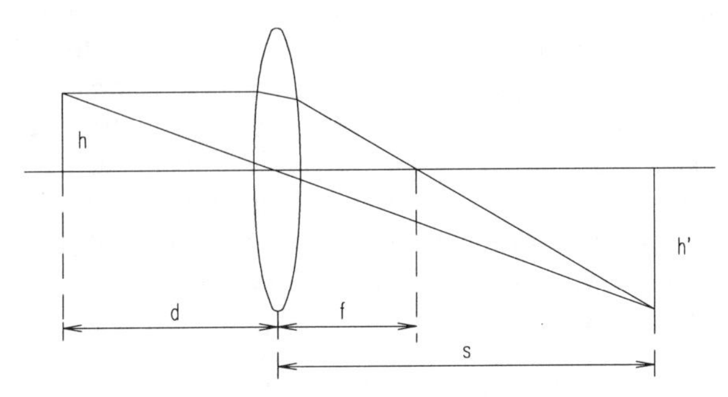

Figure 7-3. The object and image positions (h) and (h') are shown with respect to the focal plane for a thin lens (adapted from Brain, 1979).

The fundamental technical challenge in the above procedure is how to best determine the point of optimal focus; the variety of techniques reported can generally be divided into two broad classes: 1) dedicated in-focal-plane *image analysis* hardware and 2) external signal analysis of conventional composite video output.

A very common approach in the latter case involves examining the high-frequency spatial content of the video signal. The *optical transfer function (OTF)* of a lens describes how each spatial frequency component in a viewed object is attenuated by the lens as it forms an image. Severe aberration (i.e., departure of the wavefront from its ideal spherical form) due to defocusing will significantly reduce the high-frequency portion of the OTF (Krotkov & Martin, 1986). The obvious visual effect is a blurring of the resulting image. More importantly, the loss of high-frequency components in the video signal can be electronically detected to form the basis of a *focus criterion function.*

Deriving distance information from camera focus has an advantage over stereoscopic ranging in the sense that there is no need to solve the computationally intensive and sometimes error-prone correspondence problem. The principle disadvantage is the technique applies only to one observed point in the scene at any given time. Other limitations arise from measurement errors associated with the quantification of both f and s, spatial quantitization of the detector, the performance of the focus criterion function, the validity of the thin lens model describing a compound lens, and the nonlinearity of the lens equation itself (Krotkov & Martin, 1986).

7.2.1 Honeywell Autofocus Systems

The first practical autofocus system for lens-shutter cameras was developed by the Honeywell Visitronic Group in 1976. The system represents a variation of the stereoscopic ranging technique, nicely optimized for low-cost implementation through the development of a special purpose integrated circuit (IC) for autocorrelation. Two five-element photosensitive arrays were located at each end of the IC, measuring about 0.1 by 0.25 inches in size (Stauffer & Wilwerding, 1982). Figure 7-4 shows a pair of mirrors reflecting the incoming light from two viewing windows at either end of the camera housing onto these arrays. One of these images remained fixed while the other was scanned across its respective array through the mechanical rotation of the associated mirror. The angular orientation of the moving mirror at the precise instant that the IC indicated the two images were matched was directly related to the range to the subject and used to position the camera lens.

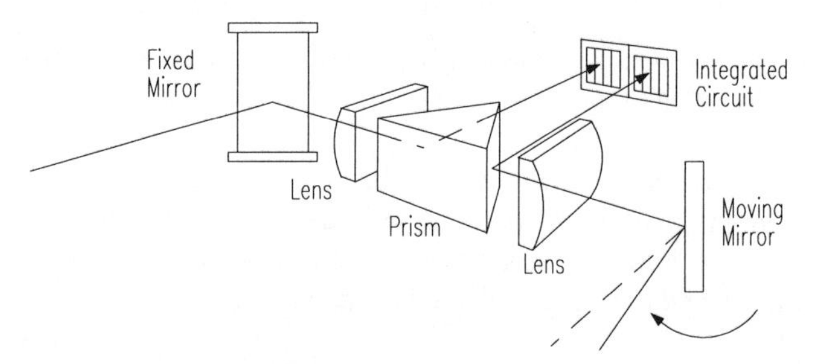

Figure 7-4. The original Visitronic *Autofocus System* employed a pair of five-element detector arrays to establish correlation (courtesy Honeywell Corp.).

The photocurrents from corresponding elements in each array were passed through a string of diodes on the IC and thus converted to voltages proportional to the log of the current. The resulting pair of voltages was then fed to a differential amplifier, which produced a difference signal proportional to the ratio of the two light intensities as seen by the respective detectors (Stauffer & Wilwerding, 1982). For four of the five-element array pairs, the absolute values of these difference signals were summed and the result subtracted from a reference voltage to yield the correlation signal. The better the scene match, the lower the differential signal for each array pair, and the higher the correlation signal.

The peak value of the correlation signal corresponded to the best scene match. An operational amplifier on the IC performed a continuous comparison between the correlation output and the previous highest value stored in a capacitor. The output from this comparator was high as long as the correlation signal was lower than the previous peak value. The last low-to-high transition represents the mirror angle corresponding to the highest peak. A potentiometer on the moving mirror produced a voltage that varied as a linear function of mirror position. The output of this potentiometer was sampled and stored when the IC indicated the peak correlation signal was present. A similar potentiometer coupled to the camera lens positioning mechanism was used to stop the lens travel when its output matched the stored voltage signifying mirror position at best focus.

The Honeywell *Through-the-Camera-Lens (TCL)* autofocus system, a second-generation refinement of the *Visitronic System*, compared the signatures of light passing through two different sectors of the camera lens as opposed to two separate viewing windows. Instead of five, there were 24 pairs of detectors arranged in an array about 5 millimeters long. Two complete arrays were provided to accommodate camera lenses with different aperture sizes (Stauffer & Wilwerding, 1982). Light from any given point in the field of view of a camera passes through all sectors of the camera lens, and subsequently arrives at the image plane from many different angles. If the lens is in focus, these components all converge again to a single point in the image plane. If the lens is not in focus, these components are displaced from one another, and the image becomes fuzzy.

Similarly, light from every point in the scene of interest passes through each sector of the lens. Thus, each sector of the lens will contribute a recognizable signature of light to the image plane, in keeping with the image viewed. (Early pinhole cameras made use of this principle; essentially there was only one sector, and so there was only one image, which was always in focus.) Practically speaking, these signatures are identical, and if the lens is in focus, they will be superimposed. As the lens moves out of focus, the signatures will be displaced laterally, and the image blurs. The Honeywell *TCL* system detects this displacement for two specific sectors (A and B) located at opposite sides of the lens as shown in Figure 7-5.

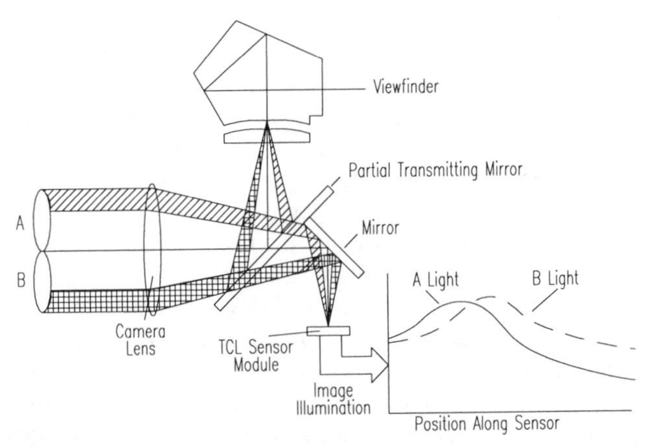

Figure 7-5. Light from two separate sectors of the same lens is compared to determine the position of best focus in the TCL Autofocus System (courtesy Honeywell Corp.).

Light from these two sectors falls upon a series of 24 microlenses mounted on the surface of the integrated circuit in the camera image plane. An array of sensors is positioned within the IC at a specified distance behind the image plane in such a fashion that light incident upon the row of microlenses and their associated image sampling apertures will diverge again to isolate the respective components arriving from each of the two lens sectors (Figure 7-6). Within each aperture image in the detector plane are two detectors, one for each of the two sectors (A and B). Output of all 24 of the A-detectors is used to construct the A-signature; the 24 B-detectors are read to form the B-signature (Stauffer & Wilwerding, 1982).

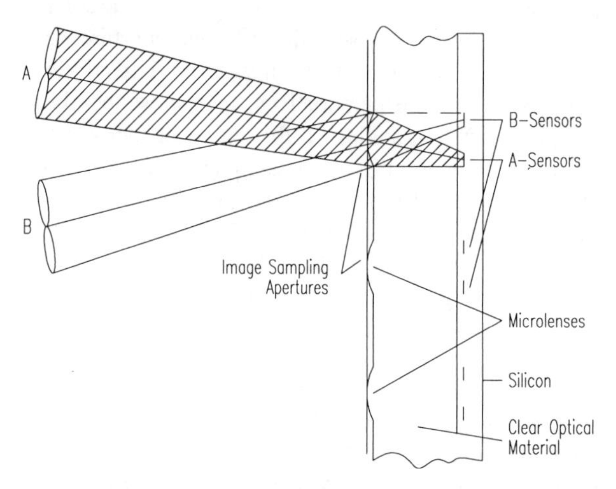

Figure 7-6. A row of microlenses focuses light on the pairs of detectors, forming two separate signatures for comparison (courtesy Honeywell Corp.).

The signatures of light passing through the two camera lens sectors can then be compared and analyzed. The distance between these lens sectors is the base of triangulation for determining range to the subject. Which signature appears to be leading the other and to what degree indicates how far and in what direction the lens must be moved to bring the images into superposition. The output of the CCD detector array is fed to a CMOS integrated circuit which contains the CCD clock circuitry and an A/D converter that digitizes the analog output for further processing by a dedicated-logic algorithm processor.

The Honeywell *TCL* circuitry operates on a 5-volt power supply, and the sensor and companion ICs together draw less than 60 milliwatts. The *TCL* system can sense if the image is in focus to where the plane of the image is within 0.05 millimeter of the position of correct focus, and could potentially provide a passive sensing capability for robotic applications provided there is adequate ambient illumination and scene contrast. The detector pairs in the *TCL* system can discriminate light differences of one part in 100, whereas the human eye is limited to one part in 10 (Stauffer & Wilwerding, 1982).

7.2.2 Associates and Ferren Swept-Focus Ranging

The swept-focus ranging technique uses a conventional video camera with a single lens of very shallow depth of field to produce an image in which only a narrow interval of object space is in focus at any given time. By means of a computer-controlled servo drive (Figure 7-7), the lens can be positioned with great accuracy over a series of discrete stops to view different range "slices." (Some systems operate with a fixed-location lens, and vary the position of the detector element to achieve the same effect.) The distance between the lens and the image plane at the detector is related to the range at which the camera is focused in accordance with the thin lens equation. Thus, if the lens is mechanically positioned to bring the desired object into focus, then the range to that object can be derived from the position of the lens.

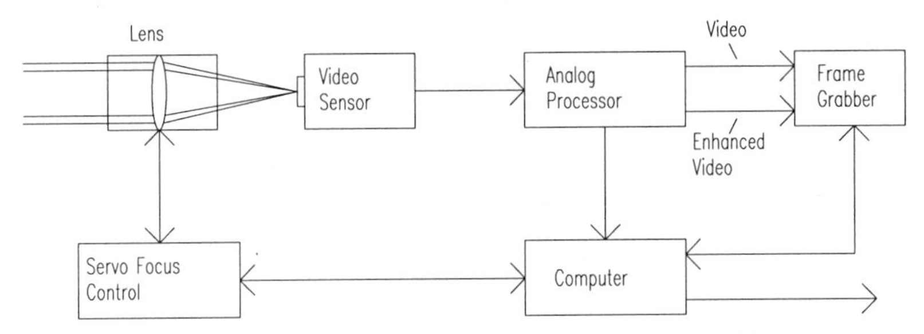

Figure 7-7. Block diagram of a typical swept-focus three-dimensional vision system (courtesy Associates and Ferren, Inc.).

An analog signal processor filters the video signal from the camera to obtain only the high-frequency portion representing information that changes rapidly across the scene, such as in-focus edges or textured material (Figure 7-8). The out-of-focus portions of an image do not contribute to the high-frequency information. This filtered signal is integrated during each video field time.

To perform ranging, the lens is successively positioned at a multitude of discrete precalculated positions, reading and storing the integrated high-frequency data as it becomes available at each position before moving to the next. At the end of this process, the resultant profile of high-frequency response with range is processed to reduce noise effects and then analyzed to determine the locations of all significant peaks. Each peak in high-frequency response represents the best-focus location of a target. The distance to each target can be found simply by reading from a look-up table the object range corresponding to the lens position where the peak occurred.

The swept-focus vision system developed by Associates and Ferren, Wainscott, NY, (Figure 7-9) was specifically intended to address the collision avoidance needs of a mobile robotic platform (Farsaie, et al., 1987; Ferren, 1986). The design therefore employed special optical preprocessing techniques to minimize the onboard computational requirements for image understanding. The system consists of a swept-focus sensor mounted on a robotic vehicle, in communication with a remote host computer and frame grabber. To determine the range to objects in the sensor's field of view, the lens is swept through hundreds of discrete focal positions, remaining at each position for $^{1}/_{60}{}^{th}$ of a second, or one video field time. During this time, the analog signal processor integrates the high-frequency response in that field. This summation is a measure of the amount of edge information in the associated range slice and representative of the relative degree of focus.

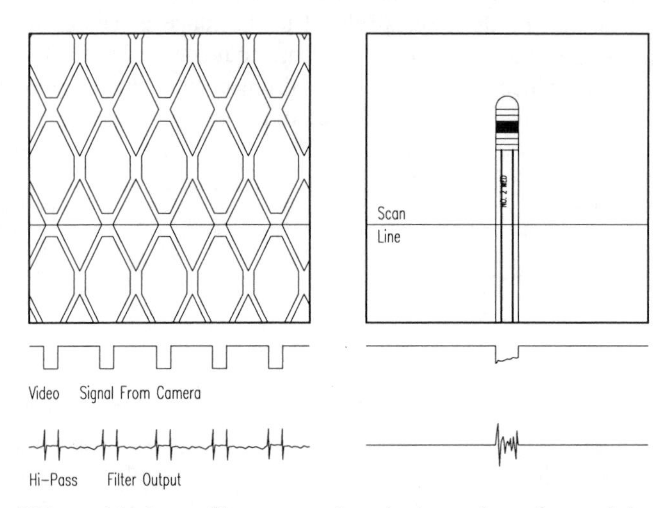

Figure 7-8. Video and high-pass filter output when viewing a piece of expanded metal and a pencil (courtesy Associates and Ferren, Inc.).

Good accuracy (about 1 inch) and repeatability are obtained with a 600-position scan over a 25-foot range interval, which takes approximately 12 seconds (50 millimeter/Fl.0 lens). Accuracy and resolution vary with range and are greatest at closer range, using the current exponential scan profile. Ranging accuracy and the ability to separate targets closely spaced in range are bounded by the physical constraints of the lens. The greater the desired accuracy and resolution, the shorter the required depth of field, which can be achieved by using a lens of longer focal length or larger aperture as illustrated by the following equation (Krotkov & Martin, 1986):

$$DOF = \frac{2afcd(d-f)}{a^2f^2 - c^2(d-f)^2}$$

where:

DOF = depth of field
a = aperture diameter
f = focal length
c = smallest dimension of detector photoreceptor
d = distance to object.

The tradeoffs involved are reduced field of view and increased size and weight. In practice, the two lenses found to be most useful are a 50 millimeter/F1.0 and a 105 millimeter/F1.8, both good quality photographic lenses. The longer lens offers better ranging accuracy and resolution but has a narrower field of view than the shorter lens.

Figure 7-9. Swept-focus camera system developed for passive three-dimensional vision applications (courtesy Associates and Ferren, Inc.).

The swept-focus vision system described has been used as the primary sensor for a mobile robot with good success (Ferren, 1986). The main factor limiting the speed of this technique is the standard 60-Hz video field rate. The system supplies accurate range data and can generate a floor-plan map of its environment that is used in map-based path planning. For such an imaging task, a quick full-range scan could be executed to find the gross location of a target. The lens could then be scanned through the identified range space at smaller increments, saving the entire video field at each position in a large bit-mapped model. A three-dimensional representation of the edge-enhanced object could thus be generated and stored in memory.

During subsequent motion of the robotic vehicle, the onboard video camera can be used as a visual proximity detector by positioning the lens at a fixed focus and monitoring the change in high-frequency content of the scene as the robot travels. A significant rise in this high-frequency information is indicative of a target coming into focus at the range that the lens is imaging. When this condition arises, the robot pauses until it can determine whether or not a collision is imminent. In this application, the 50-millimeter lens has been most useful. The accuracy of the l05-millimeter lens is superior, but its 110-degree field of view is too restrictive.

The use of optical preprocessing in the swept-focus sensor gives it some advantages over other sensing techniques. There is no *missing parts* problem since there is only a single lens, and periodic mechanical alignment is not necessary. The preprocessing action of the short depth-of-field lens also allows for ranging that is not computationally intensive. The system operates passively under normal ambient lighting conditions, responding well to all target objects except those which present a flat field, such as painted walls with no visible texture or markings.

Swept focus has acceptable accuracy for most applications, will locate multiple targets at different ranges, is not computationally intensive, does not suffer from the *missing parts* problem, and operates passively provided there is sufficient ambient light. For these reasons, the swept-focus vision system can be a good primary sensor for mobile robot applications, provided power consumption is not a critical problem. However, the addition of redundant sensors is recommended to ensure the detection of objects which are out of the camera's field of view at close range.

7.2.3 JPL *Range-from-Focus* System

A program to develop a semi-autonomous navigation system for use on a planetary rover has been underway at NASA's Jet Propulsion Laboratory (JPL) since late 1988. The initial testbed vehicle used a passive vision-based navigation technique that required a great deal of computation. To reduce the computational overhead, researchers at JPL are working an alternate approach in the form of a *range-from-focus* optical system (Figure 7-10). The goal is to minimize the necessary computation so that navigation of the rover can be practically performed on board rather than remotely from earth (Wilcox, 1990).

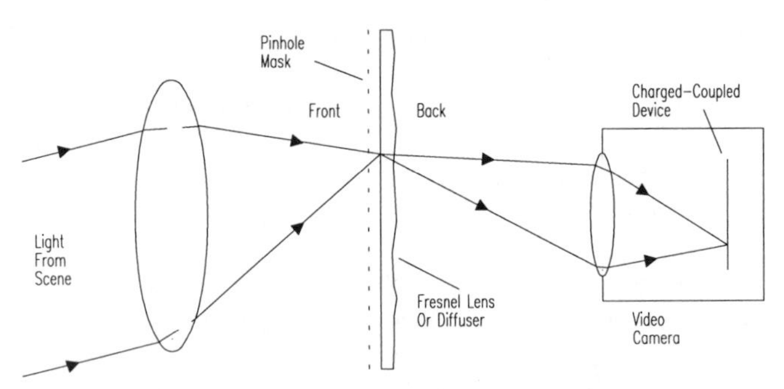

Figure 7-10. Diagram of the JPL *Range-From-Focus* system (courtesy Jet Propulsion Laboratory).

This focus-based ranging system uses a large-aperture short-focal-length lens with a pinhole mask at the prime focus. This mask is transparent only in an array of pinholes at or near the diffraction-limiting spot size of the lens. A diffuser or

Fresnel lens is placed behind the mask to direct the light coming through the holes back toward a CCD camera. The camera is focused on the mask such that there is a one-to-one correspondence between each pixel and pinhole. When successive frames from the CCD camera are differenced and the magnitude of that difference averaged, the only significant signal remaining will be in those parts of the image where the terrain is in focus.

The initial prototype will use a 75-millimeter, F1.9 lens. By way of example, if the lens is focused at a range of 10 meters, the corresponding focal distance is 75.567 millimeters; at 10.5 meters it is 75.540 millimeter. The difference in focal distance is 27 microns, which leads to a 14-micron circle of confusion for a point source at 10.5 meters. The pinhole array is focused for 10 meters, and the diffraction limiting spot is approximately 2 microns. Even a highly textured surface at 10.5 meters will not produce strong difference values between successive frames (assuming the image moves less than 14 microns across the array), whereas an object at 10 meters will produce a 100-percent contrast change with only 2 microns of image motion.

Two or three different range planes could be mixed on different video scan lines in the same sensor. To accomplish this, the pinhole array could be corrugated so that alternate scan lines represent different range distances. A practical implementation would be made from layers of photographic film, with stripes of clear film alternating with the pinhole arrays. For robotic collision avoidance purposes, it is generally not required to have a range map as dense as a standard video image (approximately 500 by 500); several pixels can be averaged horizontally, assuming they will be at approximately the same range. This approach yields two 250- by 250-pixel range maps at two different ranges from the single sensor.

Another variation would incline the image pinhole array to match the flat-earth ground plane. Using the corrugated approach previously described, one could mix the resulting images to produce a single video image depicting elevation deviations. This method would allow obstacle detection from brightness changes in the video image alone; no postprocessing would be required.

7.3 Return Signal Intensity

Ranging techniques involving *return signal intensity* determine the distance to an object based on the amplitude of energy (usually light) reflected from the object's surface. The inverse square law for emitted energy states that as the distance from a point source increases, the intensity of the source diminishes as a function of the square of the distance. If Lambertian surfaces are assumed (see Chapters 8 and 9), then this principle results in a computationally simple algorithm for range calculation. Numerous prototypes have been developed using this technique, but few have found their way to commercial products. Ctein (1982), however, reports that Kodak incorporated an active near-infrared rangefinder in the *Kodamatic*

980L instant camera that measures the brightness of the reflected flash to estimate distance for focusing purposes.

7.3.1 Programmable Near-Infrared Proximity Sensor

A custom-designed near-infrared proximity sensor was developed for use on ROBART I (Everett, 1982) to gather high-resolution geometric information in support of navigational routines as will be discussed in Chapter 10. The primary purpose of this head-mounted sensor was to provide precise angular location of prominent vertical edges such as door openings. An improved programmable version was incorporated on ROBART II (Figure 7-11) to complement range information provided by a Polaroid sonar sensor (Flynn, 1985). Adams, et al. (1990) report additional follow-up work of a similar nature at the Oxford University.

Figure 7-11. The near-infrared proximity sensor developed for ROBART II used a programmable array of four high-powered LED emitters with a variable threshold receiver. The PIN-photodiode detector is situated at the focal point of the parabolic reflector.

An astable multivibrator produces a square wave of short-duration pulses, driving high-power XC-880-A gallium-aluminum arsenide LEDs that emit energy in the near-infrared spectrum. The system uses an array of adjacent LEDs for increased range and sensitivity, with reflected energy focused on the lens of a TIL413 photodiode by a parabolic reflector. The output of this photodiode is passed through an L/C differentiator network, amplified, and fed to four separate follow-on threshold detector stages (Figure 7-12). The receiver sensitivity is broken into four discrete levels by these individually adjustable threshold comparators. A strong return will cause all four channels to go low, whereas a weak return will cause only the most sensitive channel to indicate detection.

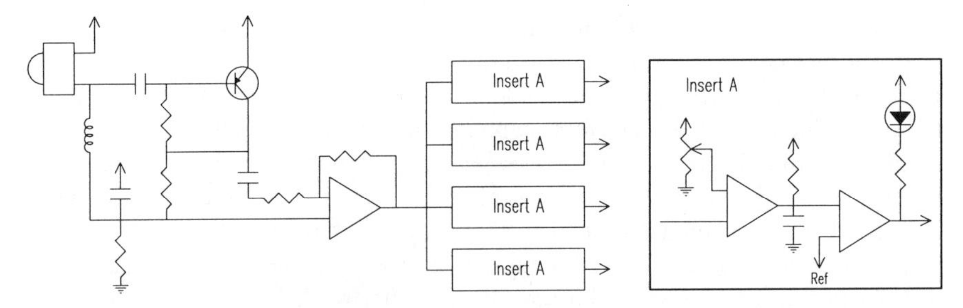

Figure 7-12. Schematic diagram of the receiver portion of the programmable near-infrared proximity sensor.

Effective range is controlled by firing combinations of LEDs, thereby emitting regulated amounts of energy (i.e., the more LEDs illuminating the scene, the farther the detection range). The number of LEDs in the array that are enabled at any given time is specified by a microprocessor, providing programmable control over the amount of transmitted energy, which in turn fixes the maximum range of the sensor (Everett & Flynn, 1986). The total number of active emitters can be any value between one and four (Figure 7-13). The robot "feels around" out to a distance of 5 or 6 feet, and notes any detected obstruction. If no reflected energy is sensed, an additional LED is activated to extend the range of the sensor a few more feet, and the area is probed again. This process is repeated as the head pans to map the entire region in terms of range discontinuities as a function of relative bearing.

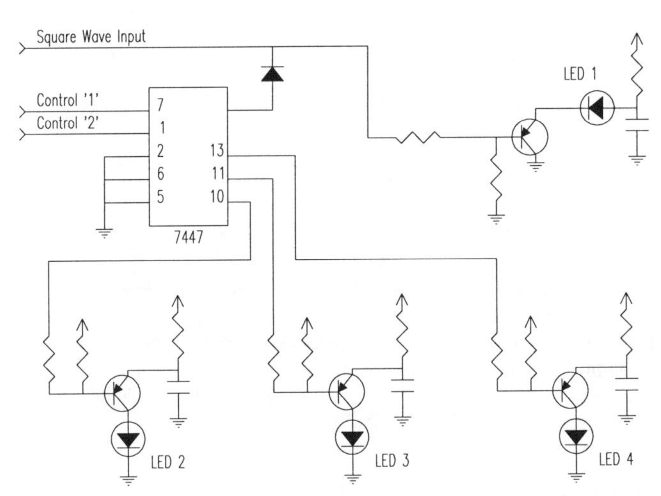

Figure 7-13. Schematic diagram of the transmitter portion of the programmable proximity detector.

The data protocol employed for communicating the information is in the form of a single byte in which the upper nibble represents the number of LEDs fired, and the lower nibble represents the number of comparators in the receiver circuitry that detected the returned energy. For example, the result "14" would indicate only one LED was needed to generate a sufficiently strong return to trigger all four threshold detectors, implying the target was relatively close. The result "41" would signify four LEDs were required, with only the most sensitive comparator responding, an indication the target was considerable distance away. A "40" would mean there was no detectable target within range.

During experimental testing the system proved capable of seeing out to an average of 6 feet with one LED active, 10 feet with two LEDs active, 13 feet with three, and a maximum average range of 15 feet attainable with all four (Everett & Flynn, 1986). Figure 7-14A shows a sonar plot of a small room generated by 256 range readings taken by the head-mounted Polaroid sensor shown in Figure 7-11, with the robot situated as shown. Note the broadened representation of the right side of the upper doorway due to the 30-degree effective beamwidth of the Polaroid sonar. Figure 7-14B shows the same data overlaid with crosses to mark the state transitions for the near-infrared proximity sensor. Examination of the recorded data shows the biggest reflectance discontinuities for the two crosses closest to the edges of door opening.

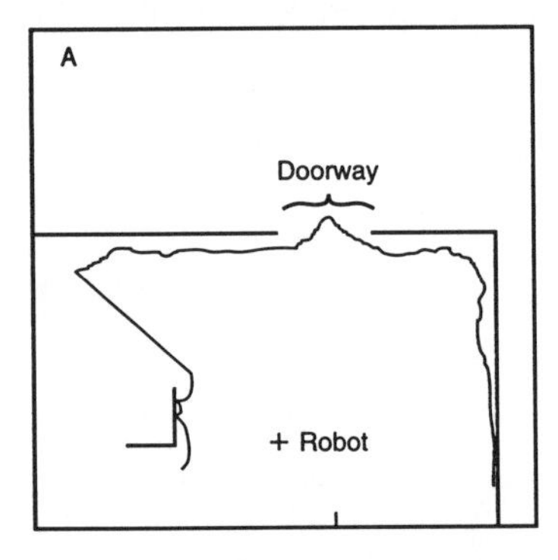

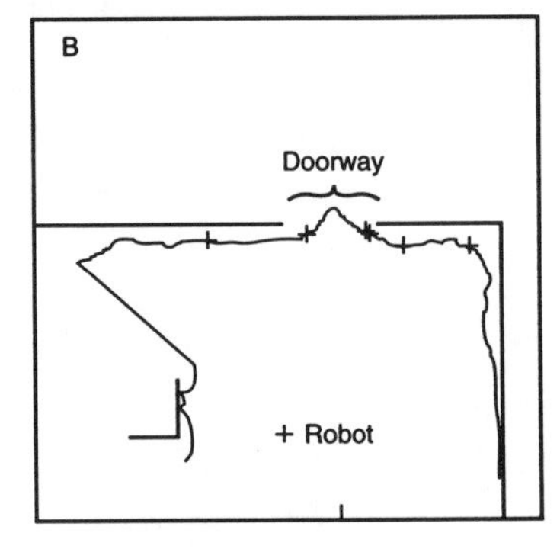

Figure 7-14. A total of 256 sonar range measurements, taken by ROBART II while stationary at the point marked by the cross, are overlaid on actual room measurements in (A). State transitions in the near-infrared sensor data marked with crosses in (B) clearly identify the open doorway boundaries at the top of the figure (courtesy MIT AI Lab).

Unfortunately, not all objects in the real world are ideally Lambertian in nature, and the varying reflectivities of typical surfaces preclude simple measurement of return signal strength from being a reliable indicator of distance under most conditions.

7.3.2 Australian National University Rangefinder

A monocular ranging technique developed at the Australian National University attempted to resolve the reflectivity problem by simultaneously measuring the return signal intensity of a pair of light sources (Jarvis, 1984). The two sources were arranged with a two-dimensional camera detector along a common optical axis that was focused on the target surface as illustrated in Figure 7-15. The displacement between the sources resulted in differing magnitudes of returned energy, each related to distance by the inverse square law. For identical collocated sources, the intensity of the return signal as sensed by the receiver should ideally be the same. However, in this configuration one emitter was closer to the scene than the other, resulting in a difference in the return signal intensity produced by the two sources. This measurable difference was exploited to yield absolute range values, and the effects of surface reflectivity (which similarly attenuates returned energy for both sources) subsequently cancel out.

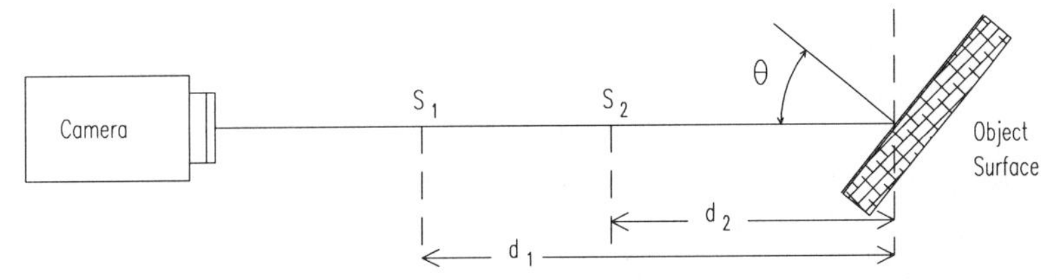

Figure 7-15. By using a pair of emitters at S_1 and S_2 with a common detector, two equations are generated with two unknowns to eliminate the influence of surface reflectivity (adapted from Jarvis, 1984).

The experimental system developed for evaluation of this technique used 35-millimeter slide projectors as the point sources. Sensitivity improved as the distance between projectors was increased. Color images of the scene (256 successive frames) were captured for three separate lighting conditions: 1) background lighting only, 2) background lighting with the far source energized, and 3) background lighting with the near source energized (Jarvis, 1984). The background illumination components were subtracted pixel by pixel from each of the actively illuminated scenes to isolate the effects of the point-source illuminators. Image data from the red, blue, and green channels were then separately processed on a VAX 11/780 to compute the range estimates for each pixel in a 128 by 128 spatial array. Of the approximately 60 seconds required for one complete ranging operation, 75 percent involved image acquisition (Jarvis, 1984). The prototype was capable of measuring range over uniform textured or colored surfaces, but encountered difficulty when observing multicolor non-planar targets.

7.3.3 MIT Near-Infrared Ranging System

A one-dimensional implementation of this ranging technique was developed by Connell at the Massachusetts Institute of Technology (MIT) Artificial Intelligence Lab. The MIT system used a pair of identical point-source LEDs positioned a known distance apart, with their incident light focused on the target surface. The emitters were individually fired in a sequential manner, with the reflected energy in each case detected by a phototransistor and digitized with an analog-to-digital converter.

By the inverse square law, the observed intensity is inversely proportional to the square of the round trip distance traveled. Furthermore, the difference in the resulting intensities caused by the offset in the distance between the LED emitters can be used to solve for the range value:

$$r = \frac{d}{\sqrt{\frac{B_1}{B_2}} - 1}$$

where:

r = the range to the target
d = the distance between emitters
B_1 = intensity of return for LED 1
B_2 = intensity of return for LED 2.

The basic assumptions made in the design are that all surfaces are Lambertian in nature and that the observed objects are wider than the field of view of the LEDs. Ambient light interference is reduced by blinking the LEDs and synchronizing the detector to look for this *on-and-off* sequence of energy returning from the observed scene.

7.3.4 Honeywell Displaced-Sensor Ranging Unit

Honeywell Visitronics has developed a prototype return-signal-intensity ranging system using a single near-infrared LED source and two displaced silicon detectors. A momentary pulse of near-infrared radiation is projected onto the target surface, while the reflected flux is simultaneously detected with two sensors that are displaced along the measurement axis (Figure 7-16). The signal from each sensor may be represented by the following:

$$S_1 \propto \frac{FR}{D^2} \quad \text{and} \quad S_2 \propto \frac{FR}{(D+d)^2}$$

where:

S_1 and S_2 are the detected signals
F = projected spot flux
R = surface reflectivity
D = distance to target
d = displacement seen by S_2.

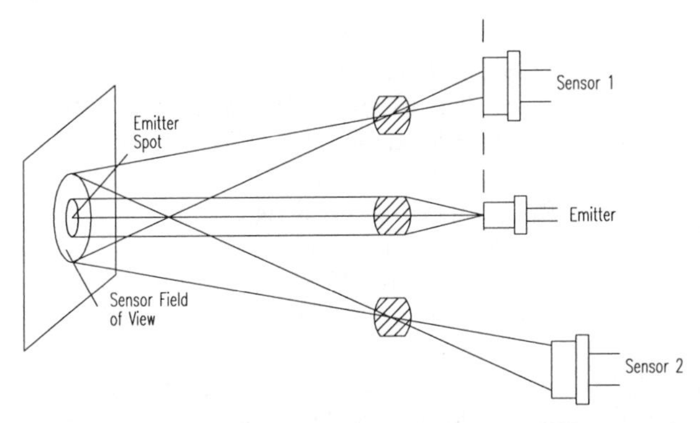

Figure 7-16. Two displaced sensors are used with a common LED source in the Honeywell return-signal-intensity prototype (courtesy Honeywell Corp.).

The detected signal intensities thus provide a means to determine range independent of the surface reflectivity. The use of twin displaced detectors as opposed to displaced emitters eliminates the need to alternately fire the LEDs and offers the advantage of matched stable response and excellent linearity. (LED emitters are temperature sensitive and their performance changes with age, thus making it difficult to maintain identical output.)

The Honeywell prototype provides a pulse-repetition frequency that is proportional to range with a 0- to 5-volt DC linear output. The system can sense objects with a surface reflectivity of 10 to 90 percent out to a distance of 5 meters, with a resolution of 6 millimeters at a distance of 1 meter. System response is less than 5 milliseconds. The prototype is packaged in an enclosure 51 by 51 by 150 millimeters with a weight of 0.65 kilograms.

7.4 References

Adams, M.D., Hu, H., Probert, P.J., "Towards a Real-Time Architecture for Obstacle Avoidance and Path Planning in Mobile Robots," IEEE International Conference on Robotics and Automation, Cincinnati, OH, pp. 584-589, May, 1990.

Beesley, M.J., *Lasers and Their Applications*, pp. 137-147, Taylor and Francis LTD, London, 1971.

Brain, A.E., "Lenses for Industrial Automation, Part One: A Brief Review of Basic Optics," SRI Technical Note 201, SRI International, Menlo Park, CA, November, 1979.

Brown, L.B., "A Random-Path Laser Interferometer System," Proceedings of the Laser Institute of America's International Congress of Applications of Lasers and Electro-Optics, San Francisco, CA, 11-14 November, 1985.

Brown, L.B., Merry, J.B., Wells, D.N., "Tracking Laser Interferometer," U.S. Patent No. 4,790,651, 30 September, 1987.

CLS, "The CMS-2000 Laser Coordinate Measuring System," Product Literature, Chesapeake Laser Systems Inc., Lanham, MD, December, 1991.

Cleveland, B.A. "An Intelligent Robotic Inspection System (IRIS)," Technical Report, MTS Systems Corporation, Minneapolis, MN, 31 March, 1986.

Conrad, D.J., Sampson, R.E., "3D Range Imaging Sensors," in *Traditional and Non-Traditional Robotic Sensors*, T.C. Henderson, ed., NATO ASI Series, Vol. F63, Springer-Verlag, pp. 35-47, 1990.

Ctein, "Autofocus Looks Sharp," *High Technology*, pp. 53-56, November-December, 1982.

Everett, H.R., "Robotics in the Navy," *Robotics Age*, pp. 6-11, November, 1985.

Everett, H.R., Flynn, A.M., "A Programmable Near-Infrared Proximity Detector for Mobile Robot Navigation", Proceedings SPIE Mobile Robots I, Cambridge, MA, pp. 221-230, October, 1986.

Everett, H.R., "Survey of Collision Avoidance and Ranging Sensors for Mobile Robots," NOSC Technical Report 1194, Naval Ocean Systems Center, San Diego, CA, March, 1988.

Farsaie, A., McKnight, T.R., Ferren, B., Harrison, C.F., "Intelligent Controllers for an Autonomous System," IEEE International Symposium on Intelligent Control, 1987.

Ferren, "3-D Computer Vision System for Robots," Final Report for Phase II, NSWC Contract No. N60921-85-D-0064, Associates and Ferren, Wainscott, New York, June, 1986.

Flynn, A.M., "Redundant Sensors for Mobile Robot Navigation," Technical Report 859, MIT Artificial Intelligence Laboratory, Cambridge, MA, October, 1985.

Goldberg, N., "Inside Autofocus: How the Majic Works," Popular Photography, pp. 77-83, February, 1982.

Jarvis, R.A., "Range from Brightness for Robotic Vision," Proceedings of 4th International Conference on Robot Vision and Sensory Controls, London, U.K., pp. 165-172, 9-11 October, 1984.

Krotkov, E., Martin, J., "Range from Focus," IEEE International Conference on Robotics and Automation, San Francisco, CA, pp.1093-1098, April, 1986.

Lau, K., et al., "Robot End Point Sensing Using Laser Tracking System," Proceedings of the NBS Sponsored Navy NAV/CIM Robot Standards Workshop, Detroit, MI, pp. 104-111, June, 1985.

Marnheim, L.A., “Autofocus: What’s it all about?” *Modern Photography*, pp. 102-178, June, 1980.

Stauffer, N., Wilwerding, D., “Electronic Focus for Cameras,” *Scientific Honeyweller*, Vol. 3, No. 1, March, 1982.

Wilcox, B.H., “Vision-based Planetary Rover Navigation,” SPIE International Conference on Visual Communications and Image Processing, Lausanne, Switzerland, October, 1990.

Wolpert, H.D., “Autoranging/Autofocus: A Survey of Systems,” *Photonics Spectra*, pp. 165-168, June, 1987.

8
Acoustical Energy

All sensors, whether active or passive, perform their function by detecting (and in most cases quantifying) the change in some specific property (or properties) of energy. Active sensors emit energy that travels away from the sensor and interacts with the object of interest, after which part of the energy is returned to the sensor. For passive sensors, the source of the monitored energy is the object itself and/or the surrounding environment. In the case of acoustical systems, it must be recognized that the *medium of propagation* can sometimes have significant influence, and such effects must be taken into account.

Sound is a vibratory mechanical perturbation that travels through an elastic medium as a longitudinal wave. For gases and liquids the velocity of wave propagation is given by (Pallas-Areny & Webster, 1992):

$$s = \sqrt{\frac{K_m}{\rho}}$$

where:

s = speed of propagation
K_m = bulk modulus of elasticity
ρ = density of medium.

Since the introduction of sonar in 1918, acoustic waves have been successfully used to determine the position, velocity, and orientation of underwater objects in both commercial and military applications (Ulrich, 1983). It therefore seems only logical we should be able to take advantage of this well-developed sonar technology for deployment on mobile robotic vehicles. This seemingly natural carry-over from underwater scenarios has been somewhat lacking, however, for a number of reasons. The speed of sound in air (assume sea level and 25°C) is 1138 feet/second, while under the same conditions in sea-water sound travels 5,034 feet/second (Bolz & Tuve, 1979). The wavelength of acoustical energy is directly proportional to the speed of propagation as shown below:

$$\lambda = \frac{s}{f}$$

where:

λ = wavelength
s = speed of sound
f = operating frequency.

This relationship means the wavelength for an underwater sonar operating at 200 KHz would be approximately 0.30 inches, while that associated with operation in air at the same frequency is in contrast only 0.07 inches. As we shall see later, the shorter the wavelength, the higher the achievable resolution. So in theory, better resolution should be obtainable with sonar in air than that associated with operation in water. In practice, however, the performance of sonar operating in air seems poor indeed in comparison to the success of underwater implementations, for several reasons.

For starters, water (being basically incompressible) is a much better conductive medium (for sound) than air. In fact, sound waves originating from sources thousands of miles away are routinely detected in oceanography and military applications. One such example involves monitoring global warming (as manifested in long-term variations in average sea-water temperature) by measuring the associated change in the speed of wave propagation over a transoceanic path. Secondly, the mismatch in acoustical impedance between the transducer and the conducting medium is much larger for air than water, resulting in reduced coupling efficiency. The high acoustic impedance of water allows for conversion efficiencies from 50 to 80 percent, depending on the desired bandwidth (Bartram, et al., 1989). In addition, underwater systems are generally looking for fairly large discrete targets in relatively non-interfering surroundings, with the added benefit of intensely powerful pulse emissions. And finally, one should keep in mind that untold millions of defense dollars have been invested over many decades in the research and development of sophisticated underwater systems that individually cost millions of dollars to procure, operate, and maintain. In contrast, most robotic designers begin to balk when the price of any sensor subsystem begins to exceed a few thousand dollars.

The range of frequencies generally associated with human hearing runs from about 20 Hz to somewhere around 20 KHz. Although sonar systems have been developed that operate (in air) within this audible range, ultrasonic frequencies (typically between 20 KHz and 200 KHz) are by far the most widely applied. It is interesting to note, however, ultrasonic frequencies as high as 600 MHz can be produced using piezoelectric quartz crystals, with an associated wavelength in air of 500 nanometers (Halliday & Resnick, 1974). (This wavelength is comparable to electromagnetic propagation in the visible light region of the energy spectrum.) Certain piezoelectric films can be made to vibrate in the gigahertz range (Campbell, 1986).

Acoustical ranging can be implemented using triangulation, time of flight (Frederiksen & Howard, 1974; Biber, et al., 1980), frequency modulation (Mitome, et al., 1984), phase-shift measurement (Fox, et al., 1983; Figueroa & Barbieri, 1991), or some combination of these techniques (Figueroa & Lamancusa, 1992). Triangulation and time-of-flight methods typically transmit discrete short-duration pulses and are effective for in-close collision avoidance needs (Chapter 10), and at longer distances for navigational referencing (Chapter 15). Frequency-modulation and phase-shift ranging techniques involving the transmission of a continuous sound wave are better suited for short-range situations where a single dominant target is present.

In addition to distance, the radial direction and velocity of a moving object can also be determined with continuous-wave systems by measuring the *Doppler shift* in frequency of the returned energy. Anyone who has ever noticed the change in siren pitch that occurs when an approaching fire truck or ambulance passes by a stationary observation point is familiar with this effect. For such a fixed observer listening to a moving source, the arriving *Doppler frequency* is expressed as (Halliday & Resnick, 1974):

$$f' = f\left(\frac{s}{s \pm v_s}\right)$$

where:

f' = Doppler frequency at observation point
f = frequency of source
s = speed of sound in air
v_s = radial velocity of source.

As the source closes on the observer, the sign of v_s is negative, resulting in a slightly higher apparent frequency. The sign of v_s becomes positive when the source is moving away from the observer, and the Doppler frequency is decreased.

Alternatively, for a moving observer listening to a fixed-location source, the observed frequency is expressed as (Halliday & Resnick, 1974):

$$f' = f\left(\frac{s \pm v_o}{s}\right)$$

where:

v_o = radial velocity of observer.

In this situation, the sign of v_o is negative if the observer is moving away from the source, resulting in a slightly lower frequency. If the observer is closing on the source, the sign of v_o is positive and the frequency is increased. Note the change in frequency for a moving source approaching a stationary observer is different from that of a moving observer approaching a fixed-location source at the same relative velocity.

In the case of a reflected wave, there is a factor of two introduced, since any change x in relative separation affects the round-trip path length by $2x$. Furthermore, in such situations it is generally more convenient to consider the *change* in frequency Δf, known as the *Doppler shift*, as opposed to the *Doppler frequency* f' (Pallas-Areny & Webster, 1992):

$$\Delta f = f_e - f_r = \frac{2 f_e v \cos\theta}{s}$$

where:

Δf = Doppler frequency shift
f_e = emitted frequency
f_r = received frequency
v = velocity of the target object
θ = relative angle between direction of motion and beam axis.

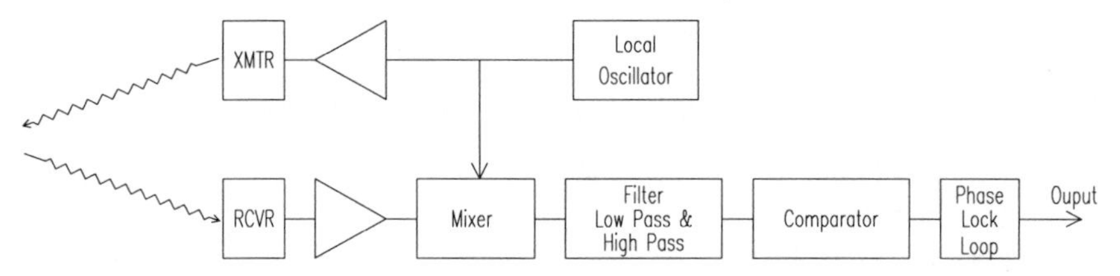

Figure 8-1. Block diagram of an ultrasonic Doppler ground speed sensor used by skiers and joggers (Milner, 1990).

8.1 Applications

Ultrasonic-based measurement systems have found broad appeal throughout the industrial community for a wide variety of purposes such as non-destructive testing (Campbell, 1986), industrial process control (Asher, 1983), stock measurement (Shirley, 1991), liquid level measurement (Shirley, 1989), safety interlocks around dangerous machinery (Irwin & Caughman, 1985), and even intrusion detection in security scenarios (Smurlo & Everett, 1993). In the recreational electronics industry major usage is seen in underwater sonar for depth and fish finding (Frederikson & Howard, 1974), and automatic camera focusing (Biber, et al., 1980). Typical robotic applications include collision avoidance (Everett, 1985), position location (Dunkin, 1985; Figueroa & Mahajan, 1994), and Doppler velocity measurements (Milner, 1990).

8.2 Performance Factors

There are three basic types of ultrasonic transducers: 1) *magnetostrictive*, 2) *piezoelectric*, and 3) *electrostatic*. The first of these categories, *magnetostrictive*, is primarily used in high-power sonar and ultrasonic cleaning applications (Campbell, 1986) and of limited utility from a mobile robotics perspective. *Piezoelectric* and *electrostatic* transducers were treated briefly in Chapter 5, but will be re-examined here from the standpoint of some unique features affecting performance.

Piezoelectric crystals change dimension under the influence of an external electrical potential and will begin to vibrate if the applied potential is made to oscillate at the crystal's resonant frequency. While the force generated can be significant, the displacement of the oscillations is typically very small, and so piezoelectric transducers tend to couple well to solids and liquids but rather poorly to low-density compressible media such as air (Campbell, 1986). Fox, et al. (1983) report using a quarter-wavelength silicon-rubber matching layer on the front face of the transducer in an attempt to achieve better coupling into air at operating frequencies of 1 to 2 MHz. There is also a mechanical inertia associated with the vibrating piezoelectric crystal. As a consequence, such transducers will display some latency (typically several cycles) in reaching full power, and tend to "ring down" longer as well when the excitation voltage is removed.

Electrostatic transducers, on the other hand, generate small forces but have a fairly large displacement amplitude, and therefore couple more efficiently to a compressible medium such as air than do piezoelectric devices. The low-inertia foil membrane allows for quicker turn-on and turn-off in comparison to the slow response of piezoelectrics, facilitating unambiguous short-duration pulses for improved timing accuracy (Campbell, 1986). Since effective operation is not limited to a unique resonance frequency, electrostatic transducers are much more broadband, but with an upper limit of several hundred kilohertz in contrast to megahertz for the piezoelectric variety.

In addition to transducer design considerations, the performance of ultrasonic ranging systems is significantly affected by target characteristics (i.e., absorption, reflectivity, directivity) and environmental phenomena, as will be discussed below.

8.2.1 Atmospheric Attenuation

As an acoustical wave travels away from its source, the signal power decreases according to the inverse square law as illustrated in Figure 8-2, dropping 6 dB as the distance from the source is doubled (Ma & Ma, 1984).

$$I = \frac{I_o}{4\pi R^2}$$

where:

I = intensity (power per unit area) at distance R
I_o = maximum (initial) intensity
R = range.

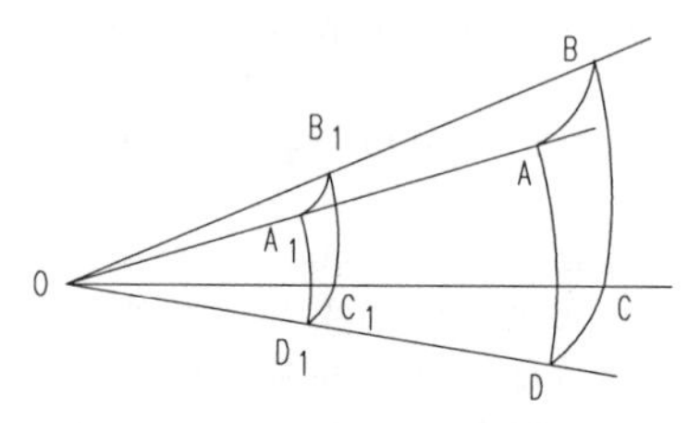

Figure 8-2. Neglecting atmospheric attenuation, the total energy flowing within the cone OABCD is independent of the distance at which it is measured, whereas the intensity per unit area falls off with the square of R (adapted from Feynman, et al., 1963).

There is also an exponential loss associated with molecular absorption of sound energy by the medium itself (Pallas-Areny & Webster, 1992):

$$I = I_o e^{-2\alpha R}$$

where:

α = attenuation coefficient for medium.

The value of α varies slightly with the humidity and dust content of the air and is a function of the operating frequency as well (higher frequency transmissions attenuate at a faster rate). The maximum detection range for an ultrasonic sensor is thus dependent on both the emitted power and frequency of operation: the lower the frequency, the longer the range.

The maximum theoretical attenuation for ultrasonic energy (Shirley, 1989) can be approximated by:

$$a_{max} = \frac{f}{100}$$

where:

a_{max} = maximum attenuation in dB/foot
f = operating frequency in KHz.

For a 20-KHz transmission, a typical absorption factor in air is approximately 0.02 dB/foot, while at 40 KHz losses run between 0.06 and 0.09 dB/foot (Ma & Ma, 1984).

Combining the above *spherical-divergence* and *molecular-absorption* attenuation factors results in the following governing equation for intensity as a function of distance R from the source:

$$I = \frac{I_o \, e^{-2\alpha R}}{4\pi \, R^2}.$$

Note that in this expression, which does not yet take into consideration any interaction with the target surface, intensity falls off with the square of the distance.

8.2.2 Target Reflectivity

The totality of all energy incident upon a target object is either reflected or absorbed, be it acoustical, optical, or RF in nature. The *directivity* of the target surface determines how much of the reflected energy is directed back towards the transducer. Since most objects scatter the signal in an isotropic fashion, the returning echo again dissipates in accordance with the inverse square law (Biber, et al., 1980), introducing an additional $4\pi R^2$ term in the denominator of the previous equation for intensity. In addition, a new factor K_r must be introduced in the numerator to account for the *reflectivity* of the target:

$$I = \frac{K_r \, I_o \, e^{-2\alpha R}}{16\pi^2 \, R^4}$$

where:

K_r = coefficient of reflection.

This *coefficient of reflection* for a planar wave arriving normal to a planar object surface is given by (Pallas-Areny & Webster, 1992):

$$K_r = \frac{I_r}{I_i} = \left(\frac{Z_a - Z_o}{Z_a + Z_o} \right)^2$$

where:

I_r = reflected intensity
I_i = incident intensity
Z_a = acoustic impedance for air
Z_o = acoustic impedance for the target object.

The bigger the impedance mismatch between the two media, the more energy will be reflected back to the source. In industrial applications, this phenomenon allows tank level measurement to be accomplished using an ultrasonic transducer in air looking down on the liquid surface, or alternatively an immersed transducer looking upward at the fluid/air interface.

Most targets are *specular* in nature with respect to the relatively long wavelength (roughly 0.25 inch at 50 KHz) of ultrasonic energy, as opposed to being *diffuse*. In the case of *specular reflection*, the angle of reflection is equal to the angle of incidence, whereas for *diffuse reflection* energy is scattered in various directions by surface irregularities equal to or larger than the wavelength of incident radiation. *Lambertian* surfaces are ideal diffuse reflectors that in theory scatter energy with equal probability in all directions (Jarvis, 1983).

To develop a more or less intuitive appreciation for this relationship to wavelength, it is perhaps helpful to consider the analogy of a pair of rubber balls impacting a hypothetical surface with the sawtooth profile shown below in Figure 8-3. Assume one ball is approximately an inch in diameter, while the other is a much larger basketball. If the sawtooth dimension d is in the same neighborhood as the diameter of the smaller ball, then there is a good chance this ball when approaching the surface at some angle of incidence θ will bounce back towards its origin as shown in Figure 8-3A. This is because on the scale of the smaller ball, the surface has a significant normal component. On the other hand, when the basketball impacts this sawtooth surface with the same angle of incidence, the surface irregularities are too small with respect to the ball diameter to be effective. The basketball therefore deflects in a specular manner as shown in Figure 8-3B.

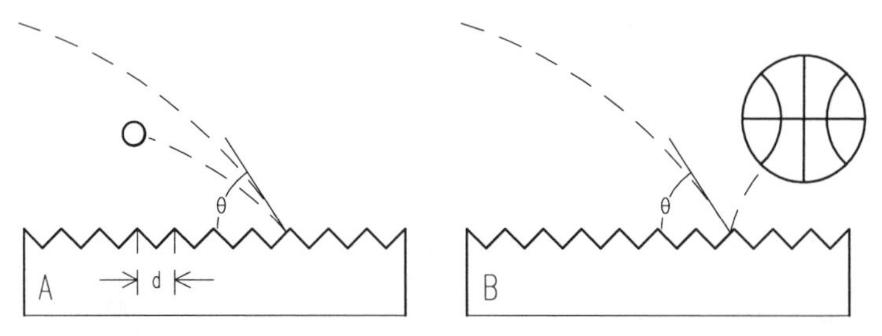

Figure 8-3. A small ball impacting a sawtooth surface as shown will generally bounce back towards its origin, whereas a ball much larger in diameter than the sawtooth dimension d will bounce away in specular fashion.

A familiar example of this effect at optical wavelengths can be seen when the beam of an ordinary flashlight is pointed towards a wall mirror at roughly a 45-degree angle. The footprint of illumination on the mirror surface is not visible, because all the light energy is deflected away in a specular fashion. In other words, you can't see the flashlight spot on the mirror itself. Now suppose the flashlight is redirected slightly towards the wall adjacent to the mirror. The spot of light shows up clearly on the wall surface, which is Lambertian in nature with respect to the wavelength of light. The wall is thus a *diffuse reflector* as opposed to a *specular reflector* for optical energy.

When the angle of incidence of a sonar beam decreases below a certain critical value, the reflected energy does not return to strike the transducer. The obvious reason for this effect is the normal component falls off as the angle of incidence becomes more shallow, as illustrated in Figure 8-4. This critical angle is a function of the operating frequency chosen and the topographical characteristics of the target surface.

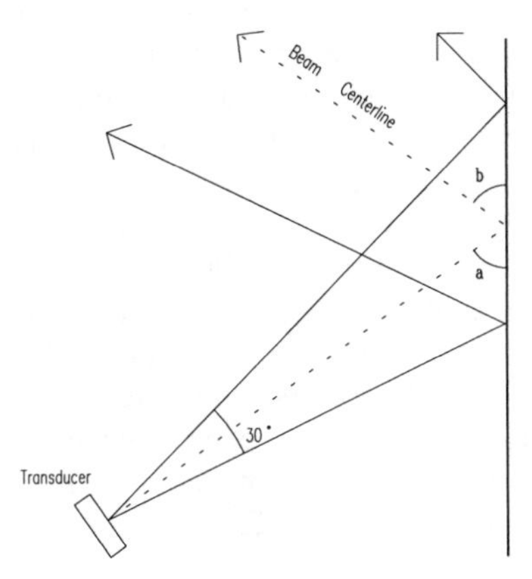

Figure 8-4. As the angle of incidence decreases below a certain critical angle, reflected energy no longer returns to the transducer.

For the Polaroid electrostatic transducers this angle turns out to be approximately 65 degrees (i.e., 25 degrees off normal) for a flat target surface made up of unfinished plywood. Transducer offset from the normal will result in either a false echo as deflected energy returns to the detector over an elongated path, or no echo as the deflected beam dissipates. In Figure 8-4 above, the ranging system would not see the wall and indicate instead maximum range, whereas in Figure 8-5 the range reported would reflect the total round trip through points A, B, and C as opposed to just A and B.

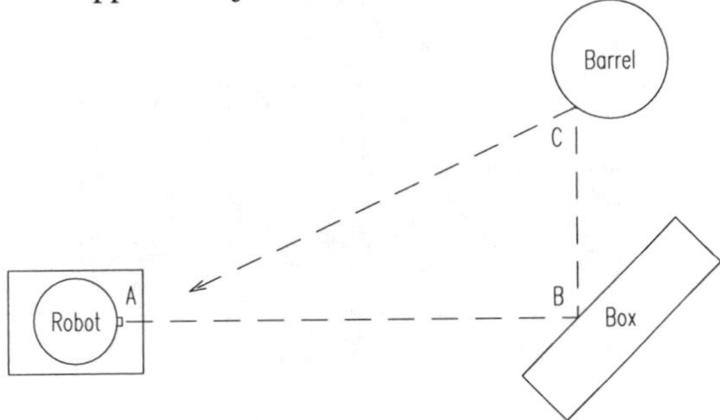

Figure 8-5. Due to specular reflection, the measured range would represent the round trip distance through points A, B, and C as opposed to the actual distance between A and B (adapted from Everett, 1985).

When the first prototype of the MDARS Interior robot was delivered to another government laboratory for formal *Technical Feasibility Testing* in early 1991, the narrow-beam collision avoidance sonar array installed by NCCOSC experienced significant problems in the form of false echo detections. These erroneous sonar readings were quickly seen to correlate with the presence of periodic expansion joints in the concrete floor surface of the test facility. The transducers in the forward-looking array were purposely installed with a 7-degree down angle to increase the probability of detection for low-lying obstructions. This approach had worked very well in our building over months of extended operations, because the smooth floors were very *specular targets* with no significant discontinuities. An overnight field change realigning the sonar beams to a horizontal orientation was required to resolve the problem.

Any significant absorption can result in a reduction of the reflected wave intensity with an adverse impact on system performance. For example, the Polaroid ultrasonic system has an advertised range of 35 feet. In testing the security module on the MDARS Interior robot (Figure 8-6), however, we found it was difficult to pick up an average size person standing upright much beyond a distance of 23 feet. Harder targets of smaller cross-sectional area, on the other hand, could be seen out to the maximum limit.

Figure 8-6. The early prototype security sensor suite on the MDARS Interior robot consisted of 24 Polaroid ultrasonic transducers, 48 passive infrared motion detectors, and six microwave motion detectors (courtesy Naval Command Control and Ocean Surveillance Center).

The amount of energy coupled into the target surface (i.e., absorbed) versus that reflected is basically determined by the difference in *acoustic impedance* (Z) between the propagation medium (air) and the target object itself. Typical values for Z are listed in Table 8-1. Maximum transmission of energy occurs in the case of a fully homogeneous medium where Z is uniform throughout. For non-homogeneous situations involving an interface between two dissimilar media, effective coupling falls off (and reflectivity subsequently goes up) as the differential in Z increases. The *coefficient of transmission* for a planar wave incident upon a planar target in a direction normal to the target surface is given by (Pallas-Areny & Webster, 1992):

$$K_t = \frac{I_t}{I_i} = \frac{4Z_a Z_o}{(Z_a + Z_o)^2}$$

where:

K_t = coefficient of transmission (absorption)
I_t = transmitted intensity
I_i = incident intensity
Z_a = acoustic impedance for air
Z_o = acoustic impedance for the target object.

Table 8-1. Typical values of acoustical impedance (Z) for various conducting media (adapted with permission from Bolz & Tuve, 1979, © CRC Press, Boca Raton, FL; and Pallas-Areny & Webster, 1992).

Medium	Z	Units
Air	4.3×10^{-4}	million Pascal-seconds/meter
Cork	1.0	million Pascal-seconds/meter
Water	1.5	million Pascal-seconds/meter
Human tissue	1.6	million Pascal-seconds/meter
Rubber	3.0	million Pascal-seconds/meter
Glass	13	million Pascal-seconds/meter
Aluminum	17	million Pascal-seconds/meter
Steel	45	million Pascal-seconds/meter
Gold	62.5	million Pascal-seconds/meter

The original Polaroid ranging module transmitted a 1-millisecond *chirp* consisting of four discrete frequencies: 8 cycles at 60 KHz, 8 cycles at 56 KHz, 16 cycles at 52.5 KHz, and 24 cycles at 49.41 KHz (Biber, et al., 1980). This technique was employed to increase the probability of signal reflection from the target, since certain surface characteristics could theoretically absorb a single-frequency waveform, preventing detection. In actual practice such frequency-dependent effects rarely arose, suggesting this aspect of the absorption problem had been somewhat overestimated. In fact, Polaroid subsequently developed an improved version of the ranging module circuit board, the *SN28827*, that operated at a single frequency of 49.1 KHz.

My daughter Rebecca compiled a significant amount of empirical data in 1993 as part of her high school science fair project entitled *Determining the Accuracy of an Ultrasonic Ranging Sensor*. One of her tests investigated the reflective properties of various target surfaces measuring 16 by 24 inches. The targets were maintained normal to a temperature-compensated Polaroid sensor (a *Digitape* ultrasonic tape measuring unit made by Houseworks) mounted 14 inches above a smooth concrete floor and 35 feet away. Starting at a point beyond the maximum range of detection, the distance between the sensor and target was decreased in 1-foot increments until a valid range reading was obtained. The following table is reproduced here with her permission:

Table 8-2. Maximum detection ranges for standardized 16-by 24-inch cross-sections of various materials.

Surface	Distance	Reading	Units
Plywood	24	24.2	feet
Towel	22	22.3	feet
Underside of rug	16	16.3	feet
Foam	13	13.3	feet
Pillow	9	9.4	feet
Blanket	8	8.4	feet
Top of rug	3	3.6	feet

8.2.3 Air Turbulence

Turbulence due to wind and temperature variations can cause bending or distortion of acoustical energy traveling through air (Shirley, 1989). Wind direction and velocity can have a noticeable push or delay effect on the wave propagation velocity, more relevant in the case of outdoor vehicles. Consideration of wind effect errors must also treat crosswind components in addition to those which travel on a parallel path either with or against the wavefront. Crosswind effects can cause the beam center to be offset from its targeted direction, diminish the intensity of returned echoes, and result in a slightly longer beam path due to deflection.

In general, little effort is made in the case of mobile robotic applications to correct for such errors. This is probably due to the fact that ultrasonic ranging is most widely employed in indoor scenarios where the effects of air turbulence are minimal, unless extreme measurement accuracy is desired. In addition, there is really no practical way to reliably measure the phenomena responsible for the interference, and so compensation is generally limited to averaging over multiple readings. This approach introduces a coordinate transformation requirement in the case of a moving platform, since the slow speed of sound limits effective update rates to roughly 2 Hz (i.e., single transducer, assuming 28 feet maximum range). Faster updates are possible if the system is range-gated to some lesser distance (Gilbreath & Everett, 1988).

8.2.4 Temperature

Recall the earlier expression for wave propagation speed (s) in a gas, as a function of density ρ and bulk modulus of elasticity K_m:

$$s = \sqrt{\frac{K_m}{\rho}}.$$

Since both these parameters change with temperature, the speed of sound in air is also temperature dependent (Pallas-Areny & Webster, 1992), and in fact directly proportional to the square root of temperature in degrees Rankine (Everett, 1985):

$$s = \sqrt{g\,k\,R\,T}$$

where:

s = speed of sound
g = gravitational constant
k = ratio of specific heats
R = gas constant
T = temperature in degrees Rankine (F + 460).

For temperature variations typically encountered in indoor robotic ranging applications, this dependence results in a significant effect even considering the short distances involved. A temperature change over the not unrealistic span of 60° to 90°F can produce a range error as large as 12 inches at a distance of 35 feet. Fortunately, this situation can be remedied through the use of a correction factor based upon the actual ambient temperature, available from an external sensor mounted on the robot. The formula is simply:

$$R_a = R_m \sqrt{\frac{T_a}{T_c}}$$

where:

R_a = actual range
R_m = measured range
T_a = actual temperature in degrees Rankine

T_c = calibration temperature in degrees Rankine.

The possibility does still exist, however, for temperature gradients between the sensor and the target to introduce range errors, since the correction factor is based on the actual temperature in the immediate vicinity of the sensor only. As in the case of air turbulence, there is generally little recourse other than averaging multiple readings. (Some industrial applications provide a temperature-stabilized column of air using a small blower or fan.)

8.2.5 Beam Geometry

Still another factor to consider is the beamwidth of the selected transducer, defined as the angle between the points at which the sound power has been reduced to half (-3 dB) its peak value (Shirley, 1989). This formal definition does not always map directly into any useful parameter in real-world usage, however. What is generally of more concern can be better described as the *effective beamwidth*, or the beam geometry constraints within which objects are reliably detected. (Reliable detection, of course, is also very much dependent on the size and shape of the object.) The width of the beam is determined by the transducer diameter and the operating frequency. The higher the frequency of the emitted energy, the narrower and more directional the beam, and hence the greater the angular resolution. Recall, however, an increase in frequency causes a corresponding increase in signal attenuation in air and decreases the maximum range of the system.

The wavelength of acoustical energy is inversely proportional to frequency as shown below:

$$\lambda = \frac{s}{f}$$

where:

λ = wavelength
s = speed of sound
f = operating frequency.

The *beam-dispersion angle* is directly proportional to this transmission wavelength (Brown, 1985):

$$\theta = 1.22\frac{\lambda}{d}$$

where:

θ = desired dispersion angle
λ = acoustic wavelength
d = transducer diameter.

The above relationship can be intuitively visualized by considering the limiting case where d approaches zero. Such a hypothetical device would theoretically function as a point source, emitting energy of equal magnitude in all directions. As d is increased, the device can be considered a planar array of point sources clustered together in circular fashion. For this configuration, the emitted energy will be in phase and at maximum intensity only along a surface normal. Destructive interference from adjacent point sources causes the beam intensity to fall off rapidly to either side up to some local minimum value as shown in Figure 8-7. Constructive interference then occurs past this minimum point, resulting in the presence of side lobes.

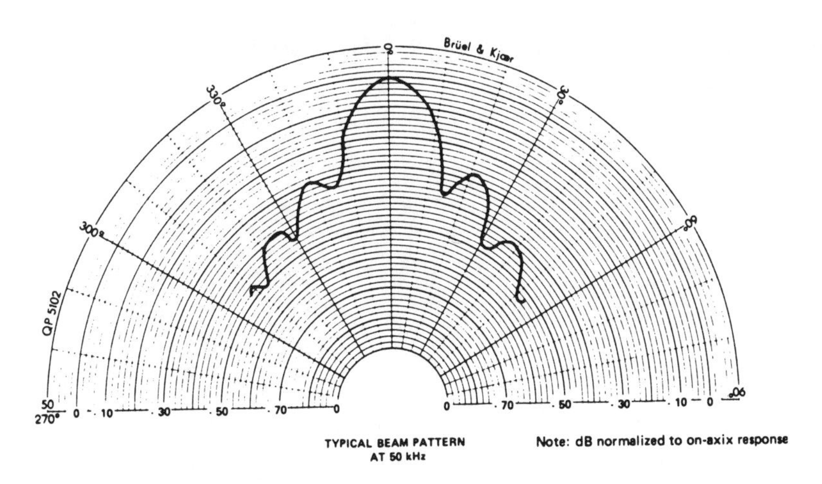

Figure 8-7. Constructive interference results in maximum power in the main lobe along the beam center axis (courtesy Polaroid Corp.).

Shirley (1989) defines the spot diameter that is *insonified* by the ultrasonic beam (i.e., footprint of the incident beam at the target surface) in terms of this beam-dispersion angle θ:

$$D = 2R \tan\frac{\theta}{2}$$

where:

D = spot diameter
R = target range.

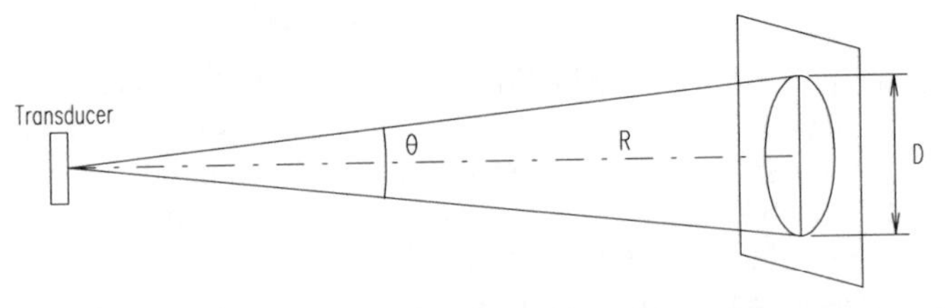

Figure 8-8. The diameter of the insonified footprint at the target surface, assuming normal incidence.

Best results are obtained when the beam centerline is maintained normal to the target surface. As the angle of incidence varies from the perpendicular, note the range actually being measured does not always correspond to that associated with the beam centerline (Figure 8-9). The beam is reflected first from the portion of the target closest to the sensor. For a 30-degree beam-dispersion angle at a distance of 15 feet from a flat target, with an angle of incidence of 70 degrees with respect to normal, the theoretical error could be as much as 10 inches. The actual line of measurement intersects the target surface at point B as opposed to point A.

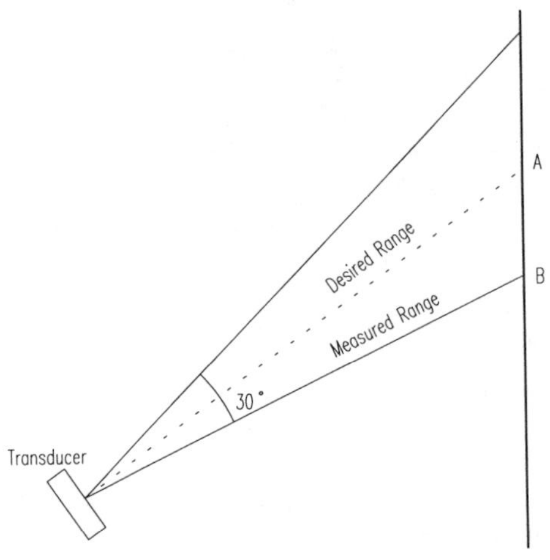

Figure 8-9. Ultrasonic ranging error due to beam divergence results in a shorter range measurement to the target surface at B instead of the desired reading to point A.

Effective beamwidth introduces some uncertainty in the perceived distance to an object from the sensor but an even greater uncertainty in the angular resolution of the object's position. A very narrow target such as a vertical pole would have a relatively large associated region of floor space that would essentially appear to the sensor to be obstructed. Worse yet, a 3-foot doorway may not be discernible at all when only 6 feet away, simply because at that distance the beam is wider than the door opening.

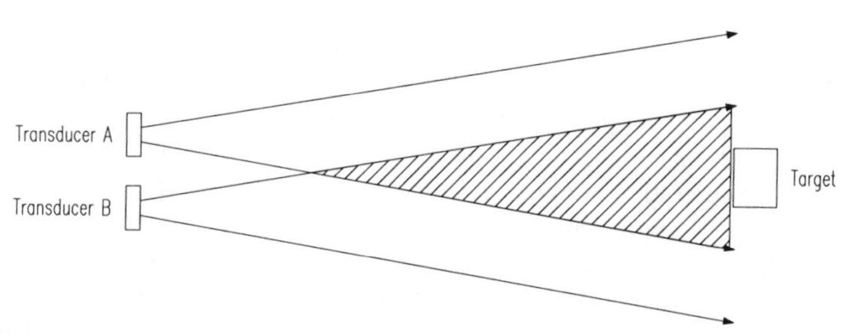

Figure 8-10. Beam-splitting techniques using two or more sensors can improve angular resolution for discrete targets (adapted from Everett, 1985).

Improved angular resolution can sometimes be obtained through beam splitting, a technique that involves the use of two or more transducers with partially overlapping beam patterns. Figure 8-10 shows how for the simplest case of two transducers, twice the angular resolution can be obtained along with a 50-percent increase in coverage area. If the target is detected by both sensors A and B, then it (or at least a portion of it) must lie in the region of overlap shown by the shaded area. If detected by A but not B, then it lies in the region at the top of the figure, and so on. Increasing the number of sensors with overlapping beam patterns decreases the size of the respective regions, and thus increases the angular resolution.

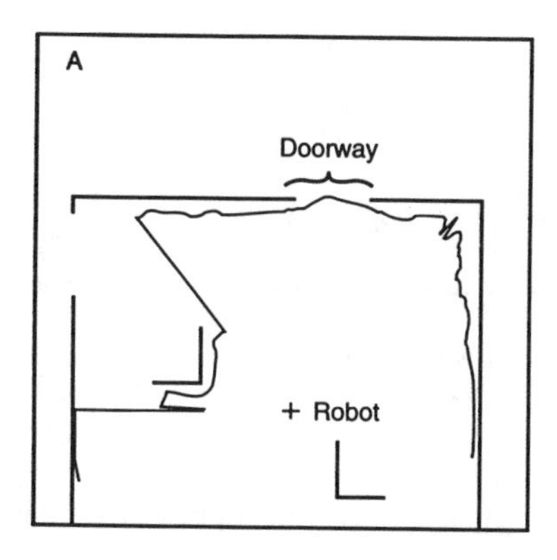

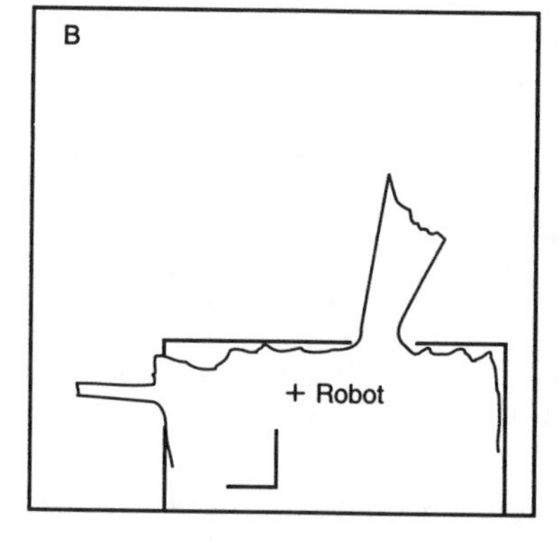

Figure 8-11. With ROBART II situated as shown in (A), the sonar beamwidth is too wide to see the open doorway; relocation of the robot as depicted in (B) allows several readings to penetrate as the head-mounted transducer is scanned left to right (plot courtesy MIT AI Lab).

It should be noted, however, that this increase in angular resolution is limited to the case of a discrete target in relatively uncluttered surroundings, such as a metal pole supporting an overhead load or a lone box in the middle of the floor. No improvement is seen for the case of an opening smaller than an individual beamwidth, such as the doorway illustrated in Figure 8-11A. The entire beam from at least one sensor must pass through the opening without striking either side in order for the opening to be detected (Figure 8-11B), and the only way to improve resolution otherwise is to decrease the individual beamwidths by increasing the operating frequency, changing transducers, or through acoustical focusing. Some designs achieve this effect through use of an attachable horn that concentrates the energy into a tighter, more powerful beam (Shirley, 1989).

Kilough and Hamel (1989) incorporated an innovative arrangement of four Polaroid transducers in a *phased-array* cluster on HERMIES as an alternative method of reducing the effective beamwidth. Figure 8-12A shows how two pairs of transducers connected in parallel are in turn wired in series to preserve the original impedance seen by the driving module. The constructive and destructive interferences resulting from the close proximity of the sensors produced the measured pattern depicted in Figure 8-12B, which agrees with the theoretical predictions for the case of a 38-millimeter center spacing and a 6.6-millimeter wavelength at 50 KHz. Note the nulls in the pattern at 5 degrees either side of centerline, and the associated side lobes. To minimize this effect, the transducers are mounted as closely spaced as possible. Measured beamwidth was reduced from 30 to 18 degrees.

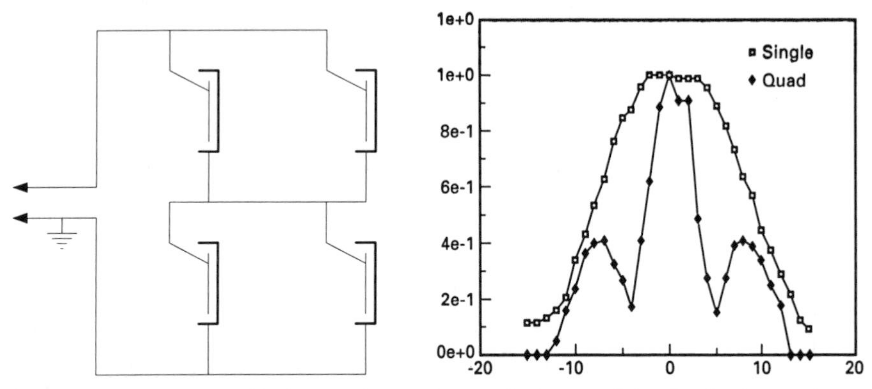

Figure 8-12. Four Polaroid electrostatic transducers wired in parallel-series fashion as shown form a phased-array configuration with the resulting beam pattern shown on the right (adapted from Kilough & Hamel, 1989).

A number of factors must be considered when choosing the optimal beamwidth for a particular application. A narrow beamwidth will not detect unwanted objects to either side, is less susceptible to background noise, and can achieve greater ranges since the energy is more concentrated (Shirley, 1989). One the other hand, for collision avoidance applications it is often desirable to detect any and all objects in front of the robot, and since extremely long ranges are not usually required, a wide-angle transducer may be a more optimal choice (Hammond, 1993). When comparing a single transducer of each type, the use of a wide beamwidth will improve chances of target detection due to the greater likelihood of some portion of the beam encountering a surface normal condition as seen in Figure 8-13. Admittedly this observation is a bit like saying the wider the beam, the more chance of hitting a target. Taken to the extreme, a hypothetical 360-degree field-of-view transducer is clearly of rather limited utility due to the total lack of azimuthal information regarding the target's whereabouts.

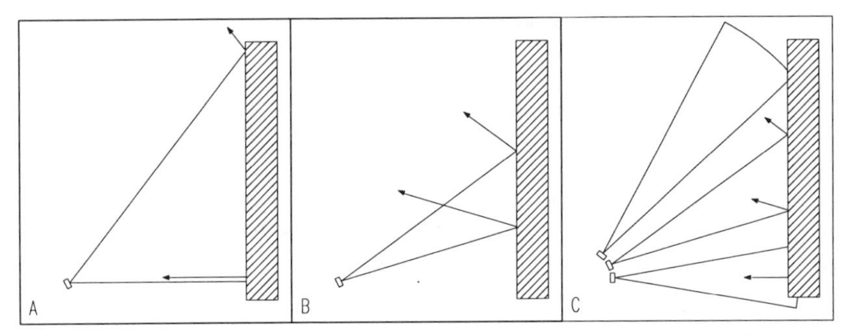

Figure 8-13. A wide-angle transducer (A) has a greater chance of encountering a surface normal condition than a single narrow-beam transducer (B), but at the expense of reduced angular resolution and effective range. A more optimal configuration is presented in (C), at a slight cost to system update rate.

Alternatively, an equivalent surface-normal condition can be realized using a cylindrical array of narrow-beam transducers to achieve the same volumetric coverage as illustrated in Figure 8-13C. This approach offers the added advantage of significantly improved angular resolution but at the expense of a slower overall update rate. The MDARS Interior robot uses a combination of wide-angle piezoelectric sonars operating at a frequency of 75 KHz for timely *obstacle detection* coverage, and a nine-element array of narrow-beam Polaroid electrostatic transducers operating at 49.4 KHz to support intelligent *obstacle avoidance*. Detection of any potential obstructions by either type of sonar causes the platform to slow to a speed commensurate with the narrow-beam update rate, whereupon the high-resolution Polaroid data is used to formulate an appropriate avoidance maneuver.

8.2.6 Noise

Borenstein & Koren (1992) of the University of Michigan Mobile Robotics Lab define three types of noise affecting the performance of ultrasonic sensors:

- *Environmental noise* resulting from the presence of external sources operating in the same space. Typical examples in industrial settings include high-pressure air blasts and harmonics from electrical arc welders.
- *Crosstalk* resulting from the proximity of other sensors in the group, which can be especially troublesome when operating in confined areas.

- *Self noise* generated by the sensor itself.

A noise-rejection measure for each of the components was developed and integrated into a single algorithm (Michigan, 1991), which was in turn combined with a fast sensor-firing algorithm. This software has been implemented and tested on a mobile platform that was able to traverse an obstacle course of densely packed 8-millimeter-diameter poles at a maximum velocity of 1 meter/second.

8.2.7 System-Specific Anomalies

A final source of error to be considered stems from case-specific peculiarities associated with the actual hardware employed. We shall again refer to the Polaroid system, in light of its widespread usage, as an illustrative example in the ensuing discussion.

Pulse Width

The 1-millisecond length of the original four-frequency Polaroid *chirp* was a potential source of range measurement error since sound travels roughly 1100 feet/second at sea level, which equates to about 13 inches/millisecond. The uncertainty and hence error arose from not knowing which of the four frequencies making up the *chirp* actually returned to trigger the receiver, but timing the echo always began at the start of the *chirp* (Everett, 1985). For the initial application of automatic camera focusing, designers were less concerned about absolute accuracy than missing a target altogether due to surface absorption of the acoustical energy. The depth of field of the camera optics would compensate for any small range errors that might be introduced due to this *chirp* ambiguity.

Even with the more recent *SN28827* ranging model operating at a single frequency of 49.1 KHz, the transmission pulse duration is 0.326 milliseconds, giving rise to a maximum theoretical error of approximately 1.7 inches. (This estimate takes into account round-trip distance, and assumes best-case echo detection after just three cycles of returned energy.) The new Polaroid *Ultrasonic Ranging Developer's Kit* allows for programmable pulse duration to alleviate this limitation in demanding applications (Polaroid, 1993).

Threshold Detection

The specific method for detection of the returned pulse can be a significant source of error in any TOF ranging system (Figueroa & Lamancusa, 1992). Kuc and Siegel (1987) point out that the intensity of a typical pulse transmission peaks in the second cycle (Figure 8-14), and so simple thresholding of the received signal can cause late detection of weak echoes. Leonard and Durrant-Whyte (1992) discuss further complications in the specific case of the integrating capacitive threshold detector employed in the Polaroid ranging module. This integrative approach was incorporated by the designers to discriminate against unwanted noise spikes (Biber, et al., 1980). Compared to strong reflections, valid but weak echo returns can take substantially longer to charge up the capacitor to the threshold level required for the comparator to change state (i.e., the *time-walk* problem identified in Chapter 5).

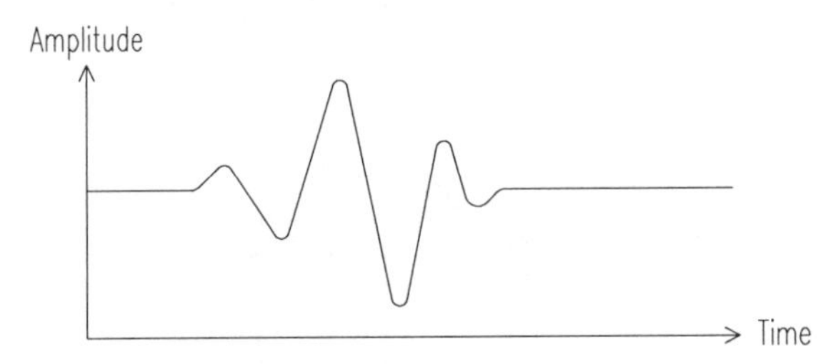

Figure 8-14. A typical pulse waveform for an electrostatic transducer can be approximated by a sinusoid that is modulated by a Gaussian envelope, peaking in intensity during the second cycle (Kuc & Siegel, 1987, © IEEE).

The effect of this charging delay is to make those targets associated with weaker returns appear further away. Ignoring the obvious worst-case scenario of a completely missed echo, maximum theoretical error is bounded by the length of the transmitted burst. The obvious question now becomes, which is more preferable: missing target detection altogether, or being alerted to target presence at the expense of range accuracy? The answer of course depends on the particular priorities of the application addressed. If the ranging sensor is being employed as a presence detector for security purposes (see Chapter 17), precise accuracy is not all that important. On the other hand, if the sensor is being used for navigational referencing (see Chapters 15 and 16), the situation may be somewhat different.

Stepped Gain

Lang, et al. (1989) experimentally confirmed error effects associated with the piecewise 16-step gain ramp employed on the earlier Polaroid *607089* ranging module. In order to precisely counter the effects of signal loss as a function of range to target (i.e., due to atmospheric attenuation and spherical divergence), the actual time-dependent gain compensation would be an exponential function inversely related to the equation presented in Section 8.2.2. A rather coarse piecewise approximation to this ideal gain curve (Figure 8-15) naturally results in a situation where the instantaneous amplifier gain is: 1) correct only for a single point in time over the period represented by a specific step value, 2) excessive prior to this point, and 3) insufficient afterwards. If the gain is too low at the time of reflected pulse train arrival, weak echoes are either missed entirely or delayed in triggering the integrating detector, resulting in an erroneous increase in the perceived range.

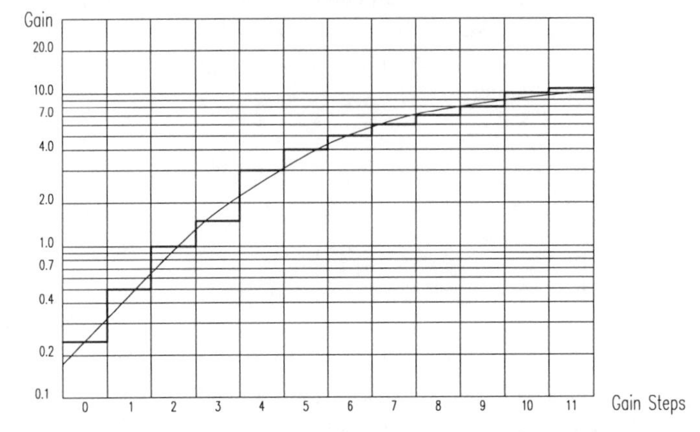

Figure 8-15. The 12-step approximation employed in the new *6500-series* receiver gain ramp results in a situation where the instantaneous gain is either above or below the ideal value for most of the step duration (adapted from Polaroid, 1993). Note the large jump in gain between steps 3 and 4.

8.3 Choosing an Operating Frequency

The operating frequency of an ultrasonic ranging system should be selected only after careful consideration of a number of factors, such as the diameter and type of transducer, anticipated target characteristics, sources of possible interference, and most importantly the nature of the intended task, to include desired angular and range resolution. Resolution is dependent on the bandwidth of the transmitted energy, and greater bandwidth can be achieved at higher frequencies but at the expense of maximum effective range. The minimum ranging distance is also a function of bandwidth, and thus higher frequencies are required in close as the distance between the detector and target decreases. Most man-made background noise sources have energy peaks below 50 KHz (Hammond, 1993), however, and so higher-frequency systems are generally preferred in acoustically noisy environments (Shirley, 1989).

8.4 Sensor Selection Case Study

The Department of Energy's Office of Technology Development has an ongoing environmental restoration effort that, among other things, seeks to develop a robotic inspection capability (Figure 8-16) for stacked 55-gallon drums of hazardous waste stored inside warehouse facilities (Byler, 1993). Aside from the obvious challenges associated with global navigation and collision avoidance, the mobile robot must be able to appropriately position itself next to a stack of barrels before activating the onboard inspection system (Heckendorn, et al., 1993). This *barrel-detection* requirement presents an interesting hypothetical case study for reviewing some of the factors discussed in this chapter that determine the applicability of an ultrasonic ranging sensor for a particular task. Careful consideration of the nature of the target surface in conjunction with the reflective-sensor performance factors (i.e., *cross-sectional area, reflectivity*, and *directivity*) can provide some helpful insights into appropriate sensor selection.

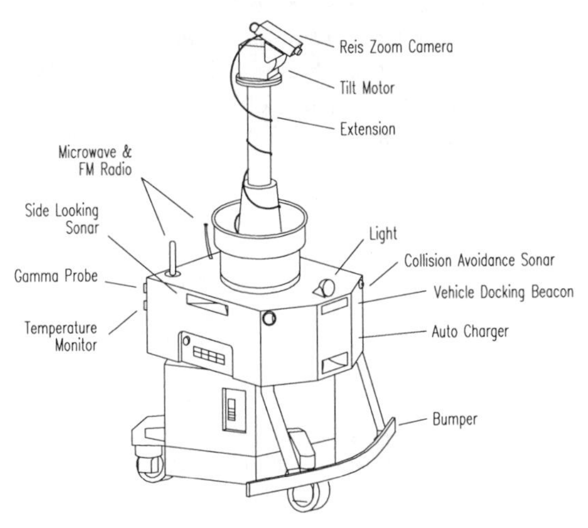

Figure 8-16. The Savannah River Site nuclear surveillance robot is equipped with an extending vertical mast illustrative of the type that could potentially be employed in a stacked barrel inspection task (adapted from Fisher, 1989).

Let's examine first the nature of the object of interest itself, namely the barrel, in terms of its target characteristics. The diameter and height of each barrel is expected to be somewhat standardized, although there may likely be some minor variations encountered. In general, however, the *cross-sectional area* of the barrel will be very large with respect to the sensor due to extremely close proximity to the robot. There is therefore no decided advantage given in this particular application to ultrasonic, RF, or optically based systems from the standpoint of *target cross-sectional area.*

The material from which the barrels are constructed is of particular interest. An all-metal barrel would suggest there might possibly be some advantage to using a low-power microwave-based ranging system to take advantage of the inherent high *reflectivity* with virtual immunity to varying surface conditions. There is no guarantee, however, that plastic barrels will not be used in some cases, perhaps for containment of corrosive agents. Plastic provides poor reflectivity for RF energy, but both plastic and metal surfaces reflect ultrasonic energy well. The color and material condition of the target surface will have little effect on either RF or ultrasonic energy, but could seriously impair performance of an active optical ranging system.

By far the most important target characteristic is the cylindrical geometry, significant in that it represents a common feature for all barrels in the warehouse and directly impacts *directivity*. A cylindrical reflector will scatter energy in all directions (i.e., as is desired in stealth applications) rather than redirect it in a concentrated fashion back towards the sensor (i.e., as in the case of a cooperative retroreflective target). This scattering reduces the strength of the return, but at the same time ensures some energy does get reflected back towards the sensor, provided an imaginary ray passing from the transducer to the cylinder axis intersects the footprint of illumination (Figure 8-17). At extremely close ranges, the resultant loss in signal strength is not a problem, and the effect of a guaranteed surface normal turns out to be highly beneficial.

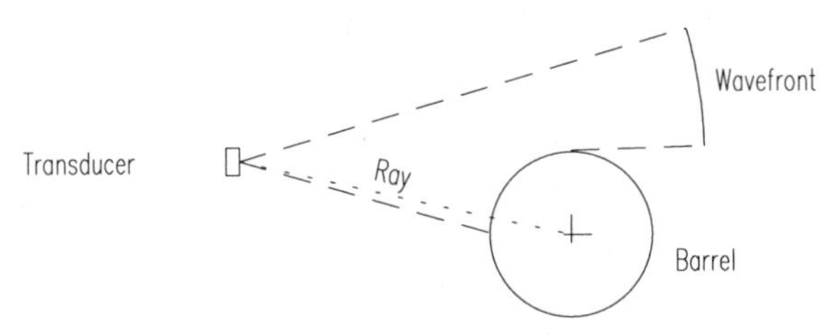

Figure 8-17. A surface normal condition exists as long as the footprint of illumination of the propagating wavefront contains a ray originating at the transducer and passing through the vertical axis of the cylindrical barrel.

The situation illustrated in Figure 8-17 above suggests the fairly wide beamwidth of an ultrasonic ranging system is actually somewhat advantageous when trying to detect a cylindrical target, as it increases the chances of a surface normal condition. In addition, it is illustrative to note that the measured range is always along the path of the ray which passes through the center of the drum, even with the transducer offset as shown. Since the barrel radius is known in advance, this observation allows for distance to be calculated to the barrel centerpoint itself. A number of such measurements made as the robot advances along a known baseline enables the precise location of the barrel to be determined through triangulation.

In summary, an ultrasonic ranging system would be expected to give good results in this particular application for the following reasons:

- Very low cost and easy to interface.
- Barrel surfaces, whether metal or plastic, provide excellent reflectivity regardless of surface color or condition.
- Fairly wide beam increases probability of detection.
- Ranges can easily be derived to barrel centerline.

8.5 References

Asher, R.C., "Ultrasonic Sensors in the Chemical and Process Industries," Journal of Physics E.: Scientific Instruments, Vol. 16, pp. 959-963, 1983.

Bartram, J.F., Ehrlich, S.L., Fredenberg, D.A., Heimann, J.H., Kuzneski, J.A., Skitzki, P., "Underwater Sound Systems," in *Electronic Engineer's Handbook,* D. Christiansen and D. Fink, eds., 3rd edition, McGraw Hill, New York, NY, pp. 25.95-25.133, 1989.

Biber, C., Ellin, S., Shenk, E., "The Polaroid Ultrasonic Ranging System," Audio Engineering Society, 67th Convention, New York, NY, October-November, 1980.

Bolz, R.E., Tuve, G.L., *CRC Handbook of Tables for Applied Engineering Science*, CRC Press, Boca Raton, FL, 1979.

Borenstein, J, Koren, Y., "Error Eliminating Rapid Ultrasonic Firing for Mobile Robot Obstacle Avoidance," IEEE International Conference on Robotics and Automation, Nice, France, May, 1992.

Brown, M.K., "Locating Object Surfaces with an Ultrasonic Range Sensor," IEEE Conference on Robotics and Automation, St. Louis, MO, pp.110-115, March, 1985.

Byler, E., "Intelligent Mobile Sensor System for Drum Inspection and Monitoring," Phase I Topical Report, DOE Contract DE-AC21-92MC29112, Martin Marietta Astronautics Group, Littleton, CO, June, 1993.

Campbell, D., "Ultrasonic Noncontact Dimensional Measurement," *Sensors*, pp. 37-43, July, 1986.

Dunkin, W.M., "Ultrasonic Position Reference Systems for an Autonomous Sentry Robot and a Robot Manipulator Arm", Masters Thesis, Naval Postgraduate School, Monterey, CA, March, 1985.

Everett, H.R., "A Multielement Ultrasonic Ranging Array," *Robotics Age*, pp.13-20 July, 1985.

Feynman, R.P., Leighton, R.B., Sands, M., *The Feynman Lectures on Physics*, Vol. 1, Addison-Wesley, Reading, MA, 1963.

Figueroa, F., Barbieri, E., "Increased Measurement Range Via Frequency Division in Ultrasonic Phase Detection Methods," *Acustica*, Vol. 73, pp. 47-49, 1991.

Figueroa, J.F., Lamancusa, J.S., "A Method for Accurate Detection of Time of Arrival: Analysis and Design of an Ultrasonic Ranging System," Journal of the Acoustical Society of America, Vol. 91, No. 1, pp. 486-494, January, 1992.

Figueroa, J.F., Mahajan, A., "A Robust Navigation System for Autonomous Vehicles Using Ultrasonics," *Control Engineering Practice*, Vol. 2, No. 1, pp. 49-59, 1994.

Fisher, J.J., "Application-Based Control of an Autonomous Mobile Robot," American Nuclear Society, Charleston, SC, Sect. 2-6, pp. 1-8, 1989.

Fox, J.D., Khuri-Yakub, B.T., Kino, G.S., "High-Frequency Acoustic Wave Measurements in Air," IEEE Ultrasonics Symposium, pp. 581-584, 1983.

Frederiksen, T.M., Howard, W.M., "A Single-Chip Monolithic Sonar System," *IEEE Journal of Solid State Circuits*, Vol. SC-9, No. 6, December, 1974.

Gilbreath, G.A., Everett, H.R., "Path Planning and Collision Avoidance for an Indoor Security Robot," SPIE Mobile Robots III, Cambridge, MA, pp. 19-27, Novemeber, 1988.

Halliday, D., Resnick, R., *Fundamentals of Physics*, John Wiley, New York, NY, 1974.

Hammond, W., "Smart Collision Avoidance Sonar Surpasses Conventional Systems," *Industrial Vehicle Technology '93: Annual Review of Industrial Vehicle Design and Engineering*, UK and International Press, pp. 64-66, 1993.

Heckendorn, F.M., Ward, C.W., Wagner, D.G., "Remote Radioactive Waste Drum Inspection with an Autonomous Mobile Robot," ANS Fifth Topical Meeting on Robotics and Remote Systems, American Nuclear Society, Knoxville, TN, pp. 487-492, April, 1993.

Irwin, C.T., Caughman, D.O., "Intelligent Robotic Integrated Ultrasonic System," Proceedings, Robots 9, Society of Manufacturing Engineers, Detroit, MI, Sect. 19, pp. 38-47, June, 1985.

Jarvis, R.A., "A Laser Time-of-Flight Range Scanner for Robotic Vision," *IEEE Transactions on Pattern Analysis and Machine Intelligence*, Vol. PAMI-5, No. 5, pp. 505-512, 1983.

Kilough, S.M., Hamel, W.R., "Sensor Capabilities for the HERMIES Experimental Robot," American Nuclear Society, Third Topical Meeting on Robotics and Remote Systems, Charleston, SC, CONF-890304, Section 4-1, pp. 1-7, March, 1989.

Kuc, R., Siegel, M.W., "Physically Based Simulation Model for Acoustic Sensor Robot Navigation," *IEEE Transactions on Pattern Analysis and Machine Intelligence*, Vol. PAMI-9, No. 6, pp. 766-778, November, 1987.

Lang, S., Korba, L., Wong, A., "Characterizing and Modeling a Sonar Ring," SPIE Mobile Robots IV, Philadelphia, PA, pp. 291-304, 1989.

Leonard, J.J., Durrant-Whyte, H.F., *Directed Sonar Sensing for Mobile Robot Navigation*, Kluwer Academic Publishers, Boston, MA, 1992.

Ma, Y.L., Ma, C., "An Ultrasonic Scanner System Used on an Intelligent Robot," IEEE IECON '84, Tokyo, Japan, pp. 745- 748, October, 1984.

Michigan, "Mobile Robotics Lab," Brochure, University of Michigan Mobile Robotics Lab, Ann Arbor, MI, 1991.

Milner, R., ""Measuring Speed and Distance with Doppler," *Sensors*, pp. 42-44, October, 1990.

Mitome, H., Koda, T., Shibata, S., "Double Doppler Ranging System Using FM Ultrasound," *Ultrasonics*, pp. 199-204, September, 1984.

Pallas-Areny, R., Webster, J.G., "Ultrasonic Based Sensors," *Sensors*, pp. 16-20, June, 1992.

Polaroid, "Polaroid Ultrasonic Ranging Developer's Kit," Publication No. PXW6431 6/93, Polaroid Corporation, Cambridge, MA, June, 1993.
Shirley, P.A., "An Introduction to Ultrasonic Sensing," *Sensors*, pp. 10-17, November, 1989.
Shirley, P.A., "An Ultrasonic Echo-Ranging Sensor for Board Inspection and Selection," *Sensors*, June, 1991.
Smurlo, R.P., Everett, H.R., "Intelligent Sensor Fusion for a Mobile Security Robot," *Sensors*, pp. 18-28, June, 1993.
Ulrich, R., *Principles of Underwater Sound for Engineers*, 1983.

9
Electromagnetic Energy

It is important to realize that *acoustical energy* and *electromagnetic energy* are two fundamentally different phenomena with some very dissimilar properties and characteristics. Perhaps most obvious is the fact that sound travels rather slowly as a longitudinal pressure wave, whereas radio and optical energy propagate as electromagnetic waves at the speed of light. Additionally, sound must be *conducted* through some type of *medium*, whereas none is required for optical or RF energy, as evidenced by their effective transmission through the vacuum of space. In fact, for electromagnetic radiation, the earth's atmosphere is not an enabling mechanism at all but rather an impediment to propagation.

The electromagnetic spectrum is depicted in Figure 9-1. All the waves represented are electromagnetic in nature, travel at the same speed c in free space, and differ only in wavelength and the type of source with which they are generated (Halliday & Resnick, 1974). The *index of refraction n* relates the speed of light in a particular medium to the speed of light in a vacuum as follows:

$$n = \frac{c}{c_m}$$

where:

n = refractive index of medium
c = speed of light in a vacuum
c_m = speed of light in medium.

The value of n for air is 1.0003; light therefore travels slightly slower in air than in the vacuum of space, and slower still in water, for which n is 1.33. Such changes in propagation velocity are responsible for the bending of light rays at an interface between different media in accordance with Snell's law (Figure 9-2):

$$n_1 \sin\theta_i = n_2 \sin\theta_t$$

where:

n_1 = index of refraction in first medium
θ_i = angle of incidence with respect to surface normal
n_2 = index of refraction in second medium
θ_t = angle of refraction with respect to surface normal.

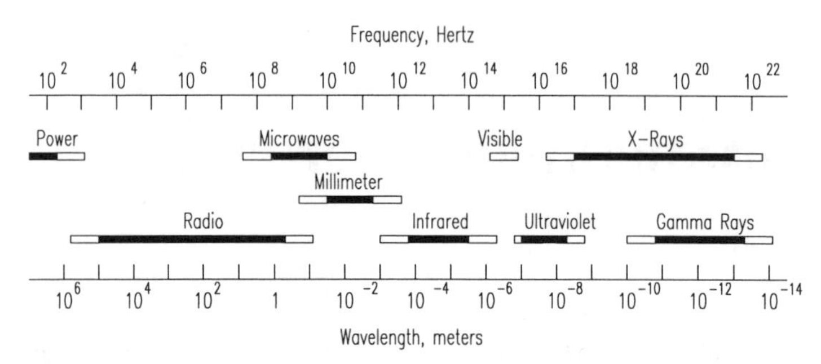

Figure 9-1. The electromagnetic spectrum is divided into several overlapping frequency intervals for which a particular energy source and detection technology exist (adapted with permission from Halliday & Resnick, 1974, © John Wiley and Sons, Inc.).

The index of refraction n is a function of the properties of atoms in the medium and the frequency of the electromagnetic radiation (Feynman, et al., 1963):

$$n = 1 + \frac{N q_e^2}{2 \varepsilon_o m (\omega_o^2 - \omega^2)}$$

where:

N = number of charges per unit volume
q_e = charge of an electron
ε_o = permittivity
m = mass of an electron
ω = frequency of the electromagnetic radiation
ω_o = resonant frequency of electron bound in an atom.

In examining the above equation, it can be seen that n slowly gets larger as the frequency increases (i.e., as ω approaches ω_o), an effect witnessed by anyone who has ever observed a rainbow in the sky. For example, the index of refraction is higher for blue light than for red. This phenomenon is known as *dispersion*, because the frequency dependence of the index n causes a prism to "disperse" the different colors into a distinctive spectrum (Feynman, et al., 1963).

Recall from the earlier discussions on acoustical energy in Chapter 8 that the *Doppler shift* for a moving observer approaching a fixed source was different from that of a moving source closing on a fixed observer at the same rate. In the case of electromagnetic energy, unlike sound, there is no *medium of transmission* relative to which the source and observer are moving (Halliday & Resnick, 1974).

There should therefore be no difference at all in the two cases under consideration: a source moving towards an observer should exhibit the same Doppler shift as an observer moving towards a source with the same relative velocity.

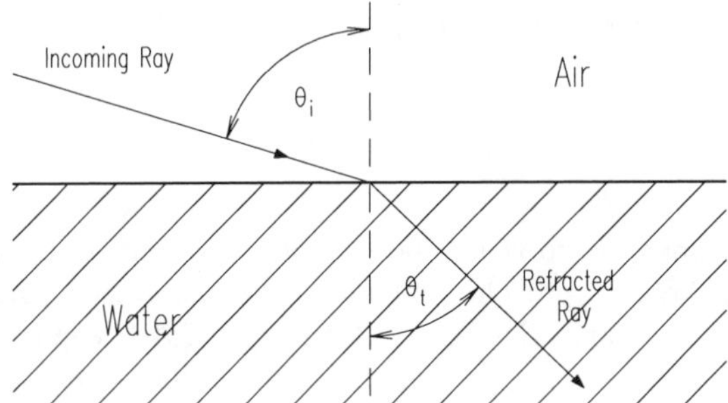

Figure 9-2. Snell's law describes the change in path angle that occurs when light passes a boundary between two media with different values of the refractive index *n*. A ray of light incident upon the surface of water as shown will appear to bend downward.

According to the theory of relativity, the observed frequency is given by (Halliday & Resnick, 1974):

$$f' = f\frac{1-\frac{v}{c}}{\sqrt{1-\left(\frac{v}{c}\right)^2}}$$

where:

f' = observed frequency
f = operating frequency of source
v = rate of separation or closure
c = speed of light.

As discussed in Chapter 8, it is usually more convenient to deal with the *Doppler shift frequency*, rather than the observed frequency, which is approximated by (Schultz, 1993):

$$f_d = \pm\frac{2vf\cos\theta}{c}$$

where:

f_d = *Doppler shift* frequency
θ = angular offset from radial path
$v \ll c$.

The sign in the above expression is positive in the case of target closure.

9.1 Optical Energy

The *optical* region of the electromagnetic spectrum is broken up into *ultraviolet*, *visible*, and *infrared* domains as shown in Figure 9-3, where the wavelength is determined by the speed of light (in a vacuum) divided by frequency:

$$\lambda = \frac{c}{f}.$$

The precise limits of the *visible* portion of the spectrum are not well defined because eye sensitivity approaches the horizontal axis in an asymptotic fashion at the upper and lower bounds (Halliday & Resnick, 1974) of approximately 400 and 700 nanometers, respectively (see Figure 9-4 later in this section). Fortunately for us humans, the emission spectrum of the sun conveniently peaks at 555 nanometers, in the very center of the visible range (Herman, et al., 1989).

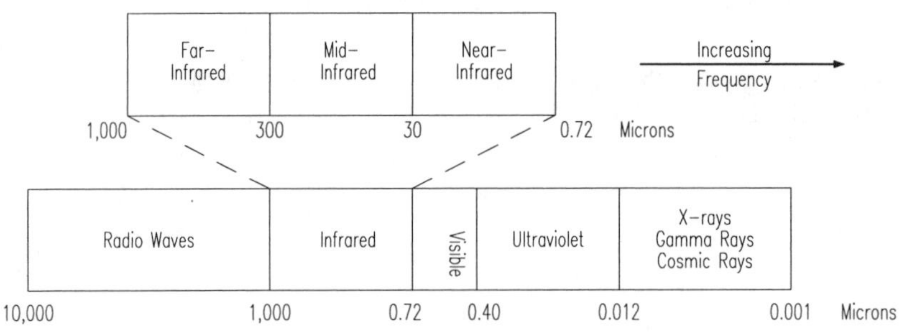

Figure 9-3. The optical portion of the electromagnetic spectrum encompasses wavelengths from 0.012 micron (ultraviolet) up to 1000 microns (infrared); the infrared region can be further subdivided into near-infrared, mid-infrared, and far-infrared (adapted from Banner, 1993, and Buschling, 1994).

The *infrared* portion of the electromagnetic spectrum encompasses wavelengths of 0.72 to 1,000 microns. All objects with an absolute temperature above 0°K emit radiant energy in accordance with the Stephan-Boltzman equation (Buschling, 1994):

$$W = \varepsilon \sigma T^4$$

where:

W = emitted energy
ε = emissivity
σ = Stephan-Boltzman constant (5.67 x 10^{-12} watts/cm^{2}K^4)
T = absolute temperature of object in degrees Kelvin.

The totality of all energy incident upon an object surface is either absorbed, reflected, or reradiated in accordance with Kirchoff's law. *Emissivity* (ε) is defined as the ratio of radiant energy emitted by a given source to that emitted by a perfect blackbody radiator of the same area at the same temperature under identical conditions (Graf, 1974). *Emissivity* is also a convenient measure of energy absorption. A hypothetical surface with an emissivity value of zero is a perfect reflector, neither absorbing nor emitting radiant energy, whereas in contrast, a theoretical *blackbody* with an ideal emissivity of one would absorb 100 percent of the supplied energy, reflecting none (Buschling, 1994).

9.1.1 Electro-Optical Sources

In 1977 the IEEE redefined a radar as "an electromagnetic means for target location and tracking" (IEEE, 1977). As pointed out by Sundaram (1979), this includes electro-optical devices such as laser radars (lidars) or laser rangefinders in general. Relative to microwave and millimeter-wave systems, electro-optical sensors are characterized by extremely short wavelengths affording much higher resolution, but suffer the greatest attenuation by atmospheric constituents. Active optical sources employed in rangefinding include broadband *incandescent*, broadband *electronic strobes*, narrowband *light emitting diodes*, *super luminescent diodes*, and coherent *lasers*. Each of these devices will be discussed briefly in the following paragraphs.

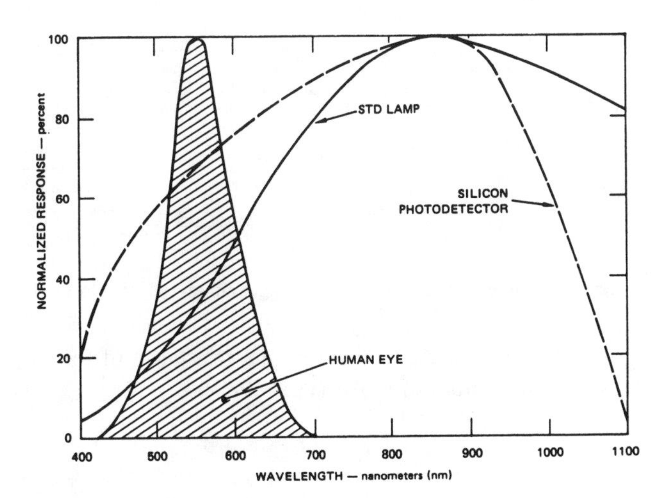

Figure 9-4. Radiation from a tungsten lamp filament is shown in comparison to the spectral sensitivity of silicon (Brain, 1979).

An *incandescent lamp* emits radiation as a consequence of the heating of a conductive filament (typically tungsten) with an electrical current. The higher the operating temperature of the filament, the greater the percentage of radiation falling in the visible portion of the energy spectrum; infrared heat lamps are simply tungsten bulbs operating at low filament temperature (Fox, 1989). The vast majority of energy given off by a conventional tungsten lamp is outside the visible region and in the infrared (Figure 9-4), which explains why fluorescent lighting, with almost no infrared component, is so much more energy efficient. Incandescent lamps transform only about five percent of their filament current into visible light (Mims, 1987). In the case of both infrared and fluorescent sources, the fundamental light emission processes occur in individual atoms, and these atoms do not operate together in a cooperative fashion (Halliday & Resnick, 1974). The resulting light output therefore is said to be *non-coherent*, with random phase variations associated with individual rays.

Electronic strobes consist of a glass or quartz flashtube filled with gas and containing two or more ionizing electrodes connected to a high-voltage DC power supply. Xenon is the preferred fill gas for most applications due to its high white-light spectral peak, but other gases such as argon, neon, krypton, and hydrogen can also be used (Fox, 1989). Figure 9-5 shows a typical trigger circuit in block diagram form. The power supply generally consists of a storage capacitor charged by the rectified output of a step-up transformer driven by an astable multivibrator. A silicon controlled rectifier (SCR) is used as an electronic switch to activate a trigger transformer when a flash is desired, coupling several thousand volts to the trigger electrode of the tube. This trigger pulse ionizes some of the xenon gas, forming a low resistance path that immediately discharges the storage capacitor through the gas-filled tube (Mims, 1987).

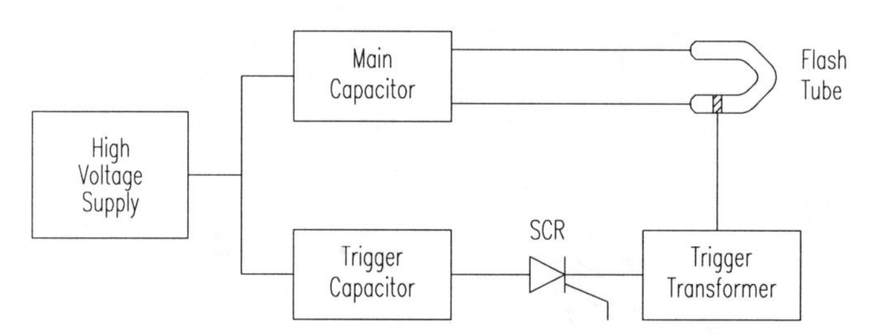

Figure 9-5. Block diagram of a typical xenon flash trigger circuit (adapted from Mims, 1987).

The length of the flash is determined by the resistance of the discharge path and the value of the capacitor as follows (Fox, 1989):

$$T_d = 3RC$$

where:

T_d = duration of flash

R = resistance in ohms
C = capacitance in farads.

Flash duration is typically very short (5 to 200 microseconds), a feature often used to advantage in freezing relative motion in inspection applications (Novini, 1985). The xenon immediately returns to its non-conductive state once the capacitor has discharged, or the flash has been terminated by interrupting the current flow using a solid-state switch (Mims, 1987). The flash can be retriggered as soon as the capacitor has recharged. The average life expectancies of ordinary xenon tubes are in the millions of flashes.

A *light emitting diode* (LED) is a solid-state P-N junction device that emits a small amount of optical energy when biased in the forward direction. LEDs produce *spontaneous emission* with a moderate spectrum (i.e., 40-100 nanometers) about a central wavelength. (As Udd (1991) points out, in sources dominated by *spontaneous emission* there is a low probability of one photon stimulating emission of another photon; such devices have important advantages in fiber-optic sensor applications, including low noise and relative immunity to optical feedback.) LEDs are attractive from the standpoint of durability, shock tolerance, low heat dissipation, small package size, and extremely long life (typically in excess of 100,000 hours).

Due to the spectral response of silicon detectors, near-infrared LEDs are the most efficient, and were the only type offered in photoelectric proximity sensors until around 1975 (Banner, 1993). Green, yellow, red, and blue versions are now readily available (Table 9-1), but near-infrared remains the most popular source in LED-based sensor applications. For example, most optical proximity detectors employ near-infrared LEDs operating between 800 and 900 nanometers. Fast cycle times allow LEDs to be used in a similar fashion to strobed illuminators for freezing motion and other image-synchronization applications, such as frame differencing between successive illuminated and non-illuminated scenes.

Table 9-1. Summary of typical light emitting diode types showing color selectivity (adapted from Fox, 1989).

LED Type	Spectral Peak	Color	Output
Gallium arsenide	540 nm	green	80 ft-lm
	900 nm	IR	10 mw
	980 nm	IR	500 mw
Gallium arsenide phosphide	560 nm	green	300 ft-lm
	610 nm	amber	200 ft-lm
	680 nm	red	450 ft-lm
Gallium aluminum arsenide	800 nm	red	1 mw
	850 nm	IR	5 mw
Silicon carbide	590 nm	yellow	150 ft-lm

Super luminescent diodes (SLDs) are a relatively new development (in the rapidly expanding field of fiber-optic communications and optical-disc technology) that can best be described as midway between the simplistic LED and the more complex coherent laser diode. The construction of all three devices is similar: a forward-biased P-N junction leads to a recombination of excess holes with electrons accompanied by emission of photon energy. While LEDs produce *spontaneous emission* only, laser diodes are physically configured so emissions in the active region oscillate back and forth several times between specially designed front and back facets. A high flux of photons past an excited state results in a high probability the excited state will be "stimulated" to radiate by a passing photon (Udd, 1991). (The principle of *stimulated emission* was first described by Einstein in 1917 and later demonstrated by T.H. Maiman in 1960 (Koper, 1987).) The SLD's characteristic laser "gain" on each forward pass results in a primary wavelength or mode of operation and what is termed a *coherent* output (Dokras, 1987), since each newly created photon has the same phase and wavelength as the stimulating photon (Udd, 1991).

LEDs have no such amplification mechanism; the output intensity simply increases with an increase in current density while the photon flux remains below the threshold required for *stimulated emission* (Udd, 1991). Surface-emitting LEDs have a wide solid-angle output beam with a Lambertian intensity distribution. Edge-emitting LEDs, on the other hand, have a waveguide mechanism built into their structure that results in a narrow Gaussian intensity pattern (Dokras, 1987). A more detailed description of the construction and operation of these two devices is presented by Udd (1991). An SLD is like an edge-emitting LED but with a single-pass gain feature similar to the laser, resulting in a combination of *spontaneous* and *stimulated emission*. The power output is therefore greater than a conventional LED, but as current density is increased, the device is unable to achieve multiple-pass gain as does a laser diode (Dokras, 1987). This limitation is assured by disabling the lasing cavity, typically through: 1) antireflection coating the end facets, 2) using proton bombardment to make one end of the cavity an absorber, or 3) mechanically destroying an end facet (Udd, 1991).

At present, the majority of active optical ranging devices employ laser sources, in equipment based on triangulation, time of flight, phase detection, interferometry, and return signal intensity. Lasers exist in a variety of types: solid-state (Sharp, 1989), liquid (Manzo, 1989), gas (Janney, 1989), and semiconductor (Shurtz, 1989). The more well-known types are gas lasers like helium-neon (HeNe) or the solid-state variety like neodymium: yttrium-aluminum-garnet (Nd:YAG). The recent advent of semiconductor-based laser diodes has had significant impact on the rangefinder instrument community (Depkovich & Wolfe, 1984). Although they typically have reduced power output and poorer spectral quality relative to other lasers, semiconductor devices are compact, rugged, reliable, and efficient, with sufficient quality of performance for most sensing needs. An often used laser of this type is the gallium arsenide (GaAs) laser diode, which emits in the near-infrared region.

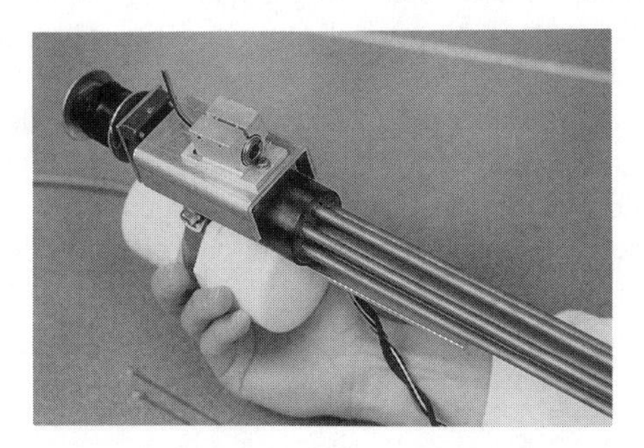

Figure 9-6. A visible-red laser diode is used as a manual sight for the pneumatic dart gun (see Chapter 1) on ROBART III.

This dynamic expansion in usage can be better understood by reviewing some of the inherent qualities of laser light (Depkovich & Wolfe, 1984). Lasers produce an intense well-collimated output, an important property in distinguishing the signal from background illumination, particularly in long-distance applications. The use of energy from the optical portion of the spectrum minimizes the specular reflectance problems encountered with acoustics, with the exception of polished surfaces (Jarvis, 1983). Furthermore, lasers generally transmit spectrally pure light of a single wavelength, void of extraneous signals and noise. This quality can be exploited by placing narrowband optical filters matching the source wavelength in front of the detector component. Filters of this type will reject ambient light, resulting in an improved signal-to-noise ratio for the system.

Along with these advantages there also exist some disadvantages that must be taken into account (Depkovich & Wolfe, 1984). Laser-based systems represent a potential safety problem in that the intense and often invisible beam can be an eye hazard. Gas lasers require high-voltage power supplies that present some danger of electrical shock. Laser sources typically suffer from low overall power efficiency. Lasing materials can be unstable with short lifetimes, resulting in reliability problems. Finally, some laser-based ranging techniques require the use of retroreflective mirrors or prisms at observed points, effectively eliminating selective sensing in unstructured surroundings.

9.1.2 Performance Factors

Atmospheric Attenuation

Attenuation of optical energy due to atmospheric effects can occur in the form of: 1) *scattering* and 2) *absorption. Scattering* is a disordered change in the direction of propagation and/or polarization of electromagnetic waves upon encountering molecules and aerosols (dust, sea salt, soot, ash, etc.) in the earth's atmosphere (Figure 9-7). *Molecular scattering*, being proportional to the fourth power of frequency (Feynman, et al., 1963), is strongly wavelength dependent but essentially negligible outside the visible portion of the spectrum; scattering by aerosols is less dependent on wavelength and can occur in both the visible and infrared regions (Herman, et al., 1985).

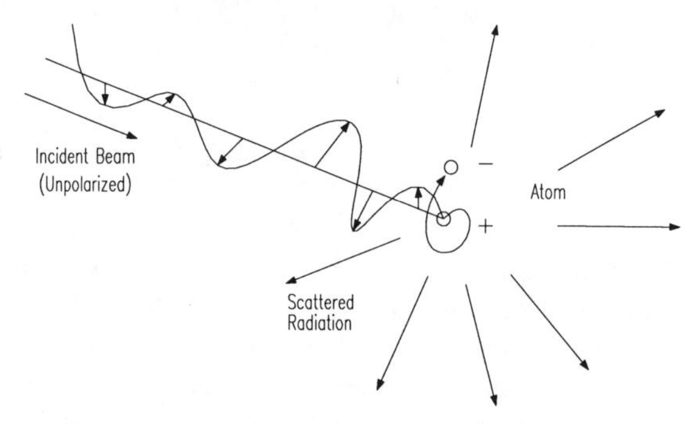

Figure 9-7. In molecular scattering, a beam of incident radiation causes the charges (electrons) in the atom to move; the moving electrons in turn radiate in various directions (Feynman, et al., 1963).

Visibility is a quantitative indicator of atmospheric attenuation. The World Meteorological Organization defines *meteorological optical range* (*MOR*) as the length of a path in the atmosphere required to reduce the luminous flux of a white light of color temperature 2700°K to five percent of its original value (Crosby, 1988). The factor 0.05 (i.e., five percent) is known as the *contrast threshold* ε. The *extinction coefficient* σ is a measure of the reduction of transmitted light due to atmospheric effects, where σ is the summation of the *absorption coefficient* α and the *scattering coefficient* β (Crosby, 1988). The *apparent contrast* C of an object against a uniform horizon sky varies exponentially with the *extinction coefficient* and range, a relationship known as Koschmieder's law:

$$C = e^{-\sigma r}$$

where:

C = apparent contrast

σ = extinction coefficient
r = range to object.

As the *apparent contrast C* approaches the *contrast threshold* ε (0.05), the above equation reduces to (Crosby, 1988):

$$r = \frac{3}{\sigma}$$

where:

r = maximum visibility range (MOR).

Solutions to the actual radiative transfer function describing the effects of atmospheric scattering involve very complex mathematical functions, and so a detailed discussion of atmospheric scattering and absorption is beyond the scope of this chapter. Herman, et al. (1985) provide a comprehensive mathematical treatment of the subject of atmospheric scattering in *The Infrared Handbook* (Wolfe & Zissis, 1985), with appropriate aerosol models and extensive tables of optical constants for various airborne particulates. LaRocca (1985) provides a similar treatment of atmospheric absorption in the same volume.

Surface Reflectivity

As discussed in Chapter 8, the parameter that governs the ratio of reflected versus transmitted sound energy at the interface between two media is the acoustical impedance Z; in the case of optical energy it is the index of refraction n. When light strikes a specular surface as illustrated in Figure 9-8, a portion of the wave is reflected as shown, with the *angle of reflection* θ_r equal to the *angle of incidence* θ_i. The angle θ_t of the *refracted* wave is given by Snell's law.

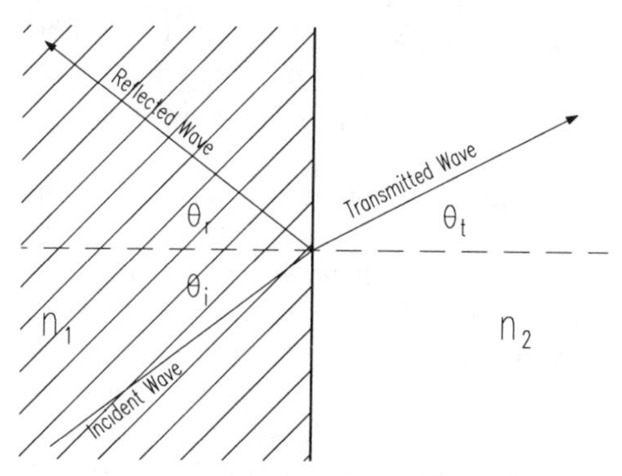

Figure 9-8. For specular surfaces, the angle of reflection is equal to the angle of incidence, while the product $n \sin \theta$ is the same for the incident and refracted beam (Feynman, et al., 1964).

For *specular* surfaces, the intensity of the reflected light depends on the *angle of incidence* and also the *polarization*. For situations where the electric field, ***E***, is perpendicular to the plane of incidence, the *reflection coefficient* is (Feynman, et al., 1964):

$$R_c = \frac{I_r}{I_i} = \frac{\sin^2(\theta_i - \theta_t)}{\sin^2(\theta_i + \theta_t)}.$$

For ***E*** parallel to the plane of incidence, the *reflection coefficient* is (Feynman, et al., 1964):

$$R_c = \frac{I_r}{I_i} = \frac{\tan^2(\theta_i - \theta_t)}{\tan^2(\theta_i + \theta_t)}.$$

For normal incidence (any polarization):

$$R_c = \frac{I_r}{I_i} = \frac{(n_2 - n_1)}{(n_2 + n_1)}$$

where:

R_c = reflection coefficient
I_r = reflected intensity
I_i = incident intensity.

Due to the shorter wavelengths involved, optical energy is in general less susceptible than microwave or millimeter-wave energy to problems associated with *specular reflection*. As was illustrated in Chapter 8 for the case of a flashlight beam aimed at a mirror, however, specular reflection will occur on highly polished or shiny surfaces. Anyone who has experienced driving at night on a wet road knows firsthand the results in the form of diminished headlight effectiveness. The normally *diffuse* road surface is transformed into a *specular* reflector by the pooling of water in tiny pits and crevices, markedly reducing the surface normal component and consequently the amount of light reflected back to the vehicle.

This last example brings up an important point: the amplitude of a surface reflection is not so much a property of the material, as is the index of refraction, but a "surface property" that depends on the specific topographical characteristics of the target (Feynman, et al., 1964). The above formulas hold only if the change in index is sudden, occurring within a distance of a single wavelength or less. Recall a similar criteria was established in Chapter 8 regarding *specular reflection* for acoustical waveforms. For light, however, the short wavelengths involved dictate that for very smooth surface conditions, the interface transition must occur over a distance corresponding to only a few atoms (Feynman, et al., 1964). For other than highly polished surfaces, the transition happens over a distance of several wavelengths, and the nature of the reflection will be *diffuse* rather than *specular*.

Glass, clear plastic, and other transparent substances with little or no reflectance properties can cause problems. In fact, the unknown reflectivity of observed targets is perhaps the most significant problem in optical range measurement and, coupled with the changing angle of incidence, causes the returned energy to vary significantly in amplitude. As a result, detection capabilities over a wide dynamic range (between 80 to 100 dB) are required, complicating the design of the receiver electronics.

Air Turbulence

Turbulence-induced pressure variations can cause random irregularities in the index of refraction of the atmosphere that will perturb an optical wavefront, causing image motion, image distortion, and added blur (Hufnagel, 1985). As the distorted wavefront continues to propagate, its local irradiance will randomly vary as a consequence of defocusing and spreading effects, a phenomenon known as *scintillation*, an example of which is seen in the twinkling of distant stars. The general consensus is the index of refraction does not vary rapidly in either space or time (Hufnagel, 1985), causing only small-angle refraction and diffraction of the radiation, with no changes in polarization (Saleh, 1967). The effects of turbulence on the index of refraction n are small, however, in comparison to those of temperature, and therefore can usually be ignored (Hufnagel, 1985). This is particularly true in the case of most robotic applications, considering the relatively short distances involved.

Temperature

In addition to wavelength and ambient pressure, the index of refraction is also dependent on temperature and humidity. Of all these parameters, temperature is by far the most significant, influencing the rate of index change as follows (Hufnagel, 1985):

$$\frac{dn}{dT} = \frac{78\,P}{T^2} \times 10^{-6}$$

where:

n = index of refraction
P = local air pressure in millibars
T = absolute temperature in degrees Kelvin.

A condition known as *thermal blooming* occurs when the radiation is strong enough to significantly heat the air along its path, but such effects are generally limited to application of very high-power laser beams (Hufnagel, 1985).

A final consideration should be given to temperature influence on componentry as opposed to the atmosphere itself. The wavelength of a laser diode can be shifted by temperature changes at the P-N junction. This shift in spectral output with temperature is approximately 0.25 nanometers per degree centigrade in the case of gallium-arsenide lasers (Gibson, 1989). A temperature increase will also radically lower the power output of semiconductor emitters, including LEDs and SLDs. Laser diodes can employ tiny closed-loop thermoelectric coolers within the emitter package to compensate for such effects (Dokras, 1987).

Beam Geometry

One of the decided advantages of optical systems relative to acoustical and RF is the ease with which the beam can be focused, using very small and inexpensive lenses. Laser beams in particular are narrow and collimated with little or no divergence, enabling a powerful spatially-selective beam of energy to be concentrated on a small spot at long distances. Typical values for very inexpensive scanning laser systems (see Chapter 6) are in the neighborhood of 0.3 to 0.5 milliradians. Wider beamwidths (i.e., in the neighborhood of 5 degrees) are sometimes employed in proximity sensor (Banner, 1993) and collision avoidance ranging applications (Moser & Everett, 1989) to increase the volumetric coverage and probability of target detection, but at the expense of range.

9.1.3 Choosing an Operating Wavelength

To be used with mobile robotic systems, an optical ranging system must function effectively under normal ambient lighting conditions, which makes the choice of the energy source somewhat critical. Some structured light systems use an incandescent lamp or xenon flash directed through a slit or patterned mask and projected onto the surface. Others use laser beams that are mechanically or electronically scanned at high rates to create the desired illumination. The major criterion for selecting a light source is to be sure that its intensity peaks at a spectral frequency other than that of the ambient energy. The camera (or detector) should be outfitted with a matching narrowband filter to complement the source and improve detection.

For example, ultraviolet light with a wavelength of 0.2 to 0.3 microns is effective outdoors because absorption by atmospheric ozone prevents the transmission of sunlight energy less than 0.3 microns in length. However, an ultraviolet source of the required power density level would be hazardous in indoor environments (not eye safe). Contrast this with infrared energy near 2.8 microns, which is better suited to indoor activities because man-made objects tend to reflect infrared energy well. Infrared loses some of its usefulness outdoors due to the inherent radiation emitted by the natural terrain, roadways, and objects (LeMoigue & Waxman, 1984). There is a strong component of near-infrared energy in sunlight and incandescent light, but ambient noise effects can be reduced by modulating the source over time, then demodulating the received energy at the detector. This technique effectively subtracts the constant illumination of the background.

The design of electro-optical systems must be optimized to extract the necessary information from ambient noise and clutter with a comfortable signal-to-noise ratio. For active systems, the actual source should be chosen according to the following guidelines (Dokras, 1987). It must produce energy:

- With sufficient intensity.
- At the required wavelength (or within an appropriate spectrum).
- With the desired radiation pattern.

9.2 Microwave Radar

The portion of the electromagnetic spectrum considered to be the useful frequency range for practical radar is between 3 and 100 GHz (Miller et al., 1985). Most modern conventional radars use microwave energy and operate in the L, S, C, or X bands (IEEE, 1976). The use of letter designations (see Table 9-2) was adopted for security reasons during World War II and persisted afterwards for sake of convenience. For the most part unaffected by fog, rain, dust, haze, and smoke, radar systems can produce astonishing accuracies in terms of target discrimination and range computation when combined with computerized signal processing (Nowogrodzki, 1983).

Table 9-2. Designated radar frequency bands (IEEE Standard 521-1976) shown in relation to VHF and UHF allocations (Barton, 1989).

Band	Frequency Range	Units
VHF	30-300	MHz
UHF	300-1,000	MHz
P	230-1,000	MHz
L	1,000-2,000	MHz
S	2,000-4000	MHz
C	4,000-8,000	MHz
X	8,000-12,500	MHz
K_u	12.5-18	GHz
K	18-26.5	GHz
K_a	26.5-40	GHz
Millimeter	> 30	GHz

Ranging is accomplished by pulsed TOF methods, CW phase detection, or CW frequency modulation. Pulsed systems can detect targets up to distances on the order of hundreds of miles, relying on the measurement of the round-trip time of a wave propagating at the speed of light. Near-field measurements (less than 100 feet) are more difficult for this type of system because the extremely sharp short-duration signals that must be generated and detected are expensive and complicated to realize, and virtually impossible for distances less than 1 foot (Schultz, 1993). CW systems, on the other hand, are effective at shorter ranges because the phase-detection or frequency-shift measurements are not dependent on the wave velocity and are also well suited for measuring the speed of moving objects by Doppler methods.

9.2.1 Applications

Microwave radars are extensively employed in both military and commercial surveillance, tracking, and navigational applications. Microwaves are ideally suited for long-range sensing because the resolution is generally sufficient, attenuation of the beams in the atmosphere is minimal, and low-mode guiding structures can be constructed. The relatively long wavelengths provide radar systems with an "all weather" capability, overcoming the absorption and scattering effects of air, weather, and other obscurants.

Microwaves are also used for shorter-range sensing needs such as tail-warning radar and ground-control radar for aircraft, typically involving distances in hundreds of feet. Other such uses include tank level indicators (Williams, 1989; Fitch, 1993), traffic control (Schultz, 1993), motion sensors (Hebeisen, 1993), presence detectors (Williams, 1991), and obstacle avoidance radars (Siuru, 1994), operating over ranges from a few feet to a few hundred yards. Microwave systems have been in the developmental stage for quite some time but only came into their prime within the last 20 years or so with the advent of inexpensive, reliable solid-state components as alternatives to the typically fragile, power-consuming thermionic devices (Nowogrodzki, 1983). Equipment for transmitting, receiving, and processing of the waveform is now widely available.

9.2.2 Performance Factors

The basic *radar equation* expresses the relationship between the signal power received at the antenna as a function of antenna size and the emitted power of the system (Ridenour, 1947):

$$S = \left[\frac{PG}{4\pi R^2}\right]\left[\frac{\sigma}{4\pi R^2}\right]\left[\frac{G\lambda^2}{4\pi}\right]$$

where:

S = signal power received
P = transmitted power
G = antenna gain
λ = wavelength
σ = radar cross-section of target
R = range to target.

The quantity in the first set of square brackets represents the power density of the incident wave at the target. For an isotropic antenna distribution radiating energy equally in all directions, the transmitted power P would simply be divided by the area of a sphere ($4\pi R^2$). Since in reality some type of directional antenna is generally employed, the radiated energy is concentrated by the associated antenna gain G.

As was the case for sonar in Chapter 8, we make the assumption here that the incident wave is reflected from the target in an isotropic Lambertian fashion, dispersing yet again in accordance with the inverse square law. This relationship is expressed in the second set of brackets in the radar equation above, where σ (the *radar cross-section*) serves as a combined representation of target *cross-sectional area*, *reflectivity*, and *directivity,* accounting for the fact the reflected energy distribution may not be purely isotropic. (The concept of *radar cross section* will be re-examined in more detail later in the section addressing millimeter-wave radar performance factors.) The first two sets of brackets together therefore give the power density of the returning wave at the receiving antenna, while the last set of brackets represents the cross-sectional area of the receiving antenna.

This most basic form of the radar equation assumes a monostatic configuration where the same antenna is used for both transmission and reception. For a more detailed explanation of terms and treatment of bistatic (separate antennae) configurations, see Blake (1990). In addition, any losses due to atmospheric scattering or absorption, which can be quite significant, are not yet taken into account. (This atmospheric attenuation aspect of radar performance will be treated in some detail later in this chapter.) Otherwise, however, the above equation very closely parallels the expression presented in Chapter 8 for the intensity of a reflected acoustical wave, with signal strength falling off (for both radar and sonar) with the fourth power of distance.

A major component consideration in the implementation of radar ranging capability is the configuration of the transmitting and receiving antennae. Conventional long-range monostatic systems typically feature a large *parabolic reflector* with the detector (or feed) positioned at the focal point of the dish (Figure 9-9A). The relationship between transmitted beamwidth and antenna diameter is expressed in the following equation:

$$\theta = 1.22\frac{\lambda}{d}$$

where:

θ = beamwidth
λ = wavelength
d = diameter of the reflector.

Increasing the diameter of the reflector results in improved range capability due to the more powerful (tighter focused) outgoing beam, and the larger antenna surface area with which to intercept the reflected energy (i.e., the gain parameters in the first and third brackets of the radar equation). Disadvantages include the need to manipulate a bulky mechanical system with high inertial loading, with a massive supporting structure required to offset the effects of vibration and wind. These factors have been major impediments to deployment of conventional radar-based ranging systems on board mobile robots.

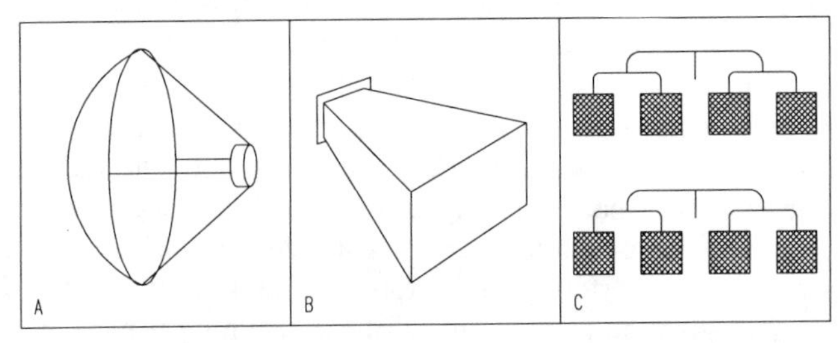

Figure 9-9. Common configurations of microwave antennae include: (A) reflective dish with feed situated at focal point, (B) conventional horn, and (C) two-dimensional microstrip arrays (adapted from Hebeisen, 1993).

To offset some of these drawbacks, many short-range commercial applications use a *horn* antenna configuration. Contrary to popular assumption, the dimensions of the horn opening are inversely proportional to beamwidth. In other words, the longer dimension of the horn results in the more narrow beam dimension, and vice versa. The antenna orientation depicted in Figure 9-9B, for example, would result in a rather broad vertical beam cross-section with a more narrow horizontal profile. Horn size tends to be reasonably small for beamwidths greater than 20 degrees, but fairly unrealistic in applications requiring less than 15 degrees (Schultz, 1993).

Phased-array antenna configurations (Figure 9-9C) present an alternative arrangement in the form of an array of multiple small antennae separated by distances of a few wavelengths. The transmissions from each radiator diverge and overlap with neighboring transmissions in a constructive and destructive fashion based on their phase relationships. By properly adjusting the phases, the antenna array can be tuned to a desired direction and intensity, as well as electronically scanned across the field of view. The small size of the individual transmitter-receivers facilitates low-profile designs and reduces problems due to wind and vibration. The resulting smaller coverage area decreases overall effectiveness, however, while the requirement for electronically variable phase control increases overall system complexity. Flat-plate dipoles and microstrip patch radiators are especially popular in applications involving ranges of 200 feet or less (Schultz, 1993).

Effective detection range is influenced by a number of factors besides antenna design and system performance, one of the more significant being atmospheric attenuation. Under fair weather conditions, clear atmosphere progresses from completely transparent all the way to 100-percent opaque as frequency is increased from 1 to 1000 GHz (Falcone & Abreu, 1979). Rain and snow can cause significant attenuation in signals above 2 GHz, as the physical size of the droplet becomes comparable to the wavelength (Hebeisen, 1989). Other issues for consideration include backscatter from airborne moisture, ground multipath interference, background clutter, reflectivity and directivity of the target surface, and natural cover such as snow or foliage.

Relatively speaking, there are a number of disadvantages associated with microwave radars from a robotics perspective:

- Microwave energy, like ultrasonic, is susceptible to specular reflections at the target surface, requiring receivers and signal processors with wide dynamic ranges.
- Available bandwidth is lower than shorter-wavelength millimeter-wave and electro-optical systems, resulting in reduced resolution.
- Longer wavelengths translate into large and bulky system components.
- Large antenna sizes are required to get narrow directional beams.
- The wide beamwidths result in problems due to side lobes, backscatter, and multipath reflections.
- Less available bandwidth, longer range capability, and wide beamwidth collectively make covert undetected operations difficult, with increased susceptibility to intercept and jamming.

Significant environmental constraints on microwave radar were identified by Barton (1977) and summarized by Johnston (1979). For all the above reasons, conventional radar systems operating in the microwave portion of the energy spectrum have less applicability to the short-range collision-avoidance needs of a mobile robotic platform.

9.3 Millimeter-Wave Radar

Millimeter waves constitute that portion of the electromagnetic spectrum with wavelengths of about 500 micrometers to 1 centimeter (30 to 300 GHz), midway between microwave and electro-optic. Millimeter waves possess several properties which differ substantially from microwave radiation. Relative to microwaves, millimeter-wave systems have significantly less range capability, primarily due to atmospheric attenuation and backscatter, but this generalized observation is given with regard to the more conventional applications of air-search and tracking radars. For the more unorthodox shorter-range needs of a mobile robotic system, this characteristic is not necessarily a limiting disadvantage.

In fact, Strom (1977) points out that for short-range applications, the performance of millimeter wave is actually superior to microwave under poor weather conditions:

> "If we compare a 3-mm radar and a 3-cm radar which are designed for short range usage and are equivalent in all other respects, we note that the equivalence in aperture provides 20 dB more gain at the short wavelength to offset atmospheric attenuation, and the equivalence in signal bandwidth (when considered as a percentage of operating frequency) yields a tenfold improvement in range resolution. Thus, the combined effect of antenna gain and range resolution is to reduce the clutter cell size by 30 dB."

The shorter wavelengths result in a narrow beam, with relatively small-sized antenna apertures for a given bandwidth. While the overall physical size of the system is reduced, the smaller apertures mean less collected energy, which again limits the effective range of the system.

Shorter wavelengths yield more accurate range and Doppler measurements. The ratio of wavelength to target size improves for higher frequencies, enabling better detection of small objects such as poles, wires, trees, and road signs. The high Doppler frequencies (conveniently in the audio range) provide good detection of slowly moving targets (Richard, 1976). The low power requirements of shorter wavelengths are attractive for robotic applications. In addition, shorter wavelengths translate into smaller componentry, which leads to reduced size and weight; the tradeoffs are that atmospheric attenuation increases, and the smaller antenna sizes lead to reduced receiver sensitivities.

Another advantageous feature is the extremely large bandwidth associated with millimeter waves; for the four major bands (35, 94, 140, and 220 GHz), the associated bandwidths are 16, 23, 26, and 70 GHz respectively. This means the entire microwave region could fit into any one of the millimeter bands (Sundaram, 1979). A bandwidth of one percent at 300 GHz is equal to all frequencies below S-band center frequency at 3 GHz (Johnston, 1978). More bandwidth translates into reduced multipath, greater resolution and sensitivity, reduced interference between mutual users of the band, and greater resistance to jamming.

9.3.1 Applications

Millimeter waves have been proposed for numerous applications (Skolnik, 1970) and have been the subject of theoretical studies (and much debate) since the early 1950s (Johnston, 1978). Unfortunately, the technology was not sufficiently developed during the earlier part of this period, and it was not until recent solid-state advances in the past two decades that practical devices could be developed and tested. Likely beneficiaries of this technology include remote environmental sensing, interference-free communications, low-angle tracking radar, high-resolution and imaging radar, spectroscopy (Senitzky & Oliner, 1970), terminal guidance, active fusing, range finding equipment, and automobile braking. Of these, the most common usage today is target tracking and designation in the military. The narrow beamwidth of millimeter-wave transmissions is highly immune to ground reflection problems when following targets at low-elevation angles, making such radars highly effective at low-angle tracking.

In addition to their limited range capabilities, millimeter-wave systems are not used for large-volume search and target acquisition because of their narrow beamwidths. Anyone who has ever tried to find a specific object of interest with a very high-power telescope is familiar with this problem. Some such optical devices have wider-angle "spotter scopes" attached on a parallel axis to assist in first locating the target. Certain military ground-to-air defense systems have adopted the same approach, employing longer-range wide-beam microwave search radars for initial acquisition, and then switching to a millimeter tracking radar for weapons system control.

Short-range, low-power millimeter-wave systems would appear to be well suited to the collision avoidance and navigational ranging needs of an exterior mobile robot. In fact, Johnston (1978) suggested automobile braking as being the largest potential commercial application of millimeter-wave radar. As Weisberg (1977) pointed out, the ability to use smaller antennae is a dominating characteristic influencing selection of millimeter wave over microwave. He cited three platforms most directly affected: satellites, missiles, and mini-RPVs. The envisioned application of millimeter ranging systems on robotic vehicles can be considered an extension of Weisberg's third category.

9.3.2 Performance Factors

Atmospheric Attenuation

All electromagnetic energy is absorbed to some degree by atmospheric gases (water, carbon dioxide, oxygen, and ozone), and from attenuation due to haze, fog, clouds, and rain (Sundaram, 1979). The absorbed energy is converted into heat and then lost to the surrounding atmosphere (Hebeisen, 1989). Frequency selective absorption takes place at the higher frequencies, due to resonances of the atmospheric gases (Van Vleck, 1964), and varies with atmospheric pressure, temperature, and relative humidity (Dyer & Currie, 1978). Two most pronounced effects are due to the magnetic interaction of oxygen and the electric polarity of the water molecule in water vapor (Koester, et al., 1976). These regions of maximum absorption are denoted as *absorption bands*, with the most significant being around 60 GHz (Sundaram, 1979). Similarly, atmospheric absorption by gases is minimal for certain frequencies (Table 9-3), appropriately called *atmospheric windows* (Sundaram, 1979).

Table 9-3. Naturally occurring atmospheric windows (adapted from Sundaram, 1979).

Window	Favored Wavelengths
Visible	0.4 to 0.7 μm
IR	3 to 5 μm, 8 to 12 μm
RF	3 and 10 cm (main microwave bands)
MMW	8.5, 3.2, 2.1, 1.4 mm
	(35, 94, 140, 220 GHz)

Relative to microwaves, millimeter waves display greater interaction with the environment. This attribute is a bonus in radiometry applications in that sensors can detect small particles and carry on frequency-selective interaction with gases. However, the resulting atmospheric attenuation limits the maximum range and restricts operation of such devices in adverse weather conditions. Millimeter waves interact with atmospheric gases, particulate matter, and suspended moisture droplets (hydrometers) through three primary mechanisms: absorption, scattering, and refraction (Dyer, et al., 1978). Attenuation can occur due to absorption by water vapor and gases, absorption from condensed water droplets in fog and rain, and scattering from water droplets in rain (Richard, 1976). Of these, rain is by far the most significant factor, with its associated attenuation being directly proportional to the size of the water droplets and the rainfall rate (Hebeisen, 1989), and to the 2.5 power of frequency (Barton, 1989). Dry snow, on the other hand (with the exception of very heavy snowfall rates), produces very little attenuation (Dyer & Currie, 1978). Because airborne particulates such as dust, smog, and smoke have dielectric constants much smaller than that of water, their associated attenuation cross section is reduced, resulting in negligible millimeter wave attenuation (Dyer & Currie, 1978).

Richard (1976) characterizes clear weather attenuation as small for 35, 70, and 94 GHz, appreciable at 140 and 240 GHz and prohibitively large at 360 GHz and above. Fair weather ranges are generally restricted to 10 to 20 kilometers (Sundaram, 1979), and typical employment with comparative all-weather performance to 3-cm microwave is more likely limited to between 5 and 10 kilometers (Strom, 1977). Johnston (1979) provides an excellent overview of early work addressing the various factors affecting millimeter-wave propagation, to include attenuation, backscatter, foliage penetration, phase variations, polarization effects, and surface phenomena. Another comprehensive survey of millimeter-wave propagation data treating attenuation, backscatter, and foliage is presented by Dyer and Currie (1978). Clear weather atmospheric as well as calculated rain attenuation curves versus frequency of operation are presented by Richard (1976) and numerous others.

In summary, atmospheric losses in the four millimeter bands, while higher than those for microwave energy, are significantly lower than that associated with optical devices, particularly in the case of rain, fog, smoke, dust, and haze. For the robotic applications envisioned (collision avoidance, navigation, communications), the range limitations impose no significant constraints.

Radar Cross-Section

Radar cross-section, a term used to quantify the ability of a particular target to return energy to the receiving antenna, is a function of three independent parameters (Schultz, 1993):

- *Geometric cross-section* — refers to the amount of incident radiation intercepted by the target's cross-sectional area and is basically influenced by target size and orientation (aspect ratio).
- *Reflectivity* — determines that portion of the incident energy reflected versus absorbed by the target and is primarily influenced by target composition.
- *Directivity* — is a measure of the reflected energy returning to the receiving antenna and is mainly influenced by target geometry.

The *coefficient of reflection* introduced earlier in this chapter in the particular case of optical energy applies to radar reflectivity as well. Recall for normal incidence:

$$R = \frac{I_r}{I_i} = \frac{(n_2 - n_1)}{(n_2 + n_1)}$$

where:

R = reflection coefficient
I_r = reflected intensity
I_i = incident intensity.

The actual value of R depends on target surface conditions, material composition, and the frequency of the incident wave, but in general is greater for lossy materials possessing: 1) higher permittivity (dielectric constant), 2) higher conductivity, and 3) lower permeability (Hebeisen, 1989). In other words, lossless dielectric materials such as plastic and glass reflect little energy, whereas conductive materials such as aluminum or steel make good radar targets. The more magnetic the conductive material, the less its reflectivity, and so one would expect aluminum to reflect more energy than steel. Typical values of permittivity, conductivity, and permeability for several representative materials are listed in Table 9-4.

Table 9-4. Typical values of relative permittivity, conductivity (s/m), and relative permeability for common materials, for low frequencies at room temperature (adapted from Hebeisen, 1989).

Material	Relative Permittivity	Conductivity	Relative Permeability
Air	1.0	-	-
Bakelite	5.0	-	-
Glass	4-10	-	-
Oil	2.3	10^{-11}	-
Dry soil	3-4	10^{-5}	-
Distilled water	80	2×10^{-4}	-
Silver	-	6.17×10^{7}	0.99998
Copper	-	5.80×10^{7}	0.99999
Aluminum	-	3.54×10^{7}	1.000021
Iron	-	10^{7}	4,000

To get an intuitive appreciation for the values listed in Table 9-4 above, try heating a small glass container of baby oil in a microwave oven for a set time interval, and then heat the same amount of water for an identical length of time. The temperature of the oil shows negligible increase, whereas the container of water can easily be brought to a boil. Water, being a better conductor than oil, absorbs more of the RF energy, which is then converted into heat. Non-conductive materials that have a high moisture content, therefore, make relatively good targets (Figure 9-10).

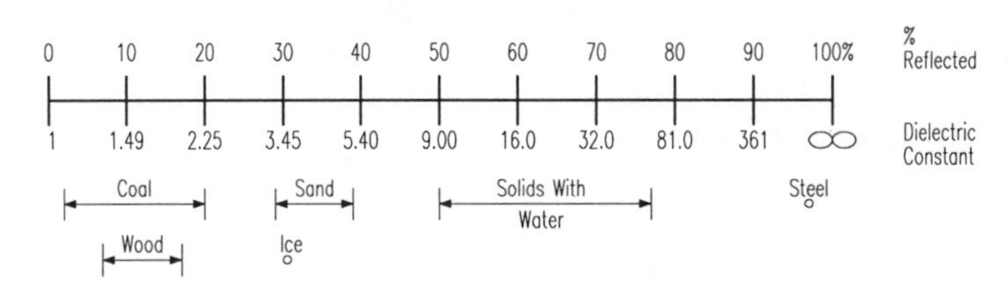

Figure 9-10. The percentage of reflection of microwave energy off various solids, showing the relative reflectivity for high-moisture-content solids in relationship to steel (adapted from Hebeisen, 1993).

All other factors being equal, a general rule of thumb is the radar cross-section of a particular object will increase with size and also with the frequency of incident radiation, up until the point where the object's size is much larger than the operating wavelength, after which little additional improvement is seen (Hebeisen, 1989). This relationship is the reason multipath reflections in RF communications are so much more of a problem at higher frequencies; the radiated energy is more likely to reflect from the various objects in the immediate environment. The longer the wavelength (i.e., the lower the frequency), the easier it is for the energy to penetrate through surrounding structure for non-line-of-sight operation.

Multipath Reflections

In addition to the desired signal return from discrete targets, multipath reflections from intermediate surfaces must also be taken into account in ground-based scenarios. (Multipath reflections are generally not a problem in surface-to-air radar tracking applications.) When a horizontal beam is projected from a mobile robotic platform, there is a good chance the lower part of the beam will illuminate a portion of the ground as well as the target of interest (Figure 9-11). If the ground-reflected signal is not suppressed by the directional characteristics of the receiving antenna, this leads to constructive and destructive interference, causing fades. Ground multipath and clutter, however, do not usually limit the performance of millimeter ground-to-ground systems as seriously as foliage obscuration, background clutter, and terrain masking (Richard, 1976).

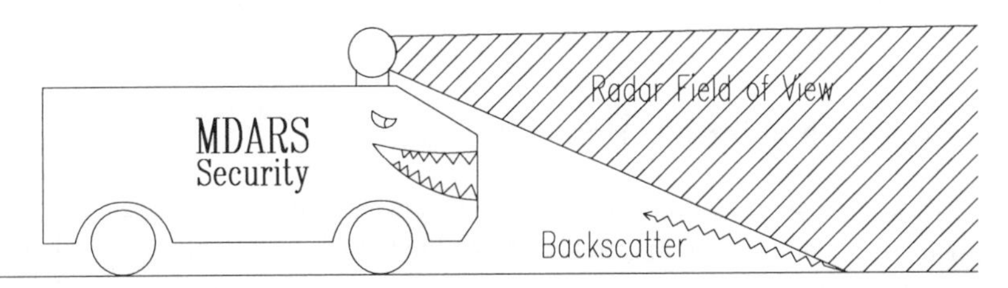

Figure 9-11. Ground backscatter at low elevation angles can be a problem in robotic collision-avoidance radar applications.

When scattered energy is returned in the direction of the receiver (i.e., backscatter), it can seriously contribute to the noise level, sometimes lowering the signal-to-noise ratio to the point target detection is difficult or impossible. Backscatter effects from rain are dependent upon the rainfall rate and the frequency of operation, and appear most significant in the region of 20 to 40 GHz, which includes the 35-GHz K_a band of frequencies (Richard, 1976). Backscatter from fog and clouds is more than two orders of magnitude less than that associated with rain (Lo, et al., 1975). In addition to causing backscatter along the path, an accumulation of snow on the reflecting surface can effectively mask the target altogether. The best way to reduce backscatter is to limit the size of the resolution cell (i.e., the beamwidth). Other techniques include narrow-beam antennae, short pulse durations for TOF systems (Richard, 1976), wide-band frequency modulation (McGee, 1968), frequency optimization, and circular polarization. The use of circular polarization, for example, can reduce rain return by 10 dB or more (Strom, 1977).

Temperature

The speed of light, quite unlike the speed of sound, is not significantly influenced by ambient temperature variations. Temperature differentials and humidity can, however, generate time-varying refractions, producing beam wander that results in short-term fades (Strom, 1977). This topic was previously treated in Section 9.1.2.

Beam Geometry

The shorter wavelength of millimeter-wave radiation results in narrower beamwidths relative to microwave. By way of illustration, a millimeter-wave antenna with a 12-cm-diameter aperture provides a 1.8-degree beamwidth at 94 GHz, as opposed to 18 degrees at 10 GHz (Sundaram, 1979). Narrow beams mean increased range, greater angular resolution, reduced noise and interference, minimized side lobe returns, fewer multipath problems, and less chance of detection and jamming.

9.3.3 Choosing an Operating Frequency

Tradeoffs in the selection of an operating frequency involve antenna size, atmospheric attenuation properties, and available component technology (Koester, et al., 1976). Strom (1977) provides a very comprehensive overview of frequency optimization that takes into account the effects of attenuation, clutter, and system parameters. These factors must be collectively considered from a systems integration point of view when establishing design parameters.

For example, due to narrower beamwidth, side lobes and ground multipath effects are much reduced for 220 GHz. Since operation at that frequency band would also entail the smallest size components, it might seem like the logical choice for most applications. However, Weisberg (1977) points out that on a hot humid day, the required power for a 3-kilometer target acquisition system jumps from a fraction of a watt at 94 GHz to tens of kilowatts if operated at 220 GHz.

9.4 References

Banner, *Handbook of Photoelectric Sensing*, Banner Engineering Corp., Minneapolis, MN, 1993.

Barton, D.K., "Philosophy of Radar Design," in *Radar Technology*, E. Brookner, ed., ARTECH House Books, Dedham, MA, 1977.

Barton, D.K., "Radar Principles," in *Electronic Engineer's Handbook,* D. Christiansen and D. Fink, eds., 3rd edition, New York, McGraw Hill, pp. 25.2-25.53, 1989.

Blake, L., "Prediction of Radar Range," in *The Radar Handbook*, Chapt. 2, Skolnik, M., ed., 2nd edition, McGraw Hill, New York, NY, 1990.

Buschling, R., "Understanding and Applying IR Temperature Sensors," *Sensors*, pp. 32-37, October, 1994.

Crosby, J.D., "Visibility Measurement: An Assessment of Two Techniques," *Sensors*, pp. 32-40, October, 1988.

Depkovich, T., Wolfe, W., "Definition of Requirements and Components for a Robotic Locating System," Final Report No. MCR- 83-669, Martin Marietta Denver Aerospace, Denver, CO, February, 1984.

Dokras, S., "Active Components in Fiber-Optic Sensors," *Sensors*, pp. 20-23, April, 1987.

Dyer, F.B., Currie, N.C., "Environmental effects on Millimeter Radar Performance," AGARD Conference Proceedings, CP 245, *Millimeter and Submillimeter Wave Propagation and Circuits*, pp. 2.1 - 2.9, 1978.

Falcone, V.J., Abreu, L.W., "Atmospheric Attenuation of Millimeter and Submillimeter Waves," IEEE EASCON-79 Conference Record, Vol. 1, pp. 36-41, 1979.

Feynman, R.P., Leighton, R.B., Sands, M., *The Feynman Lectures on Physics*, Vol. 1, Addison-Wesley, Reading, MA, 1963.

Feynman, R.P., Leighton, R.B., Sands, M., *The Feynman Lectures on Physics*, Vol. 2, Addison-Wesley, Reading, MA, 1964.

Fitch, F.M., "Measuring Level with Radar Technology," *Sensors*, pp. 40-41, April, 1993.

Fox, C.S. "Lamps, Luminous Tubes, and Other Noncoherent Electric Radiation Sources," *Electronic Engineer's Handbook,* D. Christiansen and D. Fink, eds., 3rd edition, New York, McGraw Hill, pp. 11.4-11.11, 1989.

Gibson, S.B., "Application of Semiconductor Lasers," *Electronic Engineer's Handbook,* D. Christiansen and D. Fink, eds., 3rd edition, New York, McGraw Hill, pp. 11.37-11.41, 1989.

Graf, R.F., *Dictionary of Electronics*, Howard W. Sams, Indianapolis, IN, 1974.

Halliday, D., Resnick, R., *Fundamentals of Physics*, John Wiley, New York, NY, 1974.

Hebeisen, S., "Target and Environmental Characteristics Which Affect the Performance of Microwave Sensing Systems," Sensors Expo International, Cleveland, OH, September, 1989.

Hebeisen, S., "Microwave Proximity Sensing," *Sensors*, pp. 22-27, June, 1993.

Herman, B., LaRocca, A.J., Turner, R.E., "Atmospheric Scattering," in *The Infrared Handbook*, Wolfe, W.L., Zissis, G.J., eds., pp. 5.1-5.131, 1985.

Hufnagel, R.E., "Propagation Through Atmospheric Turbulence," in *The Infrared Handbook*, Wolfe, W.L., Zissis, G.J., eds., pp. 6.1-6.56, 1985.

IEEE, "IEEE Standard Letter Designations for Radar Bands," IEEE Standard 521-1976, November 30, 1976.

IEEE, "IEEE Standard Radar Definitions," IEEE Standard 686-1977, November, 1977.

Janny, G.M., "Gas Lasers," *Electronic Engineer's Handbook,* D. Christiansen and D. Fink, eds., 3rd edition, New York, McGraw Hill, pp. 11.27-11.31, 1989.

Jarvis, R.A., "A Laser Time-of-Flight Range Scanner for Robotic Vision," IEEE Transactions on Pattern Analysis and Machine Intelligence, Vol. PAMI-5, No. 5, pp. 505-512, September, 1983.

Johnston, S.L., "Some Aspects of Millimeter Wave Radar," Proceedings International Conference on Radar, Paris, France, December 4-8, pp. 148-159, 1978.

Johnston, S.L., "A Radar Engineer Looks at Current Millimeter- Submillimeter Atmospheric Propagation Data," IEEE EASCON-79 Conference Record, Vol. 1, pp. 27-35, 1979.

Koester, K.L., Kosowsky, L., and Sparacio, J.F., "Millimeter Wave Propagation," Appendix A to "Millimeter Wave Radar Applications to Weapons Systems," V.L. Richards, pp. 77-105, June, 1976.

Koper, J.G., "A Three-Axis Ring Laser Gyroscope," *Sensors*, pp. 8-21, March, 1987.

LaRocca, A.J., "Atmospheric Absorption," in *The Infrared Handbook*, Wolfe, W.L., Zissis, G.J., eds., pp. 5.1-5.131, 1985.

Le Moigue, J., Waxman, A.M., "Projected Light Grids for Short Range Navigation of Autonomous Robots," Proceedings of 7th IEEE Conference on Pattern Recognition, Montreal, Canada, pp. 203-206, July - August, 1984.

Lo, L.T., Fannin, B.M., Straiton, A.W., "Attenuation of 8.6 and 3.2 mm Radio Waves by Clouds," *IEEE Transactions on Antennae and Propagation*, Vol. AP-23, No. 6, November, 1975.

Manzo, P.R., "Liquid Lasers," *Electronic Engineer's Handbook,* D. Christiansen and D. Fink, eds., 3rd edition, New York, McGraw Hill, pp. 11.25-11.27, 1989.

McGee, R., "Multipath Suppression by Swept Frequency Methods," Ballistic Research Laboratories Memorandum Report No 1950, November, 1968.

Miller, D.L. et al., "Advanced Military Robotics," Interim Report No. R84-48603-001, Martin Marietta Denver Aerospace, Denver CO, 26 July, 1985.

Mims, F.M., Forrest Mims' Circuit Scrapbook II, Howard W. Sams, Indianapolis, IN, p. 95, 1987.

Moser, J., Everett, H.R., "Wide-Angle Active Optical Triangulation Ranging System," SPIE Vol. 1195, Mobile Robots IV, Philadelphia, PA, November, 1989.

Novini, A., "Fundamentals of Machine Vision Lighting," Penn Video, December, 1985.

Nowogrodski, M., "Microwave CW Radars in Industrial Applications," *Electro, 1983 Conference Record*, Vol. 8, pp. 1-7, 1983.
Richard, V.W., "Millimeter Wave Radar Applications to Weapons Systems," U.S. Army Ballistic Research Labs Report No. 2631, June, 1976.
Ridenour, L.N., *Radar Systems Engineering*, MIT Radiation Laboratory Series, McGraw Hill, pp. 143-147, 1947.
Saleh, A.A.M., "An Investigation of Laser Wave Depolarization Due to Atmospheric Transmission," *IEEE Journal of Quantum Electronics*, Vol. QE-3, pp. 540-543, November, 1967.
Schultz, W., "Traffic and Vehicular Control Using Microwave Sensors," *Sensors*, pp. 34-49, October, 1993.
Senitzky, B., Oliner, A.A., "Submillimeter Waves - A Transition Region," *Submillimeter Waves*, Microwave Research Institute Symposia Series, Fox, J., ed., Polytechnic Press of the Polytechnic Institute of Brooklyn, NY, 31 March - 2 April, 1970.
Sharp, E.J., "Solid Optically Pumped Lasers," *Electronic Engineer's Handbook*, D. Christiansen and D. Fink, eds., 3rd edition, New York, McGraw Hill, pp. 11.18-11.24, 1989.
Shurtz, R.R., "Semiconductor Lasers and LEDs," *Electronic Engineer's Handbook,* D. Christiansen and D. Fink, eds., 3rd edition, New York, McGraw Hill, pp. 11.31-11.36, 1989.
Siuru, B., "The Smart Vehicles Are Here," *Popular Electronics*, Vol. 11, No. 1, pp. 41-45, January, 1994.
Skolnik, M.I., "Millimeter and Submillimeter Wave Applications," *Submillimeter Waves*, Microwave Research Institute Symposia Series, Fox, J., ed., Polytechnic Press of the Polytechnic Institute of Brooklyn, New York, pp. 9-26, March 31 - April 2, 1970.
Strom, L.D., "The Unresolved Problems and Issues," 6th DARPA Tri-Service Millimeter Wave Conference, pp. 10-27, 1977.
Sundaram, G.S., "Millimetre Waves - The Much Awaited Technology Breakthrough?", *International Defense Review*, part of the Jane's Information Group, Vol. 11, No. 2, February, 1979.
Udd, E., "Fiber Optic Sensors Based on the Sagnac Interferometer and Passive Ring Resonator," in *Fiber Optic Sensors: An Introduction for Engineers and Scientists*, E. Udd, ed., John Wiley, New York, pp. 233-269, 1991.
Van Vleck, J.H., "Theory of Absorption by Uncondensed Gases," in D.E. Kerr (ed.), *Propagation of Short Radio Waves*, Boston Technical Publishers, pp. 646-664, MA, 1964.
Weisberg, L.R., "Millimeter Waves - The Coming of Age," 6th DARPA Tri-Service Millimeter Wave Conference, pp. 4-9, 1977.
Williams, H.S., "Proximity Sensing with Microwave Technology," *Sensors*, pp. 6-15, June, 1989.
Williams, H.S., "The Basic Principles of Microwave Sensing," *Sensors*, pp. 26-28, May, 1991.

Wolfe, W.L., Zissis, G.J., eds., *The Infrared Handbook*, Environmental Research Institute of Michigan, Ann Arbor, MI, 1985.

10
Collision Avoidance

Truly autonomous control implies the ability of a free-roaming platform to travel anywhere so desired, subject to nominal considerations of terrain traversibility. Many potential applications await an indoor mobile robot that could move in a purposeful fashion from room to room, with the intelligence to avoid objects and if necessary choose alternative routes of its own planning. *Navigational referencing sensors* typically require high angular and/or range resolution over fairly long distances and will be extensively treated in several follow-on chapters. *Collision avoidance sensors*, on the other hand, usually operate over shorter ranges with less resolution required. The field of view should provide sufficient coverage for a turning vehicle, and allow enough time to stop or alter course.

The various proximity and non-contact ranging techniques presented in Chapters 3 through 9 all represent potential candidate solutions for meeting the collision avoidance needs of a moving platform. We shall examine in this chapter some of the issues involved in incorporating such sensors on indoor robotic systems, and then consider a few special problems that arise in more complex outdoor scenarios.

10.1 Navigational Control Strategies

A number of different navigational control strategies have been adopted by various parties, and lumping them together here under a common heading is a bit like mixing apples and oranges. Accordingly, it is perhaps advantageous to first make a distinction between high-level *global navigation* (i.e., planning an optimal path to some desired goal location in world coordinates), and *localized navigation* (i.e., piloting the robot around unexpected obstructions). We shall address only the latter category in this chapter, from the dual perspective of: 1) the required sensors and 2) the interpretation of data collected by those sensors for purposes of collision avoidance.

Six general approaches for such localized collision avoidance will be discussed:

- "Wander" routine.
- Circumnavigation.
- Potential fields.
- Certainty grids.
- Motor schema.
- Vector field histogram.

The above candidates can be subdivided into three basic categories: 1) *reactive control*, 2) *representational world modeling*, and 3) some combination of both.

10.1.1 Reactive Control

Reactive control for our purposes refers to a behavior-based strategy that directly couples real-time sensory information to motor actions, without the use of intervening symbolic representations that attempt to model in an absolute sense all or part of the robot's operating environment. Arkin (1992a) lists the following general characteristics of reactive control:

- It is typically manifested by a decomposition into primitive behaviors.
- Global representations are avoided.
- Sensor decoupling is preferred over sensor fusion.
- It is well suited for dynamically changing environments.

Wander

The most simplistic *reactive* collision avoidance capability for an autonomous mobile robot is perhaps illustrated by the basic *wander routine* implemented by several research groups (Everett, 1982; Brooks, 1986; Arkin, 1987; Anderson & Donath, 1988). The term *wander* is used here to describe a behavioral primitive that involves traveling more or less in a straight line until an obstacle is encountered, altering course to avoid impact, then resuming straight-line motion. Such a capability can be simply hard-coded, rule-based, or inherent in a more sophisticated layered *subsumption architecture* (Brooks, 1986).

By way of illustration, the *wander routine* employed on ROBART I was based on a six-level scheme of proximity and impact detection using the following sensor inputs (see Figure 10-1):

- A positionable near-infrared proximity scanner mounted on the head.
- A forward-looking LM-1812 sonar mounted 20 inches above the floor.
- Ten near-infrared proximity detectors to sense close (< 18 inches) obstructions.
- Projecting "cat-whisker" tactile sensors to detect pending (< 6 inches) collisions.
- Contact bumpers to detect actual impact.

- Drive motor current sensors to monitor for overload condition indicative of a stall.

The first two categories were loosely classified as non-contact ranging sensors that looked out ahead of the robot for planning purposes, while the next three were considered close-in proximity and tactile sensors requiring immediate action. Drive motor overload was a last-resort internal sensor in the event contact with an object was for whatever reason not detected by any of the above.

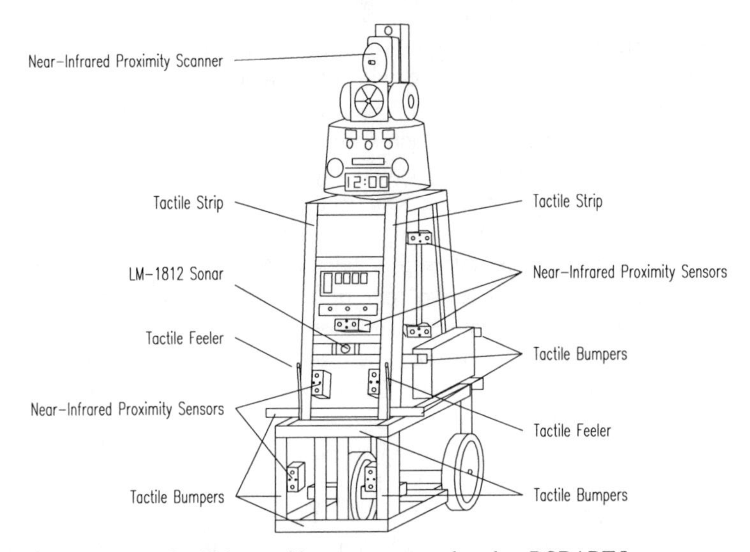

Figure 10-1. Location of collision avoidance sensors employed on ROBART I.

In some ways the software implementation was similar to Brooks' subsumption approach (1986), in that it was a bottom-up design with two distinctly separate hierarchical layers: 1) a *low-level interrupt-driven layer* and 2) an *intermediate-level polling layer* in the main program loop. This layering was basically just an algorithmic differentiation of software categories running on a single processor, however, and limited in actual embodiment to only two layers, although a future higher-level expansion was suggested (Everett, 1982). Brooks, on the other hand, developed a much more powerful and versatile *subsumption architecture,* wherein multiple layers are implemented as additional finite-state machines to support progressively intelligent control.

Those sensors monitoring ROBART's close-in environment (i.e., proximity detectors, feeler probes, bumpers, drive current overload) were considered high priority and consequently read by a maskable interrupt request (IRQ) routine. Unless deactivated by the main program loop, the IRQ routine continuously monitored the sensor output states, and would redirect the motion of the robot in accordance with preprogrammed responses specifically tailored to the individual sensors in alarm. By way of illustration, a preprogrammed response for a right-front bumper impact would consist of the following steps:

- Stop all forward travel.
- Turn steering full right.
- Back up for *x* number of seconds while monitoring rear bumper.
- Stop and center steering.
- Resume forward travel.

Multiple behaviors were incorporated within both the IRQ routines and the main code, arbitrated in accordance with preassigned priorities in the event of conflict. For example, the collision avoidance interrupt service routine would poll all potential inputs to determine which specific device had triggered the interrupt request. Those inputs representing actual impact with an obstacle were naturally ranked higher in terms of polling sequence than inputs associated with "cat-whisker" probes, which in turn had precedence over near-infrared proximity detectors, and so forth. The interrupt service code would initiate the associated canned response for the first active condition discovered by the polling routine, thereby ensuring the higher concern situations received priority attention. The canned avoidance responses in turn would also poll the other inputs to ensure appropriate reaction as depicted in the above example (i.e., monitoring rear bumper while backing).

Whenever a close-in collision avoidance sensor triggered an interrupt, execution of the intermediate level software was temporarily suspended as the controlling microprocessor switched to the interrupt service routine. The low-level avoidance maneuvers would then dominate until the robot was clear of the detected obstruction (quite unlike the subsumption architecture approach). The intermediate level software, on the other hand, had the ability to disable IRQ interrupts associated with collision avoidance sensors, or otherwise suppress or inhibit the lower level behaviors.

A typical example involving suppression as well as complete disabling is illustrated for the case where the robot is homing in on its battery recharging station. The intermediate-level docking software would set a flag that the low-level interrupt service routine always checked before executing canned avoidance maneuvers. If a potential obstacle were encountered during a docking procedure, the normal avoidance response would be suppressed. The robot would instead back up a short distance, then move forward for a predetermined interval at fixed offset from the charger. This action effectively caused the robot to travel along a circular arc to a new relative position before reattempting to dock, thus clearing the intervening obstruction (Everett, 1982). Obstacles are still avoided in this fashion, but without losing sight of the higher-level goal to recharge. Once the robot closed to within a predetermined distance from the charger, all collision avoidance behavior is deactivated to permit physical contact with the charger contacts.

To facilitate somewhat more intelligent movement than the purely reactionary "bump-and-recover" IRQ routines, the intermediate-level software would repeatedly poll the sonar and head-mounted near-infrared scanner on each pass through the main loop. These longer-range sensors were tasked with monitoring the area in front of the robot out to a distance of approximately 5 feet and storing a suitable representation of detected targets in a relative world model as illustrated in Figure 10-2. The *wander algorithm* reacted to the information in the model by choosing the least obstructed direction for continued transit in the event the forward path became blocked. Since all zones were equally weighted in a binary fashion (i.e., either blocked or clear), the least obstructed direction was taken to be that opening defined by the largest number of adjacent clear zones. The inherent simplicity of this modeling scheme enabled real-time on-the-fly response, without the robot having to "stop and think" before continuing on its way.

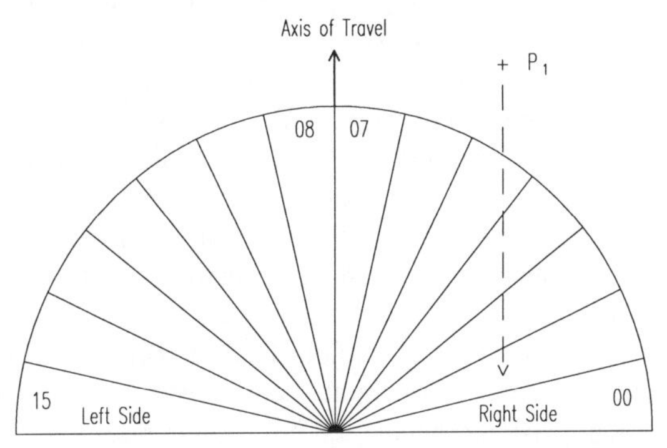

Figure 10-2. The world model employed on ROBART I consisted of sixteen wedge-shaped zones relative to the direction of travel (Everett, 1982).

A fundamental deficiency with this simplistic world representation arises from the fact that a polar model cannot be continuously updated over time to develop a set of certainty values reflecting an increasingly more accurate probability of zone occupancy. For example, a detected obstacle located at point P_1 in the above diagram would transition due to robot motion from *Zone 05* to *Zone 00*, crossing through all zones in between. Repetitive sightings would likely not be associated with the same zone number. As a result, reactions are always made to "snapshot" sensor information subject to numerous sources of potential error, usually resulting in jerky or erratic vehicle movement. Borenstein and Koren (1990a; 1990b) solve this problem by deriving the polar model in real time from a certainty grid representation, as will be discussed in a later section.

Circumnavigation

The term *circumnavigation* describes a collision avoidance behavior in which the robot deflects laterally to "sidestep" an obstruction, while still attempting to move in the general direction of the goal. When the onboard sensors indicate the perceived object is no longer a threat, normal transit is resumed along the desired path. In a sense, *circumnavigation* can be thought of as a *wander* behavior that reverts to a *goal-seeking* behavior when clear, instead of simply resuming straight-line motion.

A good example of a *circumnavigation* collision avoidance behavior is that implemented by Cybermotion, Inc. on their *K2A Navmaster* autonomous vehicle. In normal operation, the *K2A* drive controller calculates a motion vector from its perceived position to some downloaded *X-Y* goal location. This vector is recomputed on the fly as the robot moves so that vehicle heading is continuously reset in accordance with the vector orientation. If a threatening obstacle is detected in the forward path of the robot, speed of advance is reduced and a fixed bias is added to the heading command. The sign of the bias is such that the platform veers away in the direction of free space. Once the obstruction is cleared, the steering bias is removed, and the robot closes on the goal location.

The *circumnavigation* approach has the obvious advantages of simplicity and speed of execution without any requirement for significantly more complex processing power. The technique is obviously limited, however, in that only relatively minor incursions into the intended path can be surmounted in this fashion. Any obstacle that significantly blocks the desired route can potentially deflect the vehicle too far from its intended trajectory for normal path resumption to occur. In addition, circumnavigation techniques must always progress forward in the general direction of the goal without backtracking. No provision is made for choosing alternate routes if the original path is completely blocked.

In addition to being specific behaviors, both *wander* and *circumnavigation* can also be considered as stand-alone *collision avoidance control strategies*. (We're touching now upon that apples and oranges problem I mentioned earlier.) The following sections deal with additional examples of control strategies for collision avoidance (and other purposes) that are capable of implementing not only *wander* and *circumnavigation* but various other behaviors as well.

Potential Fields

The concept of *potential fields* was introduced by Krogh (1984) for simulations of localized mobile robot control, and by Khatib (1985) for manipulator control using Cartesian as opposed to joint coordinates (Tilove, 1990). The classical approach involves an artificial force acting upon the robot, derived from the vector summation of an attractive force representing the goal and a number of repulsive forces associated with the individual known obstacles (Tilove, 1990):

$$\vec{F}_t(x) = \vec{F}_o(x) + \vec{F}_g(x)$$

where:

$\vec{F}_t(x)$ = resultant artificial force vector

$\vec{F}_o(x)$ = resultant of repulsive obstacle forces

$\vec{F}_g(x)$ = attractive goal force.

The attractive goal force can be classically represented as (Tilove, 1990):

$$\vec{F}_g(x) = Q_{goal} \frac{\vec{x} - \vec{x}_g}{|\vec{x} - \vec{x}_g|}$$

where:

Q_{goal} = a positive constant (i.e., the "charge" of the goal).

The *classical potential field* is the summation of the attractive goal force and the repulsive force contributions from those directions defined by the various fields of view of the obstacle detection sensors. The individual repulsive forces are aligned away from their respective obstacles and towards the robot, falling off with the *k-th* power of separation distance (Tilove, 1990). For example, an early MIT implementation on the robot depicted in Figure 10-3 treated each detected sonar target as the origin of a repulsive force decaying as the square of the indicated range (Brooks, 1986). The desired vehicle heading was represented as an attractive force. The resultant of all such virtual forces acting on the robot, if greater than a predetermined threshold, was used to compute the instantaneous drivemotor commands for steering and velocity, effectively moving the platform away from obstacles and in the general direction of the goal.

Figure 10-3. Brooks (1986) applied the *classical potential field* method to an autonomous mobile robot equipped with a ring of 12 Polaroid ranging sensors (courtesy MIT AI Lab).

Alternatively, Arkin (1992b) uses a repulsive force magnitude that is a linear function of obstacle range:

$$O_m = G\,\frac{S-d}{S-R} \quad \text{for } R < d \leq S$$

where:

O_m = magnitude of repulsive force associated with obstacle
G = gain constant
S = sphere of influence from center of obstacle (i.e., $O_m = 0$ for $d > S$)
d = distance from center of obstacle to robot
R = radius of obstacle.

In both of the preceding examples, since the resulting field depends only upon the relative positions of nearby obstacles with respect to the robot, it is possible for repulsive forces to be generated by objects that in fact do not lie along the intended path of travel. Such a situation is illustrated in Figure 10-4A below.

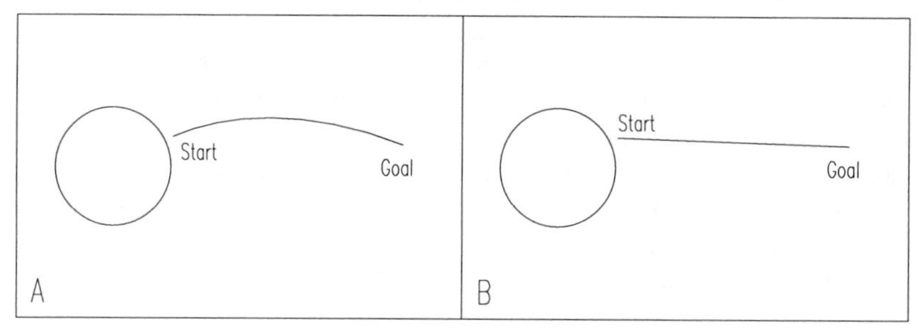

Figure 10-4. (A) The *classical potential field* method considers only separation distance, causing the robot to deviate from a straight-line path segment even though moving away from the circular obstacle; (B) the *generalized potential field* method (see below) considers relative velocity in addition to separation distance (adapted from Tilove, 1990, © IEEE).

In recognition of the above concerns, Krogh (1984) had introduced the concept of *generalized potential fields,* wherein the potential field intensity is a function of not only relative position with respect to obstacles but also the robot's instantaneous velocity vector at that position. The *generalized potential* is the inverse of what Krogh calls the *reserve avoidance time.* Consider a robot approaching a stationary object at some constant velocity V_o as illustrated in Figure 10-5. There is some maximum allowable deceleration rate a_{max} that will bring the robot to a halt in the shortest possible length of time t_1. Similarly there is some minimum deceleration rate a_{min} that will cause the robot to stop just before impact over some longer time interval t_2. *Reserve avoidance time* is simply the difference in time required to stop for the two cases of maximum-allowed versus minimum-required decelerations, (i.e., t_2 - t_1). The *generalized potential field* is thus sensitive to time to impact as opposed to separation distance (Figure 10-4B), and approaches infinity as the *reserve avoidance time* approaches zero (Tilove, 1990).

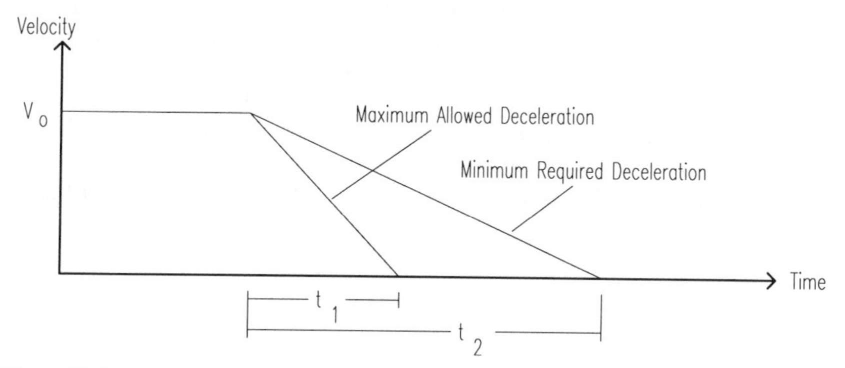

Figure 10-5. Krogh (1984) defines the *generalized potential* as the inverse of *reserve avoidance time*, which is the difference in stopping times associated with *maximum-allowed* and *minimum-required* decelerations (i.e., $t_2 - t_1$).

The principle limitation of the potential field approach is its vulnerability to becoming boxed in or "trapped" by intervening obstacles as illustrated in Figure 10-6. This problem was predicted by Culbertson (1963) for the more general case of "memoryless robots" that react to current stimuli in a deterministic fashion without taking into consideration the results of previous behavior under similar conditions. The likely occurrence of cyclic behavior as well as local maxima and minima make any system that relies solely on the potential-field navigation approach somewhat unreliable (Arkin, 1992a). To get around this problem, Thorpe (1984a; 1984b) employed a grid-based search to find a good low-cost path to the goal, adjusted the path off grid to further minimize costs, then executed the path with a variant of potential fields to keep the vehicle on the path. Krogh & Thorpe (1986) discuss the integration of a generalized potential field collision avoidance scheme with a global path planner based on *certainty grids* (to be discussed in a later section) for optimal route planning and trap recovery.

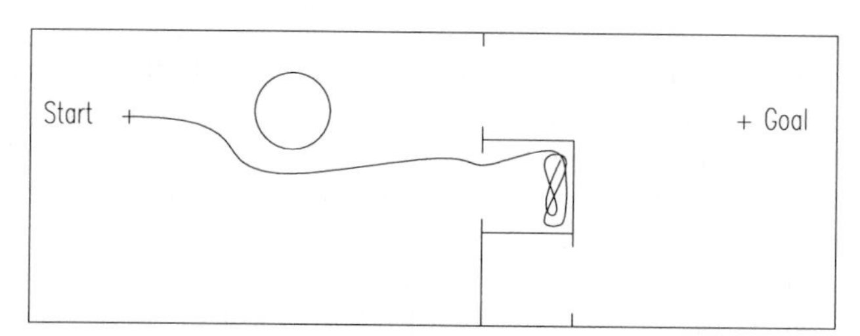

Figure 10-6. The robot successfully negotiated the first obstruction but has become trapped by the U-shaped structure of the closet and is unable to reach the goal in the next room.

Motor Schema

The *motor-schema* navigational approach (Arkin, 1989) developed at the Georgia Institute of Technology, Atlanta, GA, employs a collision avoidance strategy that is very much analogous to potential fields. *Schemas* are basically behavioral primitives, or more specifically, parameterized motor behaviors and their associated perceptual strategies (Arkin & Murphy, 1990), which in turn are denoted *motor schemas* and *perceptual schemas*, respectively. The *Motor-Schema Manager* (Figure 10-7) orchestrates the appropriate interaction of the various *schemas* to achieve intelligent goal-driven actions in a coordinated fashion.

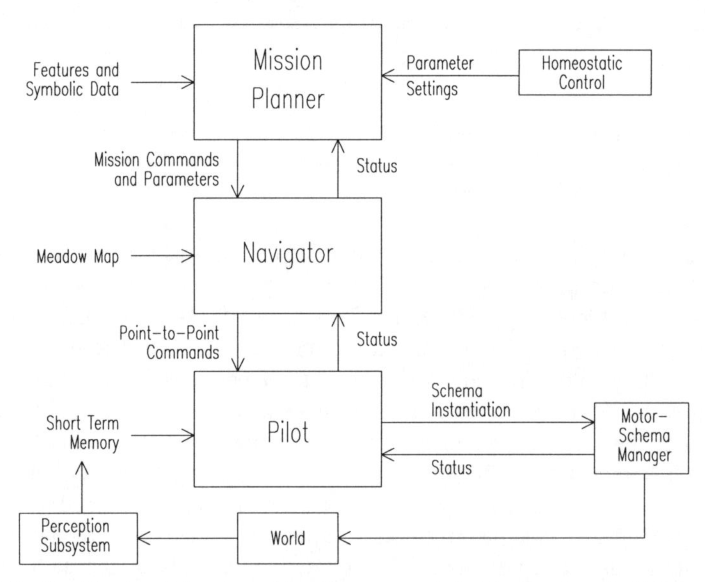

Figure 10-7. The Planning Subsystem of the *Autonomous Robot Architecture (AuRA)* developed by the Georgia Institute of Technology incorporates both a hierarchical planner (Mission Planner, Navigator, and Pilot) and a distributed control plan executor known as the Motor-Schema Manager (Arkin & Murphy, 1990).

Lyons (1986) defines a *schema* as "a generic specification of a computing agent." Each schema represents a general behavior that is instantiated when a copy of the generic specification is parameterized and activated. A collection of such schemas provide the potential family of actions for control of the robot. For example, initial schemas implemented at Georgia Tech on a *Denning Research Vehicle* included: 1) *move-ahead*, 2) *avoid-static-obstacle*, 3) *move-to-goal*, and 4) *stay-on-path* (Arkin & Murphy, 1990). The output of a schema is a single velocity vector reflecting the resolution of all potential field influences experienced by the robot at any given location, and this single vector is used to compute the desired trajectory in real time.

Referring again to Figure 10-7, the *Pilot* implements the desired path generated by the *Navigator* in a piecewise fashion by passing the appropriate selections of both sensing strategies and motor behaviors to the *Motor-Schema Manager* for instantiation. As the robot executes the path, the cartographer builds up a model of surrounding obstacles as perceived by the assigned sensors. If any detected obstacle threatens traversal of a specified path segment, the *Pilot* and *Schema Manager* coordinate in an attempt to avoid the obstruction and resume safe transit along the desired route. Should the resulting path trajectory deviate substantially from the originally specified path, the *Navigator* will be reinvoked to compute a new global path that takes into account the recently acquired sensor data and associated updates to the world model. This fall-back feature also eliminates the common tendency for conventional potential-field approaches to become boxed in or cyclically unstable.

When a suspected obstruction is detected, the instantiated motor schema associated with that particular obstacle begins to produce a repulsive force that deflects the robot away. The magnitude of the deflection vector is a direct function of the certainty of the obstacle's existence. The commanded velocity and direction of the robot is derived from the normalized vector addition of the individual output vectors for all of the active schema instantiations. No arbitration between competing behaviors is involved (Arkin, 1992a). The simplicity of this summation approach and the fact that each schema is a distributed computing agent (preferably running on separate processors on a parallel machine) combine to enable real-time robust performance while the platform is in motion (Arkin, 1989). A learning capability is realized by allowing the system to determine its own schema gains and parameter values, within preset bounds (Clark, et al, 1992).

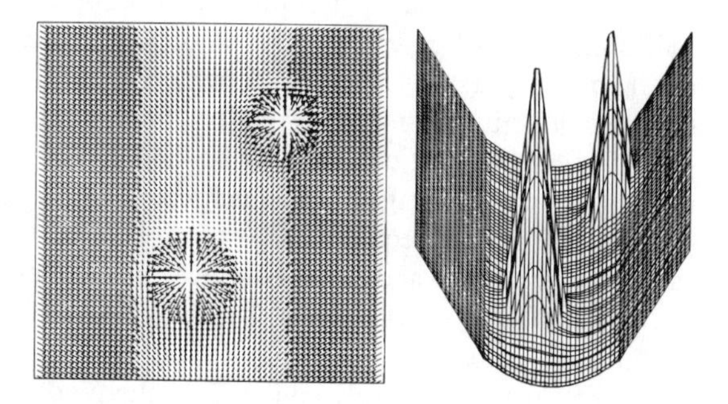

Figure 10-8. Two *avoid-obstacle* plus one *stay-on-path* plus one *move-to-goal* schema instantiations are depicted (left) in 2-D vector representation and (right) in 3-D analog (Arkin, 1989). The entire potential field (shown here for clarity) is not calculated, only each behavior's individual contributions.

10.1.2 Representational World Modeling

Representational world modeling involves the coupling of appropriate non-contact ranging sensors with some type of absolute world modeling capability. Traditional efforts reflected in the literature can generally be decomposed into the following subelements:

- Collecting sensor data on surrounding objects.
- Building an appropriate world model.
- Planning the desired path.
- Avoiding obstacles while en route.

The principle advantage of such an approach over purely reactive strategies is that finding a clear path to the goal is guaranteed, provided such a path exists.

One of the simplest absolute world model representations is a two-dimensional array of cells, where each cell in the array corresponds to a square of fixed size in the region being mapped. Free space is indicated with a cell value of zero; a non-zero cell value indicates an object. The most compact form of a cell map consists of one bit per cell, and thus indicates only the binary presence or absence of an object. By using multiple bits per cell, additional descriptive information can be represented in the map, including the probability of a given square being occupied. This feature is useful when the precise location of an object is unknown, and will be discussed in more detail in the following subsection. Memory usage is independent of map content, so cluttered surroundings are not a problem. However, the resolution of the map is only as good as the square size, and doubling the resolution quadruples the memory requirements.

A slightly more sophisticated and elegant approach is to use a quadtree representation (Fryxell, 1988). Each map begins as a square that is subdivided into four smaller squares. Each of these squares is in turn recursively subdivided (down to the map resolution if necessary) until the region occupied by the square is homogeneous (all object or all free space). For an uncluttered environment, a substantial savings in memory usage is achieved, with a decrease in find-path execution time, since the effective map size is smaller. In highly inhomogeneous environments however, memory usage can increase beyond that of the simple cell map, thus negating the primary advantage of the quadtree. Octrees can be used if a three-dimensional representation is required.

A third technique uses polyhedra and curved surfaces or geometric primitives to represent objects in the workspace (Lozano-Perez & Wesley, 1979; Brooks & Lozano-Perez, 1983). Such maps are quite compact, and with no inherent *grid*, the locations of the objects can be precisely entered into the model. These maps are also easily extended into three dimensions, in contrast to the *cell map* where memory cost would be prohibitive. However, updating the map with real-world data is difficult, as it is hard to accurately glean polygonal information from inexpensive sensors mounted on a mobile robot (Gilbreath & Everett, 1988). Statistical uncertainty of the existence of objects is difficult to implement as well.

Regardless of the particular map representation employed, target distance information must be acquired and entered into the world model as the robot is moving. This seemingly trivial operation turns out to be somewhat difficult due to problems associated with the operation of ultrasonic ranging systems in air. These problems include temperature dependence, which has a significant impact on range accuracy, and beam dispersion, which contributes to angular uncertainty. Specular reflections from target surfaces can cause additional problems. Adjacent sensor interaction requires the transducers in the array be individually fired in sequence rather than simultaneously. Finally, the slow speed of sound in air yields marginal update rates, resulting in significant displacements due to robot motion during the sequential firing of all transducers in the array. Consequently, effective interpretation of range data is critical to achieve a reasonably accurate representation of surrounding obstacles.

Certainty Grids

Moravec and Elfes (1985) of CMU describe a scheme for mapping imprecise sonar range returns into certainty grids using probability distribution functions. For each sensor reading, the assigned probability of an object being at the exact indicated range and bearing decreases radially from a maximum value at that point, according to a specified distribution function. In addition, a second distribution function characterizes the "emptiness" of cells between the sensor and the returned range. Points near the sensor have a high probability of being unoccupied, with decreasing probability for those points closer to the indicated range value or displaced from the beam centerline. The CMU technique is applied to a map where the state of occupancy for all cells is initially marked as unknown. The robot is moved to various vantage points in the room; several sonar readings are taken at each point and averaged to create the probability map. The robot thus creates its own map in an exploratory fashion but must stop periodically to take sensor readings.

Fryxell (1988) also uses a probability scheme for mapping sonar data, modeling the sonars as rays and taking several range readings from different vantage points in the robot's operating environment. Two arrays are constructed in memory, one observing the number of times each cell was "hit" and the other observing each time a cell was "missed." A voting procedure combining both maps is then used to create the final representation, where each cell is marked as either occupied or unoccupied.

Beckerman and Oblow (1988) use a similar method but model the sonar beam as a cone subtending an angle of 18 degrees. The reduced effective beamwidth (18 degrees versus 30 degrees) is achieved by employing the four-transducer phased-array system developed by Kilough and Hamel (1989) for use on the robot HERMIES-IIB. The head-mounted array is sequentially repositioned in azimuth to achieve the desired coverage. As with Fryxell (1988), the robot is moved to various points in the room to make sonar observations. These data are saved in auxiliary buffers and used to update a cumulative map, with each cell labeled as conflicting, unknown, occupied, or empty. (A conflicting cell occurs when one or more sonar readings intersect such that one marks the cell as occupied while the other marks it as empty.) After all the non-conflicting data has been integrated into the cumulative map, the original saved data for each observation point are used to resolve the status of the remaining conflicting cells through pattern analysis. This technique generates maps similar to those created by Fryxell's method, but with better resolution even though fewer sonar readings are taken.

A faster and less computationally expensive variation of these procedures was implemented by Gilbreath (1988) on ROBART II. By using a simplified probability scheme and range-gating a fixed array of sonar sensors, the mapping process can take place in real time while the robot is in motion. When entering range data into the model during actual execution of a path segment, only the center seven transducers in the lower sonar array are activated. If a given sensor reading indicates an object is within 5 feet, the cell at the indicated location is twice incremented (up to a specified maximum). The probability value assigned to each of the eight neighboring cells is incremented once, to partially take into account uncertainties arising from the 30-degree dispersion angle of the ultrasonic beam. (Borenstein and Koren (1990b) carry this simplification one step further by eliminating the probability distribution altogether and incrementing only a single cell on the beam centerline at the indicated range.)

In addition, each time a sonar range value is processed, all the cells within a cone 10-degrees wide and 4-feet long (or less if an object appears within 4 feet) have their assigned values decremented by one. This procedure erodes objects no longer present and also serves to refine their representation as the robot approaches. If the probability is reduced to zero, the cell is again regarded as free space. *Transient objects* are erased from the map at a slightly slower rate than they are entered, so the system tends to err on the side of avoiding obstructions. As with object addition, *permanent objects* and *growth* are left untouched.

Early bit-mapped collision avoidance approaches involved the development of a second localized *relative* map which represented the relative locations of objects detected in front of the robot by onboard sensors while traversing a path segment (Crowley, 1985; Harrington & Klarer, 1987; Everett, et al., 1988). When range to an obstacle fell below a critical threshold, robot motion was halted and a path around the obstacle was planned, using the smaller relative map. In this approach, however, the relative map is very transitory in nature, created at the beginning of each move and discarded at the end. The only information in the map is that obtained from range sensors while the robot is in motion. Since there is no memory of previously encountered obstacles, no learning curve exists, and several avoidance maneuvers may be required to complete the path if the area is congested.

The real-time mapping procedure employed on ROBART II, however, encodes the position of newly detected transient objects into the original *absolute* map while a path is being executed. This scheme has the advantage that all previous information about the environment is also available. All collision avoidance sensor information is statistically represented, based on the number of times something was seen at a given cell location. Figure 10-9 shows a three-dimensional bar-chart depiction of such a map created by a second-generation version of the collision avoidance software that was ported over to the MDARS Interior robot (Everett, et al., 1994). The height of each vertical bar is proportional to the probability that the given cell is occupied.

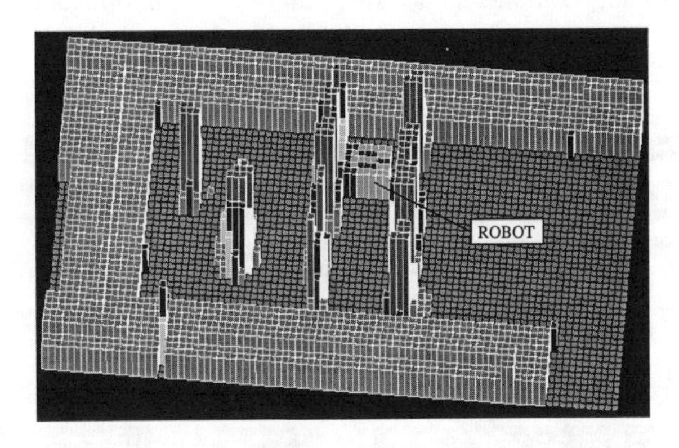

Figure 10-9. Three-dimensional probability distribution resulting from maze traversal during formal collision avoidance technical feasibility testing of the MDARS Interior robot (courtesy Naval Command Control and Ocean Surveillance Center).

The distinction between *permanent* and *transient* objects is an important feature largely responsible for the robust nature of the modeling scheme. *Permanent* objects remain in the model as a baseline from which to restart if the model for some reason becomes overly congested and must be flushed; only the *transient* objects are deleted. Only relatively immobile objects such as walls, desks, filing cabinets, etc. are recorded during the initial map generation procedure. *Permanent objects* are created under human supervision and cannot be later erased by the robot during path execution. *Transient objects* (i.e., chairs, trash cans, carts) are not recorded during the original map-building evolution and present a problem during actual path execution (hence the need for an effective collision avoidance capability).

Each object in the map is automatically *grown* by half the width of the robot in order to model the robot as a dimensionless point during subsequent find-path operations (Lozano-Perez & Wesley, 1979). The path planner will always avoid permanent objects and their associated growth, whereas the algorithm can "eat through" temporary growth surrounding *transient* objects in an attempt to find a path. This ability was found to be necessary, as in congested environments the growth operation often closes off feasible paths due to inaccuracies inherent in the range data. The cost of traversing *transient* growth increases linearly in the direction of the associated object to minimize chances of a collision.

In the original implementation of this collision avoidance strategy on ROBART II, sonar range data was transmitted in real time over an RF datalink to the world modeling software running on an *80386*-based desktop PC. The MDARS Interior program requirement to monitor twelve or more robots from a common host console (Everett, et al., 1993) made this scheme somewhat impractical. As a consequence, the concept of operation was modified slightly to eliminate continuous transfer of sonar data from the various remote platforms to the host. Instead, position-stamped range and bearing information collected over the last 10 feet of travel are stored in a circular buffer on board each MDARS platform.

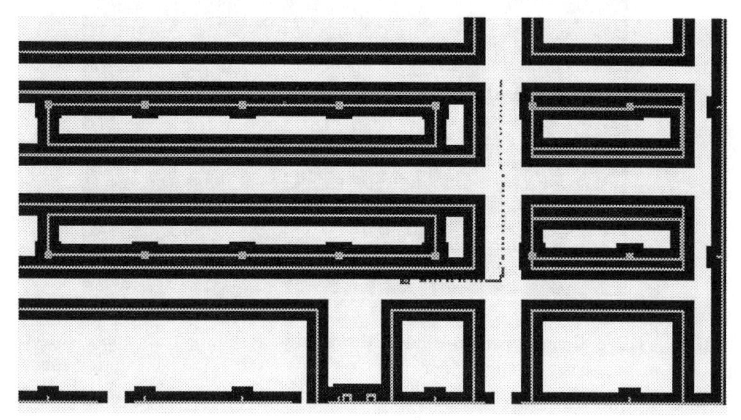

Figure 10-10. Screen dump from the MDARS Planner after an upload of historical collision avoidance data from the robot's circular buffer. Dots corresponding to position stamps for collected data can be used to recreate the path of approach (courtesy Naval Command Control and Ocean Surveillance Center).

If the on-board software determines a threatening object has entered the protected envelope in the direction of travel, the robot is halted and a "blocked" status is reported to the host control architecture. A planner resource is then assigned by the host to resolve the problem. The historical sonar and proximity sensor data are uploaded from the platform after the fact and used to update the world model for path planning purposes. The resultant avoidance maneuver is then downloaded to the robot for execution. This approach allows for a limited number of planner resources to be shared by a large number of robots but is somewhat less than optimal in the sense that the robot must stop and wait while an avoidance maneuver is being generated. One method for overcoming this inconvenience is discussed in the next section.

10.1.3 Combined Approach

In addition to the *reactive control* and *representational world modeling* schemes discussed in the preceding sections, there are also some interesting and innovative implementations employing combinations of both methodologies to achieve more robust operation in dynamic environments.

Vector Field Histogram

As was implied earlier, the *vector field histogram (VFH)* technique developed at the University of Michigan (Borenstein & Koren, 1990a; 1990b) is a combination of both Cartesian and polar representations. This hybrid approach allows for accumulated versus "snapshot" data interpretation in building the model, while exploiting the "on-the-fly" response capability of a polar representation. The VFH method does not reduce the perceived obstacle field to a single resultant vector, and thus allows object distribution to be taken into account in choosing the appropriate avoidance maneuver (Wolfensberger & Wright, 1993). This feature allows the robot to enter narrow passages that would otherwise be inaccessible with a *potential field* collision avoidance strategy.

Raschke and Borenstein (1990) claim more accurate representation of surrounding obstructions is achieved using a *histogramic probability* scheme as opposed to alternative methods employing an assumed probability function. Only a single cell (on the beam centerline at the indicated range) is incremented for each sonar reading (i.e., no probability distribution function is applied to neighboring cells). The rapid sampling of each sonar sensor while the vehicle is moving, however, creates a temporal or *histogramic probability* distribution, wherein recurring sightings cause occupied cells to achieve high certainty values over some finite time frame (Borenstein & Koren, 1990b). The individual *certainty values* represent an assigned *level of evidence* for the existence of an obstacle at that particular grid location.

The *certainty values* from the X-Y *histogram grid* (which is absolute and does not translate with robot motion) are repeatedly mapped over into a *polar histogram* that subsequently reflects the probability of obstruction as a function of relative bearing (Figure 10-11). To simplify this procedure, a smaller window of n-by-n cells symmetrically overlays the vehicle centerpoint, defining a dynamic subset of the *histogram grid* known as the *active region*. Only those cells contained within this *active region* (appropriately denoted as *active cells*) can influence the avoidance behavior of the robot.

A 360-degree *polar histogram* comprised of an integral number of identical pie-shaped sectors is generated around the vehicle's current location, with each sector assigned a variable representing the *polar obstacle density* in that particular direction. To establish the individual values of this circular array, the *certainty value* associated with each *active cell* is treated as an *obstacle vector* oriented along an imaginary line running from the midpoint of the cell to the vehicle centerpoint. The magnitude of this *obstacle vector* is calculated using the equation (Borenstein & Koren, 1990a):

$$m_{i,j} = (c^*_{i,j})^2 \left[a - b d_{i,j}\right]$$

where:

$m_{i,j}$ = magnitude of obstacle vector at cell (i,j)
$c^*_{i,j}$ = certainty value of active cell (i,j)
$d_{i,j}$ = distance between active cell (i,j) and vehicle centerpoint
a, b = positive constants.

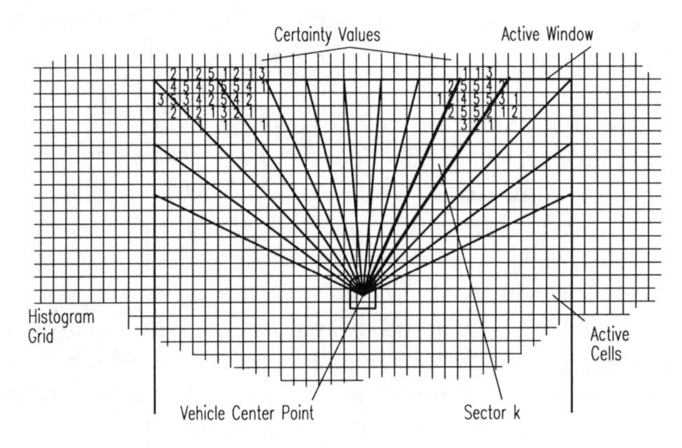

Figure 10-11. *Active cells* from a 33-by-33-cell *active region* of the Cartesian *histogram grid* are mapped into the *polar histogram* to facilitate real-time response (adapted from Borenstein & Koren, 1990a, © IEEE)

The intensity of the avoidance reaction is thus tailored to the square of the magnitude of the *level of evidence* that a perceived object is actually present (Borenstein & Koren, 1990b) and falls off linearly with increasing distance *d*. Clusters of neighboring non-zero cell values are summed together to yield a composite *obstacle cluster strength* that presents an even stronger measure of occupancy for multicell representations indicative of real targets (Borenstein & Koren, 1990a):

$$h_k = \sum_{i,j} m_{i,j}$$

where:

h_k = polar obstacle density for sector *k*.

Any random noise appearing as unreinforced sightings (i.e., a single unclustered cell) is basically ignored by the avoidance algorithm. Those sectors for which the associated *polar obstacle densities* fall below an assigned threshold are suitable candidate headings for robot motion, with the logical choice being the heading most closely matching the direction of the desired goal. The magnitude of *polar obstacle density* in the direction of forward travel is indicative of anticipated congestion and can be used accordingly to establish an appropriate speed of advance.

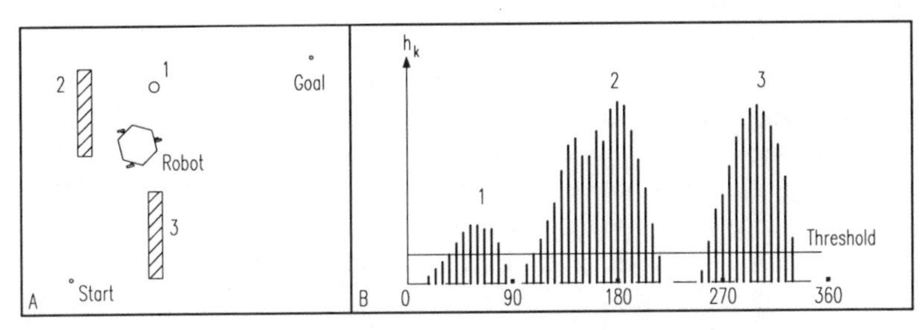

Figure 10-12. Three detected obstacles (A) consisting of two partitions and a vertical pole produce the histogramic probability distribution shown in (B) plotted as a function of relative bearing (adapted from Borenstein and Koren, 1990a).

The *VFH* algorithms were initially implemented at the University of Michigan on a modified Cybermotion *K2A* platform equipped with an onboard *80386*-based PC-compatible computer. An add-on ring of 24 Polaroid ultrasonic transducers (Figure 10-13) provides 360-degree coverage out to 200 centimeters, with a system update rate of around 6 Hz (i.e., 160 milliseconds required to update all sensors). The following steps occur on each pass through the control loop (Borenstein & Koren, 1990a):

- The most recent sonar data is read from the ranging modules.
- The *Histogramic Certainty Grid* is updated.
- A new *Polar Histogram* is created.
- The free sectors and steering direction are determined.
- The maximum allowable speed command is calculated.
- Speed and steering commands are passed to the low-level drive controller.
- Vehicle navigation parameters (X, Y, θ) are received from drive controller.

The above actions repeat every 27 milliseconds, enabling robust real-time operation. Consistent results were achieved in successfully avoiding obstructions in cluttered environments at platform speeds of up to 0.78 meters per second (Borenstein and Koren, 1990a; 1990b).

The success of the *VFH* approach in moving the robot to the desired goal position, however, is situationally dependent. Wolfensberger and Wright (1993) cite four sensitivity factors that influence robust performance:

- *Goal position* — The likelihood of becoming trapped when a valid path exists is sensitive to the relative location of the goal with respect to the robot and any obstructions. In other words, the attractive goal force can in certain situations draw the robot into an intermediate trap as illustrated in Figure 10-14A.

- *Threshold level* — The value assigned to the polar histogram threshold can influence the tendency to enter a trap situation (Figure 10-14B), or potentially preclude the robot from reaching the desired goal as illustrated in Figure 10-14C.
- *Sensor range* — The effective range of the sensors determines whether or not a trap situation can be fully assessed. For example, if the sensor cannot see the back wall of a trap of the type illustrated in Figure 10-14A, the algorithm will not realize there is an eventual obstruction, and will subsequently allow the robot to enter.
- *Map scale* — The relative dimensions of the map features (i.e., with respect to sensor range) likewise influence the ability to assess a trap situation. If the trap illustrated in Figure 10-14 is large enough in terms of depth, the sensors cannot perceive the back wall prior to entry.

Figure 10-13. CARMEL, developed by the University of Michigan, incorporates a Cybermotion *K2A Navmaster* base outfitted with a ring of 24 Polaroid ultrasonic rangefinders (courtesy University of Michigan).

As was the case with potential fields, a limitation of the *vector field histogram* approach is its inherent vulnerability to becoming caught in dead-end trap scenarios. In a series of 100 simulation runs conducted by Wolfensberger & Wright (1993), the *VFH* method became trapped and was unable to reach the goal 30 percent of the time. (It should be noted the potential-field method failed in 79 percent of the tests.) This susceptibility to trapping is not an issue if the *VFH* collision avoidance algorithm is used in conjunction with a higher level global planner as was in fact intended (Borenstein & Koren, 1991). The problem becomes a simple matter of detecting that a trap situation has occurred. (Zhao and BeMent (1990) define a *VFH* trap situation as any time the robot's heading exceeds an angle of 90 degrees with respect to an imaginary line drawn from the current position to the desired goal.) The *VHF* algorithm does not attempt to resolve the problem, but instead automatically passes control to the global path planning, which in turn is better suited to the task of optimal trap recovery. This integrated approach provides a very powerful and robust method for real-time response while ensuring an optimal path for goal attainment.

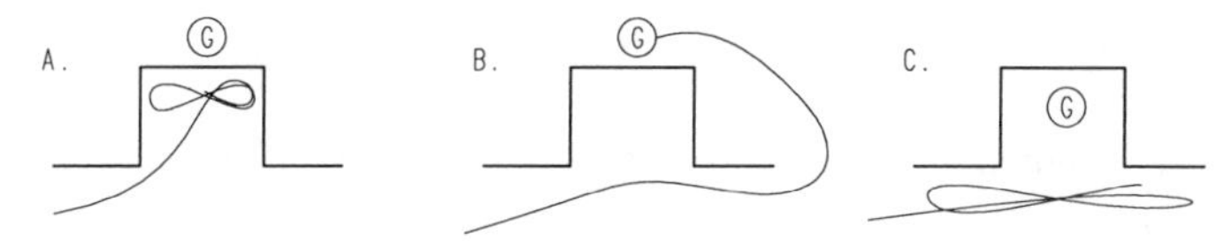

Figure 10-14. The location of the goal in (A) is such that the robot is drawn into an intermediate trap situation, while in (B) the *polar histogram threshold* has been decreased so the proximity of the back wall is above the threshold value. This lower threshold can preclude the robot from reaching the relocated goal in (C) (adapted from Wolfensberger & Wright, 1993).

10.2 Exterior Application Considerations

The collision avoidance problem for exterior applications is much more complex than for interior, even in the case of relatively structured scenarios (i.e., as opposed to open-terrain cross-country operations in unfamiliar territory). An example of such a lower risk category is illustrated by the MDARS Exterior vehicle, which operates almost exclusively on existing facility roadways, for which the associated collision avoidance needs can be subdivided into a number of specific scenarios:

- Fixed obstructions blocking part or all of the roadway.
- Moving obstructions on the roadway.
- Moving obstructions at an intersection (crossroads or train track).
- Potholes or washouts in the roadway.
- Obstructions along the sides of the roadway.
- Hazards (open ditches, lakes, mud bogs, etc.) along the sides of the roadway.

One of the most obvious concerns in fielding an autonomous exterior navigation capability at an industrial site is the need to deal with railroad crossings and roadway intersections. The remote platform must anticipate arrival at such locations in order to “stop and look” before proceeding. The most practical means for detection of oncoming traffic would seem to be Doppler radar and video motion detection.

The added variable which significantly complicates matters in outdoor settings (relative to indoor) is terrain traversability. Wilcox (1994) uses the terminology *non-geometric hazards* to describe pitfalls that cannot be characterized solely by shape, but rather by their properties (such as friction and density, for example, that in turn could adversely impact tire slippage or sinkage). In indoor environments, the floor surface is known in advance and permanent in nature, with the only significant hazards being drop-offs along loading docks and stairwells. Outdoors, this is not the case. Road surfaces can undergo day-to-day as well as seasonal variations in drivability, and hard to detect but potentially hazardous conditions can coexist in close proximity along either side.

This situation introduces two fundamental problems:

- The potential hazard must be detected in time to suitably alter the vehicle's course.
- Some representation of terrain traversability must be encoded within the world model for consideration by the path planning algorithms.

These issues are of little concern in indoor warehouse environments, where it is generally assumed that any areas of potential danger will be readily detected by onboard sensors before an accident can occur. Even if the remote platform is sufficiently disoriented with respect to its true absolute position and orientation, it is generally physically bounded by some type of structure. If the robot wanders too far from its intended location, it will eventually encounter an easily detectable wall or shelf and be halted by the onboard collision avoidance system.

In outdoor environments, however, there is no such bounding structure. Accumulated dead reckoning errors could result in a large enough offset between actual and perceived platform position and heading to where the vehicle could stray off the roadway. Detection of roadway limits is extremely difficult under all weather conditions likely to be encountered, and there is a very real possibility the platform could wind up in a ditch.

Automatic execution of any avoidance maneuver must also consider the fact that other vehicles may be operating on the road section. It is highly probable that conditions will be encountered where an obstacle blocks all or part of the right side of the road, and the required avoidance maneuver by necessity crosses the roadway centerline. Some reliable means of checking for oncoming traffic must precede any automatic execution of the unrestricted path. As in the case of railroad crossings and roadway intersections above, Doppler radar and image processing are strong contenders for this technological need. It must be realized, however, that humans address this issue with the most sophisticated sensors (eyes) and processing resources (brain) in existence, coupled with an extensive database of learned experiences, yet still on occasion make fatal mistakes.

Possible candidates for broad-area first-alert coverage include stereo vision and microwave radar, and a mix of the two is highly desirable from the standpoint of increased likelihood of target detection. Microwave radar is the preferred choice for *intelligent vehicle highway systems* (Siuru, 1994), due to the high reflectivity associated with metal targets (i.e., other vehicles), and the ability to see through obscurants such as fog, rain, or snow. Higher-resolution mapping of target location can also be addressed by stereo vision, with complimentary back-up from narrow-beam ultrasonic sensors (assuming vehicle speed is sufficiently reduced upon initial detection of a potential problem).

In the automotive industry, BMW incorporates four ultrasonic transducers on both front and rear bumpers in its *Park Distance Control* system, but maximum range in this parking-assist application is limited to about 5 feet (Siuru, 1994). This system uses piezoceramic transducers sealed in a membrane for environmental protection, which very likely means ranges much in excess of 5 feet are impractical. Environmental-grade Polaroid transducers have been incorporated in some exterior applications, however, with supposedly good survivability even under adverse conditions, although signal quality and hence performance degrade somewhat in the presence of wind and rain. Alternate possibilities worthy of consideration for the high-resolution mapping task are scanning-laser and structured-light range-finding approaches.

10.3 Navigational Re-referencing

Although reactionary control strategies are unsurpassed in terms of real-time localized control for collision avoidance, purely reactionary control schemes have somewhat limited appeal outside of the laboratory environment. Practical applications generally require some type of global representation to ensure logical and timely attainment of goals. A coupling of global and localized schemes has been suggested and in some cases demonstrated by a number of research groups (Krogh & Thorpe, 1986; Arkin, 1992a; Borenstein & Koren, 1991), and probably represents the optimal approach for most real-world situations for a number of reasons.

The integrity of any world model constructed and refined as the robot moves about its workspace is unfortunately directly dependent upon the accuracy of the robot's perceived location and orientation. Accumulated dead-reckoning errors soon render the information entered into the model invalid, in that the associated geographical reference point for the acquired data is incorrect. As the accuracy of the model degrades, the ability of the robot to successfully navigate and avoid collisions diminishes rapidly, until it fails altogether. For largely this very reason, only a very small handful of autonomous mobile robots have been successfully fielded in real-world applications, despite millions of dollars in developmental efforts over the last several decades.

In a nutshell, except for the case of highly structured environments (such as hospital corridors, for example), it is very difficult to keep from getting lost when operating in the real world. Some reliable and routine means of periodically resetting the navigational parameters (X, Y, and θ) is therefore required if continuous unattended operation is to be sustained. Such methods and their associated sensor needs are addressed next in Chapters 11 through 16.

10.4 References

Anderson, T.L., Donath, M., "Synthesis of Reflexive Behavior for a Mobile Robot Based Upon a Stimulus-Response Paradigm," SPIE Mobile Robots III, Vol 1007, W. Wolfe, Ed., Cambridge, MA, pp. 198-211, November, 1988.

Arkin, R.C., "Motor-Schema-Based Navigation for a Mobile Robot: An Approach to Programming by Behavior," IEEE International Conference on Robotics and Automation, Raleigh, NC, 1987.

Arkin, R.C., Murphy, R.R., "Autonomous Navigation in a Manufacturing Environment," *IEEE Transactions on Robotics and Automation*, Vol. 6, No. 4, pp. 445-454, August, 1990.

Arkin, R.C., "Motor-Schema-Based Mobile Robot Navigation," *International Journal of Robotics Research*, Vol. 8., No. 4, pp. 92-112, August, 1989.

Arkin, R.C., "Behavior-Based Robot Navigation for Extended Domains," *Adaptive Behavior*, Vol. 1, No. 2, MIT, Cambridge, MA, pp. 201-225, 1992a.

Arkin, R.C., "Homeostatic Control for a Mobile Robot: Dynamic Replanning in Hazardous Environments," *Journal of Robotic Systems*, Vol. 9, No. 2, pp. 197-214, 1992b.

Beckerman, M., Oblow, E.M., "Treatment of Systematic Errors in the Processing of Wide Angle Sonar Sensor Data for Robotic Navigation," Oak Ridge National Laboratory Technical Memo, CESAR-88/07, February, 1988.

Borenstein, J., Koren, Y., "High-Speed Obstacle Avoidance for Mobile Robots," IEEE Symposium on Intelligent Control, Arlington, VA, pp. 382-384, August, 1988

Borenstein, J., Koren, Y., "Real-Time Obstacle Avoidance for Fast Autonomous and Semi-autonomous Mobile Robots," American Nuclear Society, Third Topical Meeting on Robotics and Remote Systems, Charleston, SC, CONF-890304, Section 4-4, pp. 1-6, March, 1989a.

Borenstein, J., Koren, Y., "Real-Time Obstacle Avoidance for Fast Mobile Robots," *IEEE Transactions on Systems, Man, and Cybernetics*, Vol. 19, No. 5, pp. 1179-1187, September/October, 1989b.

Borenstein, J., Koren, Y., "Real-Time Obstacle Avoidance for Fast Mobile Robots in Cluttered Environments," IEEE International Conference on Robotics and Automation, Vol. CH2876-1, Cincinnati, OH, pp. 572-577, May, 1990a.

Borenstein, J., Koren, Y., "Real-Time Map Building for Fast Mobile Robot Obstacle Avoidance," SPIE Vol. 1388, Mobile Robots V, Boston, MA, November, 1990b.

Borenstein, J., Koren, Y., "The Vector Field Histogram - Fast Obstacle Avoidance for Mobile Robots," *IEEE Journal of Robotics and Automation*, Vol. 7, No. 3, pp. 278-288, June, 1991.

Brooks, R.A., "A Robust Layered Control System for a Mobile Robot," *IEEE Journal of Robotics and Automation*, Vol. RA-2, No. 1, pp. 14-20, 1986.

Brooks, R.A., Lozano-Perez, T., "A Subdivision Algorithm in Configuration Space for Findpath with Rotation," International Joint Conference on Artificial Intelligence, Karlsruhe, Germany, 1983.

Clark, R.J., Arkin, R.C., Ram, A., "Learning Momentum: On-Line Performance Enhancement for Reactive Systems," IEEE International Conference on Robotics and Automation, Nice, France, pp. 111-116, May, 1992.

Crowley, J.L., "Navigation for an Intelligent Mobile Robot," *IEEE Journal of Robotics and Automation*, Vol. RA-1, No. 1, March, 1985.

Culbertson, J., *The Minds of Robots: Sense Data, Memory Images, and Behavior in Conscious Automata*, University of Illinois Press, p. 50, Chicago, IL, 1963.

Everett, H.R., "A Microprocessor Controlled Autonomous Sentry Robot", Masters Thesis, Naval Postgraduate School, Monterey, CA, October, 1982.

Everett, H.R., "A Multi-Element Ultrasonic Ranging Array", *Robotics Age*, pp. 13-20, July, 1985.

Everett, H.R., Gilbreath, G.A., Bianchini, G.L., "Environmental Modeling for a Mobile Sentry Robot", NOSC Technical Document 1230, Naval Ocean Systems Center, San Diego, CA, January, 1988.

Everett, H.R., Gilbreath, G.A., Heath-Pastore, T.A., Laird, R.T., "Coordinated Control of Multiple Security Robots," SPIE Mobile Robots VIII, Vol. 2058, Boston, MA, pp. 292-305, September, 1993.

Everett, H.R., Gilbreath, G.A., Heath-Pastore, T.A., Laird, R.T., "Controliing Multiple Security Robots in a Warehouse Environment," NASA Conference on Intelligent Robotics in Field, Factory, Service, and Space, Vol. 1, Houston, TX, pp. 93-102, March, 1994.

Fryxell, R.C., "Navigation Planning Using Quadtrees," SPIE Mobile Robots II, Cambridge, MA, pp. 256-261, November, 1987.

Gilbreath, G.A., Everett, H.R., "Path Planning and Collision Avoidance for an Indoor Security Robot," SPIE Mobile Robots III, Cambridge, MA, pp. 19-27, Novemeber, 1988.

Harrington, J.J., Klarer, P.R., "SIR-1: An Autonomous Mobile Sentry Robot," Technical Report SAND87-1128, UC-15, Sandia National Laboratories, May, 1987.

Khatib, O., "Real-Time Obstacle Avoidance for Manipulators and Mobile Robots," IEEE Conference on Robotics and Automation, pp. 500-505, March, 1985.

Kilough, S.M., Hamel, W.R., "Sensor Capabilities for the HERMIES Experimental Robot," American Nuclear Society, Third Topical Meeting on Robotics and Remote Systems, Charleston, SC, CONF-890304, Section 4-1, pp. 1-7, March, 1989.

Krogh, B.H., "A Generalized Potential Field Approach to Obstacle Avoidance Control," Proceedings, SME Conference, *Robotics Research: The Next Five Years and Beyond*, Bethlehem, PA, August, 1984.

Krogh, B.H., Thorpe, C.E., "Integrated Path Planning and Dynamic Steering Control for Autonomous Vehicles," IEEE International Conference on Robotics and Automation, San Francisco, CA, pp. 1664-1669, April, 1986.

Lozano-Perez, T., Wesley, M.A., "An Algorithm for Planning Collision-Free Paths Among Polyhedral Obstacles," *Communications of the ACM*, Vol. 22, No. 10, pp. 560-570, 1979.

Lyons, D., "RS: A Formal Model of Distributed Computation for Sensory-Based Robot Control," Ph.D. thesis, COINS Technical Report 86-43, University of Massachusetts, Amherst, MA, 1986.

Moravec, H. P., Elfes, A., "High Resolution Maps from Wide Angle Sonar," Proceedings of the 1985 IEEE International Conference on Robotics and Automation, St. Louis, MO, pp. 116-121, March, 1985.

Raschke, U., Borentstein, J., "A Comparison of Grid-Type Map Building Techniques by Index of Performance," IEEE International Conference on Robotics and Auutomation, Cincinnati, OH, May, 1990.

Siuru, B., "The Smart Vehicles Are Here," *Popular Electronics*, Vol. 11, No. 1, pp. 41-45, January, 1994.

Thorpe, C., "FIDO: Vision and Navigation for a Mobile Robot," PhD Thesis, Carnegie Mellon University, Pittsburgh, PA, November, 1984a.

Thorpe, C. "Path Relaxation: Path Planning for a Mobile Robot," Proceedings, AAAI-84, National Conference on Artificial Intelligence, University of Texas, Austin, TX, August, 1984b.

Tilove, R.B., "Local Obstacle Avoidance for Mobile Robots Based on the Method of Artificial Potentials," IEEE International Conference on Robotics and Automation, Vol. 1, Cincinnati, OH, pp. 566-571, May, 1990.

Wilcox, B.H., "Non-Geometric Hazard Detection for a Mars Microrover," Conference on Intelligent Robotics in Field, Factory, Service, and Space, NASA Conference Publication 3251, Houston, TX, Vol. 1, pp. 675-684, March, 1994.

Wolfensberger, M., Wright, D., "Synthesis of Reflexive Algorithms with Intelligence for Effective Robot Path Planning in Unknown Environments," SPIE Vol. 2058, Mobile Robots VIII, Boston, MA, pp. 70-81, September, 1993.

Zhao, Y., BeMent, S.L., "A Heuristic Search Approach for Mobile Robot Trap Recovery," SPIE Vol. 1388, Mobile Robots V, Boston, MA, pp. 122-130, November, 1990.

11
Guidepath Following

One of the simplest forms of autonomous platform control is *guidepath following*, where the vehicle reacts to the sensed position of some external continuous-path reference track. *Automated guided vehicles (AGVs)* have found extensive use in factories and warehouses for material transfer, in modern office scenarios for supplies and mail delivery, and in hospitals for distribution of meals and medication to nursing stations. In 1989 about 500 AGVs were sold in the United States, compared to approximately 3000 vehicles purchased by European companies, and an estimated 5000 more by Japanese corporations in the same time period (Cahners, 1990).

Advantages of guidepath control in material handling applications are seen primarily in the improved efficiency and reduction of manpower that arise from the fact that an operator is no longer required to perform the delivery function. AGV systems offer a much more flexible and cheaper alternative to conventional hard automation. Large numbers of vehicles can operate simultaneously without getting lost or disoriented, scheduled and controlled by a central computer that monitors overall system operation and remote vehicle flow. Communication with individual vehicles can be over RF links, modulated near-infrared light beams, or other means.

From a manufacturing or industrial point of view, guidepath following represents a tremendous improvement over hard automation through significantly increased flexibility, in terms of both route modifications and floor space usage. From an autonomous mobile robot perspective, however, the fundamental disadvantage of guidepath control is the lack of flexibility in the system: a vehicle cannot be commanded to go to a previously unserviced location unless the guidepath is first modified. While this requirement is certainly an inconvenience for any desired changes to product flow lines, it represents a significant drawback in the case of a security robot, for example, that must investigate a potential break-in at some arbitrary remote location.

The most common guidepath following schemes in use today involve some type of stripe or wire guidepath permanently installed on the floor of the operating area. Specialized sensors mounted on the front of the platform are used to servocontrol the steering mechanism, causing the vehicle to follow the intended route. For purposes of this discussion, such guidance schemes can be divided into three general categories: 1) those which sense and follow the audio or RF field from a closed-loop wire embedded in the floor, 2) those which optically sense and follow some type of stripe affixed to the floor surface, and 3) those which sense a permanent-magnet pathway. A very general comparison of relevant features is provided in Table 11-1 below.

Table 11-1. General comparison (there will be some exceptions) of common guidepath schemes (adapted with changes from Kamewaka & Uemura, 1987).

Salient Feature	Embedded Wire	Optical Stripe	Magnetic Tape
Relative installation cost	High	Low	Low - medium
Maintenance cost	Low	High	Low - medium
Flexibility	None	High	Medium
Communication channel	Yes	No	No
Longitudinal markers	Intersections only	Yes	Yes
Branching	Yes	Yes	Yes
Passing	No	No	No
Breakage tolerant	No	Yes	Yes
Works outdoors	Yes	No	Yes

11.1 Wire Guided

Wire-guided systems represent a practical and proven technology for those AGV applications that enjoy long-term stability in terms of both product flow and equipment location, with higher positioning accuracies (< 0.25 inches typical) than generally achievable using other methods. An inherent safety characteristic is the fact the vehicle will stop automatically should the guidance signal disappear, a feature often used to ensure AGVs do not enter a particular area or load handling station until the appropriate time. By combining off-wire dead-reckoning capabilities with fixed wire guidepaths, a high degree of flexibility can be achieved. Figure 11-1 shows a representative *wire-guided* system installed by Control Engineering, Harbor Springs, MI, for delivery of food and medical supplies in hospital environments.

The principle of operation is illustrated in Figure 11-2 below. A pair of inductive pick-up coils is arranged on the bottom of the vehicle chassis on either side of centerline. The signal amplitude induced in these sense coils due to the alternating magnetic field generated by current flow through the wire guidepath is an inverse function of the coil-to-wire separation distance squared. Any difference in amplitude between the left and right sensor signals can thus be used to servo vehicle heading in order to maintain a balanced output condition. Multiple adjacent wire runs excited at different operating frequencies (generally within the 2- to 10-KHz range) are sometimes employed to facilitate path branches. AGV Products, for example, uses 2575, 3433, 4390, and 5240 Hz in their four-frequency path installations (AGV, 1993). No inherent provisions are readily available for resolving longitudinal position along the route from the guidewire itself, other than detecting cross wires at intersections, and so most systems rely on complementary optical or magnetic sensors for this purpose if required.

Figure 11-1. This wire-guided system is used to deliver meals and medical supplies to nursing stations at Balboa Naval Hospital (courtesy Naval Medical Center, San Diego, CA).

As an added benefit, RF-modulated information and vehicle commands can be relayed to the AGV over the wire itself. (Since no realistic capability exists to similarly support AGV-to-host communications, this technique is not generally employed in the more sophisticated multivehicle installations, which rely instead on full-duplex RF or optical datalinks.)

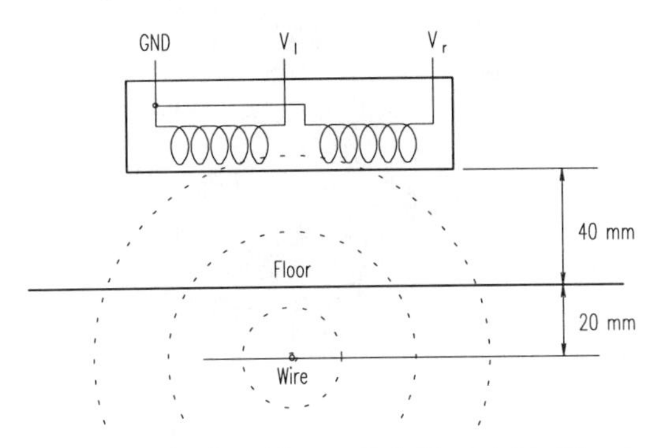

Figure 11-2. Use of a pair of inductive pick-up coils is illustrated in this schematic drawing of the *ANT10* guidepath antenna manufactured by AGV Products, Inc. (courtesy AGV Products, Inc.).

For repetitive AGV operations that involve few if any changes to the guidepath, wire-guided technology is hard to beat from the standpoint of reliability. Initial installation is expensive and then difficult to alter, but the individual vehicle costs are relatively low due to the simplicity of required onboard equipment. Cracks in the flooring due to settling has been known to cause wire breakage, however, which is both disruptive and expensive to repair (Guidoni, 1992). Current installation techniques usually incorporate a styrofoam rope between the wire and epoxy or grout sealer to allow freedom of motion in the event of slab expansion or settling, thus eliminating this problem.

Though traditionally used indoors, there have been some very successful applications of wire-guided technology in rather demanding exterior settings. Barrier Systems, Inc., Carson City, NV, has successfully applied the concept in the control of a large automated vehicle that repositions concrete highway barriers along the 1.7-mile bridge connecting San Diego with Coronado Island (Figure 11-3). Moving at speeds up to 5 miles per hour, the patented system picks up and laterally displaces 2,500 of these barrier sections a distance of 6 feet (four times a day) to optimize lane usage during rush-hour traffic (Murray, 1994). The 25-ton front- and rear-steered hydrostatic-drive vehicle relies on wire guidance to ensure proper placement of the mechanically linked barrier sections, with a path-tracking accuracy of 2.5 centimeters. A similar wire-guided system is in operation along a 10-mile section of the R.L. Thornton Freeway (Interstate 30) in Dallas, TX.

Figure 11-3. The *Quickchange® Moveable Barrier* system uses buried-wire guidepath control to automatically relocate concrete barrier sections on San Diego's Coronado Bridge for optimal lane usage during rush-hour traffic (courtesy Barrier Systems, Inc.).

11.2 Optical Stripe

Optical stripe-following schemes have evolved over the past several decades as favorable alternatives to wire-guided schemes for those less structured applications requiring more flexibility in terms of path additions and/or modifications. Various implementations of the concept exist, including the most elementary case of a high-contrast (dark-on-light, light-on-dark) line, a retroreflective tape illuminated by an onboard light source, and a chemical stripe that glows when irradiated by ultraviolet energy. A number of marker and barcode schemes are employed to yield localized longitudinal position and branching information along the path of travel. Petriu (1991) describes a unique although somewhat impractical implementation employing three parallel tracks (Figure 11-4) to provide continuous absolute position encoding over the full length of the path.

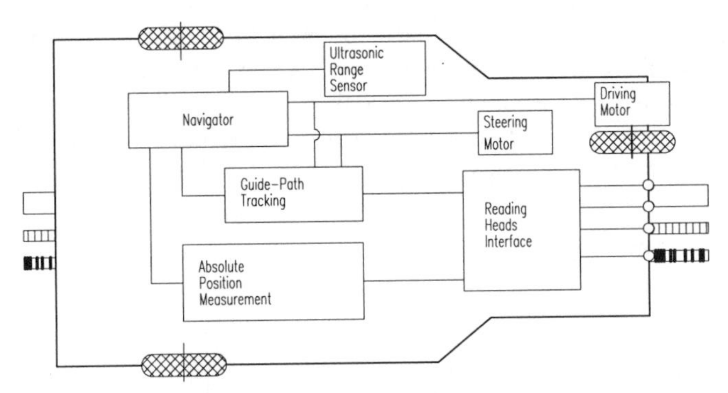

Figure 11-4. Four optical sensors on this experimental AGV follow a three-stripe guidepath that yields continuous absolute position information along the route (adapted from Petriu, 1991, © IEEE).

Most optically based systems have the ability to follow either the left edge or the right edge of the stripe on command, a capability that facilitates branches down alternate paths. Installation problems arise in the case of curved pathways when conventional retroreflective tape is employed as the guidepath stripe. (Retroreflective tape comes on rolls and has little lateral flexibility.) Tsumura (1986) proposed the requirement for curved tape sections of varying radii as a solution in place of piecewise approximation. This problem is of course avoided in systems that follow a painted guidepath.

Optical guidepaths are much easier to install and more flexible in terms of subsequent changes to the route layouts than embedded-wire schemes. Since the guidepath is passive, there is essentially no limitation on route length. Unlike wire-guided systems, any small breaks or discontinuities in an optical path are of little significance, but the tradeoff is recurring installation and maintenance costs due to the exposed nature of the stripe. Some users claim the inherent visibility of the stripe is an advantage from the standpoint of alerting humans to the route, and as a consequence fewer obstructions are inadvertently left to block the vehicle. The same effect could be achieved, of course, by painting lane markings on the floor for embedded-wire runs, but it involves an extra expense. Another disadvantage associated with optical stripe followers (relative to wire-based schemes) is seen in the potential for path occlusion by minor debris inadvertently left on the floor (i.e., packing materials, trash, paint spills, metal shavings).

11.2.1 ModBot Optical Stripe Tracker

The prototype stripe follower developed for use on the ModBot is based on a near-infrared analog sensor module (P/N C5-1DN05) manufactured by Banner Engineering (Banner, 1993a, 1993b), tracking a 1-inch-wide retroreflective tape (P/N BRT-THG-1-100). The sensor head consists of four of these modules arranged in an array to yield a 4-inch footprint of illumination on the floor as shown in the block diagram of Figure 11-5. All sensor modules are active when the system is in the acquisition mode looking for the stripe, but only the center two are used in the tracking mode once the stripe has been located and the robot is centered on the guidepath.

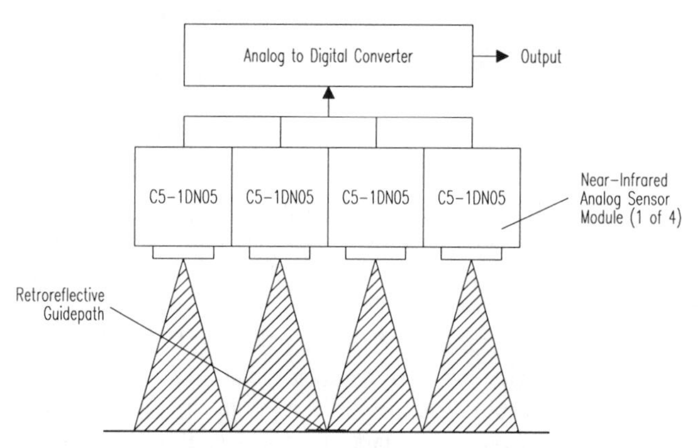

Figure 11-5. The prototype stripe following module used on the ModBot tracks a 1-inch-wide retroreflective tape affixed to the floor.

The initial design called for two additional sensors with fan-shaped beam patterns to provide a binary indication of stripe presence out to 4 inches on either side of the coverage area depicted in Figure 11-5 above. This additional capability turned out to be unnecessary, however, as the fast update rate associated with the four analog sensors proved more than sufficient for real-time tracking, and platform motion was incorporated to expand the window of acceptance during initial stripe acquisition. In the event a stripe segment was not acquired within 6 inches of its expected position (see Section 11.5), the robot would stop and pivot in place, turning 90 degrees to the left and right of its original heading while monitoring for stripe detection. With the particular geometry of the ModBot base (Figure 11-6), this action results in a 14-inch radius of curvature for the search arc.

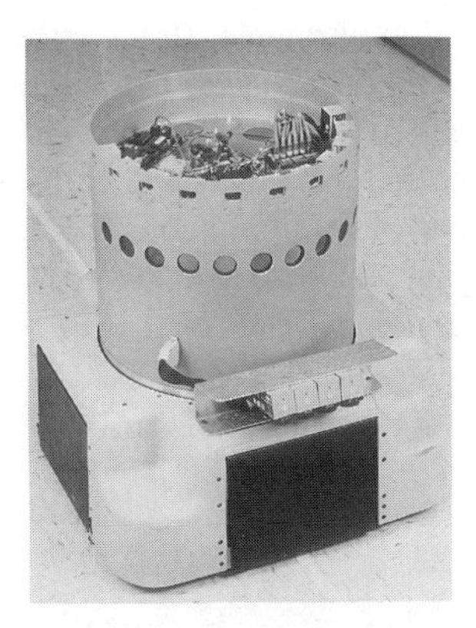

Figure 11-6. The prototype retroreflective guidepath tracking sensor employed on the ModBot was based on four Banner Engineering analog retroreflective sensors (courtesy Naval Command Control and Ocean Surveillance Center).

To assist in making a conventional straight-line transition from off-path to guidepath control, the binary stripe-status information (i.e., present, not present) is monitored for each of the four sensor heads to establish the intercept angle with respect to the path (Figure 11-7). For example, if the robot's search trajectory crosses the guidepath in a perpendicular fashion, all four channels will detect the tape at approximately the same time. If the intercept angle is obtuse, the sensors will respond in sequential fashion from left to right, and vice versa if the angle is acute. The rate of sequential detection determines the magnitude of the angle. This information is used by the control software to determine how far forward the robot should continue after initial acquisition and in which direction it should then turn to center itself over the stripe and shift to tracking mode.

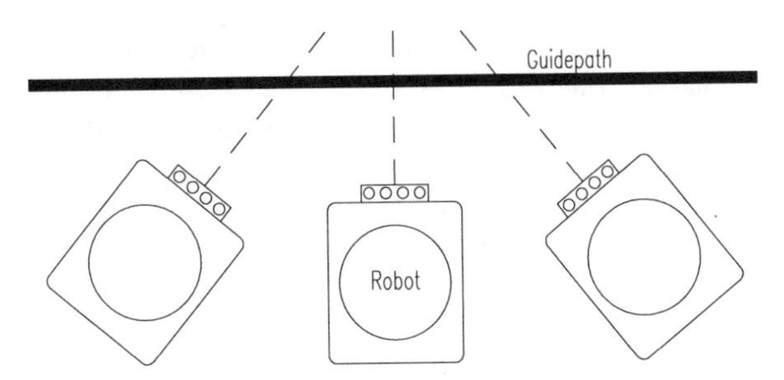

Figure 11-7. The rate and direction of individual sensor detection determine the magnitude and direction of the guidepath intercept angle.

11.2.2 U/V Stimulated Emission

Probably the most versatile optical stripe following system in widespread use today is based on the principle of *stimulated emission* of an essentially invisible chemical stripe when exposed to ultraviolet (UV) radiation. This technology was developed in the early 1970s by Lear-Siegler Corporation, Grand Rapids, MI, for use on their *Mailmobile-Series* AGVs, with patent rights subsequently sold to Bell and Howell Corporation, Zeeland, MI. Bell and Howell sold the patent rights in the early 1980s to Litton Industrial Automation, Holland, MI, in a bid to expand from office applications (Figure 11-8) into industrial material handling markets. Litton implemented the technology into its *UltraFlex®* guidepath scheme in support of the company's *Integrator®* series of AGVs (Litton, 1989). Litton's Holland-based operations were eventually sold in 1994 to Saurer Automation.

Figure 11-8. The fluorescent glow of the stimulated chemical guidepath is clearly visible beneath this Bell and Howell *Mailmobile* (see also Figure 3-27), which has been reliably picking up and delivering office mail in the main administration building at NCCOSC for over eight years (courtesy Naval Command Control and Ocean Surveillance Center).

The guidepath stripe is composed of fluorescent particles suspended in a water-based hardener that permeates the floor surface for extended durability, drying in approximately eight hours (Litton, 1992). As in the previously discussed case of retroreflective tape, the passive guidepath is unaffected by minor scuffs, abrasions, or discontinuities, and is essentially unlimited in terms of overall length. An active ultraviolet light source irradiates the guidepath, causing the embedded fluorescent particles to reradiate with a fairly narrow spectral output in the blue-green visible spectrum as seen in Figure 11-8. The lateral position sensor that detects this stimulated response is insensitive to ultraviolet wavelengths, and thus sees a fairly high signal-to-noise ratio associated with the fluorescent stripe relative to the surrounding floor surface. The Bell and Howell *Mailmobile* uses three discrete photodetectors to track the glowing guidepath, with a resultant path-tracking accuracy of ±0.25 inches (B&H, 1985), while Litton incorporated a mirror-scanned photodetector into their *UltraFlex* sensor design with a ±0.3-inch tracking accuracy (Litton, 1992). Saurer Automation has recently introduced a microprocessor-based guidepath sensor that employs a fixed array of solid-state photodetectors with a 4-inch effective field of view.

One potential drawback of the stimulated-emission guidepath scheme is the exposed nature of the path, which requires periodic maintenance upgrades to compensate for wear. Unlike the retroreflective-tape schemes discussed earlier, however, there is no need to first remove the existing chemical guidepath; the new stripe can simply be painted down right on top of the old one. The chemical solution is easily applied with a special applicator to almost any floor surface, including tile, masonry, wood, steel, and most carpeting, and dries quickly to an almost invisible finish (Litton, 1991). To minimize performance degradation due to interfering fluorescent residuals found in some carpet cleaning solutions, Bell and Howell offers a specially formulated industrial carpet shampoo free of optical brighteners (B&H, 1990). There have been some isolated instances where the guidance system was confused by the glow of fluorescing salt crystals tracked in from icy sidewalks by pedestrians in cold winter climates.

While this fluorescent guidepath approach is very well suited to AGVs operating over fairly smooth floor surfaces, the additional ground clearance generally required in typical mobile robotic applications introduces some practical concerns that can adversely affect performance. One of the biggest problems is potential interference from direct sunlight. In conventional AGV applications the guidepath sensors are routinely tucked up under the vehicle chassis to minimize exposure to such effects. Bell and Howell in fact offers an optional add-on *Sun Skirt* for use with their *Mailmobile* that effectively eliminates most ambient-light access to the protected sensor region, for improved operation in areas where patches of bright sunlight are routinely encountered (B&H, 1990). With an exposed drive configuration such as the Cybermotion *K2A Navmaster* robot, however, adequate shielding of external light sources is virtually impossible. Saurer's new sensor head features improved circuitry and algorithms that minimize ambient light interference, thereby eliminating the need for physical shielding.

11.3 Magnetic Tape

Magnetic-tape guidepath schemes involve a strip of flexible magnetic material affixed to or buried beneath the floor surface, sensed by an array of fluxgate or Hall-effect sensors (see also Chapter 12) on board the vehicle. Kamewaka & Uemura (1987) describe a belt-like magnet made of barium ferrite compounded with a flexible resin to form a 50-millimeter-wide and 0.8-millimeter-thick strip. A vertical magnetic field component of 6 Gauss is generated by this material at the nominal sensor stand-off distance of 40 millimeters. Tsumara (1986) mentions the use in Japan of a ferrite magnetic powder mixture painted directly onto the floor, but no details as to sensitivity, effectiveness, or identity of the system manufacturer are provided. Advantages include extremely low maintenance, immunity to path discontinuities, tolerance of minor debris along the path, and unlimited route lengths. Primary disadvantages are relatively high initial installation costs, potential magnetic interference from surrounding sources, and the short stand off distance required between the guidepath and sensor head.

11.3.1 Macome Magnetic Stripe Follower

Macome Corporation, Tokyo, Japan has developed a magnetic guidance unit for AGVs that employs an array of saturable-core fluxgate sensors (see Chapter 12). Two analog output signals generated by a pair of coils 80 millimeters apart provide for differential servo-control of vehicle heading. The coils are driven in and out of saturation by a square wave pulse train produced by a local oscillator as shown in Figure 11-9. The excitation voltage is divided across the inductance of the coil I_s in series with a fixed resistance R_s. Since the coil inductance changes as a function of the externally applied magnetic field, the measured voltage across R_s will be indicative of field strength and hence the tape-to-coil separation distance. This signal is detected and smoothed by a half-wave rectifier and filter circuit and compared to a similar signal generated by identical circuitry associated with the other coil. The output signals from this simple interface are sufficient to support a 40-millimeter tape stand-off distance without additional amplification (Kamewaka & Uemura, 1987).

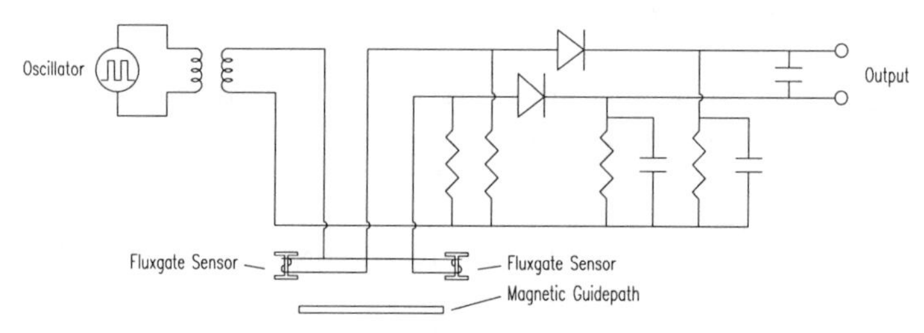

Figure 11-9. Schematic diagram of the interface circuitry associated with the saturable-core fluxgate sensors used to sense tape position (adapted from Kamewaka & Uemura, 1987).

In addition to the two analog tracking sensors, three groups of binary-output sensors furnish coarse zone (left, right, or center) and branch detection information as shown in Figure 11-10. The outputs of all binary sensors within a particular group are logically ORed together to facilitate broader coverage. The three zone signals (indicating whether the vehicle is positioned over the left side zone, right side zone, or center zone of the designated transit lane) are in turn

ORed together to produce a safety signal that stops the vehicle immediately if the guidepath is lost.

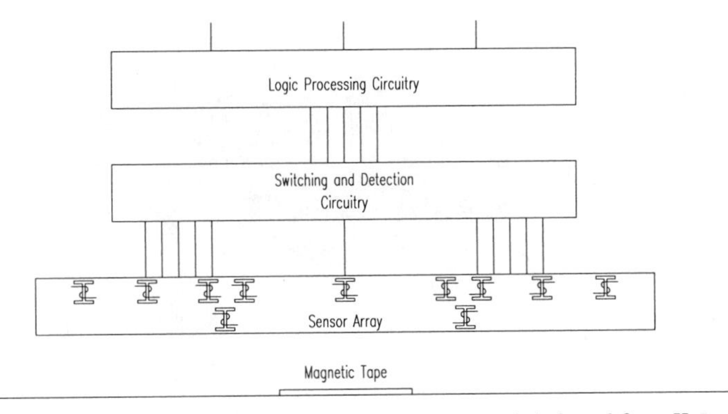

Figure 11-10. Block diagram of Macome magnetic tape sensor unit (adapted from Kamewaka & Uemura, 1987).

The left and right zone signals are logically ANDed together to generate a branch detection flag indicating the vehicle has come to a Y in the path (Figure 11-11). When a branch is detected, the vehicle is shifted in the desired direction by applying an appropriate offset voltage to the output of the steering comparator (Kamewaka & Uemura, 1987).

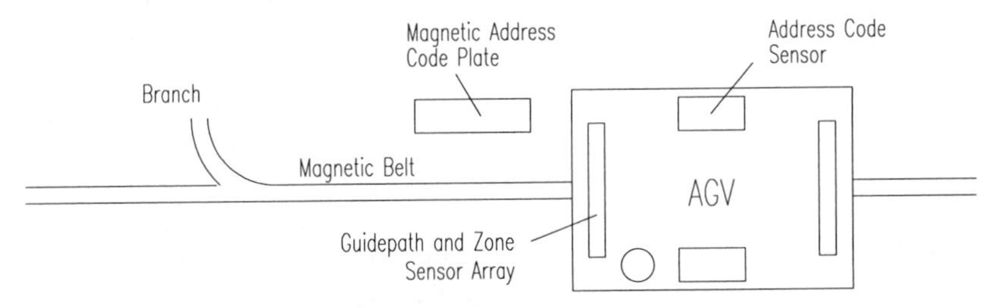

Figure 11-11. Layout of the various components of the Macome system (adapted from Kamewaka & Uemura, 1987).

11.3.2 Apogee Magnetic Stripe Follower

Apogee Robotics, Fort Collins, CO, sells a stand-alone microprocessor-based sensing head (designed for use on their *Orbitor 750* AGV) that employs an array of Hall-effect sensors for tracking a magnetic tape guidepath (Figure 11-12). Effective stand-off distance for a 0.5-inch-wide tape 0.062 inches thick is between 1 and 2 inches, with a corresponding magnetic field strength varying from 9 to 2.5 Gauss over that range. Slightly increased stand-off is possible through use of a thicker tape.

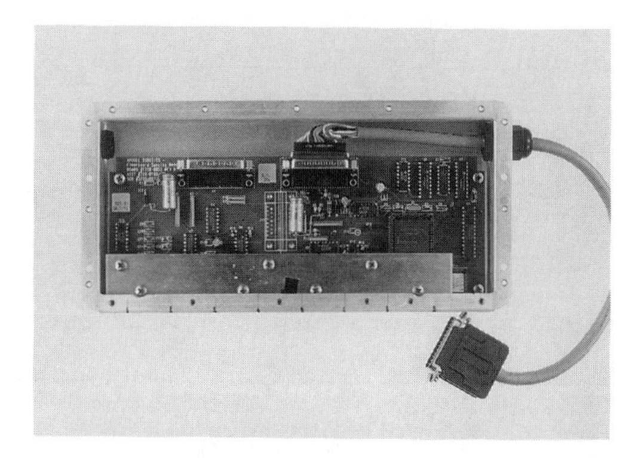

Figure 11-12. Photo of Apogee magnetic tape sensor head with cover removed (courtesy Naval Command Control and Ocean Surveillance Center).

Eight Hall-effect sensors (Microswitch *91SS12-2*) are mounted approximately 1 inch apart to yield two 4.75-inch detection zones, one centered on the vehicle axis for tracking the guidepath and the other located immediately adjacent on the right side for detecting magnetic barcode markers. Barcodes are comprised of discreet magnetic markers, with a logic "0" located 4.0 inches from the center of the guidepath, and a logic "1" displaced 5.5 inches from center. Start and stop bits are indicated by a collocation of a "0" and a "1" to yield a double-wide marker.

Allowed manufacturing tolerances for the Hall-effect devices are roughly equal to the signal level being measured. The sensor head must therefore be carefully calibrated in place after installation on the AGV, after which an automated null-offset routine is incorporated into the software to ensure maximum sensitivity. The influence of external magnetic fields caused a significant problem when we attempted to interface one of these units to the MDARS Interior robot, because unlike conventional AGVs, the rotating nature of the Cybermotion *Navmaster K2A* turret creates a constantly changing vehicle magnetic signature.

The Apogee sensor head measures 3.25 inches high by 9.5 inches wide and 1 inch deep, and requires 12 volts DC at approximately 200 milliamps. Communication with the vehicle electronics is by an RS-232 serial interface.

11.3.3 3M/Honeywell Magnetic Lateral Guidance System

The 3M Traffic Control Materials group in conjunction with the Honeywell Technology Center, Minneapolis, MN, has developed a prototype system that incorporates a magnetic guidepath into conventional retroreflective lane marking material for highway usage. Initial feasibility tests were conducted using an off-the-shelf flexible magnetic strip (*Plastiform® type B-1033*) manufactured by Arnold Engineering, Norfolk, NB. The 4-inch wide tape, consisting of barium ferrite particles suspended in a nitrile rubber binder, exhibits a remnant magnetization (B_r) of about 2500 Gauss (Stauffer, et al., 1995). Three-foot sections of this material were aligned end to end, with every other section inverted to produce an alternating (i.e., north-up, south-up, north-up) vertical field polarity, and then covered with 3M *Scotchlane™ 620-Series* pavement marking tape.

A Honeywell magnetoresistive two-axis magnetometer (see Chapter 3) in bridge configuration is used to sense the lateral position of the tape (Lenz, et al., 1990), with an effective vertical standoff of 9 inches. First-round feasibility tests using an existing three-axis sensor showed reliable marker detection with as much as 6 feet of lateral offset between the sensor and guidepath. A downward looking video camera mounted on the test vehicle adjacent to the magnetometer was used to precisely quantify actual guidepath offset for post-processing of the sensor data output. A total of 26 runs were made along a 100-foot section of marker material at speeds of 15 to 20 miles per hour, with data recorded at a 100-Hz rate. Magnetic deviation was observed to closely agree with control video deviation (in terms of repeatability) to about the same order as the video instrumentation accuracy of 0.25 inches (Stauffer, 1995).

Initially developed as a lateral position control scheme for use in *intelligent vehicle highway system* scenarios, the 3M/Honeywell approach has tremendous potential as an external guidepath for AGVs and robotic vehicles. Efforts are now underway to evaluate a prototype for navigational guidance of the MDARS Exterior program. Other near-term applications include automatic snowplow guidance and automotive safety equipment to warn drivers of impending departure from the traffic lane.

11.4 Heat and Odor Sensing

Australian researchers at the University of Monash report an unusual approach to tracking a temporary pathway similar to the technique used by ants to mark a trail back to their nest. Early work was based on a short-lived heat trail laid down by a lead robot equipped with a quartz-halogen projector bulb configured to raise the floor temperature beneath the unit during the course of transit (Russell, 1993; Kleeman & Russell, 1993). The idea was that additional robots lacking the navigational sophistication of the first could simply follow the temporary trail of the leader in convoy fashion.

A standard *pyroelectric* sensor as is commonly used for passive motion detection in security applications (see Chapter 17) was modified as shown in Figure 11-13 to sense the residual thermal energy imparted to the floor. Two thin-plate lithium-tantalate capacitor sense elements are often employed in sensors of this type for common-mode noise rejection. One capacitor generates a positive voltage across its plates when heated by incident radiation, while the other produces a negative voltage. These two voltages cancel each other for any thermal inputs that are uniform across the full field of view. A special lens is employed in security applications so the lateral motion of an intruder affects first one sensor element and then the other to produce a differential output. To eliminate this common-mode-rejection feature, a brass shim was installed as shown in the diagram to completely shield one capacitor from incoming radiation.

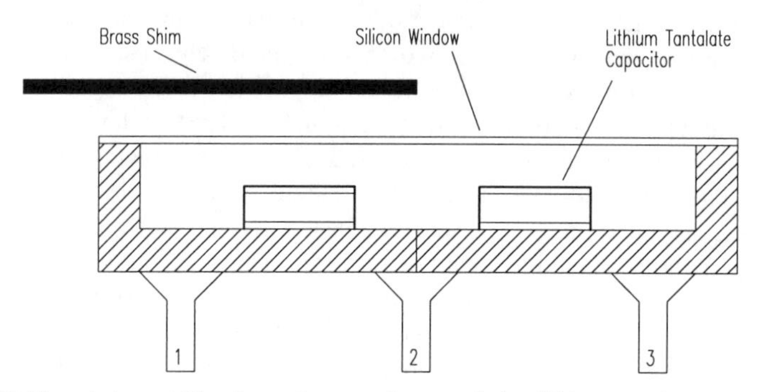

Figure 11-13. A brass shim is used to mask one of the lithium-tantalate capacitors in a conventional *pyroelectric* sensor to defeat common-mode rejection (Russell, 1993, reprinted with permission of Cambridge University Press).

Current-leakage paths in the sensor structure and associated electronics cause the voltage generated across the capacitor plates to decay with time (Russell, 1993), and so the output signal is really proportional to the magnitude of the change in incident thermal radiation. As a consequence, some means of shuttering the energy input was required between readings. In addition, two sensor readings were necessary at points 5 centimeters apart (Kleeman & Russell, 1993) to achieve the differential signal required for vehicle heading control. To meet these needs and eliminate device-specific dependencies, the same sensor was mechanically repositioned back and forth between the left and right sense apertures every 1.2 seconds by a servo-controlled linkage arm (Figure 11-14). A center "reset" position was provided to allow the sensor to settle while screened from all external radiation before reading the next aperture. The servomotor itself is mounted external to the sensor housing as illustrated for thermal isolation.

The sensor package was mounted on the front of a small tracked robot manufactured by Robotnik P/L of Melbourne, Australia. A differential drive correction δ_k was calculated as follows (Russell, 1993):

$$\delta_k = k(t_r - t_l)$$

where:

k = control system proportional gain constant
t_r = reading from right sensor position
t_l = reading from left sensor position.

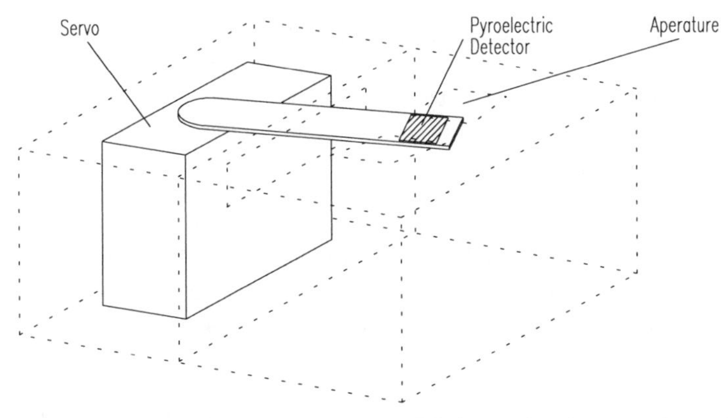

Figure 11-14. Shown inverted for clarity, a modified *pyroelectric* sensor is mechanically positioned back and forth between two sense apertures to track a heat trail on the floor (adapted from Kleeman & Russell, 1993, © IEEE).

Russell reports the optimal value of the proportional gain constant k in the above equation was found to be a function of the age of the heat trail being followed. The resultant correction factor δ_k is applied to left and right track displacements as indicated below:

$$x_l = d + \delta_k$$
$$x_r = d - \delta_k$$

where:

x_l = displacement of left track
x_r = displacement of right track
d = basic translational distance of robot along path.

The magnitude of δ_k is constrained to the range of $\pm d$ so the tracks never move backwards. An average speed of 0.3 meters per minute can be attained while following thermal trails up to 10 minutes old on vinyl floor tile (Russell, 1993). The thermal-path approach was eventually abandoned, however, due to interference problems associated with hot water and heating pipes embedded in the floor, as well as localized hot spots created by shafts of sunlight. In addition, the 70-watt halogen heater placed considerable energy demands on the limited storage capacity of the onboard battery.

The current solution involves laying down a 1-centimeter wide trail of camphor for the slaves to detect and follow (Deveza, et al, 1994). The camphor is dissolved in alcohol and applied with a felt-tip applicator; the alcohol evaporates in seconds, leaving a faint trail of camphor particles on the floor. The choice of camphor was based on its ease of detection and the fact that it is inoffensive to humans, slowly subliming over a period of several hours into a harmless vapor

(Bains, 1994). The use of a marking agent that dissipates with time is desirable, as otherwise some means must be provided to "clean up" the trail after it has served its purpose (Kleeman & Russell, 1993).

The camphor sensors (Figure 11-15) are based on the gravimetric microbalance technique, employing a quartz crystal coated with Silicone OV-17 (Russell, et al., 1994). Camphor molecules are absorbed into the crystal coating, adding to the effective mass and thus lowering the resonant frequency in direct proportion to odor concentration. Two such sensors are positioned 5 millimeters above the floor and 50 millimeters apart. The outputs from their respective oscillator circuits are compared with a reference oscillator to generate a differential signal for steering control.

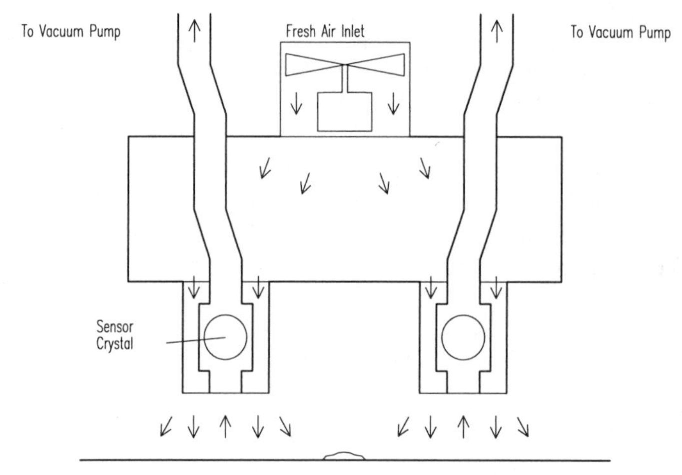

Figure 11-15. A positive pressure air curtain is used to block transient odors that may be carried by drafts into the area of the twin sensing units used to sense the trail of camphor (adapted from Russell, et al., 1994, © IEEE).

To improve tracking performance, a constant intake flow of 200 milliliters per minute is maintained by a modified aquarium pump to draw air into the sensor housing for sampling. Nevertheless, turbulence created by ventilation systems and the movement of doors and people was found to cause significant variations in sensor output, requiring time-consuming averaging for reliable operation. To overcome these effects, an over-pressure air curtain was generated by a small fan to shield extraneous inputs and focus the sensitivity on that area of the floor directly beneath the intake. At room temperature, a typical camphor/air mixture concentration of 236 PPM caused a subsequent frequency drop of 244 Hz (Russell, et al, 1994). The robot is able to follow camphor trails up to 30 minutes old.

11.5 Intermittent-Path Navigation

There is a growing trend throughout the AGV industry to move away from fixed-path guidance schemes altogether due to their inherent lack of flexibility, and more towards what has been loosely termed *off-wire* or *autonomous* guidance. (Both terms are somewhat misleading: *off-wire* overlooks optical and magnetic alternatives, and many conventional guidepath-following vehicles are already in fact fully autonomous.) Gould (1990) cites three production environments where such *free-ranging* (for lack of a better term) navigation schemes are likely to find initial application:

- Where production requirements are in a continuous state of evolution.
- In clean-room operations, where embedding a wire or painting the floor cannot be tolerated.
- In scenarios where the floor itself may be expensive and any alteration would be inappropriate.

A number of these *free-ranging* guidance schemes will be discussed later in subsequent chapters, but first let's take a quick look at a hybrid concept that incorporates some of the best features associated with both *guidepath* and *unrestricted path* navigation.

From a navigational perspective, the fundamental advantage of a fixed guidepath is the simple fact that there is essentially no way to get lost. The big disadvantage of course is the vehicle can only go where the guidepath goes. If an object temporarily obstructs a pathway, everything comes to an inconvenient halt until the obstacle is physically removed. *Unrestricted path navigation*, on the other hand, allows for free traversal around obstacles to basically any desired location for which a clear path indeed exists, but accumulated dead-reckoning errors will eventually cause the system to get lost. It seemed like a good opportunity for synergism, and so the *hybrid navigation scheme* was conceived and subsequently developed on one of the ModBots using the optical stripe follower unit previously shown in Figure 11-5.

Interestingly, while most commercial off-path navigation schemes arose from the need to momentarily leave the guidepath to maneuver around some obstacle, the ModBot implementation came about in exactly the opposite fashion. A free-ranging unrestricted path planning ability had been previously developed by Gilbreath (1988) for use on ROBART II, but suffered from eventual platform disorientation due to dead-reckoning inaccuracies. Periodic traversal of predefined tape guidepaths was therefore incorporated to routinely re-reference the platform, without giving up the ability to move freely to any desired location within the map floor plan (Everett, et al., 1990). Gilbreath's implementation of this *hybrid navigation scheme* within the existing path planner was rather elegant in its simplicity: the predefined guidepath segments were treated as "zero-cost" cells in the modified A* search algorithm, thereby giving them preferred status in the route generation process.

11.5.1 MDARS Interior Hybrid Navigation

It should be noted the *hybrid navigation scheme* is not necessarily restricted to the use of fixed guidepaths for referencing purposes. For example, the MDARS Interior robot employs a derivative of the ModBot approach that was further enhanced under a Cooperative Research and Development Agreement between Cybermotion and NCCOSC (Holland, et al., 1990). This concept combines Cybermotion's *virtual path* navigation with *unrestricted path planning* software developed by NCCOSC for improved collision avoidance. *Virtual paths* provide an effective means for correcting cumulative dead-reckoning position errors, while *unrestricted path planning* allows for transit to any desired location, with the ability to generate avoidance maneuvers around obstacles that may block a *virtual path* segment.

Figure 11-16. The MDARS Interior robot normally traverses *virtual paths* generated by the Cybermotion *Dispatcher*, switching to free-ranging off-path operation in the event of a perceived obstruction.

11.5.2 Free Ranging On Grid

As another example of the industry's desire to get away from even occasional use of guidepath segments, Frog Systems (a subsidiary of Industrial Contractors, Utrecht, The Netherlands) markets a free-ranging AGV that uses a passive floor grid to update the dead-reckoning solution in real time (Gould, 1990). An *a priori* two-dimensional map of the operating area describing the absolute location of walls, obstructions, waypoints, loading stations, and desired routes is maintained in an onboard computer, along with the precise coordinates of the grid components. Considerable flexibility is available in the *FROG (Free Ranging On Grid)* system through use of a variety of potential grid sensors, to include magnetic, optical, and RF. The vehicle's dead-reckoned position is repeatedly reset each time a known grid location is identified.

The grid itself can consist of either line segments or node points as shown in Figure 11-17. Examples of possible *magnetic line-grid* implementations include strips of metal attached to the floor with adhesive, or embedded beneath the surface as in the case of existing steel borders around removable concrete floor sections. The metal line segments are detected with an inductive proximity sensor (Section 3.2.2). An *optical line-grid* configuration has been installed at an Apple Computer factory in Singapore, using a CCD camera to detect the high-contrast intersections of a two-color checkerboard pattern in the floor tiles. Positional accuracy in this case is about 3 millimeters for the 30- by 30-centimeter grid resolution established by the chosen tile size (Van Brussel, et al., 1988).

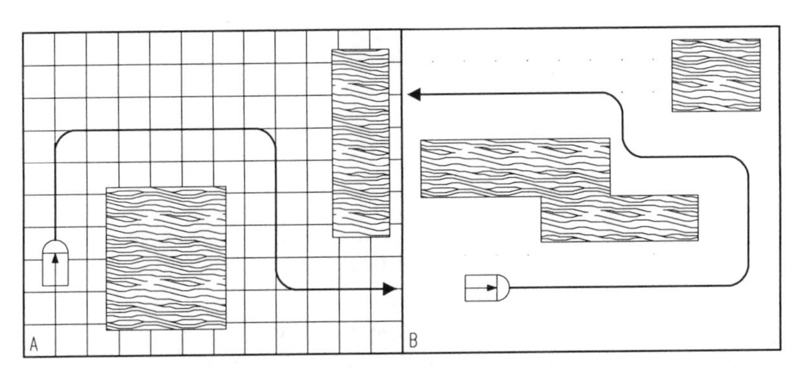

Figure 11-17. The passive grid used by FROG can consist of line segments (A) or node points (B) connected by imaginary lines (Van Brussel, et al., 1988).

Alternatively, the grid can be comprised of specific nodes in the form of metal or optical markers, or even passive RF transponder tags similar to the type used for product identification in automated inventory applications (see Chapter 17). The use of such interactive tags is attractive from the standpoint that each grid node can be individually distinguishable from all others by virtue of its own unique ID response. Such a scheme is also insensitive to adverse conditions likely to be encountered in industrial environments such as dust, dirt, and spillage, which could adversely impact the reliability of an optically based floor referencing system.

Conventional passive RF transponder tags used for personnel and product identification derive power for operation of their onboard circuitry directly from the transmitted energy emitted by the interrogator. An off-the-shelf tag of this type was modified for the *FROG* application to retransmit an easily identified "position" signal derived by dividing the incoming carrier frequency by 2^n. This intentional change in frequency facilitates detection of the weak response signal in the presence of the very powerful emissions of the vehicle's transmitting antenna (Van Brussel, et al., 1988). A "figure-eight" antenna configuration as illustrated in Figure 11-18 listens for this reply, indicative of the presence of a transponder tag at some predefined grid node location.

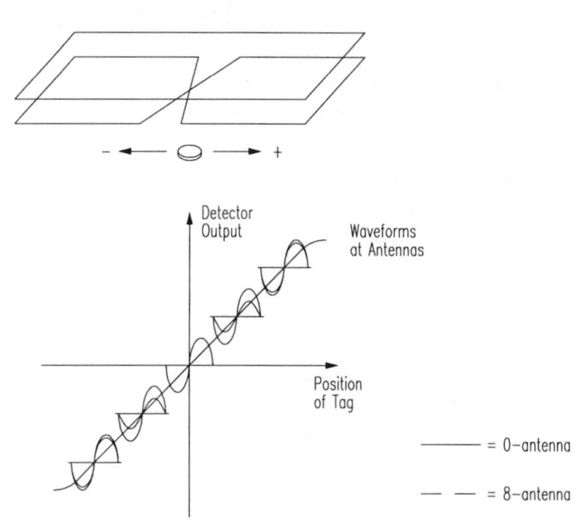

Figure 11-18. The phase of the signal received by the figure-eight antenna relative to that transmitted by the rectangular antenna indicates tag position left or right of centerline, while the signal magnitude is a linear function of lateral offset (adapted from Van Brussel, et al., 1988).

The foldover pattern in this receiving antenna generates signals in each loop that are 180 degrees out of phase, and as a consequence cancel each other completely if of equal magnitude. Such a situation would occur with the transponder tag symmetrically located with respect to the two loops (i.e., along the vehicle centerline). As the tag location shifts laterally away from symmetry, however, the loop in closer proximity begins to dominate, and a net antenna signal appears at the receiver input amplifier. The magnitude of this signal grows as a linear function of distance from centerline (within the limits of antenna coverage), while its phase in relationship to the transmitted signal determines the direction (left or right) of displacement. For a vehicle wheelbase of 1 meter, a 28- by 45-centimeter antenna situated 20 centimeters above the floor resulted in a resultant 1-centimeter positional accuracy, linear over a range of ±15 centimeters from centerline.

11.6 References

AGV, "CB20 Technical Manual," Rev. 1.0, AGV Products, Inc., Charlotte, NC, April, 1993.

Bains, S., "Robots Mark Their Territory," *SPIE OE Reports*, No. 124, pp. 1, 9, April, 1994.

Banner, Product Catalog, Banner Engineering, Minneapolis, MN, 1993a.

Banner, "Handbook of Photoelectric Sensing," Banner Engineering, Minneapolis, MN, 1993b.

B&H, "Discover Mailmobile®," Product Brochure #PM-601, Bell and Howell Mailmobile Company, Zeeland, MI, 1985.

B&H, "Mailmobile® Accessories," Product Brochure #MM-001, Bell and Howell Mailmobile Company, Zeeland, MI, 1990.

Cahners, "AGVs in America: An Inside Look," *Modern Materials Handling*, Cahners Publishing Co., a Division of Reed Elsevier, Ink., September, 1990.

Deveza, R., Russell, R.A., Thiel, D, Mackay-Sim, A., "Odour Sensing for Robot Guidance," *International Journal of Robotics Research*, 1994.

Everett, H.R., Gilbreath, G.A., Tran, T., Nieusma, J.M., "Modeling the Environment of a Mobile Security Robot," Technical Document 1835, Naval Command Control and Ocean Surveillance Center, San Diego, CA, June, 1990.

Gilbreath, G.A., Everett, H.R., "Path Planning and Collision Avoidance for an Indoor Security Robot," SPIE Mobile Robots III, Cambridge, MA, pp. 19-27, Novemeber, 1988.

Gould, L., "Is Off-Wire AGV Guidance Alive or Dead?", *Managing Automation*, pp. 38-40, May, 1990.

Guidoni, G., "SGVs Play Active Role in Chrysler's Future," *Plant, Canada's Industrial Newspaper*, Vol. 51, No. 11, 17 August, 1992.

Holland, J.M., Everett, H.R., Gilbreath, G.A., "Hybrid Navigational Control Scheme," SPIE Vol. 1388, Mobile Robots V, Boston, MA, November, 1990.

Kamewaka, S., Uemura, S., "A Magnetic Guidance Method for Automated Guided Vehicles," *IEEE Transactions on Magnetics*, Vol. MAG-23, No. 5, September, 1987.

Kleeman, L., Russell, R.A., "Thermal Path Following Robot Vehicle: Sensor Design and Motion Control," Proceedings, IEEE/RSJ International Conference on Intelligent Robots and Systems, Yokohama, Japan, July, 1993.

Lenz, J.E., et al., "A High-Sensitivity Magnetoresistive Sensor," IEEE Solid-State Sensors and Actuators Workshop, Hilton Head, SC, June, 1990.

Litton, "Litton Integrator® Automated Guided Vehicles," Product Brochure #8/89 3M M05-1303, Litton Industrial Automation, Holland, MI, August, 1989.

Litton, "Litton Integrator® Automated Guided Vehicles," Product Brochure #2/91 4M M05-1303, Litton Industrial Automation, Holland, MI, February, 1991.

Litton, "UltraFlex® Optical Guidepath from Litton," Product Brochure #1/92 5M M05-1305, Litton Industrial Automation, Holland, MI, January, 1992.

Murray, C.J., "Hydraulic Lifesaver for Highways," *Design News*, pp. 68-74, 25 April, 1994.

Petriu, E.M., "Automated Guided Vehicle with Absolute Encoded Guidepath," *IEEE Transactions on Robotics and Automation*, Vol. 7, No. 4, pp. 562-565, August, 1991.

Russell, R.A., "Mobile Robot Guidance Using a Short-Lived Heat Trail," *Robotica*, Vol. 11, Cambridge Press, pp. 427-431, 1993.

Russell, R.A., Thiel, D., Mackay-Sim, A., "Sensing Odour Trails for Mobile Robot Navigation," Proceedings, IEEE International Conference on Robotics and Automation, San Diego, CA, Vol. 3, pp. 2672-2677, May, 1994.

Stauffer, D., Lenz, J., Dahlin, T.J., "A Magnetic Lateral Guidance Concept Using Continuous Magnetic Marking," submitted for publication to *IEEE Control Systems Journal*, 1995.

Tsumura, T., "Survey of Automated Guided Vehicles in Japanese Factories," IEEE CH2282, pp. 1329-1334, 1986.

Van Brussel, H., Van Helsdingen, C.C., Machiels, K., "FROG - Free Ranging on Grid: New Perspectives in Automated Transport," Proceedings, 6th International Conference on Automated Guided Vehicle Systems, Brussels, Belgium, pp. 223-232, October, 1988.

12
Magnetic Compasses

Vehicle heading is the most significant of the navigational parameters (X, Y, and θ) in terms of its influence on accumulated dead reckoning errors. For this reason, sensors which provide a measure of absolute heading or relative angular velocity are extremely important in solving the real-world navigational needs of an autonomous platform. The most commonly known sensor of this type is probably the magnetic compass.

The terminology normally used to describe the intensity of a magnetic field is *magnetic flux density* B, measured in Gauss (G). Alternative units are the Tesla (T) and the gamma (γ), where:

$$1 \text{ Tesla} = 10^4 \text{ Gauss} = 10^9 \text{ gamma}$$

The average strength of the earth's magnetic field is 0.5 Gauss and can be represented as a dipole that fluctuates both in time and space, situated roughly 440 kilometers off center and inclined 11 degrees to the planet's axis of rotation (Fraden, 1993). This difference in location between *true north* and local *magnetic north* is referred to as *variation* (also known as *declination*), and varies with both time and geographical location. Corrective values are routinely provided in the form of *variation* or *declination tables* printed directly on the maps or charts for any given locale.

Instruments which measure magnetic fields are known as *magnetometers*. For application to mobile robot navigation, only those classes of *magnetometers* which sense the magnetic field of the earth are of interest. Such geomagnetic sensors, for purposes of this discussion, will be broken down into the following general categories:

- Mechanical magnetic compasses.
- Fluxgate compasses.
- Magnetoinductive compasses.
- Hall-effect compasses.
- Magnetoresistive compasses.
- Magnetoelastic compasses.

12.1 Mechanical Magnetic Compasses

The first recorded use of a magnetic compass was in 2634 BC, when the Chinese suspended a piece of naturally occurring magnetite from a silk thread and used it to guide a chariot over land (Carter, 1966). Much controversy surrounds the debate over whether the Chinese or the Europeans first adapted the compass for marine applications, but by the middle of the 13th century such usage was fairly widespread around the globe. William Gilbert (1600) was the first to propose that the earth itself was the source of the mysterious magnetic field that provided such a stable navigational reference for ships at sea.

The early marine compasses were little more than magnetized needles floated in water on small pieces of cork. These primitive devices evolved over the years into the reliable and time proven systems in use today, which consist of a ring magnet or pair of bar magnets attached to a graduated mica readout disk. The magnet and disk assembly floats in a mixture of water and alcohol or glycerin, such that it is free to rotate around a jeweled pivot. The fluid acts to both support the weight of the rotating assembly and to dampen its movement under rough conditions.

The sealed vessel containing the compass disk and damping fluid is typically suspended from a two-degree-of-freedom *gimbal* to decouple it from the ship's motion. This *gimbal* assembly is mounted in turn atop a floor stand or binnacle. Situated on either side of the binnacle are massive iron spheres that, along with adjustable permanent magnets in the base, are used to compensate the compass for surrounding magnetic anomalies that alter the geomagnetic lines of flux. The error resulting from such external influences (i.e., the angle between indicated and actual bearing to magnetic north) is known as compass *deviation*. A correction must be made as well for the local *variation* in order to derive true heading. *Variation* and *deviation* are usually expressed in degrees with an E or W suffix indicating which way *true north* lies from *magnetic north* or from *compass heading*. The corrections are made in sequence from left to right according to a mnemonic known to old mariners: *Can Dead Men Vote Twice*:

Compass + Deviation = Magnetic   *Magnetic + Variance = True.*

Another potential source of error which must be taken into account is *magnetic dip*, a term arising from the "dipping" action observed in compass needles attributed to the vertical component of the geomagnetic field. The dip effect varies with latitude, from no impact at the equator where the flux lines are horizontal to maximum at the poles where the lines of force are entirely vertical. For this reason, many swing-needle instruments have small adjustable weights that can be moved radially to balance the needle for any given local area of operation. In addition to gimbaled mounting, marine compasses ensure alignment in the horizontal plane by floating the magnet assembly in an inert fluid.

While gimbal and fluid suspension techniques are fairly effective for marine applications where the period of pitch-and-roll disturbances is fairly long, land-based vehicles often encounter significant acceleration effects due to rough or uneven terrain. Foster (1985) expresses the measurement error due to resulting sensor tilt in terms of the local dip angle of the earth's magnetic field:

$$\theta_{TE} = \arctan(\sin\alpha \tan\beta)$$

where:

θ_{TE} = heading error due to vehicle tilt
α = vehicle tilt displacement in north/south direction
β = dip angle of Earth's field.

12.1.1 Dinsmore *Starguide* Magnetic Compass

An extremely low-cost configuration of the mechanical magnetic compass with electronic readout is seen in a product recently announced by the Dinsmore Instrument Company, Flint, MI. The heart of the *Starguide* compass is the Dinsmore *Model 1490* digital sensor (Dinsmore, 1991), which consists of a miniaturized permanent-magnet rotor mounted in low-friction jeweled bearings. The sensor is internally damped such that if momentarily displaced 90 degrees, it will return to the indicated direction in 2.5 seconds, with no overshoot.

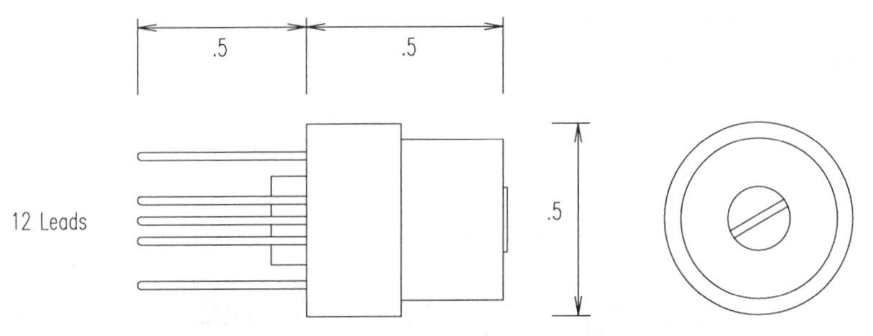

Figure 12-1. The *Model 1490* sensor used in the *Starguide* mechanical compass uses four Hall-effect sensors for electronic readout (courtesy Dinsmore Instrument Co.).

Four Hall-effect switches corresponding to the cardinal headings (N, E, W, S) are arranged around the periphery of the rotor and activated by the south pole of the magnet as the rotor aligns itself with the earth's magnetic field. Intermediate headings (NE, NW, SE, SW) are indicated through simultaneous activation of adjacent cardinal-heading switches. The Dinsmore *Starguide* is not a true *Hall-effect* compass (see Section 12.3), in that the *Hall-effect* devices are not directly sensing the geomagnetic field of the earth, but rather the angular position of a mechanical rotor.

The *Model 1490* digital sensor measures 0.5 inches in diameter by 0.63 inches high and is available separately from Dinsmore for around $12. Current consumption is 30 milliamps, and the open-collector NPN outputs can sink 25 milliamps per channel. Grenoble (1990) presents a simple circuit for interfacing the device to eight indicator LEDs. An alternative analog sensor (*Model 1525*) with a ratiometric sine-cosine output is also available for around $35. An improved analog sensor (*Model 1655*) will be introduced in the summer of 1995, identical in size and shape but with a larger output voltage swing of 1.5 volts (the output swing of the *Model 1525* is 0.8 volts). All three sensors may be subjected to unlimited magnetic flux without damage.

12.2 Fluxgate Compasses

Until most recently there was no practical alternative to the popular *fluxgate compass* for portability and long missions (Fenn, et al., 1992). (New developments in *magnetoelastic* and *magnetoinductive* magnetometers will be presented in following sections.) The term *fluxgate* is actually a trade name of Pioneer Bendix for the *saturable-core magnetometer*, derived from the gating action imposed by an AC-driven excitation coil that induces a time varying permeability in the sensor core. Before discussing the principle of operation, it is probably best to review briefly the subject of magnetic conductance, or *permeability*. The *permeability* μ of a given material is a measure of how well it serves as a path for magnetic lines of force, relative to air, which has an assigned permeability of one. Some examples of high-permeability materials are listed in Table 12-1 below.

Table 12-1. Permeability ranges for selected materials; values vary with proportional makeup, heat treatment, and mechanical working of the material (reprinted with permission from Bolz & Tuve, 1979. © CRC Press, Boca Raton, FL).

Material	Relative Permeability μ/μ_o
Supermalloy	100,000 - 1,000,000
Pure iron	25,000 - 300,000
Mumetal	20,000 - 100,000
Permalloy	2,500 - 25,000
Cast iron	100 - 600

Permeability is the magnetic circuit analogy to electrical *conductivity*, and relates *magnetic flux density* to the *magnetic field intensity* as follows:

$$B = \mu H$$

where:

B = magnetic flux density
μ = relative permeability
H = magnetic field intensity.

Since the *magnetic flux* ϕ in a magnetic circuit is analogous to *current i* in an electrical circuit, it follows that *magnetic flux density B* is the parallel to *electrical current density*.

A graphical plot of the above equation is known as the *normal magnetizing curve*, or *B-H* curve, and the *permeability* μ is the slope. An example plot is depicted in Figure 12-1 for the case of mild steel. In actuality, due to hysteresis, μ depends not only on the current value of H, but also the history of previous values and the sign of dH/dt, as will be seen later. The important thing to note at this point in the discussion is the *B-H* curve is not linear, but rather starts off with a fairly steep slope and then flattens out suddenly as H reaches a certain value. Increasing H beyond this "knee" of the *B-H* curve yields little increase in B; the material is effectively *saturated*, with a near-unity value of *permeability* (i.e., approaching that of air).

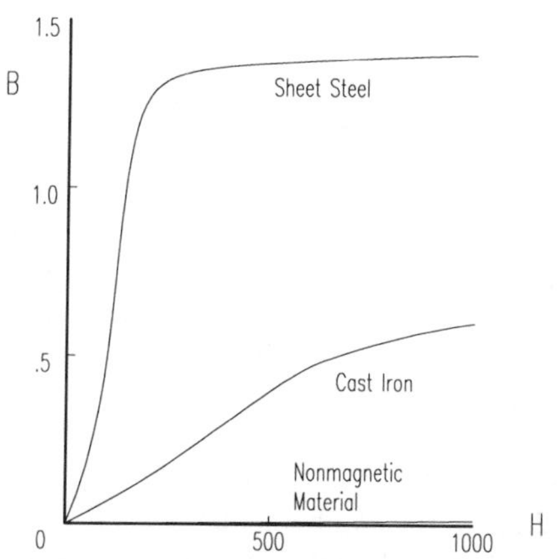

Figure 12-2. The slope of the *B-H* curve, shown here for cast iron and sheet steel, describes the *permeability* of a magnetic material, a measure of its ability (relative to air) to conduct a magnetic flux (adapted from Carlson & Gisser, 1981).

When a highly permeable material is introduced into a uniform magnetic field, the lines of force are drawn into the lower resistance path presented by the material as shown in Figure 12-3A. However, if the material is forced into saturation by some additional magnetizing force H, the lines of flux of the external field will be relatively unaffected by the presence of the saturated material, as indicated in Figure 12-3B. The fluxgate magnetometer makes use of this saturation phenomenon in order to directly measure the strength of a surrounding static magnetic field.

Various core materials have been employed in different fluxgate designs over the past 60 years, with the two most common being *permalloy* (an alloy of iron and nickel) and *mumetal* (iron, nickel, copper, and chromium). The permeable core is driven into and out of saturation by a gating signal applied to an excitation coil wound around the core. For purposes of illustration, let's assume for the moment a square-wave drive current is applied. As the core moves in and out of saturation, the flux lines from the external *B* field to be measured are drawn into and out of the core, alternating in turn between the two states depicted in Figure 12-3. (This is somewhat of an oversimplification, in that the *B-H* curve does not fully flatten out with zero slope after the knee.)

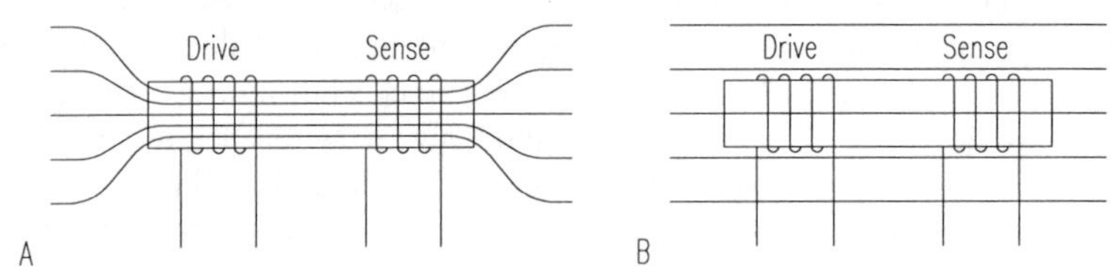

Figure 12-3. External lines of flux for: A) an unsaturated core, and B) a saturated core (adapted from Lenz, 1990, © IEEE).

These expanding and collapsing flux lines will induce positive and negative emf. surges in a sensing coil properly oriented around the core, in accordance with Faraday's law of induction. The magnitude of these surges will vary with the strength of the external magnetic field and its orientation with respect to the axis of the core and sensing coil of the fluxgate configuration. The fact that the permeability of the sensor core can be altered in a controlled fashion by the excitation coil is the underlying principle which enables the DC field being measured to induce a voltage in the sense coil. The greater the differential between the saturated and unsaturated states (i.e., the steeper the slope), the more sensitive the instrument will be.

An idealized *B-H* curve for an alternating *H*-field is shown in Figure 12-4A. The permeability (i.e., slope) is high along the section b-c of the curve, and falls to zero on either side of the saturation points H_s and $-H_s$, along segments c-d and a-b, respectively. Figure 12-4B shows a more representative situation: the difference between the left and right hand traces is due to hysteresis caused by some finite amount of permanent magnetization of the material. When a positive magnetizing force H_s is applied, the material will saturate with flux density B_s at point P_1 on the curve. When the magnetizing force is removed (i.e., $H = 0$), the flux density drops accordingly but does not return to zero. Instead, there remains some residual magnetic flux density B_r, shown at point P_2, known as the *retentivity*.

A similar effect is seen in the application of an *H*-field of opposite polarity. The flux density goes into saturation at point P_3, then passes through point P_4 as the field reverses. This hysteresis effect can create what is known as a *zero offset* (i.e., some DC bias is still present when the external *B*-field is zero) in fluxgate magnetometers. Primdahl (1970) provides an excellent mathematical analysis of the actual gating curves for fluxgate devices.

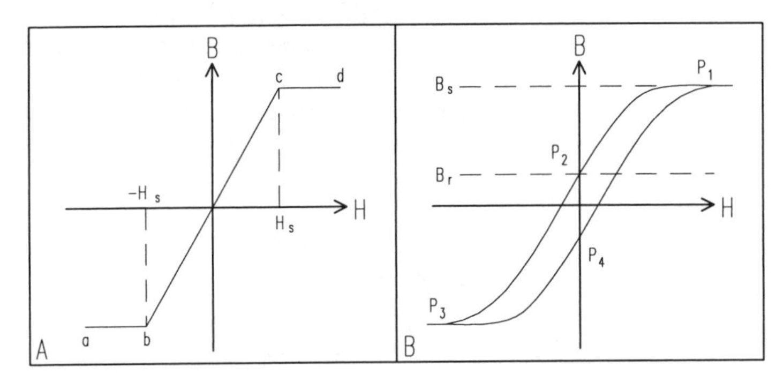

Figure 12-4. In contrast with the ideal *B-H* curve (A), minor hysteresis in the actual curve (B) results in some residual non-zero value of *B* when *H* is reduced to zero, known as the *retentivity* (adapted from Halliday & Resnick, 1974, © John Wiley and Sons, Inc.; Carlson & Gisser, 1981).

The *effective permeability* μ_a of a material is influenced to a significant extent by its geometry; Bozorth and Chapin (1942) showed how μ_a for a cylindrical rod falls off with a decrease in the length-to-diameter ratio. This relationship can be attributed to the so-called *demagnetization factor* (Hine, 1968). When a ferrous rod is coaxially aligned with the lines of flux of a magnetic field, a magnetic dipole develops in the rod itself. The associated field introduced by the north and south poles of this dipole opposes the ambient field, with a corresponding reduction of flux density through the rod. The lowered value of μ_a results in a less sensitive magnetometer, in that the "flux-gathering" capability of the core is substantially reduced.

Consider again the cylindrical rod sensor presented in Figure 12-3, now in the absence of any external magnetic field B_e. When the drive coil is energized, the lines of flux generated by the excitation current are as shown in Figure 12-5 below. Obviously there will be a strong coupling between the drive coil and the sense coil, an undesirable situation indeed since the output signal is supposed to be related to the strength of the external field only.

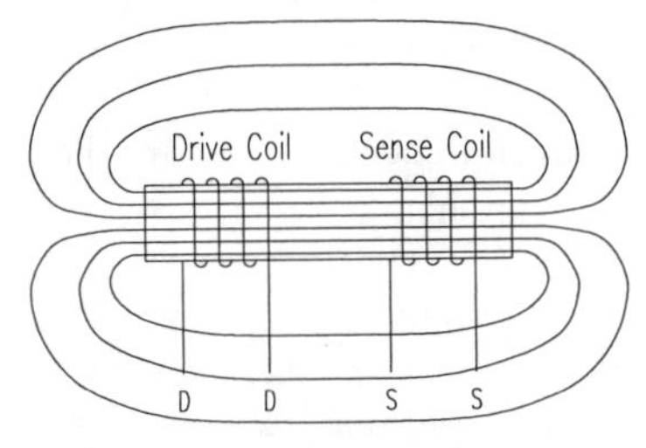

Figure 12-5. Strong coupling between the drive coil and the sense coil in this solonoidal design can interfere with the measurement of an external magnetic field.

One way around this problem is seen in the Vacquier configuration developed in the early 1940s, where two parallel rods collectively form the core, with a common sense coil as illustrated in Figure 12-6 (Primdahl, 1979). The two rods are simultaneously forced into and out of saturation, excited in antiphase by identical but oppositely wound solonoidal drive windings. In this fashion, the magnetization fluxes of the two drive windings effectively cancel each other, with no net effect on the surrounding sense coil.

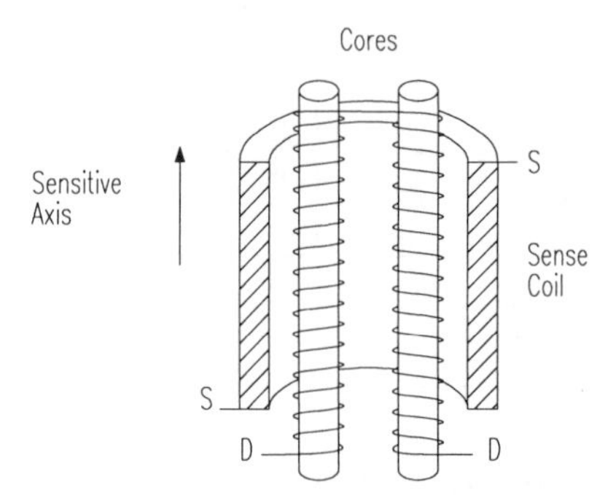

Figure 12-6. Identical but oppositely wound drive windings in the Vacquier configuration cancel the net effect of drive coupling into the surrounding sense coil, while still saturating the core material (adapted from Primdahl, 1979).

Bridges of magnetic material may be employed to couple the ends of the two rods together in a closed-loop fashion for more complete flux linkage through the core. This configuration is functionally very similar to the ring-core design first employed in 1928 by Aschenbrenner and Goubau (Geyger, 1957). In practice, there are a number of different implementations of various types of sensor cores and coil configurations as described by Stuart (1972) and Primdahl (1979). These implementations are generally divided into two classes, parallel and orthogonal, depending on whether the excitation *H*-field is parallel or perpendicular to the external *B*-field being measured. Alternative excitation strategies (sine wave, square wave, sawtooth ramp) also contribute to the variety of implementations seen in the literature. Hine (1968) outlines four different classifications of saturable inductor magnetometers based on the method of readout (i.e., how the output emf is isolated for evaluation):

- Fundamental frequency.
- Second harmonic.
- Peak output.
- Pulse-difference.

Unambiguous 360-degree resolution of the earth's geomagnetic field requires two sensing coils at right angles to each other. The ring-core geometry lends itself to such dual-axis applications in that two orthogonal pick-up coils can be configured in a symmetrical fashion around a common core. Because the drive field follows a circular path around the toroid, from the perspective of either sense coil, the flux up one side of the ring precisely cancels the flux running down the other side (Ramsden, 1994). Since there are no distinct poles in a closed-ring design, *demagnetization effects*, although still present (Stuart, 1972), are less severe. The use of a ring geometry also leads to more complete flux linkage throughout the core, implying less required drive excitation for lower power operation, and the *zero offset* can be minimized at time of manufacture by rotating the circular core (Primdahl, 1979). For these reasons, along with ease of manufacture, toroidal ring-core sensors are commonly employed in many of the low-cost fluxgate compasses available today.

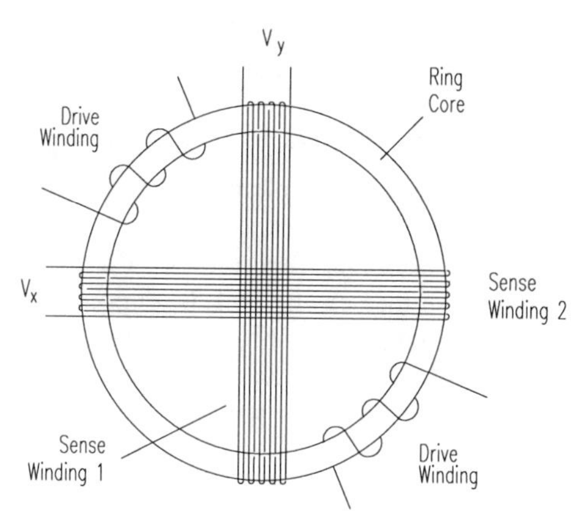

Figure 12-7. The two-channel ring core with toroidal excitation, extended by Gordon and Lundsten (1970) from the Geyger (1962) dual-drive configuration shown above, is the most popular design today in low-cost fluxgate compasses (adapted from Acuna & Pellerin, 1969).

The integrated DC output voltages V_x and V_y of the orthogonal sensing coils vary as sine and cosine functions of θ, where θ is the angle of the sensor unit relative to the earth's magnetic field. The instantaneous value of θ can be easily derived by performing two successive A/D conversions on these voltages and taking the arctangent of their quotient:

$$\theta = \arctan \frac{V_x}{V_y}.$$

Another popular two-axis core design is seen in the *Flux Valve* magnetometer developed by Sperry Corporation (Figure 12-8A). This three-legged spider configuration employs three horizontal sense coils 120 degrees apart, with a common vertical excitation coil in the middle (Hine, 1968). Referring to Figure 12-8B, the upper and lower "arms" of the sense coil S are excited by the driving coil D, with a magnetizing force H_x developed as indicated by the arrows. In the absence of an external field H_e, the flux generated in the upper and lower arms by the excitation coil is equal and opposite due to symmetry.

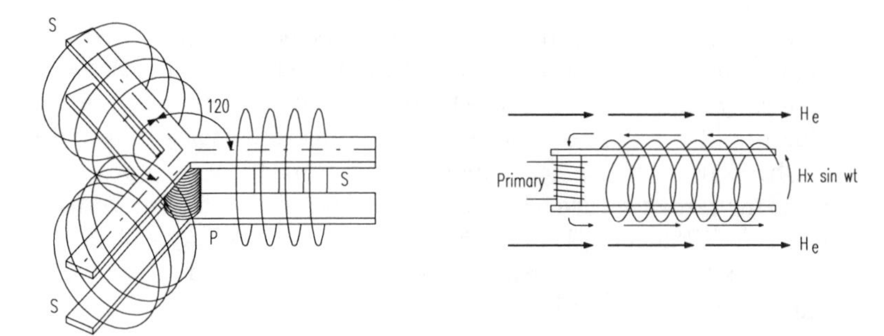

Figure 12-8. The Sperry *Flux Valve* consisted of a common drive winding P in the center of three sense windings symmetrically arranged 120 degrees apart (adapted from Hine, 1968).

When this assembly is placed in an axial magnetic field H_e, however, the instantaneous excitation field H_x complements the flow in one arm, while opposing the flow in the other. This condition is periodically reversed in the arms, of course, due to the alternating nature of the driving function. A second-harmonic output is induced in the sensing coil S, proportional to the strength and orientation of the ambient field. By observing the relationships between the magnitudes of the output signals from each of the three sense coils (Figure 12-9), the angular relationship of the *Flux Valve* with respect to the external field can be unambiguously determined.

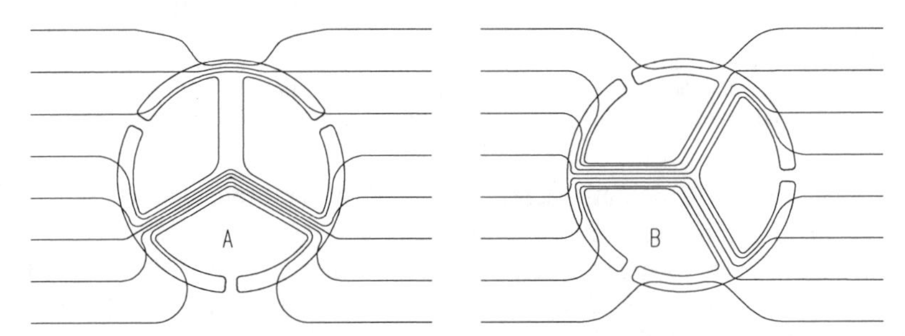

Figure 12-9. The *Flux Valve* magnetometer uses a spider-core configuration to unambiguously resolve the sensor's angular relationship to the geomagnetic field (adapted from Lenz, 1990, © IEEE).

When maintained in a level attitude, a two-axis fluxgate compass will measure the horizontal component of the earth's magnetic field, with the decided advantages of low power consumption, no moving parts, intolerance to shock and vibration, rapid start-up, and relatively low cost. If a ground vehicle is expected to operate over uneven terrain, the sensor coil is often *gimbal-mounted* and mechanically dampened to prevent serious errors introduced by the vertical component of the geomagnetic field. At latitudes encountered in the United States, a 1-degree tilt condition can result in as much as a 5-degree error in heading (Dahlin & Krantz, 1988).

Since gimbal mounting is not always effective under high-*G* operating conditions, some manufacturers have resorted to a three-axis strap-down sensor configuration, and perform a coordinate-transform from sensor coordinates to the horizontal plane of the Earth based on measured vehicle attitude (pitch and roll). Non-gimbaled strap-down sensors are advantageous as well from the standpoint of autocalibration routines that attempt to compensate for surrounding vehicle magnetic anomalies. If the sensor orientation is allowed to change with respect to the vehicle, such automatic compensation is generally not very effective.

12.2.1 Zemco Fluxgate Compasses

One of the first sensors actually employed on ROBART II for navigational referencing purposes was a fluxgate compass manufactured by Zemco Electronics, San Ramon, CA, model number DE-700. This very low-cost (around $40) unit featured a rotating analog dial and was originally intended for 12-volt DC operation in automobiles.

Figure 12-10. The Zemco *DE-700* fluxgate compass was used on ROBART II to resolve the 180-degree ambiguity in a ceiling-mounted heading reference (courtesy Naval Command Control and Ocean Surveillance Center).

A system block diagram is presented in Figure 12-11. The sensor consists of two orthogonal pick-up coils arranged around a toroidal excitation coil, driven in turn by a square-wave oscillator. The outputs V_x and V_y of amplifier channels A and B are applied across an air-core resolver to drive the display indicator. The standard resolver equations (ILC, 1982) for these two voltages are:

$$V_x = K_x \sin\theta \sin(\omega t + a_x)$$
$$V_y = K_y \cos\theta \sin(\omega t + a_y)$$

where:

θ = the resolver shaft angle
$\omega = 2\pi f$, where f is the excitation frequency
K_x and K_y are ideally equal transfer function constants
a_x and a_y are ideally zero time phase shifts.

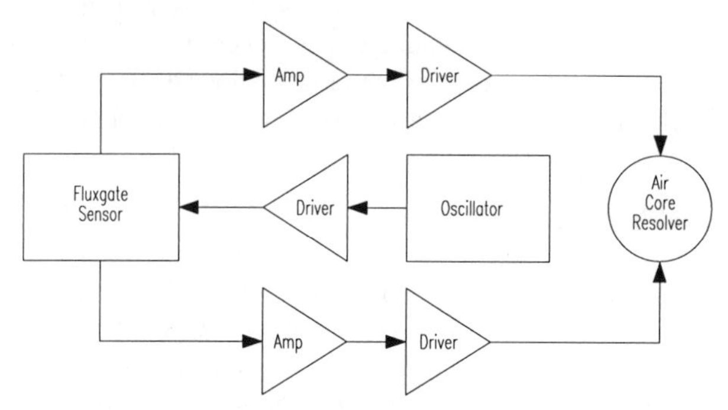

Figure 12-11. Block diagram of ZEMCO Model *DE-700* fluxgate compass (courtesy ZEMCO, Inc.).

Thus, for any static spatial angle θ, the equations reduce to:

$$V_x = K_x \sin\theta$$
$$V_y = K_y \cos\theta$$

which can be combined to yield:

$$\frac{V_x}{V_y} = \frac{\sin\theta}{\cos\theta} = \tan\theta .$$

Magnetic heading θ therefore is simply the arctangent of V_x over V_y, as previously stated.

Problems associated with the use of this particular fluxgate compass on ROBART II, however, included a fairly high current consumption (250 milliamps), and stiction in the resolver reflecting back as load into the drive circuitry, introducing some error for minor changes in vehicle heading. In addition, the sensor itself was affected by surrounding magnetic anomalies, some that existed on board the robot (i.e., current flow in nearby cable runs, drive and head positioning motors), and some present in the surrounding environment (metal desks, bookcases, large motors, etc.).

The most serious interference turned out to be the fluctuating magnetic fields due to power cables in close proximity (on the order of 12 inches) to the fluxgate sensor. As various auxiliary systems on the robot were powered up when needed and later deactivated to save power, the magnetic field surrounding the sensor would change accordingly. Significant errors could be introduced as well by minor changes in the position of cable runs, which occurred as a result of routine maintenance and trouble shooting. These problems were minimized by securing all cable runs with plastic tie-downs, and adopting a somewhat standardized protocol regarding which auxiliary systems would be activated when reading the compass.

There is no ready solution, however, for the interference effects of large metallic objects within the operating environment, and deviations of approximately 4 degrees were observed when passing within 12 inches of a large metal cabinet, for example. A final source of error was introduced by virtue of the fact the fluxgate compass had been mounted on the robot's head, as far away as possible from the effects of the drive motors and power distribution lines discussed above. The exact head position could only be read to within 0.82 degrees due to the limited resolution of the 8-bit A/D converter. In any event, an overall system error of ±10 degrees was typical and grossly insufficient for reliable dead-reckoning calculations, but the compass was not originally intended for this use (see Section 16.3.1).

This analog compass was later replaced by a newer digital version produced by Zemco, model *DE-710*, which cost approximately $90. The system block diagram is shown in Figure 12-12. This unit contained a built-in *ADC0834* A/D converter to read the amplified outputs of the two sensor channels, and employed its own *COP 421-MLA* microprocessor, which drove a liquid crystal display (LCD). All communication between the A/D converter, microprocessor, and display driver was serial in nature, with a resulting slow update rate of 0.25 Hz. The built-in LCD simulated an analog dial with an extremely coarse resolution of 20 degrees between display increments, but provision was made for serial output to an optional shift register and associated three-digit numerical display.

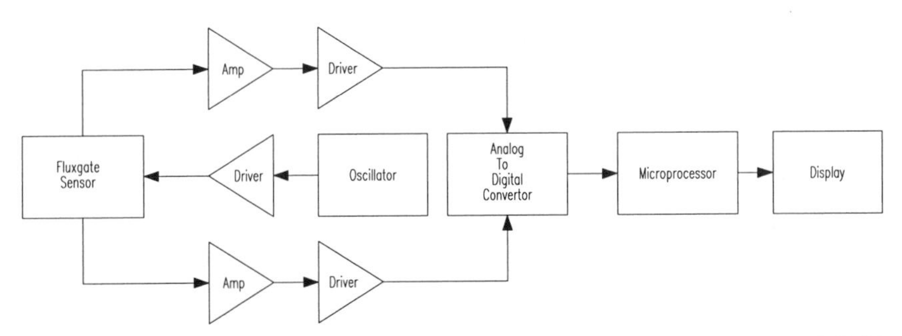

Figure 12-12. Block diagram of ZEMCO model *DE-710* fluxgate compass (courtesy ZEMCO, Inc.).

All things considered, it was determined to be more practical to discard the built-in microprocessor, A/D converter, and LCD display, and interface an external A/D converter directly to the amplifier outputs as before with the analog version. This approach resulted in a decrease in supply current from 168 to 94 milliamps. Power consumption turned out to be less of a factor when it was discovered the circuitry could be powered up for a reading, and then deactivated afterwards with no noticeable effect on accuracy.

Overall system accuracy for this configuration was typically ±6 degrees, although a valid comparison to the analog version is not possible since the digital model was mounted in a different location to minimize interference from nearby circuitry. The amount of effort put into calibration of the two systems must also be taken into account; the calibration procedure as performed was an iterative process not easily replicated from unit to unit with any quantitative measure.

12.2.2 Watson Gyro Compass

A combination fluxgate compass and solid-state rate gyro package (part number FGM-G100DHS-RS232) is available from Watson Industries, Eau Claire, WI. The system contains its own microprocessor to integrate the information from both the rate gyro and the compass for a more stable output less susceptible to interference, with an update rate of 40 Hz. The piezoelectric tuning-fork gyro (see Chapter 13) serves to filter out the effects of magnetic anomalies in the surrounding environment, while the compass counters the long-term drift of the gyro. The toroidal ring-core fluxgate sensor is internally gimbal-mounted for improved accuracy. An overall block diagram is presented in Figure 12-13.

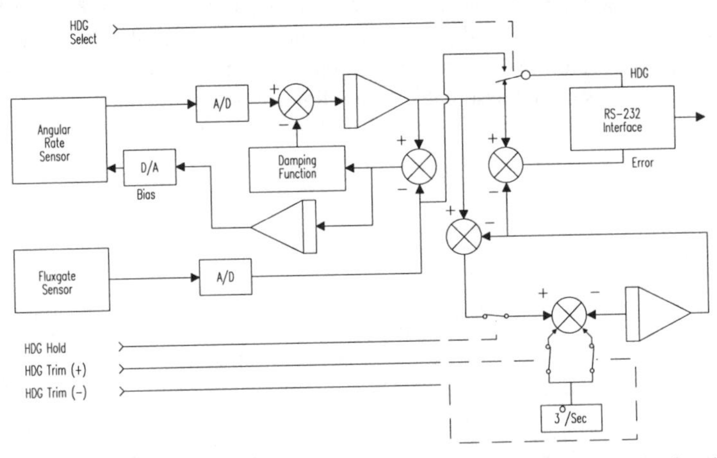

Figure 12-13. Block diagram of Watson fluxgate compass and rate gyro combination (courtesy Watson Industries, Inc.).

The Watson unit measures 2.5 by 1.75 by 3.0 inches, and weighs only 10 ounces. This integrated package is a much more expensive unit ($2500) than the low-cost Zemco fluxgate compass but is advertised to have higher accuracy (±2 degrees). Power supply requirements are 12-volts DC at 200 milliamps, and the unit provides an analog voltage output as well as a 12-bit digital output over a

2400-baud RS-232 serial link. Extensive testing of the Watson compass/gyro package on the ModBot (Figure 12-14) confirmed a fairly repeatable accuracy of about ±2 degrees in an indoor warehouse environment with planar floor surfaces.

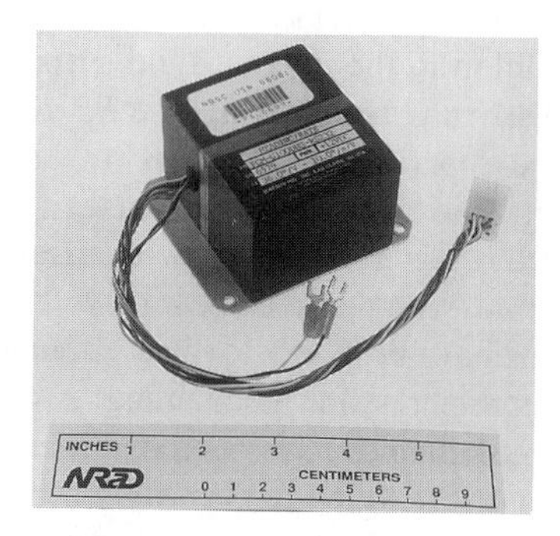

Figure 12-14. The combination fluxgate compass and solid-state rate gyro package from Watson Industries was tested on the ModBot as a potential navigational sensor for semi-structured warehouse operations (courtesy Naval Command Control and Ocean Surveillance Center).

12.2.3 KVH Fluxgate Compasses

KVH Industries, Inc., Middletown, RI, offers a complete line of fluxgate compasses and related accessories, ranging from inexpensive units targeted for the individual consumer up through sophisticated systems intended for military applications (KVH, 1993). The *C100 Compass Engine* is a versatile low-cost (less than $700) developers kit that includes a microprocessor-controlled stand-alone fluxgate subsystem based on a two-axis toroidal ring-core sensor.

Figure 12-15. The *C-100 Compass Engine* incorporates a gimbaled two-axis toroidal fluxgate design (courtesy KVH Industries, Inc.).

Two different sensor options are offered with the *C-100*: 1) the *SE-25* sensor, recommended for applications with a tilt range of ±16 degrees and 2) the *SE-10* sensor, for applications requiring up to ±45 degrees. The *SE-25* sensor provides internal gimbaling by floating the sensor coil in an inert fluid inside the lexan housing. The *SE-10* sensor provides an additional two-degree-of-freedom pendulous gimbal in addition to the internal fluid suspension. The *SE-25* sensor mounts on top of the sensor PC board, while the *SE-10* is suspended beneath it. The sensor board can be separated as much as 48 inches from the detachable electronics board with an optional cable if so desired.

The resolution of the *C100* is ±0.1 degree, with an advertised accuracy of ±0.5 degrees (after compensation, with the sensor card level) and a repeatability of ±0.2 degrees. Accuracy claims have been substantiated in fielded robotic systems by a number of users, but problems in maintaining a steady horizontal gimbal orientation due to vehicle dynamics have been reported (Rahim, 1993). Separate ±180-degree adjustments are provided for *variation* as well as *index offset* (in the event the sensor unit cannot be mounted in perfect alignment with the vehicle's axis of travel). System damping can be user-selected, anywhere in the range of 0.1 to 24 seconds settling time to final value.

An innovative automatic compensation algorithm employed in the C100 is largely responsible for the high accuracy obtained by such a relatively low-priced system. This software routine runs on the controlling microprocessor mounted on the electronics board and continually corrects for magnetic anomalies associated with the host vehicle. Three alternative user-selectable procedures are offered:

- *Eight-Point Autocompensation* — Starting from an arbitrary heading, the platform turns full circle, pausing momentarily at approximately 45-degree intervals. No known headings are required.
- *Circular Autocompensation* — Starting from an arbitrary position, the platform turns slowly through a continuous 360-degree circle. No known headings are required.
- *Three-Point Autocompensation* — Starting from an arbitrary heading, the platform turns and pauses on two additional known headings approximately 120 degrees apart.

Correction values are stored in a look-up table in non-volatile EEPROM memory. The automatic compensation routine also provides a quantitative indicator of the estimated quality of the current compensation and the magnitude of any magnetic interference present (KVH, 1993).

The *C100* configured with an *SE-25* coil assembly weighs just 2.25 ounces and draws 40 milliamps at 8 to 18 volts DC (or 18 to 28 volts DC). The combined sensor and electronics boards measure 1.8 inches wide by 4.5 inches long. RS-232 (300-9600 baud) and NMEA-0183 digital outputs are provided, as well as linear and sine/cosine analog voltage outputs. Display and housing options are also available, and two variants of the fluxgate design can be special ordered:

- *C-100-x1* — A 3-D *Helmholtz cage* is mounted around the fluxgate sensor to neutralize hard-iron field anomalies in applications where the normal hard-iron compensation is inadequate.
- *C-100-x2* — This version provides the directions and magnitudes of sensor tilt relative to the housing.

In addition, the company also markets a low-cost sensor assembly with analog sine/cosine outputs (*Model AC75*), featuring a separate hermetically sealed fluxgate that can be remotely located in the most favorable magnetic environment.

12.2.4 Applied Physics Systems Miniature Orientation Sensor

Applied Physics Systems, Mountain View, CA, offers a miniature three-axis (pitch, roll, and yaw) angular orientation sensor in a completely self-contained package measuring 0.75 by 0.75 by 4.5 inches and weighing only 50 grams (Figure 12-16). The *Model 544* sensor contains a three-axis accelerometer along with a non-gimbaled three-axis fluxgate magnetometer to produce pitch, roll, and yaw angles that are output in digital format over a bidirectional serial datalink (TTL or RS-232 compatible).

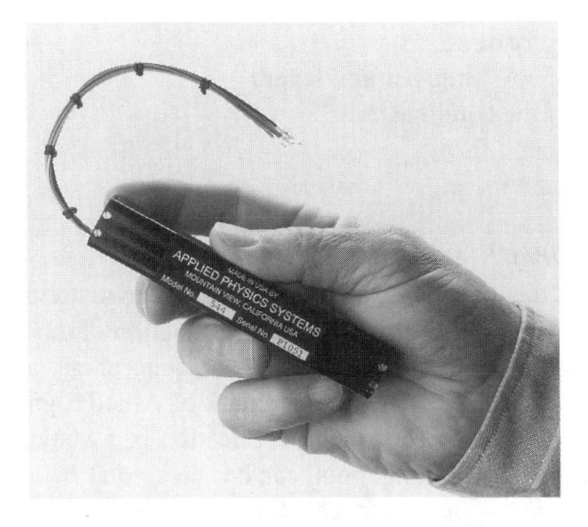

Figure 12-16. The rounded edges of the *Model 544* angular orientation sensor allow the self-contained package to fit inside a 1-inch inside-diameter cylinder (courtesy Applied Physics Systems).

Alternatively, individual acceleration and magnetic field values can be read directly for external postprocessing, with scale factors of about 4 volts per Gauss for the magnetometer, and 2 volts per *G* for the accelerometer readings. An internal *68HC11* microprocessor converts the analog sensor inputs to digital output format with an overall device accuracy of about ±0.5 degrees for each axis (pitch, roll, yaw), providing second-order temperature compensation over a range of 0 to 70°C. Maximum update rate is about 5 Hz if all six output values are read. Power consumption is approximately 50 milliamps at 5 volts DC and 30 milliamps at -5 volts DC. Various configurations of the system are available to suit different applications.

12.3 Magnetoinductive Magnetometers

A relatively new development, *magnetoinductive* sensors employ a single solenoidal winding for each axis and consume roughly an order of magnitude less power than conventional fluxgates (PNI, 1994). The sense coil serves as the inductive element in a low-power L/R relaxation oscillator, with its effective inductance being influenced by the ambient magnetic field component running parallel to the coil axis. For such a magnetic-core solenoid, the effective inductance *L* can be shown to be (Kim & Hawks, 1989):

$$L = \mu_o n^2 V \frac{dB}{dH}$$

where:

L = coil inductance
μ_o = permeability of air
n = number of coil turns per unit length
V = volume of the core material
B = total magnetic flux
H = magnetizing force.

Recall that *dB/dH*, the slope of the *B-H* curve, is simply the relative permeability μ of the core material. For a typical core material such as *METGLAS 2705M*, the relative permeability μ varies as a function of the magnetizing force *H* as shown in Figure 12-17. (*METGLAS* is an amorphous alloy of iron, boron, silicon, and carbon, and a registered trademark of Allied Signal Corporation.) By biasing the magnetizing force through application of a static DC current flow in the solenoidal coil, the operating point can be centered at H_o in the linear region of the curve as shown. The presence of an external magnetic field H_e adds or subtracts to the bias field H_o, shifting the operating point accordingly, with an associated change in the effective permeability $\mu(H)$.

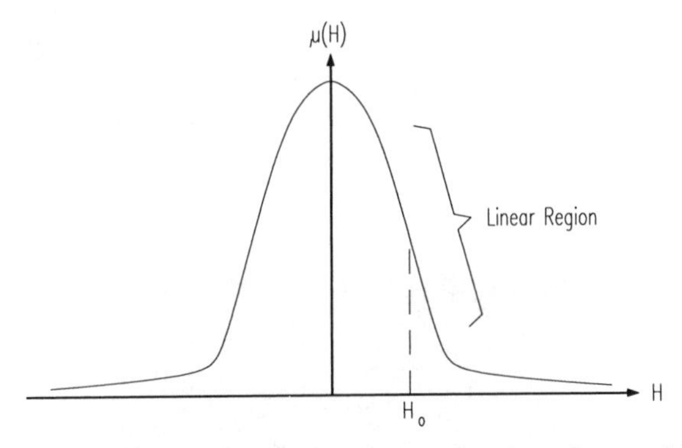

Figure 12-17. Plot of relative permeability *μ(H)* as a function of magnetizing force *H* for METGLAS core material (adapted from Kim & Hawks, 1989).

The period of the oscillator output is proportional to *L/R*, where *L* has been shown to be directly proportional to *μ(H)*. Thus the greater contribution from the ambient magnetic field H_e, the lower the effective inductance *L*, resulting in an associated increase in oscillator frequency. As a result of this direct relationship, the observed frequency shift can be as much as 100 percent as the sense coil is rotated from a parallel to antiparallel orientation with respect to the geomagnetic field (Kim & Hawks, 1989). Since there is no need to drive the core into and out of saturation as in the case of fluxgate designs, the required electronics is much simpler and power consumption greatly reduced. An additional benefit is seen in the inherently digital nature of the output signal, eliminating the cost and complexities of additional signal conditioning and an analog-to-digital interface.

12.3.1 Precision Navigation *TCM Magnetoinductive Compass*

Precision Navigation, Inc., Mountain View, CA, offers an integrated electronic solution to the problems introduced by conventional mechanical gimbaling under conditions of high dynamic loading typically experienced by ground vehicles operating on uneven terrain. The *TCM* compass shown in Figure 12-18 employs a three-axis *magnetoinductive* magnetometer to measure the *X-Y-Z* components of the geomagnetic field, along with a two-axis electrolytic inclinometer to measure vehicle attitude. The integral microprocessor uses the pitch-and-roll information to automatically correct the magnetometer outputs for tilt, providing a temperature-compensated heading solution up to 16 times each second, accurate to within ±1 degree.

A block diagram illustrating the inherent simplicity of the sense coil interface is shown in Figure 12-19. The solenoidal inductor is connected in series with resistor R_2 to form a relaxation oscillator in conjunction with the *LM339* comparator. R_2 can be used to adjust the DC coil bias current to establish the desired operating point H_o, while R_3 sets the center frequency (approximately 200

KHz) and current swing of the oscillator circuit (Kim & Hawks, 1989). The square-wave oscillator output of each of three identical channels is fed directly to the onboard microprocessor without the need for complicated interface circuitry.

Figure 12-18. The *TCM* compass employs a three-axis strap-down magnetometer in conjunction with a two-axis tilt sensor to compensate for variations in vehicle attitude (courtesy Precision Navigation, Inc.).

Automatic distortion-detection algorithms are incorporated that raise a warning flag when magnetic disturbances (i.e., close-proximity metallic objects or electrical cabling) are compromising compass accuracy. Pitch-and-roll outputs are available for external use with 0.1-degree resolution at an accuracy of ±0.2 degrees. Ambient temperature information is also provided over a range of -20 to +70°C, with an accuracy of ±0.5 degrees. Both digital outputs (RS-232 or NMEA-0183) and linear quadrature analog outputs (0-5 volts) are standard.

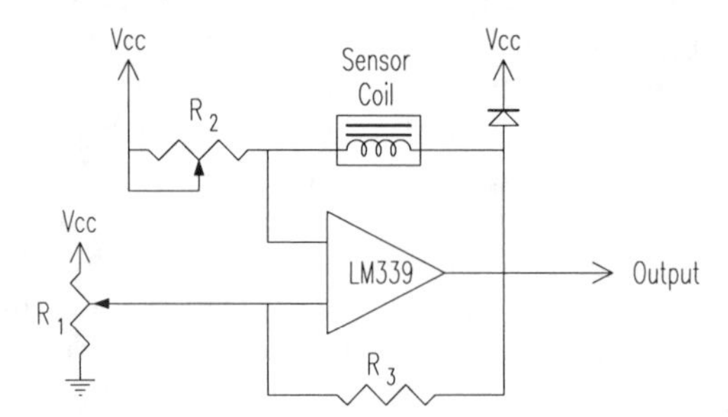

Figure 12-19. Block diagram of a single-axis sense element as implemented on the *TCM* compass (adapted from Kim & Hawks, 1989).

Power requirements for the *TCM* compass are 5 to 25 volts DC at 6 to 12 milliamps, depending on user configuration. The OEM circuit board measures 2.5

by 2 inches wide by 1.1 inches high, weighs 1.6 ounces, and with a tilt range of ±25 degrees costs only $700. (Additional tilt ranges of 60 and 90 degrees are also available at slightly higher cost.) The moderate price, extremely low power consumption, elimination of gimbal-induced measurement errors, small size and weight, plus availability of pitch, roll, and ambient temperature outputs make the *TCM-Series* a strong contender for mobile robotic applications. Field performance evaluations are currently underway for early prototypes installed on both ROBART III and the MDARS Interior robot (Chapter 1).

An extremely low-cost ($50) two-axis electronic compass without tilt compensation, the *Vector-2X*, is available as well for less demanding applications, with an overall accuracy of ±2 degrees.

12.4 Hall-Effect Compasses

Recall from Section 3.2.1 that *Hall-effect* sensors in the presence of an external magnetic field develop a DC voltage across a semiconductor region that is proportional to the magnetic field component at right angles to the direction of current flow (Wood, 1986). One advantage of this technology (i.e., relative to the fluxgate) is the inherent ability to directly sense a static flux, resulting in much simpler readout electronics. Early Hall magnetometers could not match the sensitivity and stability of the fluxgate (Primdahl, 1979), but the sensitivity of Hall devices has improved significantly. The more recent indium-antimonidide devices have a lower sensitivity limit of 10^{-3} Gauss (Lenz, 1990).

The Navy in the early 1960s showed considerable interest in a small solid-state Hall-effect compass for low-power extended operations in sonobuoys (Wiley, 1964). A number of such prototypes were built and delivered by Motorola for evaluation. The Motorola compass employed two orthogonal Hall-effect devices for temperature-nulled non-ambiguous resolution of the geomagnetic field vector. Each sensor element was fabricated from a 2- by 2- by 0.1-millimeter indium-arsenide-ferrite sandwich and inserted between two wing-like *mumetal* flux concentrators as shown in Figure 12-20. It is estimated the 2-inch magnetic concentrators increased the flux density through the sensing elements by two orders of magnitude (Wiley, 1964). The output of the Motorola unit was a variable-width pulse train, the width of the pulse being proportional to the sensed magnetic heading. Excellent response linearity was reported down to flux densities of 0.01 Gauss (Wiley, 1962).

Maenaka, et al. (1990) report on the development of a monolithic silicon magnetic compass at the Toyohashi University of Technology in Japan, based on two orthogonal Hall-effect sensors. Their use of the terminology "magnetic compass" is perhaps an unfortunate misnomer, in that the prototype device was tested with an external field of 1,000 Gauss. Contrast this field strength with that of the earth's magnetic field, which varies from only about 0.1 Gauss at the equator to about 0.9 Gauss at the poles. Silicon-based Hall-effect sensors have a lower sensitivity limit of around 10 Gauss (Lenz, 1990). It is likely the Toyohashi University device was intended for other than geomagnetic applications, such as remote position sensing of rotating mechanical assemblies.

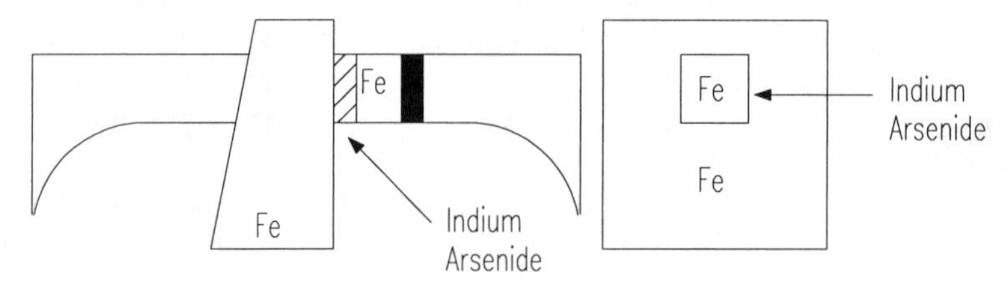

Figure 12-20. A pair of indium-arsenide-ferrite Hall-effect sensors (one shown) are positioned between flux concentrating wings of *mumetal* in this early Motorola prototype (adapted from Wiley, 1964).

This prototype Hall-effect magnetometer is still of interest in that it represents one of the first fully self-contained implementations of a two-axis magnetometer in integrated circuit form. Two vertical Hall cells (Maenaka, et al, 1987) are arranged at right angles as shown in Figure 12-21 on a 4.7-millimeter square chip, with their respective outputs coupled to a companion signal processing IC of identical size. (Two separate chips were fabricated for the prototype instead of a single integrated unit to enhance production yield.) The sensor and signal processing ICs are interconnected (along with some external variable resistors for calibration purposes) on a glass-epoxy printed circuit board.

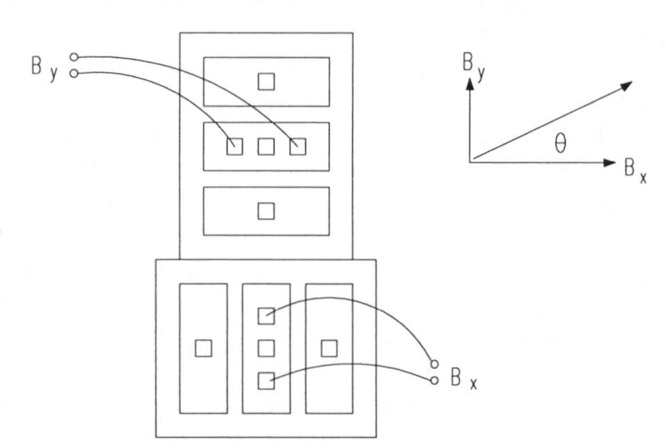

Figure 12-21. Two vertical Hall cells are arranged at right angles on a 4.7-millimeter-square chip in this two-axis magnetometer developed at the Toyohashi University of Technology in Japan (adapted from Maenaka, et al., 1990).

The dedicated signal-processing circuitry converts the *B*-field components B_x and B_y measured by the Hall sensors into an angle θ by means of the analog operation (Maenaka, et al, 1990):

$$\theta = \arctan \frac{B_x}{B_y}$$

where:

θ = angle between B-field axis and sensor
B_x = x-component of B field
B_y = y-component of B field.

The analog output of the signal-processing IC is a DC voltage that varies linearly with vector orientation of the ambient magnetic field in a plane parallel to the chip surface. Reported test results show a fairly straight-line response (i.e., ±2 percent full scale) for external field strengths ranging from 8,000 Gauss down to 500 Gauss; below this level performance begins to degrade rapidly (Maenaka, et al., 1990). A second analog output on the IC provides an indication of the absolute value of field intensity.

While the Toyohashi "magnetic compass" prototype based on silicon Hall-effect technology is incapable of detecting the earth's magnetic field, it is noteworthy nonetheless. A two-axis monolithic device of similar nature employing the more sensitive indium-antimonide Hall devices could potentially have broad appeal for low-cost applications on mobile robotic platforms. For increased sensitivity, an alternative possibility would be to use magnetoresistive sensor elements, to be discussed in the next section.

12.5 Magnetoresistive Compasses

The general theory of operation for *anisotropic magnetoresistive (AMR)* and *giant magnetoresistive (GMR)* sensors as used in short-range proximity detection was previously presented in Chapter 3. Recall three properties of the *magnetoresistive* magnetometer make it well suited for application as a geomagnetic sensor: 1) high sensitivity, 2) directionality, and 3) in the case of AMR sensors, the characteristic "flipping" action associated with the direction of internal magnetization.

AMR sensors have an open-loop sensitivity range of 10^{-2} to 50 Gauss (which easily covers the 0.1- to 1.0-Gauss range of the earth's horizontal magnetic field component), and limited-bandwidth closed-loop sensitivities approaching 10^{-6} Gauss (Lenz, 1990). Excellent sensitivity, low power consumption, small package size, and decreasing cost make both AMR and GMR sensors increasingly popular alternatives to the more conventional fluxgate designs used in robotic vehicle applications.

12.5.1 Philips AMR Compass

One of the earliest magnetoresistive sensors to be applied to a magnetic compass application is the *KMZ10B* offered by Philips Semiconductors BV, The Netherlands (Dibburn & Petersen, 1983; Kwiatkowski & Tumanski, 1986; Petersen, 1989). The limited sensitivity of this device (approximately 0.1 mV/A/m with a supply voltage of 5V DC) in comparison to the earth's maximum horizontal magnetic field (15 A/m) means that considerable attention must be given to the error-inducing effects of temperature and offset drift (Petersen, 1989).

One way around these problems is to exploit the "flipping" phenomenon (Chapter 3) by driving the device back and forth between its two possible magnetization states with square-wave excitation pulses applied to an external coil (Figure 12-22). This switching action toggles the sensor's axial magnetic field as shown in Figure 12-22A, resulting in the alternating response characteristics depicted in Figure 12-22B. Since the sensor offset remains unchanged while the signal output due to the external magnetic field H_y is inverted (Figure 12-22A), the undesirable DC offset voltages can be easily isolated from the weak AC signal.

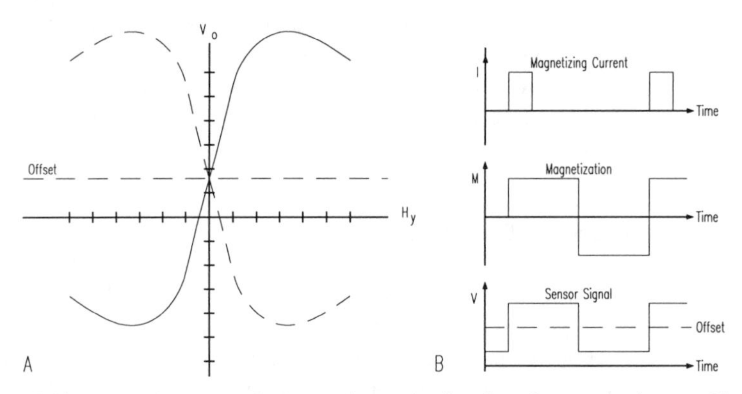

Figure 12-22. External current pulses set and reset the direction of magnetization, resulting in the "flipped" response characteristics shown by the dashed line. Note the DC offset of the device remains constant, while the signal output is inverted (adapted from Petersen, 1989).

A typical implementation of this strategy is shown in Figure 12-23. A 100-Hz square-wave generator is capacitively coupled to the external excitation coil L which surrounds two orthogonally mounted magnetoresistive sensors. The sensors' output signals are amplified and AC-coupled to a synchronous detector driven by the same square-wave source. The rectified DC voltages V_{H1} and V_{H2} are thus proportional to the measured magnetic field components H_1 and H_2. Determination of applied field direction is dependent on the ratio as opposed to absolute values of these output signals, and so as long as the two channels are calibrated to the same sensitivity, no temperature correction is required (Fraden, 1993).

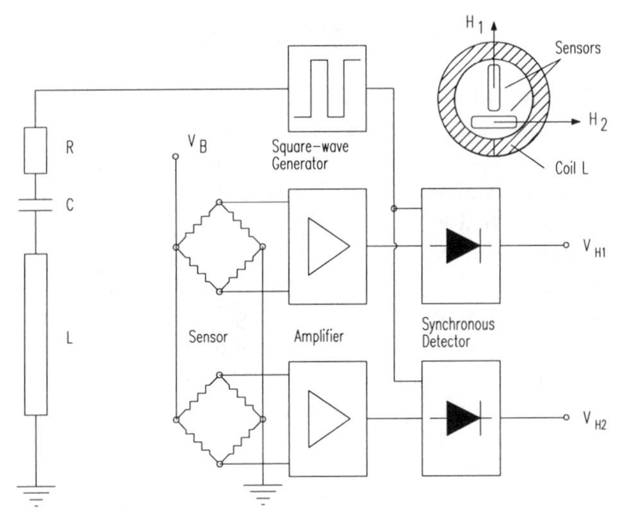

Figure 12-23. Block diagram of a two-axis magnetic compass system based on a commercially available anisotropic magnetoresistive sensor such as the Philips *KMZ10B* (Petersen, 1989).

12.5.2 Space Electronics AMR Compass

The Space Electronics *Micro-Mag* sensor introduced in Chapter 3 (SEI, 1994; Lao, 1994) can be configured as shown in Figure 12-24 to function as an anisotropic magnetoresistive (AMR) compass. The integral 350-ohm temperature compensation resistor (RTD) is connected in the lower arm of a Wheatstone bridge in series with a 100-ohm 10-turn trimming resistor. Two identical channels are required, with their associated AMR sensors mounted in an orthogonal fashion to yield output voltages proportional to the sine and cosine of magnetic field azimuth.

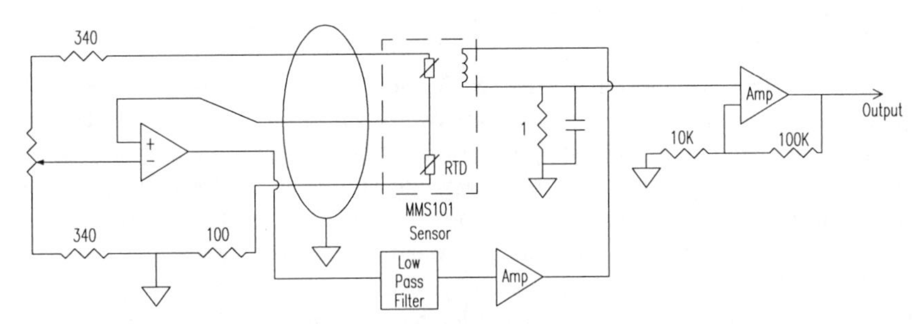

Figure 12-24. Typical application circuit for the SEI *MMS101 MicroMag* that provides an output voltage proportional to the cosine of magnetic azimuth for a gimbaled sensor in the horizontal plane (courtesy Space Electronics, Inc.).

12.5.3 Honeywell *HMR Series Smart Digital Magnetometer*

The *Honeywell Magnetoresistive (HMR) Series* of magnetometers incorporates three orthogonal sensor axes, each consisting of a *permalloy* thin-film Wheatstone bridge configuration deposited on a silicon substrate as discussed in Chapter 3 (Honeywell, 1994b). Changes in bridge resistance are converted to a digital output signal (prespecified RS-232 or RS-485) by internal A/D converters and a dedicated microprocessor, with 12-bit output resolution (11 bits plus sign). A switching technique is employed to "flip" the sensor characteristics back and forth between the two possible magnetic states (see again Chapter 3), thus canceling the DC offset and past magnetic history of the *permalloy* bridges, in addition to any offset introduced by the sensor electronics (Honeywell, 1994a). The unit is packaged in a compact rectangular enclosure measuring 1.12 by 1.75 by 3 inches as shown in Figure 12-25.

Figure 12-25. The Honeywell *HMR-Series Smart Digital Magnetometer* is a three-axis magnetoresistive magnetometer with a sensitivity of 1 milliGauss over a measurement range of ±1 Gauss (courtesy Honeywell Solid State Electronics Center).

Output values for the three axes (X, Y, and Z) are transmitted in two-byte hexadecimal format upon request from the external host processor, where they can be combined with externally supplied information regarding vehicle attitude to calculate a tilt-compensated magnetic heading solution. At 38.4 kilobaud, the maximum update rate is 54 Hz. The current bridge temperature reading is also made available with 8-bit resolution. The magnetometer has a measurement range of ±1 Gauss (each axis) with a sensitivity level of 1 milliGauss and provides a digital resolution of 0.5 milliGauss per least-significant bit. Overall accuracy is ±1 percent of full scale. Power requirements are 12 to 15 volts DC (single supply) at 40 milliamps. An *HMR Series* Development Kit is now available from the Honeywell Solid State Electronics Center, Plymouth, MN, that includes the magnetometer, power supply, cabling, operating manual, and IBM-compatible PC software.

12.6 Magnetoelastic Compasses

A number of researchers have recently investigated the use of *magnetoelastic* (also known as *magnetostrictive*) materials as sensing elements for high-resolution magnetometers. The principle of operation is based on the changes in Young's modulus experienced by magnetic alloys when exposed to an external magnetic field. The *modulus of elasticity E* of a given material is basically a measure of its stiffness, and directly relates stress to strain as follows:

$$E = \frac{\sigma}{\varepsilon}$$

where:

E = Young's modulus of elasticity
σ = applied stress
ε = resulting strain.

Any ferromagnetic material will experience some finite amount of strain (expansion or shrinkage) in the direction of magnetization due to this *magnetostriction* phenomenon. It stands to reason that if the applied stress σ remains the same, strain ε will vary inversely with any change in Young's modulus *E*. In certain amorphous metallic alloys, this effect is very pronounced.

Barrett, et al. (1973) propose a qualitative explanation, wherein individual atoms in the crystal lattice are treated as tiny magnetic dipoles. The forces exerted by these dipoles on one another depend upon their mutual orientation within the lattice; if the dipoles are aligned end to end, the opposite poles attract, and the material shrinks ever so slightly. The crystal is said to exhibit a negative *magnetostriction constant* in this direction. Conversely, if the dipoles are rotated into side-by-side alignment through the influence of some external field, like poles will repel, and the result is a small expansion.

It follows the strength of an unknown magnetic field can be accurately measured if suitable means is employed to quantify the resulting change in length of some appropriate material displaying a high *magnetostriction constant*. There are currently at least two measurement technologies with the required resolution allowing the magnetoelastic magnetometer to be a realistic contender for high-sensitivity low-cost performance: 1) *interferometric* displacement sensing and 2) *tunneling-tip* displacement sensing.

Lenz (1990) describes a *magnetoelastic* magnetometer which employs a Mach-Zender fiber-optic interferometer to measure the change in length of a *magnetostrictive* material when exposed to an external magnetic field. A laser source directs a beam of light along two optical fiber paths by way of a beam splitter as shown in Figure 12-26. One of the fibers is coated with a material (nickel iron was used) exhibiting a high *magnetostrictive* constant. The length of this fiber therefore is stretched or compressed in conjunction with any *magnetoelastic* expansion or contraction of its coating. The output beam from this fiber-optic cable is combined in a light coupler with the output beam from the uncoated reference fiber and fed to a pair of photodetectors.

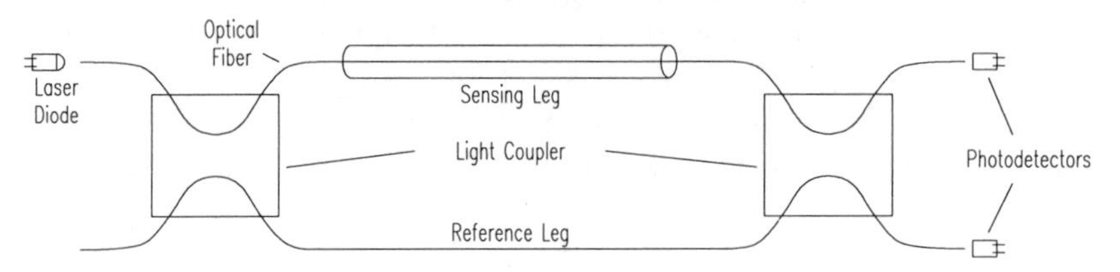

Figure 12-26. Fiber-optic magnetometers, basically a Mach-Zender interferometer with one fiber coated or attached to a magnetoelastic material, have a sensitivity range of 10^{-7} to 10 Gauss (adapted from Lenz, 1990, © IEEE).

Constructive and destructive interferences caused by differences in path lengths associated with the two fibers will cause the final output intensity as measured by the photodetectors to vary accordingly. This variation is directly related to the change in path length of the coated fiber, which in turn is a function of the magnetic field strength along the fiber axis. The prototype constructed by Lenz (1990) at Honeywell Corporation measured 4 inches long by 1 inch wide and was able to detect fields ranging from 10^{-7} Gauss up to 10 Gauss.

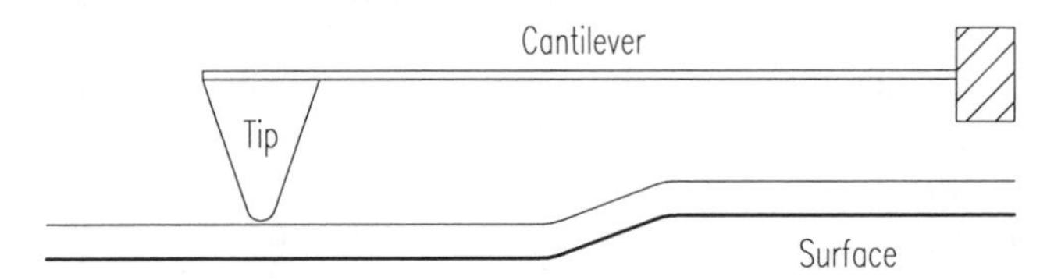

Figure 12-27. *Scanning tunneling microscopy*, invented at IBM Zurich in 1982, uses quantum mechanical tunneling of electrons across a barrier to measure separation distance at the gap (courtesy T.W. Kenny, NASA JPL).

Researchers at the Naval Research Laboratory (NRL) have developed a prototype *magnetoelastic* magnetometer capable of detecting a field as small as 6 x 10^{-5} Gauss using the *tunneling-tip-transducer* approach (Brizzolara, et al., 1989). This new displacement sensing technology, invented in 1982 at IBM Zurich, is based on the measurement of current generated by quantum mechanical tunneling of electrons across a narrow gap (Figure 12-27). An analog feedback circuit compares the measured tunnel current with a desired setpoint and outputs a drive signal to suitably adjust the distance between the tunneling electrodes with an electromechanical actuator (Kenny, et al., 1991). The instantaneous tunneling current is directly proportional to the exponential of electrode displacement. The most common actuators employed in this role are piezoelectric and electrostatic, the latter lending itself more readily to silicon micromachining techniques.

The active sense element in the NRL magnetometer is a 10-centimeter metallic-glass ribbon made from *METGLAS 2605S2*, annealed in a transverse magnetic field to yield a high magnetomechanical coupling (Brizzolara, et al., 1989). The magnetoelastic ribbon elongates when exposed to an axial magnetic field, and the magnitude of this displacement is measured by a tunneling transducer as illustrated in Figure 12-28.

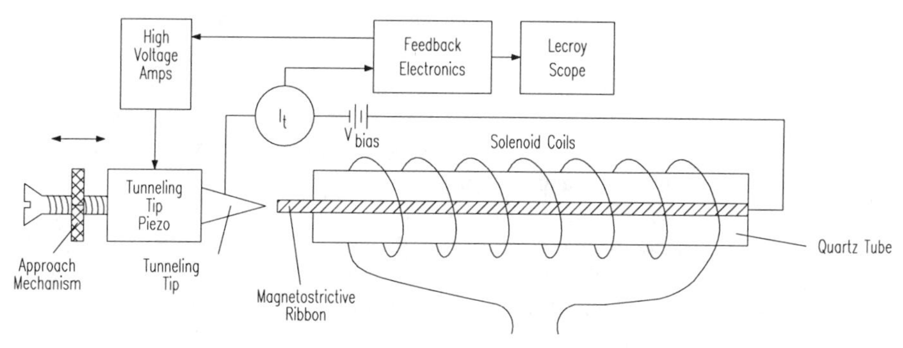

Figure 12-28. The NRL tunneling-transducer magnetometer employed a 10-centimeter magnetoelastic ribbon vertically supported in a quartz tube (Brizzolara, et al., 1989).

An electrochemically etched gold tip is mounted on a tubular piezoelectric actuator and positioned within about 1 nanometer of the free end of the *METGLAS* ribbon. The ribbon and tip are electrically biased with respect to each other, establishing a tunneling current that is fed back to the piezo actuator to maintain a constant gap separation. The degree of magnetically induced elongation of the ribbon can thus be inferred from the driving voltage applied to the piezoelectric actuator. The solenoidal coil shown in the diagram supplies a bias field of 0.85 oersted to shift the sensor into its region of maximum sensitivity.

The NRL group in collaboration with the Jet Propulsion Laboratory, Pasadena, CA, has more recently developed an alternative magnetic sensor that uses a tunneling transducer to measure the induced torque on a suspended magnet due to low-frequency field changes (DiLella, et al., 1995). The sensor consists of two micromachined silicon wafers assembled into a structure measuring approximately 1 inch by 1 inch by 0.1 inch (Figure 12-29). The upper wafer includes a permanent magnet attached to a rectangular support suspended from a pair of torsion beams. The underside of the magnet faces the tunneling tip and serves as both the tunneling counter electrode and one of two rotation control electrodes. The lower component consists of the other rotation control electrode and the tunneling tip as illustrated below.

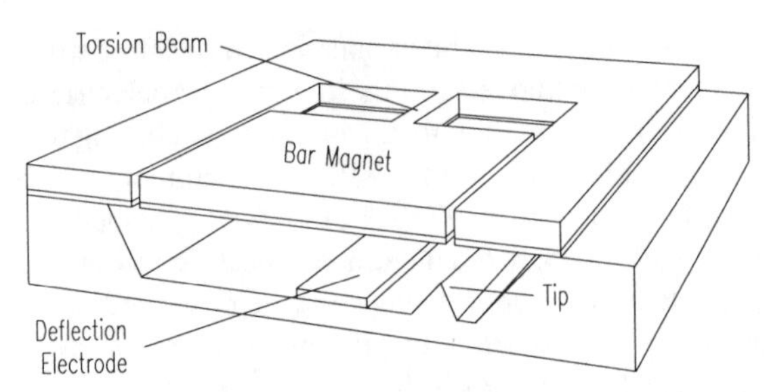

Figure 12-29. Cross-sectional diagram of the NRL/JPL micromachined magnetic-field sensor based on an electron-tunneling displacement transducer (courtesy Naval Research Lab).

Because of the offset placement of the lower rotation control electrode with respect to the longitudinal axis of the torsion beams, an electrostatic torque is generated by the voltage difference between the electrodes, rotating the magnet assembly into tunneling range of the tip. This electrostatic torque about the torsion-beam axis is balanced by the resulting torsional stress in the beams and a magnetically induced torque generated by the ambient magnetic field acting upon the permanent-magnet dipole. Once the desired tunneling current is established and maintained by a simple feedback control circuit, any subsequent change in electrode voltage can be attributed to variations in the ambient magnetic field. The calculated sensitivity limit of this sensor configuration based on fundamental noise sources is 0.002 nT/√Hz at 1 Hz, while the actual measured sensitivity of the prototype is 0.3 nT/√Hz at 1 Hz (DiLella, et al., 1995).

Fenn, et al. (1992) propose yet another tunneling *magnetoelastic* configuration with a predicted sensitivity of 2×10^{-11} Gauss, along the same order of magnitude as the cryogenically cooled SQUID. A small cantilevered beam of *METGLAS 2605S2*, excited at its resonant frequency by a gold-film electrostatic actuator, is centered between two high-permeability magnetic flux concentrators as illustrated in Figure 12-30. Any changes in the modulus of elasticity of the beam will directly affect its natural frequency; these changes in natural frequency can then be measured and directly related to the strength of the ambient magnetic field. The effective shift in natural frequency is rather small, however (Fenn reported only a 6-Hz shift at saturation), again necessitating a very precise method of measurement.

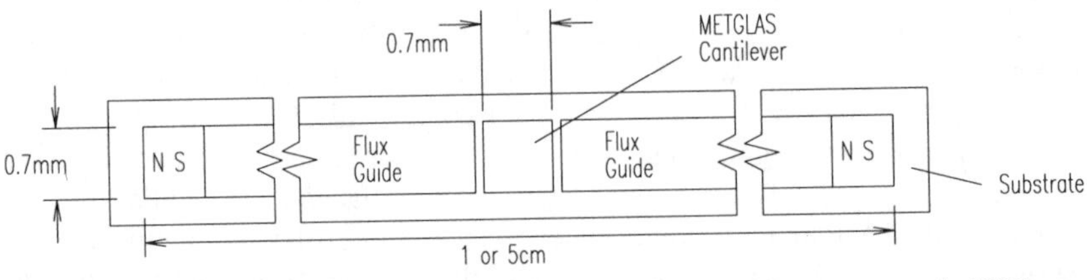

Figure 12-30. Top view of the single cantilevered design (adapted from Fenn, et al., 1992)

A second (non-magnetic) cantilever element is employed to track the displacement of the *METGLAS* reed with subangstrom resolution using tunneling-transducer displacement sensing as illustrated in Figure 12-31. A pair of electrostatic actuator plates dynamically positions the reed follower to maintain constant tunneling current in the probe gap, thus ensuring a constant lateral separation between the probe tip and the vibrating reed. The frequency of the excitation signal applied to the reed-follower actuator is therefore directly influenced by any resonant frequency changes occurring in the *METGLAS* reed. The magnetometer provides an analog voltage output which is proportional to this excitation frequency, and therefore indicative of external magnetic field amplitude.

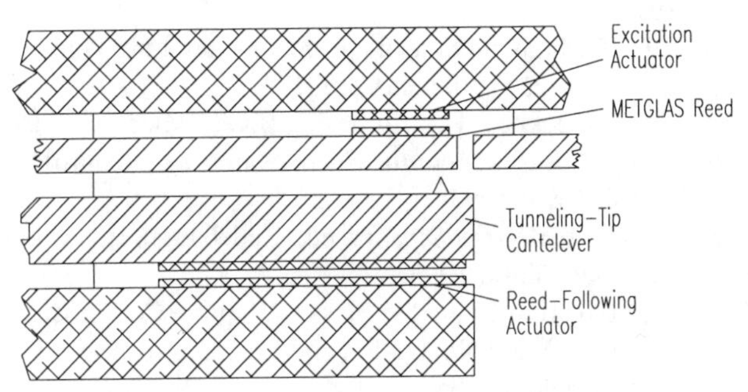

Figure 12-31. Side view of the double cantilevered design (adapted from Fenn, et al., 1992).

One anticipated problem associated with such magnetoelastic devices is that changes in Young's modulus also occur due to temperature shifts. Fenn, et al. (1992) report a 1-Hz bandwidth sensor would require a temperature stability of 10^{-7}°K during the measurement period and suggest thermal isolation using a vacuum jacket and multilayer insulation.

12.7 References

Acuna, M.H., Pellerin, C.J., "A Miniature Two-Axis Fluxgate Magnetometer," *IEEE Transactions on Geoscience Electronics*, Vol. GE-7, pp. 252-260, 1969

Barrett, C.R., Nix, W.D., Tetelman, A.S., *The Principles of Engineering Materials*, Prentice Hall, Englewood Cliffs, NJ, 1973.

Bolz, R.E., Tuve, G.L., eds., *CRC Handbook of Tables for Applied Engineering Science*, CRC Press, Boca Raton, FL, 1979.

Bozorth, R.M., Chapin, D.M., "Demagnetizing Factors of Rods," *Journal of Applied Physics*, Vol. 13, pp. 320-326, May, 1942.

Brizzolara, R.A., Colton, R.J., Wun-Fogle, M., Savage, H.T., "A Tunneling-tip Magnetometer," *Sensors and Actuators*, Vol. 20, pp. 199-205, 1989.

Carlson, A.B., Gisser, D.G., *Electrical Engineering: Concepts and Applications*, Addison-Wesley, Reading, MA, p. 644, 1981.
Carter, E.F., ed., *Dictionary of Inventions and Discoveries*, Crane, Russak, and Co., NY, 1966.
Dahlin, T., Krantz, D., "Low-Cost, Medium Accuracy Land Navigation System," *Sensors*, pp. 26-34, February, 1988.
Dibburn, U., Petersen, A., "The Magnetoresistive Sensor - a Sensitive Device for Detecting Magnetic Field Variations," *Electronic Components and Applications*, Vol. 5, No. 3, June, 1983.
DiLella, D., Colton, R.J., Kenny, T.W., Kaiser, W.J., Vote, E.C., Podosek, J.A., Miller, L.M., "A Micromachined Magnetic-Field Sensor Based on an Electron Tunneling Displacement Transducer," to be published in *Sensors and Actuators*, 1995.
Dinsmore, 1490 and 1525 Magnetic Sensors, Product Literature, Dinsmore Instrument Company, Flint, MI, January, 1991.
Fenn, R.C., Gerver, M.J., Hockney, R.L., Johnson, B.G., "Microfabricated Magnetometer Using Young's Modulus Changes in Magnetoelastic Materials," SPIE Vol. 1694, 1992.
Fraden, J., *AIP Handbook of Modern Sensors*, ed., Radebaugh, R., American Institute of Physics, New York, 1993.
Foster, M., "Vehicle Navigation Using the Plessy Adaptive Compass," RIN Conference Proceedings, Land Navigation and Location for Mobile Applications, York, England, 1985.
Geyger, W.A., *Magnetic Amplifier Circuits*, 2nd ed., McGraw-Hill, New York, 1957.
Geyger, W.A., *J. Appl. Phys.*, Vol. 33, suppl., pp. 1280-1281, 1962.
Gilbert, W., "De Magnete," 1600. (Translation: P.F. Mottelay, John Wiley, 1893.)
Gordon, D.I., Lunsten, R.H., *Rev. Phys. Appl.*, Vol. 5, pp. 175-177, 1970.
Grenoble, B., "Sensor and Logic Form Digital Compass," *Electronic Design News*, pp. 228-229, 6 December, 1990.
Halliday, D., Resnick, R., *Fundamentals of Physics*, John Wiley, New York, NY, 1974.
Hine, A., *Magnetic Compasses and Magnetometers*, Adam Hilger Ltd., London, 1968.
Honeywell, "Smart Digital Magnetometer," HMR Series Product Literature 900133, Rev. A, Honeywell Solid State Electronics Center, Plymouth, MN, August, 1994a.
Honeywell, "Permalloy Magnetic Sensors," Technical Note, 901XX, Honeywell Solid State Electronics Center, Plymouth, MN, September, 1994b.
ILC, *Synchro Conversion Handbook*, ILC Data Device Corporation, Bohemia, NY, April, 1982.

Kenny, T.W., Waltman, S.B., Reynolds, J.K., Kaiser, W.J., "Micromachined Silicon Tunnel Sensor for Motion Detection," *Applied Physics Letters*, Vol. 58, No. 1, January, 1991.

Kim, N.H., Hawks, T., "Digital Compass and Magnetometer Having a Sensor Coil Wound on a High Permeability Isotropic Core," US Patent 4,851,775, 25 July, 1989.

KVH, *C100 Compass Engine*, Product Literature, KVH Industries, Middletown, RI, April, 1993.

Kwiatkowski, W., Tumanski, S., "The Permalloy Magnetoresistive Sensors - Properties and Applications," *Journal of Physics E: Scientific Instruments*, Vol. 19, pp. 502-515, 1986.

Lao, R., "A New Wrinkle in Magnetoresistive Sensors," *Sensors*, pp. 63-65, October, 1994.

Lenz, J.E., "A Review of Magnetic Sensors," *Proceedings of the IEEE*, Vol. 78, No. 6, June, 1990.

Maenaka, K., Ohgusu, T., Ishida, M., Nakamura, T., "Novel Vertical Hall Cells in Standard Bipolar Technology," *Electronic Letters*, Vol. 23, pp. 1104-1105, 1987.

Maenaka, K., Tsukahara, M., and Nakamura, T., "Monolithic Silicon Magnetic Compass," *Sensors and Actuators*, pp. 747-750, 1990.

Petersen, A., "Magnetoresistive Sensors for Navigation," Proceedings, 7th International Conference on Automotive Electronics, London, England, pp. 87-92, October, 1989.

PNI, "TCM1 Electronic Compass Module: User's Manual," Rev. 1.01, Precision Navigation, Inc., Mountain View, CA, March, 1994

Primdahl, F., "The Fluxgate Mechanism, Part I: The Gating Curves of Parallel and Orthogonal Fluxgates," *IEEE Transactions on Magnetics*, Vol. MAG-6, No. 2, June, 1970.

Primdahl, F., "The Fluxgate Magnetometer," *Journal of Physics E: Scientific Instruments*, Vol. 12, pp. 241-253, 1979.

Rahim, W., "Feedback Limited Control System on a Skid-Steer Vehicle," ANS Fifth Topical Meeting on Robotics and Remote Systems, Knoxville, TN, Vol. 1, pp. 37-42, April, 1993.

Ramsden, E., "Measuring Magnetic Fields with Fluxgate Sensors," *Sensors*, pp. 87-90, September, 1994.

SEI, "High-Sensitivity Magnetoresistive Magnetometer," Product Literature, MMS101, Space Electronics, Inc., San Diego, CA, June, 1994.

Stuart, W.F., "Earth's Field Magnetometry, *Reports on Progress in Physics*, J.M. Zinman, Editor, Vol. 35, Part 2, pp. 803-881, 1972.

Udd, E., "Fiber Optic Sensors Based on the Sagnac Interferometer and Passive Ring Resonator," in Fiber Optic Sensors: An Introduction for Engineers and Scientists, E. Udd, Editor, John Wiley, New York, pp. 233-269, 1991.

Wiley, C.M., "Technical Review of Next Week's National Electronics Conference," *Electronics*, p. 39-41, October 5, 1962.

Wiley, C.M., "Navy Tries Solid-State Compass," *Electronics*, pp. 57-58, February 14, 1964.

Wood, T., "The Hall-Effect Sensor," *Sensors*, pp. 27-36, March, 1986.

13
Gyroscopes

Gyroscopes are for the most part insensitive to the electromagnetic and ferromagnetic anomalies that affect the accuracy of compasses and are particularly useful in applications where there is no geomagnetic field present at all (i.e., deep space), or in situations where the local field is disturbed. Two broad categories of gyroscopes will be discussed: 1) *mechanical gyroscopes* and 2) *optical gyroscopes*.

Mechanical gyroscopes operate by sensing the change in direction of some actively sustained angular or linear momentum, which in either case can be continuous or oscillatory in nature (Cochin, 1963). Probably the most well-known mechanical configuration is the *flywheel gyroscope*, a reliable orientation sensor based on the inertial properties of a rapidly spinning rotor, first demonstrated in 1810 by G.C. Bohnenberger of Germany. In 1852, the French physicist Leon Foucault showed that such a gyroscope could detect the rotation of the earth (Carter, 1966). More recently there has been considerable interest shown in a number of new products classified as *vibrating structure gyroscopes* earmarked for applications in stabilized camera optics, robotics, and intelligent-vehicle highway systems.

Optical gyroscopes have been under development now as replacements for their mechanical counterparts for over three decades. With little or no moving parts, such rotation sensors are virtually maintenance free and display no gravitational sensitivities, eliminating the need for gimbaled mounting. Fueled by a large anticipated market in the automotive industry, highly linear fiber-optic versions are now evolving that have wide dynamic range and very low projected costs.

There are two basic classes of rotation-sensing gyros, whether optical or mechanical in nature: 1) *rate gyros*, which provide a voltage or frequency output signal proportional to the turning rate and 2) *rate integrating gyros*, which indicate the actual turn angle or heading (Udd, 1991). Unlike the magnetic compass discussed in Chapter 12, however, *rate integrating gyros* can only measure relative as opposed to absolute angular position and must be initially referenced to a known orientation by some external means. One convenient way to accomplish this objective takes advantage of the earth's natural rotation..

13.1 Mechanical Gyroscopes

Mechanical gyros operate on the basis of *conservation of momentum* and can be characterized by the type of momentum sustained as follows (Cochin, 1963):

- *Continuous angular momentum* — includes the familiar spinning-rotor *flywheel gyro* and the more esoteric *electron* and *particle gyros.*
- *Oscillatory angular momentum* — employs a torsionally suspended mass oscillating back and forth at its natural frequency.
- *Continuous linear momentum* — incorporates a steady stream of fluid, plasma, or electrons, which tends to maintain its established velocity vector as the platform turns. For example, one concept uses a differential pair of hot-wire anemometers to detect the apparent lateral displacement of a flowing air column.
- *Oscillatory linear momentum* — utilizes a set of discrete masses moving back and forth (radially or in parallel) along a straight-line path. The increasingly popular *tuning-fork rate gyro*, to be discussed later in Section 13.1.3, is a good example of a radial-motion gyroscope of this type.

From a robotics perspective, the two most relevant of the above categories are the *flywheel* and *tuning-fork* gyroscopes.

13.1.1 Space-Stable Gyroscopes

The conventional *flywheel gyroscope* consists of a rapidly spinning wheel or sphere with most of its mass concentrated in the outer periphery, supported by low-friction bearings on either end of the spin axis (Cochin, 1963). Anyone who has ever ridden a bicycle has experienced (perhaps unknowingly) an interesting characteristic of this particular gyro configuration known as *gyroscopic precession.* If the rider leans the bike over to the left around its own horizontal axis, the front wheel turns left in response around the vertical axis. The effect is much more noticeable if the wheel is removed from the bike and held by both ends of its axle while rapidly spinning. If the person holding the wheel attempts to yaw it left or right about the vertical axis, a surprisingly violent reaction will be felt as the axle instead twists about the horizontal roll axis. This is due to the angular momentum associated with a spinning flywheel, which displaces the applied force by 90 degrees in the direction of spin.

The *rate of precession* Ω is proportional to the applied torque T (Fraden, 1993):

$$T = I \omega \Omega$$

where:

T = applied input torque
I = rotational inertia of rotor
ω = rotor spin rate
Ω = rate of precession.

Gyroscopic precession is a key factor involved in the operation of the *north-seeking gyrocompass*, as will be discussed in the next section.

The earth's rotational velocity at any given point on the globe can be broken up into two components: one which acts around an imaginary vertical axis normal to the surface, and another which acts around an imaginary horizontal axis tangent to the surface. These two components are known as the *vertical earth rate* and the *horizontal earth rate*, respectively. At the North Pole, for example, the component acting around the local vertical axis (*vertical earth rate*) would be precisely equal to the rotation rate of the earth, or 15 degrees per hour. As the point of interest moves down a meridian towards the equator, the *vertical earth rate* at that particular location decreases proportionally to a value of zero at the equator. Meanwhile, the *horizontal earth rate* (i.e., that component acting around a horizontal axis tangent to the earth's surface) increases from zero at the pole to a maximum value of 15 degrees-per-hour at the equator.

A typical two-axis *flywheel gyroscope* configuration is shown in Figure 13-1. The electrically driven *rotor* is suspended in a pair of precision low-friction bearings at either end of the rotor axle. The *rotor bearings* are in turn supported by a circular ring known as the *inner gimbal ring*, which in turn pivots on a second set of bearings that attach it to the *outer gimbal ring*. This pivoting action of the *inner gimbal* defines the *horizontal axis* of the gyro, which is perpendicular to the *spin axis* of the *rotor* as shown in the figure. The *outer gimbal ring* is attached to the instrument frame by a third set of bearings that define the *vertical axis* of the gyro. The *vertical axis* is perpendicular to both the *horizontal axis* and the *spin axis*.

Notice that if this configuration is oriented such that the *spin axis* points east-west, the *horizontal axis* is aligned with the north-south meridian. Since the gyro is space-stable (i.e., fixed in the inertial reference frame), the *horizontal axis* thus reads the *horizontal earth rate* component of the planet's rotation, while the *vertical axis* reads the *vertical earth rate* component. If the *spin axis* is rotated 90 degrees to a north-south alignment, the earth's rotation has no effect on the gyro's *horizontal axis*, since that axis is now orthogonal to the *horizontal earth rate* component.

Some finite amount of friction in the support bearings, external influences, and small imbalances inherent in the construction of the rotor cause even the best mechanical gyros to drift with time. Typical systems employed in inertial navigation packages by the commercial airline industry can drift as much as 0.1 degree during a six-hour flight (Martin, 1986).

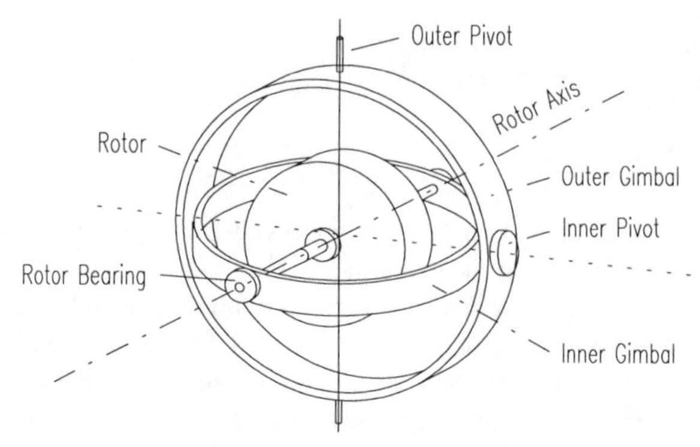

Figure 13-1. Shown here in a typical two-axis configuration, the mechanical *flywheel gyroscope* senses a change in direction of the *angular momentum* associated with a rapidly spinning rotor.

13.1.2 Gyrocompasses

The *gyrocompass* is a special configuration of the *rate integrating flywheel gyroscope*, employing a gravity reference to implement a *north-seeking* function that can be used as a *true-north* navigational reference. This phenomenon, first demonstrated in the early 1800s by Leon Foucault, was patented in Germany by Herman Anschutz-Kaempfe in 1903, and in the US by Elmer Sperry in 1908 (Carter, 1966). The US and German navies had both introduced gyrocompasses into their fleets by 1911 (Martin, 1986). The concept of operation is dependent upon four fundamental principles (Dunlap & Shufeldt, 1972):

- Gyroscopic inertia.
- Gyroscopic precession.
- Earth's rotation.
- Earth's gravitational pull.

Note the *gyrocompass* is in no way dependent upon the earth's geomagnetic field and should not be confused with the *gyromagnetic compass*, wherein a space-stable gyroscope is used to keep a conventional magnetic compass precisely aligned in the horizontal plane (Hine, 1968). The *north-seeking* capability of the *gyrocompass* is instead directly tied to the *horizontal earth rate* component of rotation measured by the gyro's *horizontal axis*. As was pointed out earlier, when the gyro *spin axis* is oriented in a north-south direction, it is insensitive to the earth's rotation, and no tilting occurs. From this it follows that if tilting is observed, the *spin axis* is no longer aligned with the meridian. The direction and magnitude of the measured tilt are directly related to the direction and magnitude of the misalignment between the spin axis and true north.

To transform a two-axis space-stable flywheel gyroscope into a *north-seeking* instrument, some type of weight can be attached to the bottom of the rotor cage in a pendulous fashion to create a simple gravity reference system. Should the gyro spin axis be caused to tilt as a consequence of misalignment with the north-south meridian as discussed above, the restoring weight will be effectively raised against the pull of gravity, creating a torque about the gyro's horizontal axis (Dunlap & Shufeldt, 1972). This resultant torque will in turn induce a perpendicular restoring force that precesses the spin axis back into alignment with the meridian. An alternative non-pendulous approach involves the use of a *mercury ballistic* consisting of one or more pairs of symmetrically arranged vials, partially filled with mercury and connected by a set of equalizing tubes (Arnold & Maunder, 1961; Cochin, 1963).

Primarily used in maritime applications, practical *gyrocompass* implementations today employ an electronic gravity reference system with sophisticated damping, and computer-controlled compensation for host platform speed as well as variations in operating latitude. Such installations are not well suited for mobile robotic applications due to their inherent size and weight, relatively high cost, long spin-up time (i.e., hours) required to initialize, shock and vibration sensitivities, control complexities, and significant power consumption.

13.1.3 Rate Gyros

The common housefly is somewhat unique in the sense that it has only a single pair of wings, instead of two wings on each side of the body as is commonly found on most other flying insects. Remnants of the hind wings, in the form of a pair of small stalks with a swelling at their ends, project outward from the thorax of the fly just behind the base of the wings (Snodgrass, 1930). Known to entomologists as "balancers," or *halteres*, these club-shaped projections consist of two small radially oscillating masses on the ends of cartilage-like fibers, effectively forming a miniature biological *tuning-fork gyro* (Cochin, 1963). If yaw is experienced, the tips of the *halteres* are subjected to *Coriolis forces*, generating muscular signals that assist the acrobatic fly in maintaining controlled flight (Arnold & Maunder, 1961).

Due to its inherent simplicity and reliability, the mechanical *tuning-fork* configuration is one of the most popular low-cost *rate gyroscopes* in use today in land-based mobile applications. Early models consisted of a permanent-magnet fork assembly, torsionally mounted in close proximity to a stationary pair of electromagnetic drive coils (Cochin, 1963). The fork tines were made to vibrate towards and away from one another at a fixed amplitude under temperature-compensated closed-loop control. Any rotation of the gyro assembly about its vertical (torsional) axis caused induced *Coriolis forces* acting on the tines (in the horizontal plane) to generate a harmonic couple about the vertical axis (Arnold & Maunder, 1961). The amplitude of torsional deflection was thus proportional to the rate of turn.

Modern solid-state implementations of the *tuning-fork rate gyro* incorporate *electrostatic* (Boxenhorn, et al., 1989) or *piezoelectric* actuation schemes (Dahlin & Krantz, 1988; Dance, 1993) in lieu of the electromechanical drive mechanisms employed in earlier designs. Variations on the vibratory element include strings, triangular and rectangular bars, cylinders, and hemispheres (Mettler & Hadaegh, 1992). The principles of operation will be addressed in more detail in the following sections describing some commercially available candidate systems.

Systron Donner *GyroChip*

The Systron Donner *GyroChip* is a solid-state angular rate sensor incorporating a micromachined quartz-crystal tuning-fork element and associated electronics sealed in a rugged and compact package as shown in Figure 13-2 (Systron Donner, undated). The double-ended tuning fork and its supporting structure are chemically etched from a single wafer of monocrystalline piezoelectric quartz (Systron Donner, 1994a). The *drive tines* at the upper end of the fork are actively driven toward and away from one another at a precisely controlled amplitude as illustrated in Figure 13-3. Each *drive tine* will experience a *Coriolis force* given by the following equation (Systron Donner, 1992):

$$F = 2m\Omega V_r$$

where:

F = Coriolis force acting on tine
m = mass of tine
Ω = input rotation rate
V_r = instantaneous linear radial velocity of tine.

Figure 13-2. The *GyroChip* solid-state gyro employs a double-ended quartz tuning fork (courtesy Systron Donner Inertial Division).

The *Coriolis force F* is perpendicular to both the input rate Ω and the instantaneous radial velocity V_r. Since the two tines are synchronously moving toward and away from each other, their respective *Coriolis forces* are perpendicular to the plane of the fork assembly, but in opposite directions. The lateral tine offset from the tuning-fork centerline (i.e., the gyro's sensitive axis) causes a torque to be generated by these forces, proportional in magnitude to the input angular velocity (Systron Donner, 1992). In that the *drive-tine* excitation is sinusoidal, the associated radial velocity V_r is also sinusoidal, which means the resultant torque produced by the *Coriolis forces* is sinusoidal as well.

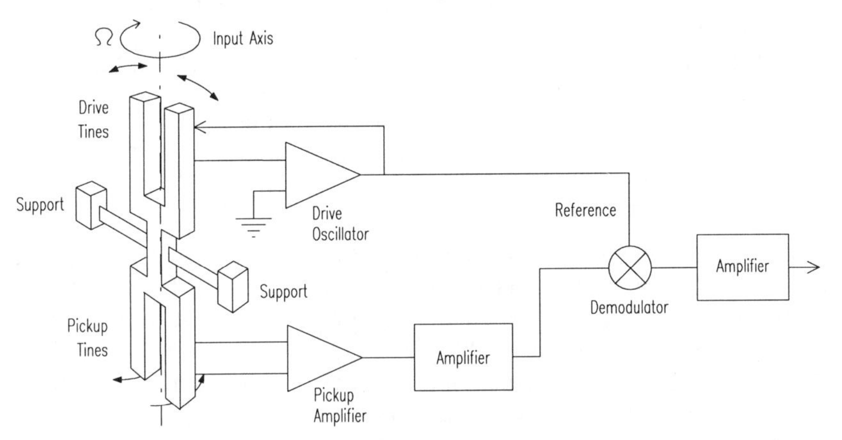

Figure 13-3. Block diagram of the *GyroChip* solid-state rate gyro (courtesy Systron Donner Inertial Division).

The *pick-up tines* react to this oscillating torque by moving in and out of plane as illustrated in Figure 13-3 above, producing a differential signal at the *pick-up amplifier* that is precisely in phase with the radial velocity of the drive tines. The output of the *pick-up amplifier* is then synchronously demodulated (based on the drive oscillator frequency) into a DC signal proportional to the angular rotation rate Ω (Systron Donner, 1992). Since only that component of angular rotation around the axis of symmetry of the tuning fork will generate (from induced Coriolis forces) an oscillating torque at the frequency of the drive tines, synchronous detection ensures input rate sensitivity is uninfluenced by off-axis components (Systron Donner, 1994a).

The *GyroChip* is characterized by a very high *mean-time-between-failure* in excess of 100,000 hours (Orlosky & Morris, 1995) and a low output-noise component that can be classified as *white noise* over a fairly wide operational bandwidth (i.e., all the way down to DC, quite unlike spinning-rotor gyro configurations). The output signal is an analog voltage that varies from 0 to ±2.5 volts DC in accordance with rotational rate and direction, with a linearity typically within less than 0.05 percent of full scale. Selected specifications are provided in Table 13-1 below. The *MotionPak* (Systron Donner, 1994c), an integrated six-DOF package consisting of three rate gyros and three accelerometers is also available for application as a low-cost *inertial navigation system (INS).*

Table 13-1. Selected specifications for the *GyroChip* solid-state rate gyro (courtesy Systron Donner Inertial Division).

Parameter	Value	Units
Range	±10 to ±1000	degrees/second
Scale factor calibration	< 1.0	percent
Linearity	< 0.05	percent full scale
Bandwidth (90 deg.)	> 60	Hz
Output noise	0.012 (typical)	deg./sec/√Hz to 100 Hz
Power	±5	volts DC
	< 0.8	milliwatts
Size	2.25 x 1.92 x .91	inches
Operating life	10 (typical)	years

The recently introduced standard-model *GyroChip II* operates from single-sided (+12 volts DC) supply; a low-noise option is also available for double-sided (±15 volts DC) operation (Systron Donner, 1994b). Selected specifications for these two devices are listed in Table 13-2. A new lower-cost version of the *GyroChip* (i.e., roughly half the current price) is expected to be available in Spring 1995 for slightly less demanding applications.

Table 13-2. Selected specifications for the *GyroChip II* (courtesy Systron Donner Inertial Division).

Parameter	Standard	Low-Noise	Units
Range	±100	±100	degrees/second
Scale factor (±2%)	15	50	millivolts/degree/second
Linearity	< 0.05	< 0.05	percent full scale
Bandwidth (90 deg.)	50	50	Hz
Output noise	< 0.05	< 0.02	degrees/second/√Hz
Power	+12 to + 18	±9 to ±18	volts DC
	35	35 (each)	milliamps
Size	2.34 x 1 x 1	2.34 x 1 x 1	inches
Operating life	> 5	> 5	years

Murata *Gyrostar* Piezoelectric Vibrating Gyroscope

An innovative single-axis *piezoelectric rate gyro* developed by Murata Electronics incorporates three *PZT-6* ceramic elements symmetrically mounted on a triangular metal bar as shown in Figure 13-4 (Nakamura, 1990). If such a bar is made to vibrate in the X direction at its natural frequency f_n, any rotation of the bar around the Z-axis introduces a Coriolis force F_c that causes vibration in the Y direction at the same frequency (Fujishima, 1991):

$$F_c = 2m\Omega\dot{y}$$

where:

F_c = Coriolis force
m = equivalent mass
Ω = angular rotation rate about Z axis.

The actual rotation rate Ω can thus be determined by measuring the amplitude of this induced vibration in the Y direction.

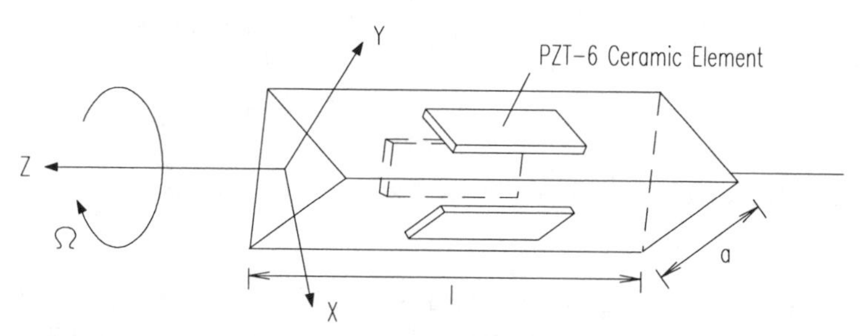

Figure 13-4. Three piezoelectric ceramic elements mounted on a metal bar of triangular cross section form an inexpensive single-axis rate gyro (courtesy Murata Electronics North America).

The flexural resonance frequency of a triangular bar as illustrated in Figure 13-4 is given by (Fujishima, et al., 1991):

$$f_n = \frac{Aa}{4\pi l^2}\sqrt{\frac{E}{6\rho}}$$

where:

A = constant
a = width of bar
l = length of bar
E = Young's modulus of elasticity
ρ = density of bar material.

To ensure good dimensional stability in the Murata gyroscope, the bar is made of *elinvar* (elastic invariable metal), a nickel-chromium-steel alloy with a very small coefficient of thermal expansion (Dance, 1993).

Referring now to Figure 13-5, the 40-millimeter *elinvar* bar is driven into oscillation by the *left* and *right* piezoelectric elements at its natural frequency of 7.85 KHz; a third piezoelectric transducer is used as a detection element to provide feedback to the drive oscillator (Fujishima, et al., 1991). The equilateral-triangle prism arrangement allows the *left* and *right* transducer elements to be located in the direction of the compound vibration mode, and consequently the same *PZT* ceramics can be used for both excitation of the bar and detection of the resultant Coriolis forces (Murata, 1994a).

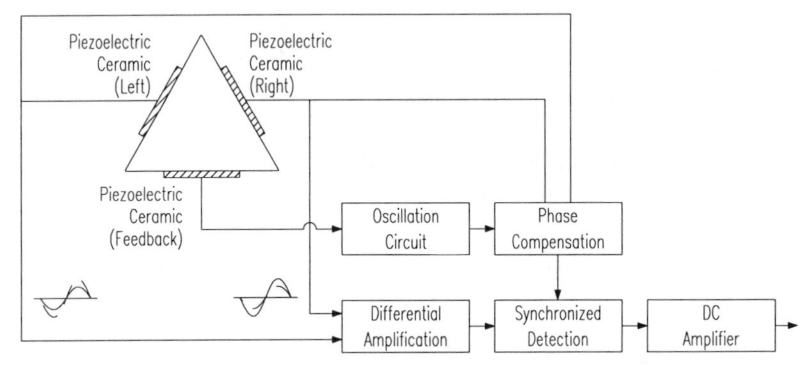

Figure 13-5. Block diagram of the *Gyrostar* piezoelectric rate gyro (courtesy Murata Electronics North America).

The gyro detects and quantifies angular rotation by subtracting the *left* and *right* piezoelectric transducer outputs from each other. This differential amplification scheme provides common-mode rejection of noise and vibration, as the left and right output signals will be equal in magnitude when Ω is zero. As the bar distorts from the effects of rotationally induced Coriolis forces, one detector output increases while the other decreases, effectively doubling the signal magnitude attributed to gyro rotation (Dance, 1993). The differential amplifier output is synchronously detected with respect to the drive oscillator signal, then further amplified to yield an analog voltage that varies linearly above and below a steady-state (no-rotation) value of 2.5 volts DC in response to changes in Ω (Murata, 1994a).

The *Gyrostar's* unique integration of piezoelectric ceramic transducers on an equilateral-triangular prism reportedly offers higher performance than conventional tuning-fork gyros for a tenth of the price (Murata, 1994a). The low power consumption (15 milliamps at 12 volts DC), small package size (roughly 1 by 1 by 2 inches), and low cost (approximately $80 in large quantities) help make the *Gyrostar* a viable and affordable option for a number of consumer as well as mobile robotic applications. On the down side, there is an inherent sensitivity to thermally induced drift (offset and scale factor), in spite of the use of the stable *elinvar* rod material. This tendency to drift is due in part to thermal gradients within the relatively large sensing element, and induced stresses from mismatches in the material thermal expansion rates at the elinvar/ceramic interfaces. Compensation techniques include high-pass filtering to block the DC component of the output signal, and repeated measurement of the offset value under static conditions of zero angular velocity (Murata, 1994a). A detailed technical manual (Murata, 1994b) of test and reliability data including thermal cycle and shock results is available from Murata Electronics North America, Smyrna, GA.

Selected specifications for the *Gyrostar Model ENV-05A* are listed in Table 13-3 below.

Table 13-3. Selected specifications for the *Gyrostar ENV-05A* (courtesy Murata Electronics North America).

Parameter	Value	Units
Range	±90 (max)	degrees/second
Scale factor	22.2	millivolts/degree/second
Scale factor stability	±5 (max)	percent (-10° to 60°C)
Linearity	±0.05 (max)	percent full scale
Hysteresis	0	
Bandwidth	7	Hz
Offset drift	200	millivolts peak-to-peak
Noise level	10 (max)	millivolts DC (rms)
Power	8 - 13.5	volts DC
	15 (max)	milliamps
Size	58 x 25 x 25	millimeters
Weight	45	grams

13.2 Optical Gyroscopes

The principle of operation of the *optical gyroscope*, first discussed by Sagnac (1913), is conceptually very simple, although several significant engineering challenges had to be overcome before practical implementation was possible. In fact, it was not until the demonstration of the helium-neon laser at Bell Labs in 1960 that Sagnac's discovery took on any serious implications; the first operational *ring-laser gyro* was developed by Warren Macek of Sperry Corporation just two years later (Martin, 1986). Navigational quality *ring-laser gyroscopes* were introduced into routine service in inertial navigation systems for the Boeing *757* and *767* in the early 1980s, and over half a million navigation systems have been installed in Japanese automobiles since 1987, many of which employ *fiber-optic gyroscopes* (Reunert, 1993). Numerous technological improvements since Macek's first prototype have made the optical gyro one of the most promising sensors likely to significantly influence mobile robot navigation in the near future.

The basic device consists of two laser beams traveling in opposite directions (i.e., counter-propagating) around a closed-loop path. The constructive and destructive interference patterns formed by splitting off and mixing a portion of the two beams can be used to determine the rate and direction of rotation of the device itself. Schulz-DuBois (1966) idealized the ring laser as a hollow doughnut-shaped mirror, wherein light follows a closed circular path. Assuming an ideal 100-percent reflective mirror surface, the optical energy inside the cavity is theoretically unaffected by any rotation of the mirror itself. The counter-propagating light beams mutually reinforce one another to create a stationary standing wave of intensity peaks and nulls as depicted in Figure 13-6, regardless of whether or not the gyro is rotating (Martin, 1986).

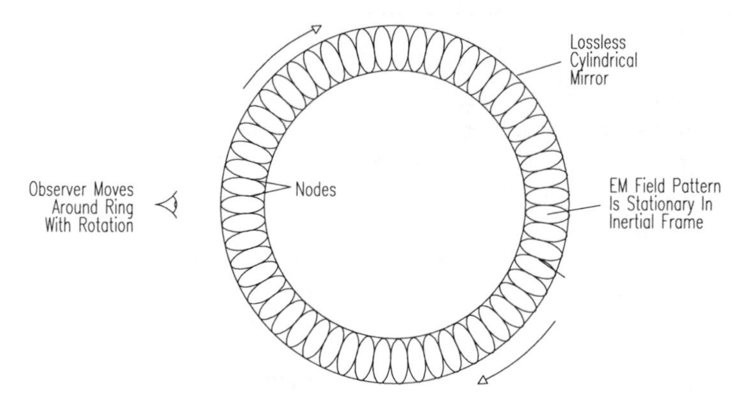

Figure 13-6. Standing wave created by counter-propagating light beams in an idealized ring-laser gyro (adapted from Martin, 1986, © IEEE).

A simplistic visualization based on the Schulz-DuBois idealization is perhaps helpful at this point in understanding the fundamental concept of operation before more detailed treatment of the subject is presented. The light and dark fringes of the nodes are somewhat analogous to the reflective stripes or slotted holes in the rotating disk of an incremental optical encoder and can be theoretically counted in similar fashion by an optical pick-off mounted on the cavity wall (Koper, 1987). (In this analogy, however, the standing-wave "disk" is actually fixed in the inertial reference frame, while the normally stationary "detector" revolves around it.) With each full rotation of the mirrored doughnut, the detector would see a number of node peaks equal to twice the optical path length of the beams divided by the wavelength of the light. For a 632.8-nanometer He-Ne wavelength in a typical 2.4-inch-diameter closed path, there are 300,000 wavelengths and hence 600,000 nodes, yielding over half a million counts per revolution (Koper, 1987).

Obviously, there is no practical way to implement this theoretical arrangement, since there is no such thing as a perfect mirror. Furthermore, the introduction of light energy into the cavity (as well as the need to observe and count the nodes on the standing wave) would interfere with the mirror performance, should such an ideal capability even exist. However, numerous practical embodiments of optical rotation sensors have been developed for use as rate gyros in navigational applications. Five general configurations will be discussed in the following subsections:

- Active optical resonators.
- Passive optical resonators.
- Open-loop fiber-optic interferometers (analog).
- Closed-loop fiber-optic interferometers (digital).
- Fiber-optic resonators.

Aronowitz (1971), Menegozzi & Lamb (1973), Chow, et al. (1985), Wilkinson (1987), and Udd (1991) provide in-depth discussions of the theory of the ring laser gyro and its fiber-optic derivatives. A comprehensive overview of the technologies and an extensive bibliography of preceding works are presented by Ezekiel and Arditty (1982) in the proceedings of the *First International Conference on Fiber Optic Rotation Sensors* held at MIT in November, 1981. An excellent treatment of the salient features, advantages, and disadvantages of *ring-laser gyros* versus *fiber-optic gyros* is presented by Udd (1985; 1991).

13.2.1 Active Ring-Laser Gyros

The *active optical resonator* configuration, more commonly known as the *ring laser gyro*, solves the problem of introducing light into the doughnut by filling the cavity itself with an active lasing medium, typically helium-neon. There are actually two beams generated by the laser that travel around the ring in opposite directions. If the gyro cavity is caused to physically rotate in the counterclockwise direction, then the counter-clockwise-propagating beam will be forced to traverse a slightly longer path than under stationary conditions. Similarly, the clockwise-propagating beam will see its closed-loop path shortened by an identical amount. This phenomenon, known as the *Sagnac effect*, in essence changes the length of the resonant cavity.

The magnitude of this change is given by the equation (Chow, et al., 1985):

$$\Delta L = \frac{4\pi r^2 \Omega}{c}$$

where:

ΔL = change in path length
r = radius of the circular beam path
Ω = angular velocity of rotation
c = speed of light.

Note that the change in path length is directly proportional to the rotation rate Ω of the cavity. Thus, to measure gyro rotation, some convenient means must be established to quantify the associated change in the optical path length.

This requirement to measure minute differences in optical path lengths is where the invention of the laser in the early 1960s provided the needed technological breakthrough that allowed Sagnac's observations to be put to practical use. For lasing to occur in a resonant cavity, the round-trip beam path must be precisely equal in length to an integral number of wavelengths at the resonant frequency. This means the wavelengths (and hence the frequencies) of the two counter-propagating beams must change, as only oscillations with wavelengths satisfying the resonance condition can be sustained in the cavity. The frequency difference between the two beams is given by the following (Chow, et al., 1985):

$$\Delta f = \frac{2fr\Omega}{c} = \frac{2r\Omega}{\lambda}$$

where:

Δf = frequency difference
λ = wavelength.

In practice, a doughnut-shaped ring cavity would be hard to realize. For an arbitrary cavity geometry, the expression becomes (Chow, et al, 1985):

$$\Delta f = \frac{4A\Omega}{P\lambda}$$

where:

A = area enclosed by the closed-loop beam path
P = perimeter of the beam path.

For single-axis gyros, the ring is generally formed by aligning three highly reflective mirrors to create a closed-loop triangular path as shown in Figure 13-7. (Some systems, such as Macek's early prototype, employ four mirrors to create a square path.) The mirrors are usually mounted to a monolithic glass-ceramic block with machined ports for the cavity bores and electrodes. The most stable systems employ linearly polarized light and minimize circularly polarized components to avoid magnetic sensitivities (Martin, 1986). The approximate quantum noise limit is due to spontaneous emission in the gain medium (Ezekiel & Arditty, 1982), representing the "best-case" scenario of the five general gyro configurations outlined in Section 12.2.2.

Dual anodes are generally incorporated as illustrated in the figure below to overcome Doppler shifts attributed to the otherwise moving medium within the laser cavity. In DC-excited plasma, the neutral atoms tend to move towards the cathode along the center of the discharge tube and towards the anode along the walls, a phenomenon known as *Langmuir flow*; the laser radiation being predominately along the tube centerline thus experiences a net motion in the medium itself (Chow, et al., 1985). The opposed dual-anode configuration introduces a reciprocity in the *Langmuir flow* which cancels the overall effect, provided the anode currents are maintained precisely equal.

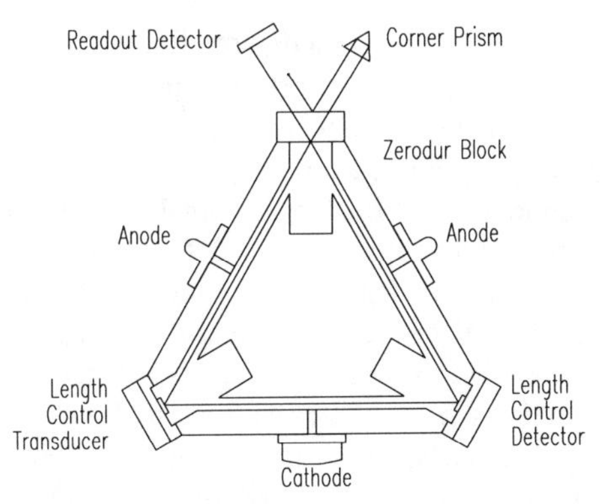

Figure 13-7. A typical three-mirror configuration of the single-axis ring-laser gyro employing dual anodes to cancel the biasing effects of induced *Langmuir flow* in the lasing medium (adapted from Udd, 1985).

The fundamental disadvantage associated with the active ring laser is a problem called *frequency lock-in,* which occurs at low rotation rates when the counter-propagating beams "lock" together in frequency (Chao, et al., 1984). This *lock-in* phenomenon is attributed to constrictions or periodic modulation of the gain medium in conjunction with the influence of a very small amount of *backscatter* from the mirror surfaces (Udd, 1985). The end result is a small deadband region (below a certain threshold of rotational velocity) for which there is no output signal as shown in Figure 13-8A. Above the lock-in threshold, output approaches the ideal linear response curve in a parabolic fashion.

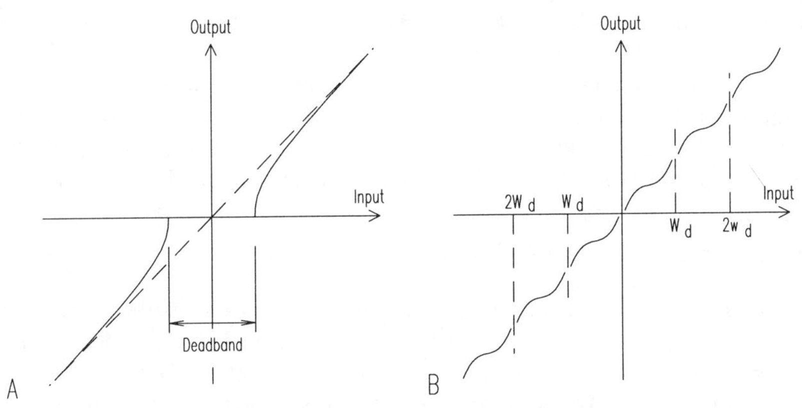

Figure 13-8. *Frequency lock-in* due to a small amount of backscatter from the mirror surfaces results in a zero-output deadband region (A) at low rotational velocities; the application of *mechanical dither* breaks the deadband region up into smaller fragments (B) that occur at input rates equal to harmonics of the dither frequency ω_d (adapted from Martin, 1986, © IEEE).

The most obvious approach to solving the *lock-in* problem is to improve the quality of the mirrors, thereby reducing the resulting *backscatter.* Again, however, perfect mirrors do not exist, and some finite amount of *backscatter* will always be present. Martin (1986) reports a representative value of 10^{-12} the power of the main beam, enough to induce *frequency lock-in* for rotational rates of several hundred degrees per hour in a typical gyro with a 20-centimeter perimeter. A more practical technique for reducing *lock-in* is to incorporate some type of biasing scheme to shift the operating point away from the deadband zone.

Mechanical dithering is the least elegant but most common and effective biasing means, introducing the obvious disadvantages of increased system complexity and reduced *mean time between failures* associated with moving parts. The entire gyro assembly is rotated back and forth about the sensing axis in an oscillatory fashion (±100 arcseconds at 400 Hz typical), with the resulting response curve shown in Figure 13-8B. State-of-the-art dithered active ring-laser gyros have a scale factor linearity that far surpasses the best mechanical gyros. Dithered biasing, unfortunately, is too slow for high-performance systems (i.e., flight control), resulting in oscillatory instabilities (Martin, 1986). Mechanical dithering can also introduce crosstalk between axes on a multi-axis system, although some of the unibody three-axis gyros employ a common dither axis to eliminate this possibility (Martin, 1986).

Buholz and Chodorow, (1967), Chesnoy (1989), Christian and Rosker (1991), as well as Dennis, et al. (1991) discuss the use of extremely short-duration laser pulses (typically 1/15 of the resonator perimeter in length) to reduce the effects of *frequency lock-in* at low rotation rates. The basic idea is to minimize the cross coupling between the two counter-propagating beams by limiting the regions in the cavity where the two pulses overlap. Wax and Chodorow (1972) report an improvement in performance of two orders of magnitude through the use of intracavity phase modulation. Other techniques based on non-linear optics have been proposed (Udd, 1985), including an approach by Litton that applies an external magnetic field to the cavity to create a directionally dependent phase shift for biasing (Martin, 1986). Yet another solution to the *lock-in* problem is to remove the lasing medium from the ring altogether, effectively forming what is known as a *passive ring resonator*, to be discussed in Section 13.2.2.

Honeywell *Modular Azimuth Position* System

The *H-726 Modular Azimuth Position System (MAPS)* developed by Honeywell's Military Avionics Division, St. Petersburg, FL, is a complete stand-alone *inertial navigation system (INS)* intended for land-based applications (Honeywell, 1992). The system can be broken down into three major subcomponents as illustrated in Figure 13-9: 1) the *Dynamic Reference Unit (DRU)*, 2) the *Control Display Unit (CDU)*, and 3) the *Vehicle Motion Sensor (VMS)*. The *DRU* consists of an *Inertial Sensor Assembly (ISA)*, an associated *Inertial Processor*, a *Navigation Processor*, interface electronics, and a low-voltage power supply. The *CDU* communicates with the *DRU* over an RS-422 datalink, providing an operator interface for mode selection, data display, and waypoint navigation. The *VMS* is essentially an incremental optical encoder that attaches to the vehicle odometer cable, providing directional information as well as 32 displacement counts per cable revolution.

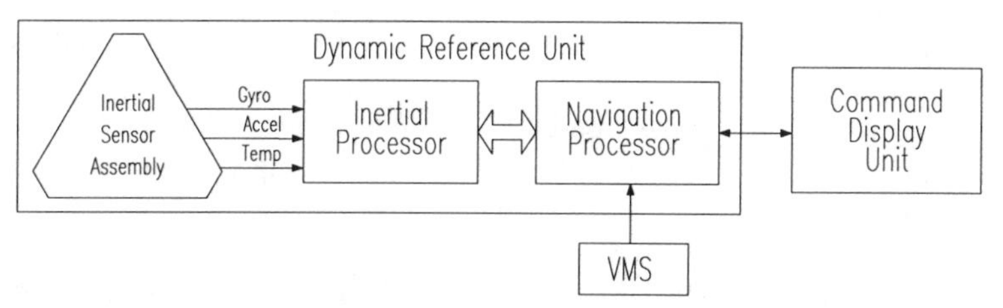

Figure 13-9. Functional block diagram of the stand-alone *H-726 Modular Azimuth Position System* (courtesy Honeywell, Inc., Military Avionics Division).

The *Inertial Sensor Assembly* is comprised of three Honeywell production-model *GG1342* ring-laser gyros and three Sundstrand *QA2000* accelerometers, mounted in a mutually orthogonal configuration as shown in Figure 13-10. The *GG1342* incorporates a three-mirror triangular lasing cavity as depicted earlier in Figure 13-7 and has a demonstrated *mean time between failure* in excess of 50,000 hours. Each of the three gyros is a completely self-contained unit including a laser block assembly, path length control transducer, readout optics, interface electronics, and mechanical dither mechanism.

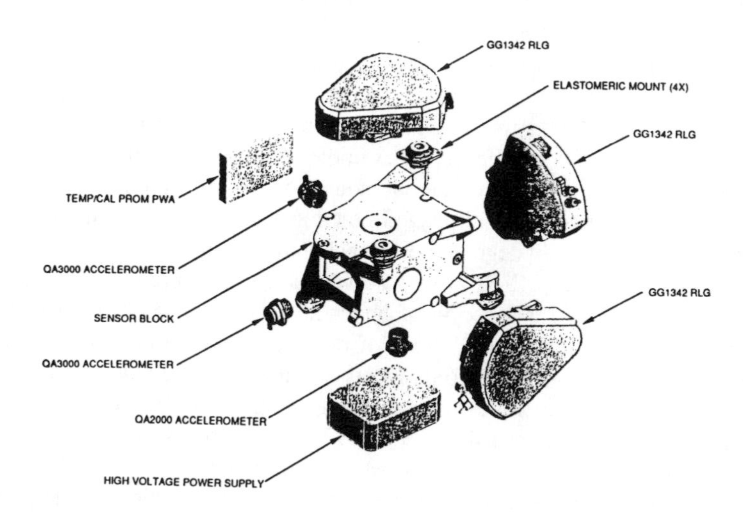

Figure 13-10. Exploded view of the *Inertial Sensor Assembly (ISA)* showing the relative orientations of the three ring-laser gyros and accelerometers (courtesy Honeywell, Inc., Military Avionics Division).

There are four basic modes of system operation:

- *Power-up* — executes a computer-controlled start-up and initialization sequence.
- *Alignment* — accepts current position data and establishes its directional reference.
- *Survey* — is the normal "run-time" operating mode.
- *Power-down* — stores current data for subsequent start-up and de-energizes system.

Upon initial start-up, the *DRU* recalls from EEPROM memory the previous location and heading of the vehicle, if available. If unknown, the current vehicle position is requested from the user in *Universal Transverse Mercator (UTM)* coordinates. A *normal alignment* can then be executed with no prior knowledge of current vehicle heading. This automatic self-alignment feature works by sensing the earth's rotation in a fashion somewhat analogous to the *north-seeking gyrocompass* previously discussed in Section 13.1.2. With the vehicle stationary, the resulting rotational components measured by the three orthogonal gyro axes can be analyzed to determine the angular orientation of the *DRU* with respect to the earth's spin axis (Huddle, 1977). The *normal alignment* process takes anywhere from 8 to 15 minutes to complete, depending on the starting latitude (Honeywell, 1992). Alternatively, a *stored heading alignment* can be performed in approximately 90 seconds using a value recalled from non-volatile memory, if the vehicle has not been moved since the DRU was last powered down.

Once initial alignment is completed, the system enters *survey mode*, ready for normal operation in either of two submodes: 1) *zero-velocity-update (ZUPT) mode* or 2) *odometer-aided mode*. The first of these options, *ZUPT mode*, is invoked only when odometry data is either not available or is of questionable validity. Sammarco (1994) reports an example of this latter situation in the case of a *MAPS* application involving autonomous navigation of a continuous mining machine, where inherent track slippage in the loose coal debris resulting from the cutting operation seriously degraded dead-reckoning results.

To re-reference under *ZUPT mode*, the vehicle must be brought to a standstill every 4 to 10 minutes for a period of about 25 seconds in order to re-establish precise alignment with the earth's axis of rotation. In addition, any differences between the measured velocity components and the known *earth-rate* values are assumed to be *DRU* velocity errors and subsequently compensated. Retroactive position-error corrections can then be made based on the perceived error vector and the duration of prior travel. The actual time interval between *ZUPTs* is a vehicle- and application-specific parameter that must be empirically determined for optimal performance (Honeywell, 1992).

In default *odometer-aided mode*, the *DRU* uses encoder-count information from the *velocity measurement system* to dampen system velocity errors as derived from the *ISA* accelerometer data. A *ZUPT* is automatically requested if a degradation is detected in dead-reckoning data (i.e., due to wheel slippage or a *VMS* component failure, for example), or if the vehicle has been in continuous motion for over an hour (Leiser, 1992). In addition, the *DRU* will continuously update its estimate of azimuth each time the vehicle stops in *odometer-aided mode* for any length of time greater than six seconds (Honeywell, 1992).

Immediately after a *ZUPT* is performed (i.e., before *DRU* velocity errors reaccumulate to any significant degree), the inertial velocity data is used to dynamically recalibrate the odometer scale factor. (Odometry calibration can change as a consequence of variations in tire pressure or vehicle loading as discussed in Chapter 2.) A 19-state *Kalman filter* in the *Navigation Processor* trims the inertial and *VMS* component coefficients during run-time operation (Leiser, 1992). Should odometer velocity not agree with the inertial measurement of velocity to within some prespecified window of acceptance, the *DRU* will set a *VMS* fault condition indicator on the *CDU* and request another *ZUPT*. The system then remains in *ZUPT mode* unless overridden by the operator.

Kearfott *Monolithic Ring-Laser Gyro*

In order to significantly reduce package size and component costs, the Kearfott Guidance and Navigation Corporation, Wayne, NJ, developed a single-block three-axis design employing only six mirrors, each mounted to the center of a cube face as shown in Figure 13-11. Within the enclosed volume of the glass-ceramic block are three mutually orthogonal and functionally independent ring-laser gyros that share a common set of mirrors. Although each mirror is used by two different gyros in the monolithic design, initial system tests in 1981 confirmed there was no interactive crosstalk between axes (Koper, 1987). The reduction in the number of required mirrors represents a considerable costs savings, in that the high-quality mirrors are one of the more costly components in an active ring-laser design. Similarly, only a single cathode and one dither mechanism are needed, instead of three each, further reducing component and assembly costs.

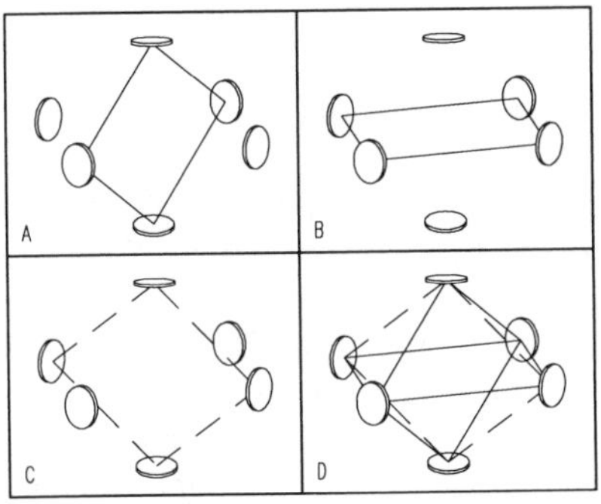

Figure 13-11. Kearfott six-mirror configuration of a three-axis ring laser gyro (adapted from Koper, 1987).

More importantly, however, the monolithic shared-mirror design optimizes the performance-to-size ratio by permitting the maximum closed-path diameter for a given volume. (Increasing the optical path length improves gyro performance, just as using a larger diameter slotted disk with more holes would increase the achievable resolution of an incremental optical encoder, to revisit the Schultz-Dubois analogy.) The medium-sized Kearfott *Monolithic Ring Laser Gyro* shown in Figure 13-12 is typically used for missile guidance and navigation; smaller low-cost units are employed in tactical weapon systems.

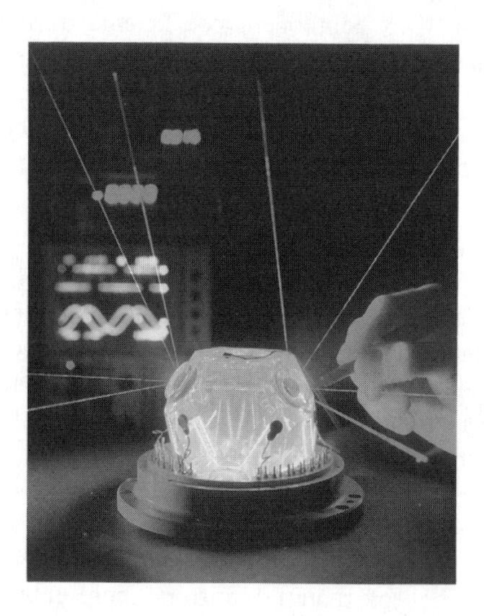

Figure 13-12. Beams of 632.8-nanometer laser light can be seen leaving the ring-laser gyro cavities through the mirrors of the compact Kearfott three-axis *Monolithic Ring-Laser Gyro*, which measures pitch, roll, and yaw in one integrated package (courtesy Kearfott Guidance and Navigation Corporation).

13.2.2 Passive Ring Resonator Gyros

The *passive ring resonator gyro* makes use of a laser source external to the ring cavity (Figure 13-13), and thus circumvents the *frequency lock-in* problem that arises when the gain medium is internal to the cavity itself. The passive configuration also eliminates problems arising from changes in the optical path length within the interferometer due to variations in the index of refraction of the gain medium (Chow, et al., 1985). The theoretical quantum noise limit is determined by *photon shot noise* and is slightly higher (i.e., worse) than the theoretical limit seen for the *active ring-laser gyro* (Ezekiel & Arditty, 1982).

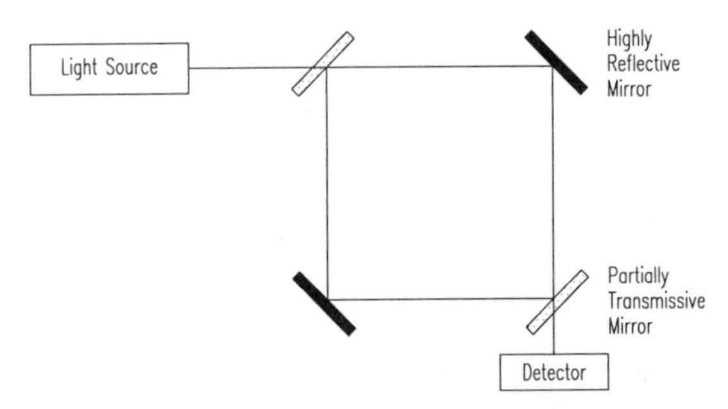

Figure 13-13. Passive ring resonator gyro with laser source external to the ring cavity (adapted with permission from Udd, 1991, © John Wiley and Sons, Inc.).

Classical implementations using mirrored optical resonators patterned after the active ring predecessors suffered from inherently bulky packaging in comparison to the newly emerging alternatives afforded by fiber-optic technology. Such fiber-optic derivatives also promised additional advantage in longer-length multiturn resonators for increased sensitivity in smaller, rugged, and less expensive packages. As a consequence, the *resonant fiber-optic gyro (RFOG)* presented later in Section 13.2.5 has emerged as the most popular of the resonator configurations (Sanders, 1992).

13.2.3 Open-Loop Interferometric Fiber-Optic Gyros

The concurrent development of optical fiber technology, spurred mainly by the communications industry, presented a potential low-cost alternative to the high-tolerance machining and clean-room assembly required for ring-laser gyros. The glass fiber in essence forms an internally reflective waveguide for optical energy, along the lines of a small-diameter linear implementation of the doughnut-shaped mirror cavity conceptualized by Schulz-DuBois (1966). The use of multiple turns of fiber means the resultant path-length change due to the Sagnac effect is essentially multiplied by a factor N equal to the integer number of turns, thereby providing significantly improved resolution (Udd, 1985). An additional advantage of the fiber-optic configuration stems from the fact that operation is not dependent on a high-finesse cavity, thereby significantly reducing manufacturing costs (Blake, et al., 1989).

Recall from Chapter 9 the *refractive index n* relates the speed of light in a particular medium to the speed of light in a vacuum as follows:

$$n = \frac{c}{c_m}$$

where:

n = refractive index of medium

c = speed of light in a vacuum
c_m = speed of light in medium.

Step-index multimode fiber (Figure 13-14) is made up of a core region of glass with index of refraction n_{co}, surrounded by a protective cladding with a lower index of refraction n_{cl} (Nolan, et al., 1991). The lower refractive index in the cladding is necessary to ensure total internal reflection of the light propagating through the core region. The terminology *step-index* refers to this "stepped" discontinuity in the refractive index that occurs at the core-cladding interface.

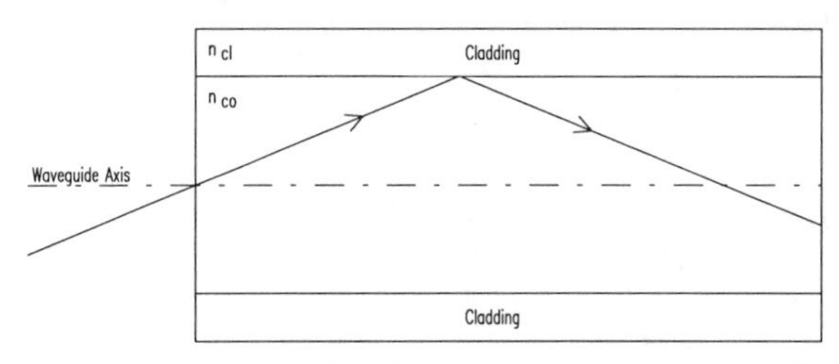

Figure 13-14. Step-index multi-mode fiber (adapted with permission from Nolan, et al., 1991, © John Wiley and Sons, Inc.).

Referring now to Figure 13-15, as long as the entry angle (with respect to the waveguide axis) of an incoming ray is less than a certain critical angle θ_c, the ray will be guided down the fiber, virtually without loss. The *numerical aperture* of the fiber quantifies this parameter of acceptance (i.e., the light-collecting ability of the fiber), and is defined as follows (Nolan, et al., 1991):

$$NA = \sin\theta_c = \sqrt{n_{co}^{\ 2} - n_{cl}^{\ 2}}$$

where:

NA = numerical aperture of the fiber
θ_c = critical angle of acceptance
n_{co} = index of refraction of glass core
n_{cl} = index of refraction of cladding.

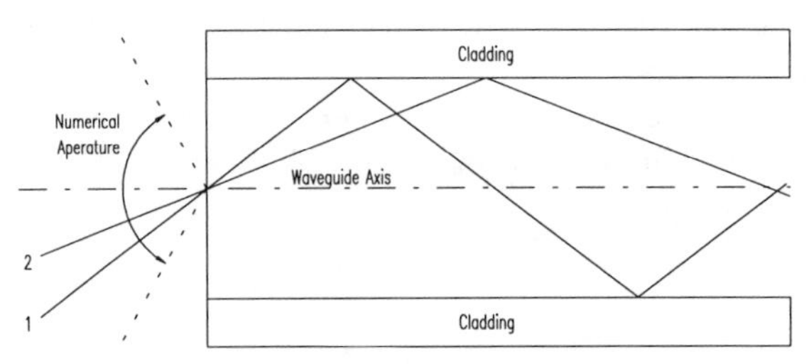

Figure 13-15. Entry angles of incoming rays 1 and 2 determine propagation paths in the fiber core (adapted with permission from Nolan, et al., 1991, © John Wiley and Sons, Inc.).

As illustrated in Figure 13-15 above, a number of rays following different-length paths can simultaneously propagate down the fiber, as long as their respective entry angles are less than the critical angle of acceptance θ_c. Multiple-path propagation of this nature occurs where the core diameter is much larger than the wavelength of the guided energy, giving rise to the term *multimode fiber*. Such multimode operation is clearly undesirable in gyro applications, where the objective is to eliminate all non-reciprocal conditions other than that imposed by the Sagnac effect itself. As the diameter of the core is reduced to approach the operating wavelength, a cutoff condition is reached where just a single mode is allowed to propagate, constrained to travel only along the waveguide axis (Nolan & Blaszyk, 1991).

Light can randomly change polarization states as it propagates through standard *single-mode fiber*. The use of special *polarization-maintaining fiber*, such as *PRSM Corning*, maintains the original polarization state of the light along the path of travel (Reunert, 1993). This is important, since light of different polarization states travels through an optical fiber at different speeds. A typical block diagram of the "minimum-reciprocal" IFOG configuration is presented in Figure 13-16. *Polarization-maintaining single-mode fiber* (Nolan & Blaszyk, 1991) is employed to ensure the two counter-propagating beams in the loop follow identical paths in the absence of rotation.

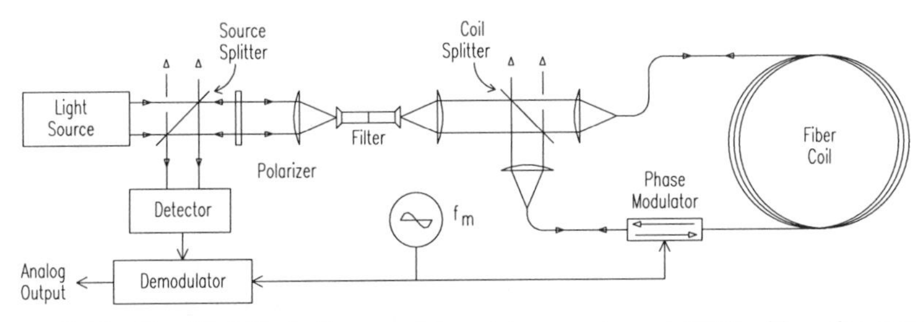

Figure 13-16. Simplified block diagram of the *minimum-reciprocal IFOG* with analog output (adapted from Ezekiel & Arditty, 1982; Lefevre, 1992).

The *Sagnac phase shift* between the two beams introduced by gyro rotation is given by (Udd, 1985):

$$Z_R = \frac{LD}{\lambda c}\Omega$$

where:

Z_R = number of fringes of phase shift due to gyro rotation
L = length of optical fiber in loop
D = diameter of loop
λ = wavelength of optical energy
c = speed of light in a vacuum
Ω = rotation rate.

The stability of the scale factor relating Z_R to Ω in the equation above is thus dependent on the stability of L, D, and λ (Ezekiel & Arditty, 1982). Practical implementations usually operate over plus or minus half a fringe (i.e., $\pm\pi$ radian of phase difference) with a theoretical sensitivity of 10^{-6} radian or less of phase shift (Lefevre, 1992). IFOG sensitivity may be improved by increasing L (i.e., adding more turns of fiber in the sensing loop), peaking at an optimal length on the order of several kilometers, after which the fiber attenuation (1 dB per kilometer typical) begins to degrade performance (Ezekiel & Arditty, 1982). This large amount of required fiber represents a rather significant percentage of overall system cost.

The two counter-propagating beams reunite at the detector, which monitors the cosinusoidal intensity changes caused by constructive and destructive interference. The peak intensity occurs as shown in Figure 13-17A at the point of zero rotation rate, where the phase shift $\Delta\phi$ between the counter-propagating beams is equal to zero. Unfortunately, there is no way to determine the direction of rotation directly from the intensity information (as can be inferred from the symmetrical nature of the plot with respect to the Y-axis), and the sensitivity of I to small changes in rotation rate is greatly reduced due to the horizontal nature of the slope (Blake, 1989).

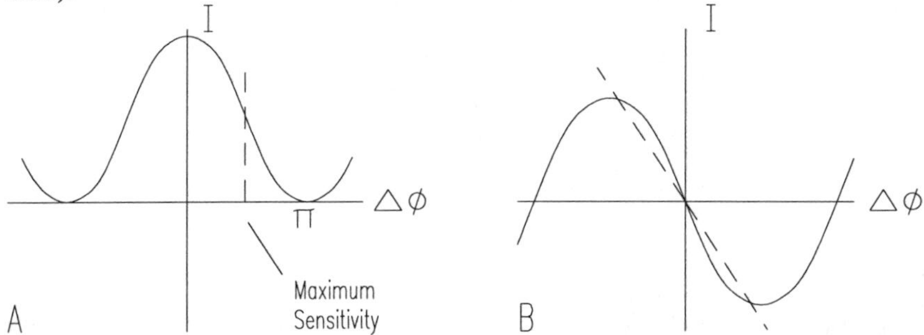

Figure 13-17. (A) Plot of detector intensity versus phase shift, and, (B) resultant demodulator output versus phase shift for the analog *open-loop IFOG* (adapted from Ezekiel & Arditty, 1982).

To overcome these deficiencies, non-reciprocal phase shifts between the two beams are introduced at an oscillatory rate ω, usually by phase modulation of the beams near one end of the interferometer coil (Udd, 1991). This phase modulation can be accomplished using a length of fiber wound around a piezoelectric cylinder and introduces a bias of $\pi/2$ to shift the operating point over into the region of maximum sensitivity on the response curve as shown in Figure 13-17A (Ezekiel & Arditty, 1982). The output of the photodetector is then synchronously demodulated and filtered to yield the sinusoidal analog representation of $\Delta\phi$ shown in Figure 13-17B. Note the direction of rotation is now easily determined from the sign of the output. Disadvantages of this open-loop approach include the non-linear relationship of the demodulated output to rotation rate Ω, and the inherent susceptibility to errors caused by variations in the light source intensity or component tolerances. As Blake (1989) points out, it is difficult to achieve good linearity in analog electronic componentry over six orders of magnitude of dynamic range.

An interesting characteristic of the *open-loop IFOG* is the absence of any narrow-band laser source (Burns, et al., 1983), the enabling technology allowing the *Sagnac effect* to reach practical implementation in the first place. A low-coherence source, such as a *superluminescent diode (SLD)*, is typically employed instead to reduce the effects of noise (Udd, 1985; Tai, et al., 1986), the primary source of which is backscattering within the fiber and at any interfaces. As a result of such backscatter, in addition to the two primary counter-propagating waves in the loop there are also a number of parasitic waves that yield secondary interferometers (Lefevre, 1992). The limited temporal coherence of the broadband SLD causes any interference due to backscattering to average to zero, suppressing the contrast of these spurious interferometers. The detection system becomes sensitive only to the interference between waves that followed identical paths (Ezekiel and Arditty, 1982; Lefevre, 1992).

The *open-loop IFOG* is attractive from the standpoint of reduced manufacturing costs, high tolerance to shock and vibration, insensitivity to gravitational effects, quick start-up, and fairly good sensitivity in terms of bias drift rate and the random walk coefficient. Coil geometry is not critical, and no path-length control is needed. Disadvantages include the long length of optical fiber required (relative to other fiber-optic gyro designs, as will be discussed later), limited dynamic range in comparison to *active ring-laser gyros*, and scale factor variations due to analog component drifts (Adrian, 1991). *Open-loop* configurations are therefore most suited to the needs of low-cost systems in applications requiring only moderate accuracy, such as gyrocompassing in automobile navigation, pitch and roll indicators, and attitude stabilization.

Hitachi Fiber-Optic Gyroscopes

Hitachi Cable, Ltd., Tokyo, Japan, offers several relatively inexpensive single-axis *open-loop IFOG* configurations intended primarily for use in automotive applications. The Hitachi *HOFG-4FT* was the original IFOG used for factory-installed vehicle navigation systems on the Toyota Mark II (a car model sold in Japan). The *HOFG-4FT* was recognized as one of the "Most Technologically Significant New Products of the Year" in 1993 by *R&D* magazine, which subsequently presented an *R&D 100* award to Hitachi. The company established a manufacturing facility in Hitachi City, Japan, with a capacity of 2,500 IFOGs per shift per month in order to meet the demands of automotive customers.

Hitachi has continued to invest in the development of lower-cost designs and manufacturing processes more in line with the needs of the automotive industry. The model *HOFG-X*, for example, is descended from the original *HOFG-4FT* design, while the *HGA-D* (Hitachi, 1994b) represents the follow-on generation of

IFOGs for in-vehicle systems. The design modifications of the *HGA-D* make it more cost effective for applications requiring serial output. Selected performance specifications are presented in Table 13-4.

Table 13-4. Selected specifications for the *HGA-D* fiber-optic gyroscope.

Parameter	Value	Units
Range	±60	degrees/second
Linearity	±2	percent scale
Random walk	1.3	degrees/√Hz
Thermal drift	±0.05	degrees/second/5°C
Update rate	10 (typical)	milliseconds
Warm up time	3 (typical)	seconds
Power	9 - 16	volts DC
	250	milliamps
Size	175 x 120 x 50	millimeters
Output	9600	baud (serial TTL)

In addition to automotive applications, Hitachi has developed IFOGs for a range of industrial and commercial uses. The *HOFG-1* has found wide employment throughout Japan in mobile robotic systems, to include industrial cleaning robots, autonomous heavy equipment, and agricultural helicopters. The system block diagram is shown in Figure 13-18. A single DB-9 connector accommodates power, reset, RS-232 serial communications, and analog output lines (Hitachi, 1994a). Selected specifications are listed in Table 13-5 below.

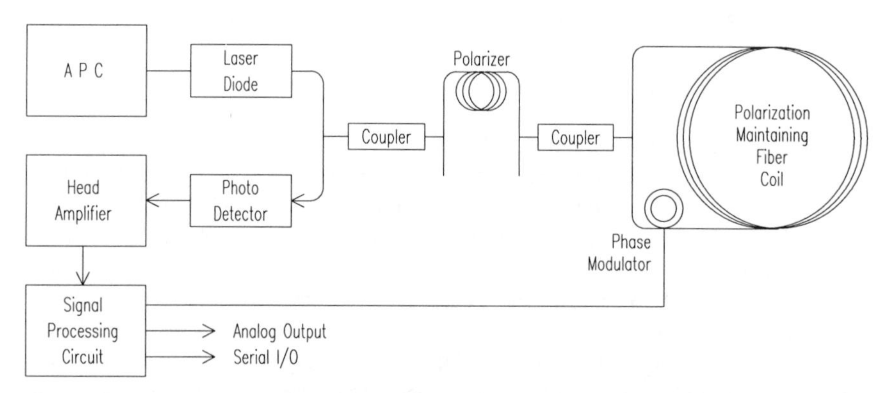

Figure 13-18. The *HOFG-1* open-loop IFOG provides a serial RS-232 as well as an analog (±2.5 volts DC) output of angle and angular rate (adapted from Hitachi, 1994a).

Table 13-5. Selected specifications for the *HOFG-1* fiber-optic gyroscope.

Parameter	Value	Units
Range	±60	degrees/second
Linearity	±1	percent scale
Random walk	<0.1	degrees/√Hz
Update rate	15 (minimum)	milliseconds
Warm up time	5 (typical)	seconds
Power	10 - 16	volts DC
	500	milliamps
Size	120 x 100 x 70	millimeters
Output	9600	baud (serial RS-232C)

13.2.4 Closed-Loop Interferometric Fiber-Optic Gyros

For applications (such as aircraft navigation) demanding higher accuracy than that afforded by the *open-loop IFOG*, the *closed-loop* configuration offers significant promise, with drifts in the 0.001 to 0.01 degree-per-hour range and scale-factor stabilities greater than 100 ppm (Adrian, 1991). Closed-loop digital signal processing is considerably more complex than the analog signal processing employed on *open-loop IFOG* configurations. Feedback into a frequency- or phase-shifting element (Figure 13-19) is employed to cancel the rotationally induced Sagnac phase shift. Since the system is always operated at a null condition where $\Delta\phi$ is equal to zero, minor variations in light-source intensity and analog component tolerances have markedly reduced effect (Ezekiel & Arditty, 1982).

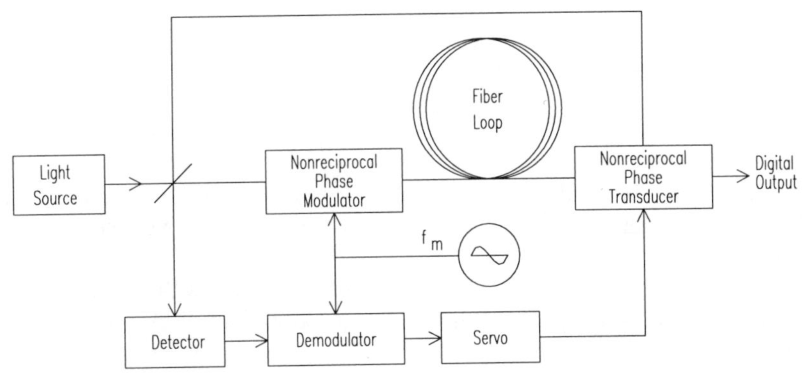

Figure 13-19. Simplified block diagram of a *closed-loop IFOG* employing a *non-reciprocal phase transducer* to null out the Sagnac phase shift $\Delta\phi$ introduced by rotation rate Ω (adapted from Ezekiel & Arditty, 1982).

Referring again to Figure 13-19, the output of the *demodulator* is passed to a servo amplifier that in turn drives a *non-reciprocal phase transducer (NRPT)*, typically an electro-optic frequency shifter placed within the fiber interferometer (Ezekiel & Arditty, 1982). The *NRPT* introduces a frequency difference between the two counter-propagating beams, resulting in an associated fringe shift at the detector given by (Udd, 1991):

$$Z_F = -\frac{\Delta f\, Ln}{c}$$

where:

Z_F = fringe shift due to frequency difference
Δf = frequency difference introduced by the *NRPT*
n = index of refraction
c = speed of light.

To null out $\Delta\phi$ at the detector, the fringe shift Z_R due to gyro rotation must be precisely offset by the fringe shift Z_F due to the relative frequency difference of the two beams:

$$\Delta\phi = Z_R + Z_F = 0 .$$

Substituting the previous expressions for Z_F and Z_R and solving for Δf yields (Ezekiel & Arditty, 1982; Udd, 1991):

$$\Delta f = \frac{4AN}{n\lambda L}\Omega = \frac{4A}{n\lambda P}\Omega = \frac{D}{n\lambda}\Omega$$

where:

A = area of fiber loop
N = number of turns in the loop
L = total length of fiber cable
P = loop perimeter
D = loop diameter.

The gyro output, being the servo-controlled frequency shift Δf imparted by the *NRPT*, is thus inherently digital, as opposed to an analog DC voltage level, and also linear.

Ezekiel and Arditty (1982) list the following advantages of the closed-loop configuration over the open-loop IFOG design previously discussed in Section 13.2.3:

- It is independent of intensity variations in the light source, since the system is operated at null.
- It is independent of individual component gains (assuming high open-loop gain maintained).

- Linearity and stability depend only on the *non-reciprocal phase transducer.*

13.2.5 Resonant Fiber-Optic Gyros

The *resonant fiber-optic gyro (RFOG)* evolved as a solid-state derivative of the *passive ring resonator gyro* discussed in Section 13.2.2. A passive resonant cavity is formed from a multiturn closed loop of optical fiber as shown in Figure 13-20. An input coupler provides a means for injecting frequency-modulated light from a laser source into the resonant loop in both the clockwise and counterclockwise directions. As the frequency of the modulated light passes through a value such that the perimeter of the loop precisely matches an integral number of wavelengths at that frequency, input energy is strongly coupled into the loop (Sanders, 1992). In the absence of loop rotation, maximum coupling for both beam directions occurs in a sharp peak centered at this resonant frequency.

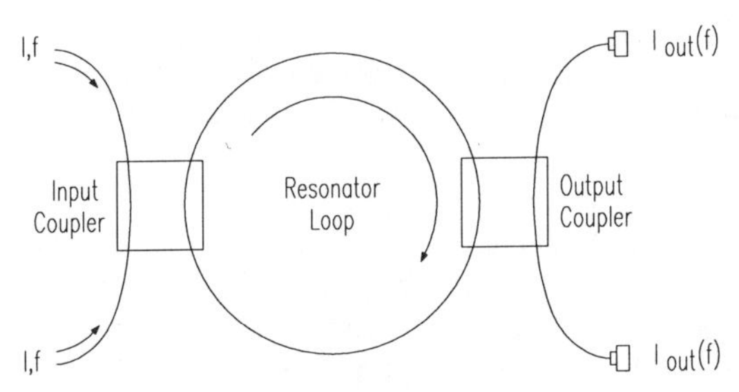

Figure 13-20. In the *resonant fiber-optic gyro (RFOG)*, maximum optical coupling into the loop occurs at that resonant frequency which yields an integral number of wavelengths corresponding to the loop perimeter (adapted from Sanders, 1992).

If the loop is caused to rotate in the clockwise direction, of course, the Sagnac effect causes the perceived loop perimeter to lengthen for the clockwise-traveling beam, and to shorten for the counterclockwise-traveling beam. The resonant frequencies must shift accordingly, and energy is subsequently coupled into the loop at two different frequencies and directions during each cycle of the sinusoidal FM sweep. An output coupler samples the intensity of the energy in the loop by passing a percentage of the two counter-rotating beams to their respective detectors as shown in the diagram. The demodulated output from these detectors will show resonance peaks as illustrated in Figure 13-21, separated by a frequency difference Δf given by the following (Sanders, 1992):

$$\Delta f = \frac{D}{\lambda n}\Omega$$

where:

Δf = frequency difference between counter-propagating beams
D = diameter of the resonant loop
Ω = rotational velocity
λ = free-space wavelength of the laser
n = refractive index of the fiber.

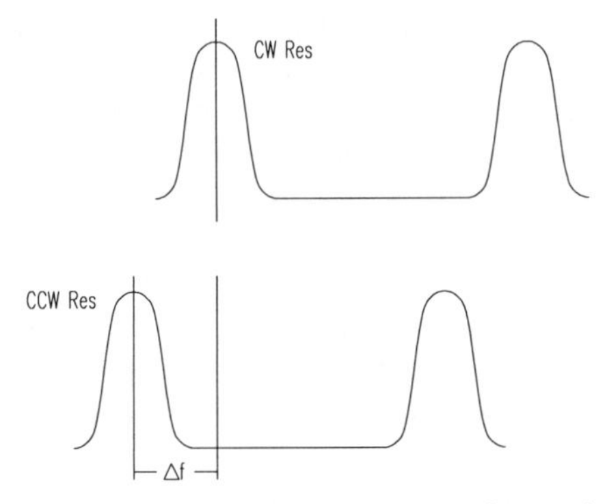

Figure 13-21. The difference (Δf) between the resonance frequencies associated with the clockwise and counterclockwise beams provides a measure of rotation rate Ω (adapted from Sanders, 1992).

In practice, the laser frequency is usually adjusted to maintain resonance in one direction, while an electro-optical frequency shifter is used to drive the other direction back into resonance. This requires a frequency shift of two times the induced Sagnac effect, since the first direction has been locked. Actual rotation rate is then determined from the magnitude of the frequency shift.

Like the IFOG, the all-solid-state RFOG is attractive from the standpoint of high reliability, long life, quick start-up, and light weight. The principle advantage of the RFOG, however, is that it requires significantly less fiber (from 10-100 times less) in the sensing coil than the IFOG configuration, while achieving the same shot-noise-limited performance (Sanders, 1992). Sanders attributes this to the fact that light traverses the sensing loop multiple times, as opposed to once in the IFOG counterpart. On the down side are the requirements for a highly coherent source and extremely low-loss fiber components (Adrian, 1991).

13.3 References

Adrian, P., "Technical Advances in Fiber-Optic Sensors: Theory and Applications," *Sensors*, pp. 23-45, September, 1991.

Arnold, R.N., Maunder, L., *Gyrodynamics and its Engineering Applications*, Academic Press, New York, NY, 1961.

Aronowitz, F., "The Ring Laser Gyro," *Laser Applications*, Vol. 1, M. Ross, ed., Academic Press, 1971.

Blake, J., Cox, J., Feth, J, Goettsche, R., "Design, Development, and Test of a 3-Inch Open Loop All Fiber Gyro," MSD-TR-89-21, 14th Biennial Guidance Test Symposium, Holloman AFB, NM, pp. 255- 266, October, 1989.

Boxenhorn, B., Dew, B., Grelff, P., "The Micromechanical Inertial Guidance System and its Application," MSD-TR-89-21, 14th Biennial Guidance Test Symposium, Holloman AFB, NM, pp. 113-131, October, 1989.

Buholz, N., Chodorow, M., "Acoustic Wave Amplitude Modulation of a Multimode Ring Laser," *IEEE Journal of Quantum Electronics*, Vol. QE-3, No. 11, pp. 454-459, November, 1967.

Burns, W.K., Chen, C.L., Moeller, R.P., "Fiber-Optic Gyroscopes with Broad-Band Sources," *IEEE Journal of Lightwave Technology*, Vol. LT-1, p. 98, 1983.

Carter, E.F., ed., *Dictionary of Inventions and Discoveries*, Crane, Russak, and Co., NY, 1966.

Chao, S., Lim, W.L., Hammond, J.A., "Lock-in Growth in a Ring Laser Gyro," Proceedings, Physics and Optical Ring Gyros Conference, SPIE Vol 487, Snowbird, UT, pp. 50-57, January, 1984.

Chesnoy, J., "Picosecond Gyrolaser," *Optics Letters*, Vol 14, No. 18, pp. 990-992, September, 1989.

Chow, W.W., Gea-Banacloche, J., Pedrotti, L.M., Sanders, V.E., Schleich, W., Scully, M.O., "The Ring Laser Gyro," *Reviews of Modern Physics*, Vol. 57, No. 1, pp. 61-104, January, 1985.

Christian, W.R., Rosker, M.J., "Picosecond Pulsed Diode Ring Laser Gyroscope," *Optics Letters*, Vol. 16, No. 20, pp. 1587-1589, October, 1991.

Cochin, I., *Analysis and Design of the Gyroscope for Inertial Guidance*, John Wiley and Sons, New York, NY, 1963.

Dahlin, T., Krantz, D., "Low-Cost, Medium Accuracy Land Navigation System," *Sensors*, pp. 26-34, February, 1988.

Dance, B., "Piezoelectric Ceramic Elements Form Compact Gyroscope," *Design News*, pp. 113-115, 20 September, 1993.

Dennis, M.L., Diels, J.M., Lai, M., "Femtosecond Ring Dye Laser: A Potential New Laser Gyro," *Optics Letters*, Vol. 16, No. 7, pp. 529-531, April 1, 1991.

Dunlap, G.D., Shufeldt, H.H., *Dutton's Navigation and Piloting*, Naval Institute Press, pp. 557-579, 1972.

Ezekiel, S., Arditty, H.J., Editors, Fiber Optic Rotation Sensors and Related Technologies, Proceedings of the First International Conference, MIT, Springer-Verlag, New York, 1982.

Fraden, J., *AIP Handbook of Modern Sensors*, Radebaugh, R., ed., American Institute of Physics, New York, 1993.

Fujishima, S., Nakamura, T., Fujimoto, K., "Piezoelectric Vibratory Gyroscope Using Flexural Vibration of a Triangular Bar," Frequency Control Symposium, 29 May, 1991.

Hine, A., *Magnetic Compasses and Magnetometers*, Adam Hilger Ltd., London, 1968.

Hitachi, "Fiber Optic Gyroscope (HOFG-1)," Specification No SP 94-28-1005, Hitachi Cable, Ltd., Tokyo, Japan, 10 August, 1994a.

Hitachi, "Fiber Optic Gyroscope (HGA-D)," Specification No SP 94-28-1006, Hitachi Cable, Ltd., Tokyo, Japan, 10 August, 1994b.

Honeywell, "H-726 Modular Azimuth Position System," Technical Description, 1192-12025, Honeywell, Inc., Military Avionics Division, St. Petersburg, FL, December, 1992.

Huddle, J.R., "The Theoretical Principles for Design of the Inertial Surveyor for Position and Gravity Determination," First International Symposium on Inertial Technology for Surveying and Geodesy, Ottawa, Ontario, Canada, 12-14 October, 1977.

Koper, J.G., "A Three-Axis Ring Laser Gyroscope," *Sensors*, pp. 8-21, March, 1987.

Lefevre, H.C., "The Interferometric Fiber-Optic Gyroscope," in *Fiber Optic Sensors*, Udd, E., ed., Vol. CR44, SPIE Optical Engineering Press, Bellingham, WA, September, 1992.

Leiser, K.E., "The Ring Laser Gyro Modular Azimuth Position System Comes of Age: First Article Test Results and Present Applications," reprinted in: *H-726 Modular Azimuth Position System*, Technical Description, 1192-12025, Honeywell, Inc., Military Avionics Division, St. Petersburg, FL, December, 1992.

Martin, G.J., "Gyroscopes May Cease Spinning," *IEEE Spectrum*, pp. 48-53, February, 1986.

Menegozzi, L.N., Lamb, W.E., "Theory of a Ring Laser," *Physical Review A*, Vol. 1, No. 4, pp. 2103-2125, October, 1973.

Mettler, E., Hadaegh, F.Y., "Space Micro-Guidance and Control: Applications and Architectures," Sensors and Sensor Systems for Guidance and Navigation II, SPIE Vol. 1694, Orlando, FL, pp. 144-158, April, 1992.

Murata, "Gyrostar Piezoelectric Vibrating Gyroscope," Product Literature, Catalog No. G-09-B, Murata Electronics North America, Inc., Smyrna, Georgia, 1994a.

Murata, "Gyrostar Piezoelectric Vibrating Gyroscope: Test and Reliability Data," Technical Manual, Catalog No. T-03-B, Murata Electronics North America, Inc., Smyrna, Georgia, 1994b.

Nakamura, T., "Vibration Gyroscope Employs Piezoelectric Vibrator," JEE, pp. 99-104, September, 1990.

Nolan, D.A., Blaszyk, P.E., Udd, E., "Optical Fibers", in *Fiber Optic Sensors: An Introduction for Engineers and Scientists*, E. Udd, ed., John Wiley, New York, pp. 9-26, 1991.

Orlosky, S.D., Morris, H.D., "A Quartz Rotational Rate Sensor," *Sensors*, pp. 27-31, February, 1995.

Reunert, M.K., "Fiber Optic Gyroscopes: Principles and Applications," *Sensors*, pp. 37-38, August, 1993.

Sagnac, G.M., "L'ether lumineux demontre par l'effet du vent relatif d'ether dans un interferometre en rotation uniforme," C.R. Academy of Science, 95, pp. 708-710, 1913.

Sammarco, J.J., "A Navigational System for Continuous Mining Machines," *Sensors*, pp. 11-17, January, 1994.

Sanders, G.A., "Critical Review of Resonator Fiber Optic Gyroscope Technology," in *Fiber Optic Sensors*, Udd, E., Editor, Vol. CR44, SPIE Optical Engineering Press, Bellingham, WA, September, 1992.

Schulz-DuBois, E.O., "Alternative Interpretation of Rotation Rate Sensing by Ring Laser," *IEEE Journal of Quantum Electronics*, Vol. QE-2, No. 8, pp. 299-305, August, 1966.

Snodgrass, R.E., *Insects: Their Ways and Means of Living*, Abbot, C.G., ed., Vol. 5 of the Smithsonian Scientific Series, Smithsonian Institution, New York, NY, 1930.

Systron Donner, "GyroChip," Product Literature, Systron Donner Inertial Division, Concord, CA, undated.

Systron Donner, "GyroChip Theory of Operation," Application Note, Systron Donner Inertial Division, Concord, CA, July, 1992.

Systron Donner, "GyroChip: Industrial Solid-State Rotation Sensor," Product Literature, Systron Donner Inertial Division, Concord, CA, February, 1994a.

Systron Donner, "GyroChip II," Product Literature, Systron Donner Inertial Division, Concord, CA, February, 1994b.

Systron Donner, "MotionPak: Solid-State 6-DOF Motion Sensor," Product Literature, Systron Donner Inertial Division, Concord, CA, June, 1994c.

Tai, S., Kojima, K., Noda, S., Kyuma, K., Hamanaka, K., Nakayama, T., "All-Fibre Gyroscope Using Depolarized Superluminescent Diode," *Electronic Letters*, Vol. 22, p. 546, 1986.

Udd, E., "Fiberoptic vs. Ring Laser Gyros: An Assessment of the Technology," in *Laser Focus/Electro Optics*, December, 1985.

Udd, E., "Fiber Optic Sensors Based on the Sagnac Interferometer and Passive Ring Resonator," in *Fiber Optic Sensors: An Introduction for Engineers and Scientists*, E. Udd, ed., John Wiley, New York, pp. 233-269, 1991.

Wax, S.I., Chodorow, M., "Phase Modulation of a Ring-Laser Gyro - Part II: Experimental Results," *IEEE Journal of Quantum Electronics*, pp. 352-361, March, 1972.

Wilkinson, J.R., "Ring Lasers," *Progress in Quantum Electronics*, Moss, T.S., Stenholm, S., Firth, W.J., Phillips, W.D., and Kaiser, W., eds., Vol. 11, No. 1, Pergamon Press, Oxford, 1987.

14

RF Position-Location Systems

RF position-location techniques can be subdivided into the two broad classes of *ground-based* systems and *satellite-based* systems. Typical non-robotic applications include marine and aircraft navigation, race car performance analysis, range instrumentation, unmanned mobile target control, mine localization, hazardous materials mapping, dredge positioning, geodetic surveys, and fleet management. Fairly low-cost localized implementation of such systems have recently found commercial application providing position location and range information for golfers (Purkey, 1994).

14.1 Ground-Based RF Systems

Ground-based RF position location systems are typically of two types: 1) *passive hyperbolic line-of-position systems* that compare the time-of-arrival phase differences of incoming signals simultaneously emitted from surveyed transmitter sites; and 2) *active radar-like trilateration systems* that measure the round-trip propagation delays for a number of fixed-reference transponders. Passive systems are generally preferable when a large number of vehicles must operate in the same local area, for obvious reasons.

14.1.1 Loran

An early example of the first category is seen in *Loran* (short for long range navigation). Developed at MIT during World War II, such systems compare the time of arrival of two identical signals broadcast simultaneously from high-power transmitters located at surveyed sites with a known separation baseline. For each finite time difference (as measured by the receiver) there is an associated hyperbolic line of position as shown in Figure 14-1. Two or more pairs of master-slave stations are required to get intersecting hyperbolic lines resulting in a two-dimensional (latitude and longitude) fix.

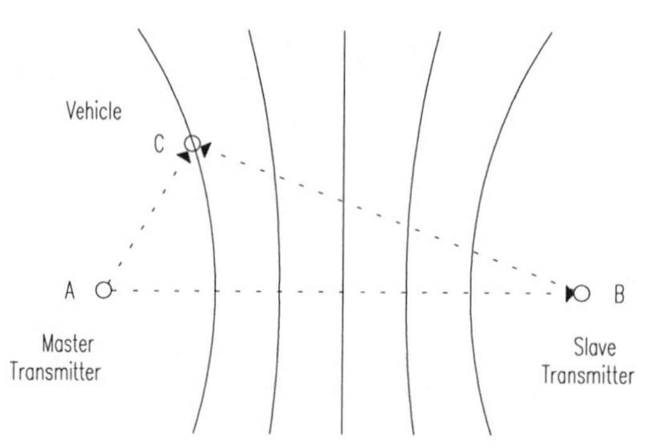

Figure 14-1. For each hyperbolic line of position, length ABC minus length AC equals some constant *K* (adapted from Dodington, 1989).

The original implementation (*Loran A*) was aimed at assisting convoys of liberty ships crossing the North Atlantic in stormy winter weather. Two 100-kilowatt slave transmitters were located about 200 miles on either side of the master station. Non-line-of-sight ground-wave propagation at around 2 MHz was employed, with pulsed as opposed to continuous-wave transmissions to aid in skywave discrimination. The time-of-arrival difference was simply measured as the lateral separation of the two pulses on an oscilloscope display, with a typical accuracy of around 1 microsecond. This numerical value was matched to the appropriate line of position on a special Loran chart of the region, and the procedure then repeated for another set of transmitters. For discrimination purposes, four different frequencies were used, 50 KHz apart, with 24 different pulse repetition rates in the neighborhood of 20 to 35 pulses per second (Dodington, 1989). In situations where the hyperbolic lines intersected more or less at right angles, the resulting (best-case) accuracy was about 1.5 kilometers.

Loran A was phased out in the early '80s in favor of *Loran C*, which achieves much longer over-the-horizon ranges through use of 5-megawatt pulses radiated from 1300-foot towers at a lower carrier frequency of 100 KHz. For improved accuracy, the phase differences of the first three cycles of the master and slave pulses are tracked by phase-lock loops in the receiver and converted to a digital readout, which is again cross-referenced to a preprinted chart. Effective operational range is about 1000 miles, with best-case accuracies in the neighborhood of 100 meters. Coverage is provided by about 50 transmitter sites to all US coastal waters and parts of the North Atlantic, North Pacific, and the Mediterranean.

14.1.2 Kaman Sciences *Radio Frequency Navigation Grid*

The Remote Control Program Group of Kaman Sciences Corporation, Colorado Springs, CO, has developed a scaled-down version of a Loran-type hyperbolic position-location system known as the *Radio Frequency Navigation Grid*

(RFNG). The original application in the late 1970s involved autonomous route control of unmanned mobile targets used in live-fire testing of the laser-guided *Copperhead* artillery round (Stokes, 1989). The various remote vehicles sense their position by measuring the phase differences in received signals from a master transmitter and two slaves situated at surveyed sites within a 100-square-kilometer area as shown in Figure 14-2. Best-case system resolution is 1.5 inches at a 20-Hz update rate, resulting in a typical positioning repeatability of ±1 meter for a 60-ton tank running at speeds up to 45 kilometers per hour.

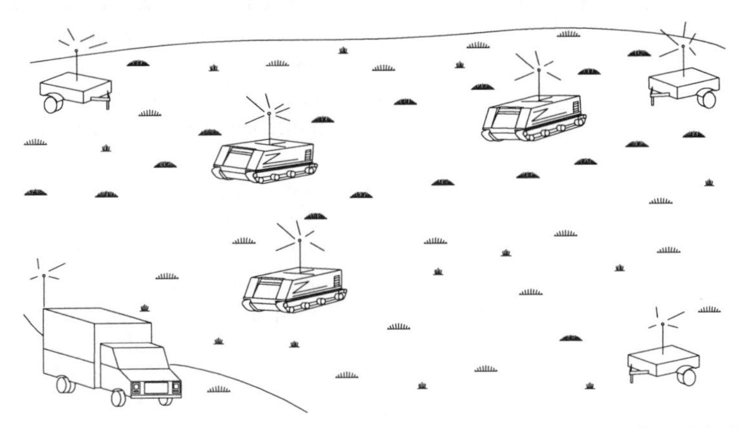

Figure 14-2. Kaman Sciences 1500-watt navigation grid is a scaled-down version of the Loran concept, covering an area 8 to 15 kilometers on a side with a position-location repeatability of ±1 meter (courtesy Kaman Sciences Corp.).

Path trajectories were originally taught by driving a vehicle over the desired route and recording the actual phase differences observed. This file was then played back at run time and compared to measured phase difference values, with vehicle steering servoed in an appropriate manner to null any observed error signal. This approach resulted in the exact replication of the recorded trail, including any changes in velocity or direction. Vehicle speeds in excess of 30 miles per hour are supported over path lengths of up to 15 kilometers (Stokes, 1989). Multiple canned paths can be stored and changed remotely, but vehicle travel must always begin from a known start point (Byrne, et al., 1992).

The *Threat Array Control and Tracking Information Center (TACTIC)* is offered by Kaman Sciences to augment the RFNG by tracking and displaying the location and orientation of up to 24 remote vehicles (Kaman, 1991). Real-time telemetry and recording of vehicle heading, position, velocity, status, and other designated parameters (i.e., fuel level, oil pressure, battery voltage) are supported at a 1-Hz update rate. The *TACTIC* operator has direct control over engine start, automatic path playback, vehicle pause/resume, and emergency halt functions. Non-line-of-sight operation is supported through use of a 23.825-MHz grid frequency in conjunction with a 72-MHz control and communications channel.

In response to requirements for column operations, an *Intelligent Collision Avoidance (ICA)* capability was added to provide automatic management of the vehicle array. The *ICA* software runs on a higher-level computer that utilizes the *TACTIC* computer to perform two-way data communication with each vehicle in the array. The *TACTIC* computer also preconditions vehicle data by computing position, speed, heading, and vehicle status for presentation to the *ICA*, which in turn computes actual vehicle spacing and location. The *ICA* module then commands (through *TACTIC*) necessary adjustments to each vehicle in order to maintain commanded column speed and vehicle spacing. Should a vehicle abort its mission for loss of data communication, off-path conditions, command abort, or other reasons, the *ICA* recognizes the abort and plans a detour around the affected vehicle. The detour is transmitted via *TACTIC* to all of the vehicles in the column. Each vehicle approaching the aborted unit then executes the detour and returns to the main trail.

Pertinent mission data are presented on the *ICA* monitor. These include a graphical display of the intended column route, and the position and status of each vehicle. Also displayed are data showing the communication status of each vehicle as well as mission elapsed time and time remaining for the chosen "reference" vehicle to reach a location of interest which could be, for example, a planned impact location. *TACTIC* logs all mission critical data as well as information provided by ancillary sensors such as exhaust gas temperature, engine rpm, etc. Other features of the enhanced *RFNG* vehicle control system include:

- Expanded operating area (100-plus square kilometers).
- Operation of vehicles from a remote location (i.e., *ICA* is used at range control centers 25 or more miles from the vehicle array).
- Remote turn on/off of grid stations where commercial power is available.
- Mission programming without having to drive the vehicle over the desired mission profile.
- Ability to operate vehicles with manual clutch and shift type transmissions.

14.1.3 Precision Technology Tracking and Telemetry System

Precision Technology, Inc., of Downsview, Ontario, has recently introduced to the automotive racing world an interesting variation of conventional phase-shift measurement approaches to position location. The company's *Precision Location* tracking and telemetry system employs a number of receive-only antennae situated at fixed locations around a racetrack to monitor a continuous sine wave transmission from a moving vehicle (Figure 14-3). By comparing the signals received by the various antennae to a common reference signal of identical frequency generated at the base station, relative changes in vehicle position with respect to each antenna can be inferred from resulting shifts in the respective phase relationships.

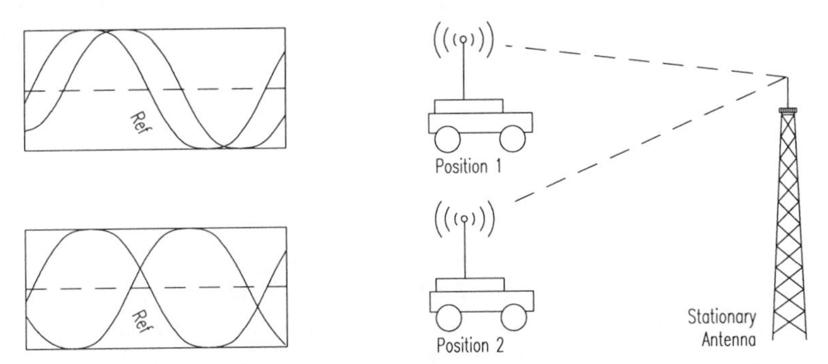

Figure 14-3. Changes in the phase relationship between the received signal and a reference sine wave generated by the base station are used to track relative movement of a race car with respect to the known location of the antenna (adapted from Duchnowski, 1992).

The system update rate for acquiring phase data from each antenna is 100 Hz. The default reporting of vehicle *X-Y* coordinate data is 10 Hz, but this information can be derived from the stored phase data if necessary at any desired sample rate up to 100 Hz. The inherent drift in the vehicle's oscillator with respect to the reference oscillator shows up in the phase data as identical offsets in the distance of the car from each of the three antennae (Figure 14-4). Although this relative bias does not affect the least-squares position determination, the accumulating drift will eventually lead to numbers that become too big for the double-precision calculations to handle. The software therefore subtracts an equivalent amount of perceived offset from each measurement to keep the numbers small.

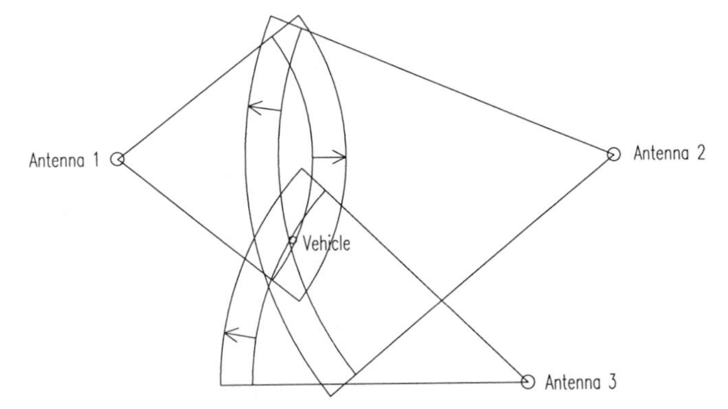

Figure 14-4. Oscillator drift will appear to increase (or decrease) the radii from all antennae by the same amount; the software minimizes the least-squares distance from the solution point to the three range arcs (courtesy Precision Technology, Inc.).

The 40.58-MHz VHF signal allows for non-line-of-sight operation to the moving vehicle, with a resulting precision of approximately 1 to 10 centimeters (Duchnowski, 1992). From a robotics perspective, problems with this approach arise when more than one vehicle must be tracked. A next-generation *Precision Location* system is currently being introduced that eliminates the need for hard-wire connections to each of the three antennae locations.

14.1.4 Motorola *Mini-Ranger Falcon*

An example of the active-transponder category of ground-based RF position-location techniques is seen in the *Mini-Ranger Falcon* series of range positioning systems offered by the Government and Systems Technology Group of Motorola, Inc., Scottsdale, AZ. The *Falcon 484* configuration depicted in Figure 14-5 is capable of measuring line-of-sight distances from 100 meters out to 75 kilometers. An initial calibration is performed at a known location to determine the turn-around delay (TAD) for each transponder (i.e., the time required to transmit a response back to the interrogator after receipt of interrogation). The actual distance between the interrogator and a given transponder is found by (Byrne, et al., 1992):

$$D = \frac{(T_e - T_d)c}{2}$$

where:

D = separation distance
T_e = total elapsed time
T_d = transponder turn-around delay
c = speed of light.

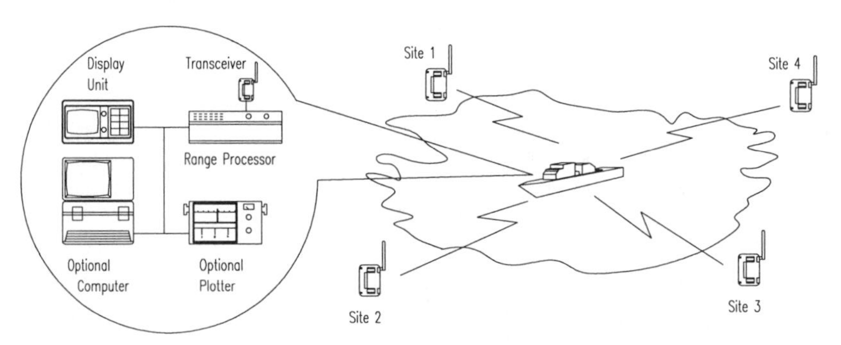

Figure 14-5. Motorola's *Mini-Ranger Falcon 484* RF position location system provides ±2 meters accuracy over ranges of 100 meters to 75 kilometers (courtesy Motorola, Inc.).

The *MC6809*-based range processor performs a least-squares position solution at a 1-Hz update rate, using range inputs from two, three, or four of 16 possible reference transponders. The individual reference stations answer only to uniquely coded interrogations and operate in C-band (5410-5890 MHz) to avoid interference from popular X-band marine radars (Motorola, undated). Up to 20 mobile users can time share the *Falcon 484* system (50 milliseconds/user maximum). System resolution is in tenths of units (meters, feet, or yards) with a range accuracy of ±2 meters probable. Power requirements for the fixed-location reference stations are 22 to 32 volts DC at 13 watts nominal, 8.5 watts standby, while the mobile range processor and its associated transmitter/receiver and display unit draw 150 watts at 22 to 32 volts DC.

14.1.5 Harris *Infogeometric* System

Harris Technologies, Inc., (HTI), Clifton, VA, is developing a ground-based RF position-location and communications strategy wherein moderately priced *infogeometric (IG)* devices cooperatively form self-organizing instrumentation and communication networks (Harris, 1994). Precision position location on the move is based on high-speed range trilateration from fixed reference devices. Each *IG* device in the network has full awareness of the identity, location, and orientation of all other *IG* devices (Figure 14-6), and can communicate in both party-line and point-to-point communication modes.

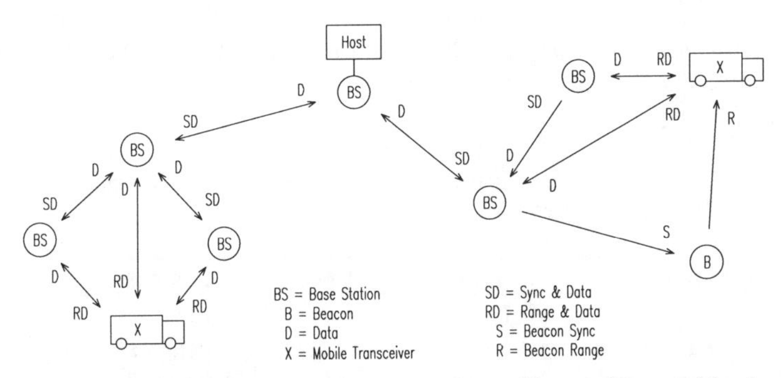

Figure 14-6. A self-organizing *infogeometric* network provides precision position location of multiple platforms through high-speed range trilateration to fixed-location *base stations* and *beacons* (courtesy Harris Technologies, Inc.).

The *IG* devices employ digital *code-division-multiple-access (CDMA)* spread-spectrum RF hardware that can provide the following functional capabilities:

- Network-level mutual autocalibration.
- Associative location and orientation tracking.
- Party-line and point-to-point data communications (with video and audio options).
- Distributed sensor data fusion.

To improve accuracy over alternative range-lateration schemes, the HTI system incorporates mutual data communications, permitting each mobile user access to the time-tagged range measurements made by fixed reference devices and all other mobile users. This additional network-level range and timing information permits more accurate time synchronization among device clocks, and automatic detection and compensation for uncalibrated hardware delays. Each omnidirectional CDMA spread-spectrum "geometric" transmission uniquely identifies the identity, location, and orientation of the transmitting source. Typically the available geometric measurement update rate is in excess of 1 MHz. Tracking and communications at operating ranges of up to 20 kilometers are supported by transmission power levels of 1 to 3 watts. Typical "raw data" measurement resolution and accuracies are cited in Table 14-1 below.

Table 14-1. Raw data measurement resolution and accuracy.

Parameter	Resolution	Biasing	Dimension
Range	1	5	meters
Bearing (Az, El)	2	2	degrees
Orientation (Az)	2	2	degrees

Enhanced tracking accuracies for selected applications can be provided as cited in Table 14-2 below. This significant improvement in performance is provided by *sensor data-fusion algorithms* that exploit the high degree of relational redundancy characteristic of infogeometric network measurements and communications.

Table 14-2. Enhanced tracking resolution and accuracies obtained through sensor data fusion.

Parameter	Resolution	Biasing	Dimension
Range	0.1 - 0.3	0.1 - 0.3	meters
Bearing	0.5 - 1.0	0.5 - 1.0	degrees
Orientation	0.5 - 1.0	0.5 - 1.0	degrees

These *data-fusion algorithms* also provide the following additional capabilities:

- Enhanced tracking in multipath and clutter, permitting precision robotic tracking even when operating indoors.
- Enhanced near/far interference reduction, for shared-spectrum operations in potentially large user networks (i.e., hundreds to thousands).

Distributed *IG* networks support precision tracking, communications, and command and control among a wide variety of potential user devices. No absolute surveyed or known location references are needed when a number of mobile units are interacting on a purely relational basis. When operating within an absolute coordinate frame as is typically the case with most semi-structured robotic applications, only a few surveyed reference locations are required.

Prototype 902- to 928-MHz FCC-compatible *infogeometric transceivers* developed for the MDARS Exterior program were field-tested in October 1994 over distances of 6 to 8 kilometers, using 1-watt transmitters equipped with 3-dB omni-directional antennae. Kinematic tracking tests under typical multipath conditions showed repeatedly achievable resolutions of 10 to 30 centimeters RMS in range, and 1 to 2 degrees RMS in bearing. The nominal communication data rate of 64 kilobits per second can be increased if necessary with a subsequent reduction in detection range and signal jamming resistance. Similarly, the 10-Hz position-update rate required by the MDARS performance specification can be increased at the expense of slightly reduced tracking resolution.

14.2 Satellite-Based Systems

14.2.1 Transit Satellite Navigation System

The Transit satellite navigation system was developed by the US Navy to provide accurate worldwide position location for ballistic missile submarines (Stansell, 1971). The first Transit satellite was launched in 1959; the system became operational in 1964 and was declassified and made available in 1967 for commercial maritime usage. By 1990, seven satellites were on line in polar orbits, accompanied by six spares (Getting, 1993).

The principle of operation for Transit is based on the observed Doppler shift of a stable CW transmitter on board a satellite as it passes through its *closest point of approach* (CPA) to the receiver. (This effect was first noticed by Navy researchers at the Applied Research Laboratory run by Johns Hopkins University, while closely monitoring the transmitter on the Soviet Sputnik 1.) The slope of the resultant Doppler frequency curve is directly related to the radial component of relative velocity, and thus indicative of the slant range from receiver (assumed to be at sea level) to satellite.

In other words, if the receiver is located in the plane of the orbit, the radial velocity component will be maximized, and slant range at CPA will be equal to the satellite altitude directly overhead. As the receiver moves away from a position directly below the orbital path (Figure 14-7), the radial component of relative velocity falls off while the tangential component increases. The overall effect shows up as a decrease in the slope of the curve representing the received Doppler frequency as a function of time during satellite passage (Figure 14-8). Based on the measured value of the slope, the actual slant range from the receiver to the known satellite position (as transmitted by the satellite) at CPA can be derived from precalculated tables. This slant range offset establishes one of two required position coordinates, while the actual time of closest approach provides the other coordinate.

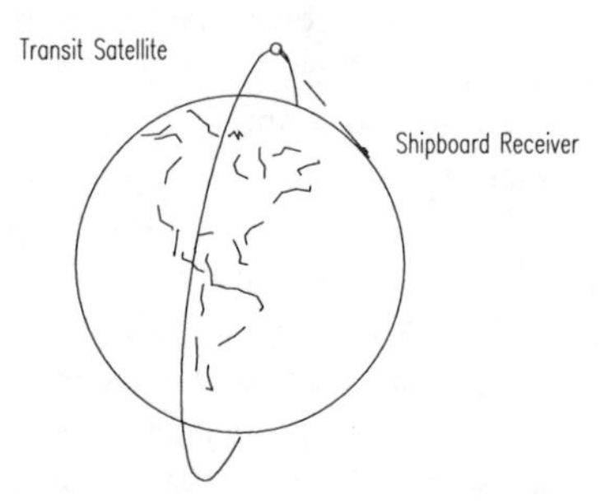

Figure 14-7. Radial component of relative velocity decreases for receiver locations further away from the orbital plane of the satellite (adapted from Dodington, 1989).

In actuality, not one but two separate CW transmissions are broadcast at 150 and 400 MHz. This technique allows the receiver to estimate the propagation delay introduced by the free-electron density of the ionosphere, based on the measured difference in time of arrival of the two signals (Getting, 1993). The Transit satellites circle the globe every hour and 45 minutes in fairly low (600 mile) polar orbits, and may be in view for as long as 20 minutes, during which time repetitive fixes can be taken every 30 seconds. Minor deviations in satellite trajectory caused by air friction and variations in the earth's magnetic field are tracked by ground stations, with updated positional information passed to the satellites for global rebroadcast. The *circular error probable* (CEP) for a single fix can approach 50 to 200 meters for a receiver mounted on a slowly moving platform with known course and speed (Dodington, 1989). For fixed-location receivers, higher accuracies approaching a few meters are possible by averaging the readings over a long period of time (Getting, 1993). In fact, before being abandoned by geodesists in favor of GPS (Section 14.2.2), Transit was providing sub-meter accuracies in certain structured scenarios.

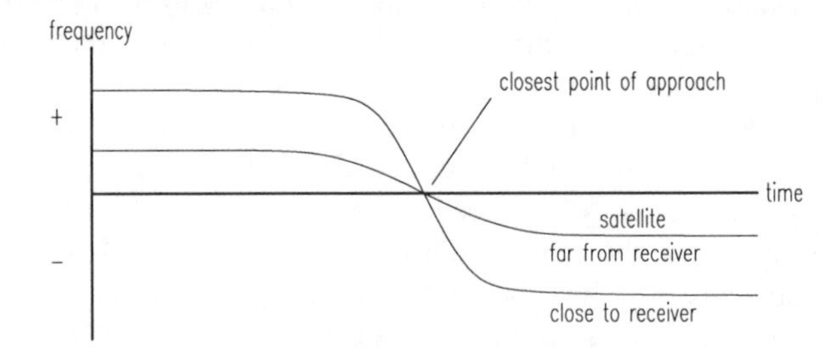

Figure 14-8. Slope of the Doppler frequency shift at time of passage is indirectly related to the slant range between receiver and satellite (Dodington, 1989).

14.2.2 Navstar Global Positioning System

The more recent *Navstar global positioning system (GPS)* developed as a Joint Services Program by the Department of Defense uses a constellation of 24 satellites orbiting the earth every 12 hours at an altitude of about 10,900 nautical miles. Four satellites are located in each of six planes inclined 55 degrees with respect to the plane of the earth's equator (Getting, 1993). The absolute three-dimensional location of any GPS receiver is determined through *trilateration* techniques based on time of flight for uniquely coded *spread-spectrum* radio signals transmitted by the satellites. Precisely measured signal propagation times are converted to *pseudoranges* representing the line-of-sight distances between the receiver and a number of reference satellites in known orbital positions. The measured distances have to be adjusted for receiver clock offset, as will be discussed later, hence the term *pseudoranges*. Knowing the exact distance from the ground receiver to three satellites theoretically allows for calculation of receiver latitude, longitude, and altitude.

Although conceptually rather simple (see Hurn (1993) for an introductory overview), this design philosophy introduces at least four obvious technical challenges:

- Time synchronization between individual satellites and GPS receivers.
- Precise real-time location of satellite position.
- Accurate measurement of signal propagation time.
- Sufficient signal-to-noise ratio for reliable operation in the presence of interference and possible jamming.

A less obvious challenge arises from the fact that, according to Einstein's theory of general relativity, time appears to run slower near a massive body like the earth than at higher altitudes. If not taken into proper consideration, this rather intriguing phenomenon can lead to position errors on the order of miles (Hawking, 1990). Having introduced this particular issue for sake of awareness, I must now defer further discussion as beyond the scope of this chapter.

The first of the previously mentioned problems is addressed through the use of sophisticated atomic clocks (relying on the vibration period of the cesium atom as a time reference) on each of the satellites to generate time ticks at a frequency of 10.23 MHz. (The Block II satellites actually carry two cesium and two rubidium atomic clocks). Each satellite transmits a periodic pseudo-random code on two different L-band frequencies (designated *L1* and *L2*) using *spread-spectrum* techniques in the internationally assigned navigational frequency band. The *L1* and *L2* frequencies of 1575.42 and 1227.6 MHz are generated by multiplying the cesium-clock time ticks by 154 and 128, respectively. The individual satellite clocks are monitored by dedicated ground tracking stations operated by the US Air Force and continuously advised of their measured offsets from official GPS time. (GPS time is kept by a virtual master clock devised through a combination of clocks at the tracking station and the individual satellites.) High precision in this regard is critical since electromagnetic radiation travels at the speed of light, roughly 1 foot per nanosecond.

There are two possible methods that can be used to establish the exact time required for signal propagation, both necessitating careful phase comparison between satellite-transmitted and locally generated waveforms. The signals involved can be: 1) the *L1* or *L2* carrier frequencies themselves or 2) the pseudo-random code modulated onto the carrier frequencies. This section will address only this second option, referred to as *code-phase tracking*, which was the original intended mode of operation for GPS. The subsequently developed and more accurate *carrier-phase tracking* approach, used primarily for static surveying applications (and more recently mobile scenarios as well), will be discussed in a later section.

In the more conventional *code-phase tracking* scheme, an identical pseudocode sequence is generated in the GPS receiver on the ground and compared to the received code from the satellite. The locally generated code is shifted in time during this comparison process until maximum correlation is observed, at which point the induced delay represents the time of arrival as measured by the receiver's clock. The problem then becomes establishing the relationship between the atomic clock on the satellite and the inexpensive quartz-crystal clock employed in the GPS receiver. Deriving this ΔT is accomplished by measuring the range to a fourth satellite, resulting in four independent trilateration equations with four unknowns (i.e., X, Y, Z, and ΔT). Details of the mathematics involved are presented by Langley (1991).

As was the case with the Transit system, precise real-time location of satellite position is determined by a number of widely distributed tracking and telemetry stations at surveyed locations around the world. Referring to Figure 14-9, all measured and received data are forwarded to a master station for analysis and referenced to GPS time. Change orders and signal-coding corrections are generated by the master station and then sent to the satellite control facilities for uploading (Getting, 1993). In this fashion the satellites are continuously advised of their current position as perceived by the earth-based tracking stations, and encode this *ephemeris* information into their *L1* and *L2* transmissions to the GPS receivers. (*Ephemeris* is the space vehicle orbit characteristics, a set of numbers that precisely describe the vehicle's orbit when entered into a specific group of equations.)

GPS time is measured in terms of 1.5-second *epochs* generated by the atomic clocks aboard the various satellites, which are maintained in effective synchronization by uplinked corrections as measured by the Control Segment. These individualized timing correction offsets are included in the navigation messages transmitted by the satellites to user receivers on the ground. The *time of week (TOW)* is defined as the number of elapsed *epochs* since the beginning of the week. The *Z-count* is a 29-bit binary number, where the 10 most-significant digits represent the GPS week number, and the 19 least-significant digits portray the current *time of week* (Mathers, 1994).

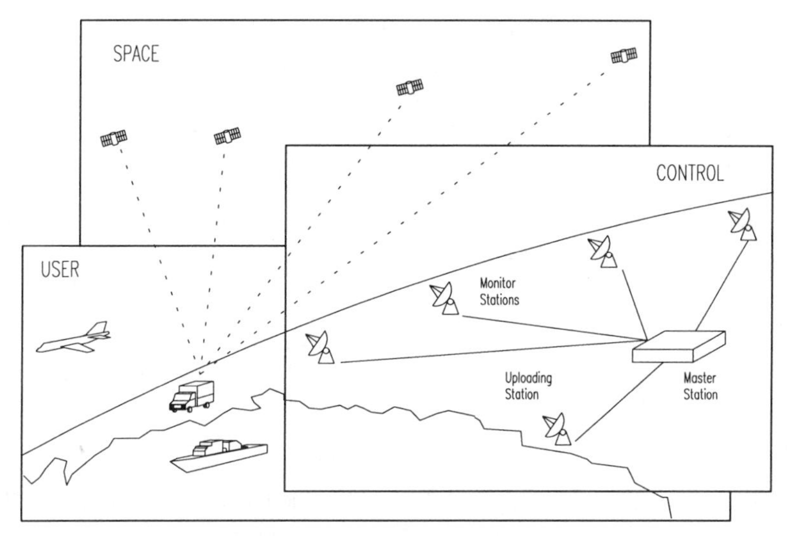

Figure 14-9. The Navstar Global Positioning System consists of three fundamental segments: Space, Control, and User (adapted from Getting, 1993, © IEEE).

In addition to its own timing offset and orbital information, each space vehicle transmits data on all other satellites in the constellation to enable any ground receiver to build up an *almanac* after a "cold start." Diagnostic information with respect to the status of certain onboard systems and expected range-measurement accuracy is also included. This collective "housekeeping" *NAV message* is superimposed on both the *L1* and *L2* signals through modulo-two addition to the pseudo-random code modulation at a very low (50-bits/second) data rate, and requires 12.5 minutes for complete downloading (Ellowitz, 1992). Timing offset and *ephemeris* information are repeated at 30-second intervals during this procedure to facilitate initial pseudorange measurements.

To further complicate matters, the sheer length of the unique pseudocode segment assigned to each individual Navstar satellite (i.e., around 6.2 trillion bits) for repetitive transmission can potentially cause initial synchronization by the ground receiver to take considerable time. For this and other reasons, each satellite broadcasts two different non-interfering pseudocodes. The first of these is called the *coarse acquisition*, or *C/A-code*, and is transmitted on the *L1* frequency to assist in acquisition. There are 1023 different *C/A-codes*, each having 1023 chips (code bits) repeated 1000 times a second (Getting, 1993) for an effective chip rate of 1.023 MHz (i.e., one-tenth the cesium clock rate). Each satellite has its own individual *C/A-code* assignment so as to enable unique identification by the ground receivers. While the *C/A-code* alone can be employed by civilian users to obtain a fix, the resultant positional accuracy is somewhat degraded due to the long chip wavelength of 300 meters. This public-usage satellite navigation service is formally referred to as *Standard Positioning Service*, or *SPS*.

The more advanced military capability known as the *Precise Positioning Service*, or *PPS*, is supported by the *Y-code,* formerly the *precision* or *P-code* prior to encryption 1 January 1994. (The *Y-code* is in fact a scrambled version of the *P-code.*) The *Y-code* is transmitted on both the *L1* and *L2* frequencies and scrambled for reception only by authorized users with appropriate cryptographic keys and equipment. In addition to limiting user access, this *antispoofing* encryption also ensures bona fide recipients will not inadvertently track false GPS-like signals generated by unfriendly forces. A special *Hand-Over-Word* is transmitted in each six-second subframe of the C/A-code *NAV message* to facilitate acquisition of the more complex Y-Code.

The major functional difference between the Y- and C/A-codes is the length of the code segment. While the C/A-code is 1023 bits long and repeats every millisecond, the Y-code is 2.35×10^{14} bits long and requires 266 days to complete (Ellowitz, 1992). Each satellite uses a one-week segment of this total code sequence; there are thus 36 unique Y-codes (31 for satellites and five spares for other purposes) each consisting of 6.18×10^{12} code bits set to repeat at midnight on Saturday of each week (ARINC, 1991). The higher chip rate of 10.23 MHz (equal to the cesium clock rate) in the precision Y-code results in a chip wavelength of 30 meters for the Y-code as compared to 300 meters for the C/A-code (Ellowitz, 1992), and thus facilitates slightly more precise time-of-arrival measurement for military purposes. For a more detailed treatment of the C/A- and Y-code signal structures, see Spilker (1978).

Brown and Hwang (1992) discuss a number of potential pseudorange error sources as summarized in Table 14-3. Positional uncertainties related to the space vehicles are clearly a factor, introducing as much as 3 meters standard deviation in pseudorange measurement accuracy. As the radiated signal propagates downward towards the earth, varying atmospheric effects introduce an uncertainty in the actual time of arrival due to changes to the speed of signal propagation. Recall from Chapter 9 that the speed of light c is only constant in a vacuum and slows in other media by a factor known as the *index of refraction n*:

$$c_m = \frac{c}{n}.$$

Electrons are liberated from the atoms found in various ionospheric gases by the ultraviolet light from the sun, thus contributing to the free-electron density, which in turn directly influences the value of n as given in the equation presented earlier in Chapter 9:

$$n = 1 + \frac{Nq_e^2}{2\varepsilon_o m\omega^2}$$

where:

N = number of charges per unit volume
q_e = charge of an electron
ε_o = permittivity of free space
m = mass of an electron
ω = frequency of the electromagnetic radiation

For free electrons, $\omega_0 = 0$ as there is no elastic restoring force (Feynman, et al., 1963).

In other words, the higher the free-electron density of the ionosphere, the more charges per unit volume (N), and the greater the subsequent slowing effect on signal propagation speed c_m. For this reason the mean *ionospheric group delay* during the day (around 50 nanoseconds) is much worse than at night (around 10 nanoseconds) when solar activity is greatly reduced (Spilker, 1978). Ionospheric effects are also very much a function of satellite elevation angle, which determines the length of the signal path through the region of influence, with the shortest possible path of course resulting when the satellite is directly overhead. At low elevation angles the above delays can easily triple (Spilker, 1978). Notice, however, the inverse square dependence on frequency ω in the above equation. This relationship can be exploited by PPS users to dynamically determine the *ionospheric group delay* effects through use of dual-frequency transmissions (i.e., *L1* and *L2*), resulting in two independent equations involving the same satellite (and therefore the same propagation path).

Tropospheric group delays caused by water vapor and other atmospheric constituents, on the other hand, are basically independent of frequency, and more pronounced due to the higher air density. Fortunately, however, tropospheric effects are fairly easy to model based on local barometric pressure, temperature, and humidity measurements (Spilker, 1978). Due to the more immediate proximity of the troposphere (9-16 kilometers) relative to the receiver, the group delay effects are even more dependent on satellite elevation than that associated with ionospheric delays as discussed above. This dependence increases exponentially below elevation angles of 15 degrees due to the stronger concentration of water vapor at altitudes below 12 kilometers (Lachapelle, et al., 1987). Local variations in the moisture content of the troposphere cause problems in eliminating this particular error component through differential-GPS techniques, as will be discussed later in this chapter.

Multipath reflections (i.e., from clouds, land masses, water surfaces) can increase the perceived time of flight beyond that associated with the optimal straight-line path (Figure 14-10). In general, the problem is much worse for marine applications than land-based scenarios, due to the higher reflectivity of water, and also more serious for low satellite elevation angles. The pseudo-random code-modulation scheme employed in the C/A- and Y-code signals inherently rejects signals outside of one chip-size of the direct pseudorange, effectively bounding the maximum possible multipath pseudorange error to 293.2 and 29.3 meters, respectively (Lachapelle, et al., 1987).

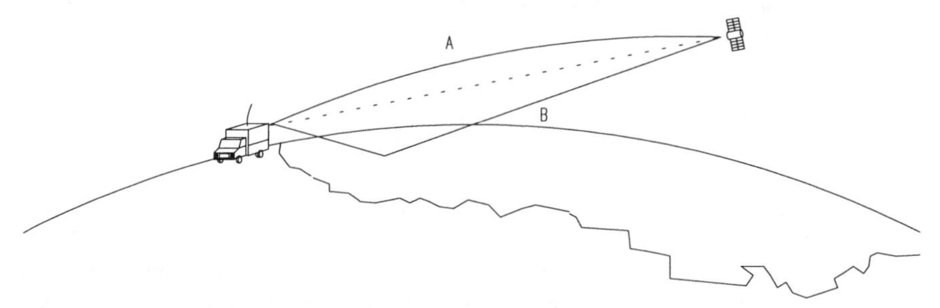

Figure 14-10. Atmospheric refraction (A) and multipath reflections (B) contribute to pseudorange measurement errors, especially at low satellite elevation angles.

In addition to the aforementioned atmospheric effects, small errors in timing and satellite position have been deliberately introduced into the C/A- and P-codes by the master station to prevent a hostile nation from using GPS in support of precision weapons delivery. This intentional degradation in SPS positional accuracy to around 100 meters (2drms) best-case is termed *selective availability (S/A)* (Gothard, 1993). *Selective availability* has been continuously in effect (with a few exceptions) since the end of Operation Desert Storm. The level of induced error was set to zero during the Persian Gulf War from August 1990 until July 1991 to improve the accuracy of commercial hand-held GPS receivers used by coalition ground forces.

There are two aspects of *selective availability*: *epsilon* and *dither*. *Epsilon* is intentional error in the superimposed navigation message regarding the exact location (*ephemeris*) of the satellite, whereas *dither* is induced error in the satellite clock that creates uncertainty in the time-stamped data. Encrypted correction parameters are incorporated into the Y-Code to allow authorized PPS users to remove the effects of both *epsilon* and *dither*. Some SPS receivers (for example, the Trimble *Ensign*) employ running-average filtering to statistically reduce the *epsilon* error over time (typically hours) to a reported value of 15 meters *spherical error probable (SEP)*.

Additional errors can be attributed to group delay uncertainties introduced by the processing and passage of the signal through the satellite electronics. Receiver noise and resolution must also be taken into account. Motazed (1993) reports fairly significant differences of ±0.02 to ±0.07 arcminutes in calculated latitudes and longitudes for two identical C/A-code receivers placed side by side. And finally, the particular dynamics of the mobile vehicle that hosts the GPS receiver plays a noteworthy role, in that best-case conditions are associated with a static platform, and any substantial velocity and acceleration will adversely affect the solution.

Table 14-3. Summary of potential error sources for measured pseudoranges (adapted with permission from Brown & Hwang, 1992, © John Wiley and Sons, Inc.).

Error Source	Standard Deviation	Units
Satellite position	3	meters
Ionospheric effects	5	meters
Tropospheric effects	2	meters
Multipath reflection	5	meters
Selective Availability	30	meters

All of the error sources listed in Table 14-3 are further influenced by the particular geometry of the four reference satellites at time of sighting. Ignoring time synchronization needs for the moment (i.e., three versus four satellites required), the most accurate three-dimensional trilateration solutions will result when the bearing or sight lines extending from the receiver to the respective satellites are mutually orthogonal. If the satellites are spaced closely together in a tight cluster or otherwise arranged in a more or less collinear fashion with respect to the receiver as shown in Figure 14-11, the desired orthogonality is lost and the solution degrades accordingly. This error multiplier, which can range from acceptable values of two or three all the way to infinity, is termed *geometric dilution of precision* (Byrne, et al., 1993).

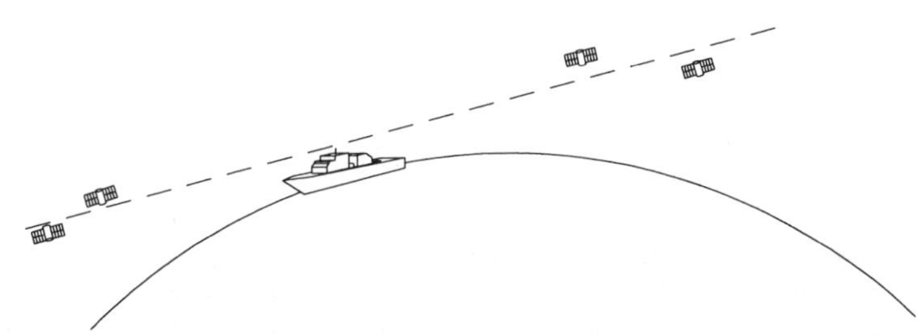

Figure 14-11. Worst case *geometric dilution of precision (GDOP)* errors occur when the receiver and satellites approach a collinear configuration as shown.

Kihara and Okada (1984) mathematically show where the minimum achievable (best-case) value for GDOP is 1.5811 and occurs when the four required GPS satellites are symmetrically located with an angle of 109.47 degrees between adjacent bearing lines as shown in Figure 14-12. In reality (since this criteria places three of the satellites just below the horizon), optimal performance is seen with three satellites equally spaced just above the horizon and one directly overhead.

Overall SPS accuracy can be significantly enhanced when combined with information from other navigational sensors, such as odometry, magnetic compasses, rate gyros, and inclinometers. For example Zexel Corporation, Farmington Hills, MI, has developed an integrated system for Oldsmobile that uses vehicle displacement information from a car's electronic speedometer, in addition to azimuthal information from the Murata piezoelectric gyro (Chapter 13), to augment the solution from a Rockwell GPS receiver (Schofield, 1994). A street-map database supplied by Navigation Technologies Corporation of Sunnyvale, CA, is employed to generate a CRT map display for the driver, with computer-synthesized speech output to advise of upcoming turns and landmarks (Lyons, 1994). The Zexel satellite navigation package was introduced as an option in Avis rental cars in the state of California in late 1994, with coast-to-coast availability expected to follow in 1995 as the associated map databases are tested and debugged.

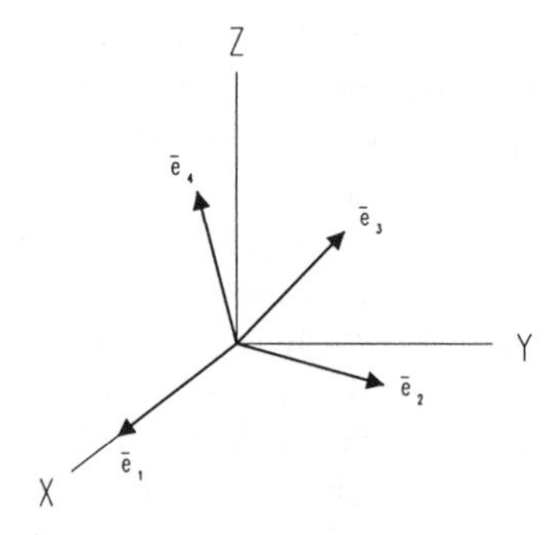

Figure 14-12. *GDOP* error contribution is minimal for four GPS satellites symmetrically situated with respect to the receiver (at origin) along bearing lines 109.47 degrees apart (Kihara & Okada, 1984).

Advertised accuracy of this combined GPS/dead-reckoning system is around 10 meters, a significant improvement over the 100 meters best case for unaided SPS. The positional accuracy needs of an exterior autonomous mobile robot, however, are a little more demanding than can be met with conventional SPS systems of this type, even when augmented by additional dead-reckoning sensors. Accordingly, most system designers are relying instead on an enhanced variation known as *differential GPS (DGPS).* With the exception of multipath (and to a certain extent tropospheric delays), the collective effects of all the error sources depicted in Table 14-3 can be essentially eliminated through use of this practice, the topic of our next section.

Code-Phase Differential GPS

The *differential GPS* concept is based on the premise that a second GPS receiver in fairly close proximity (i.e., within several tens of kilometers) to the first will experience basically the same error effects when viewing identical reference satellites, since the satellites are so far away in comparison to the receiver separation baseline. If this second receiver is fixed at a precisely surveyed location (Figure 14-13), its calculated solution can be compared to the known position to generate a composite error vector representative of prevailing conditions in that immediate locale. This differential correction can then be passed to the first receiver to null out the unwanted effects, effectively reducing position error for commercial systems to well under 10 meters. Hurn (1994) presents a concise and easy to read introduction to the basic fundamentals of DGPS. The terminology *kinematic DGPS* is often used to describe situations where the first receiver is mobile, (i.e., moving in real time), as would be the case in any robotics application.

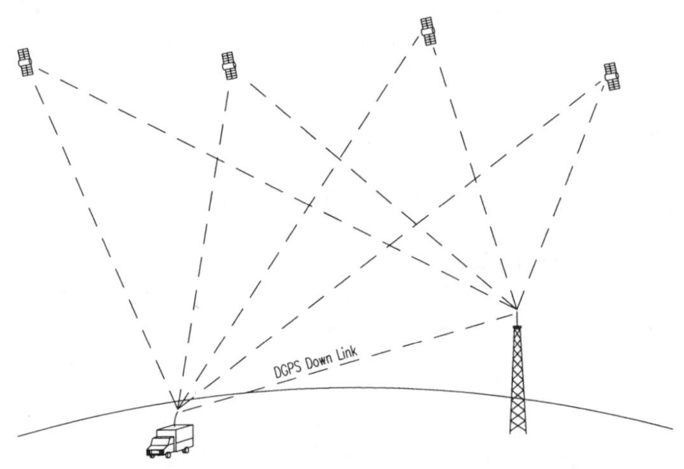

Figure 14-13. A differential GPS receiver placed at a surveyed reference site can be used to relay real-time corrections over an RF downlink to the DGPS-equipped vehicle.

In actuality, there are a number of ways this DGPS concept can be implemented, based on whether the corrections are made in the measurement domain or the solution domain, and the types of datalink employed (Blackwell, 1985). From a robotics perspective, only those procedures which make the upgraded solution available at the mobile-platform end of the datalink are of interest, thus narrowing the possibilities down to two candidate techniques. The general scenario described in the preceding paragraph, where the fixed reference receiver calculates a position correction vector, requires the following information be uplinked from the base station to the mobile unit (Blackwell, 1985):

- ΔX, ΔY, and ΔZ corrections for each user's set of satellites.
- Rate of change of above corrections
- Age of ephemeris data (each satellite) used by the base station.
- Address of the mobile platform.

The fundamental drawback to this approach is the ambiguity associated with which four satellites are used to generate the position solution. If any of the potentially numerous mobile receivers select a dissimilar combination from the available satellites in view than was selected by the reference station receiver, the differential corrections are no longer valid. Since most modern multichannel receivers employ dynamic algorithms for optimal satellite selection, this satellite coordination disconnect presents a real problem.

Accordingly, most equipment vendors have elected to support an alternative method wherein corrections are made in the measurement domain (i.e., to the pseudoranges themselves). The reference receiver basically knows where the satellites are supposed to be and how long the signal propagation should take, and compares this expected transit time to that actually required. The necessary information that must be transferred to the remote vehicle under this scheme is (Blackwell, 1985):

- Pseudorange corrections for each satellite in view.
- Rate of change of above corrections
- Age of ephemeris data (each satellite) used by the reference receiver.

This approach is advantageous in that any number of remote vehicles within the coverage envelope of the DGPS uplink can make use of this correction information simply by choosing the appropriate subset of data (i.e., that corresponding to the four specific satellites selected by their respective onboard receivers).

The fixed DGPS reference station transmits these pseudorange corrections several times a minute (depending on the data rate, message format, and number of satellites in view) to any differential-capable receiver within range. For example, a *Type 1* message format (see Table 14-4) with four satellites in view results in an average transmission rate of once every seven seconds, while 18 seconds would be required with 11 satellites in view (Mathers, 1994). The pseudorange correction is passed to the remote platform in the form of a first-order polynomial (RTCM, 1994):

$$PRC(t) = PRC(t_o) + RRC*\left[t - t_o\right]$$

where:

$PRC(t)$ = pseudorange correction at time t
$PRC(t_o)$ = correction at modified Z-count reference time t_o
RRC = range rate correction.

This correction is simply added to the appropriate pseudorange as measured by the mobile receiver:

$$PR(t) = PRM(t) + PRC(t)$$

where:

$PR(t)$ = differentially corrected measurement for time t
$PRM(t)$ = pseudorange measured by mobile receiver at time t.

Many commercial GPS receivers are available with differential capability, and most now follow the *RTCM SC-104* standard developed by the Radio Technical Commission for Maritime Services (Special Committee 104) to promote interoperability. Version 2.1 of this standard now identifies 26 formal message types (Table 14-4), each consisting of a variable number of 30-bit words (RTCM, 1994). The first two words in each message type form the message header consisting of: 1) a preamble (a fixed sequence of binary digits), 2) the message type, 3) the reference station identifier, 4) the modified Z count, 5) a sequence number that increments with each message, 6) the message length (number of words), and 7) the reference station health code (Mathers, 1994). Only the first 24 bits of each word are used for data, with the last 6 bits reserved for parity check.

Table 14-4. Format for the first 21 of 26 identified RTCM SC-104 message types; only seven are defined, nine reserved with form and content to be specified later, and the rest are undefined (Langley, 1994).

Type	Status	Explanation
1	Fixed	Differential GPS Corrections
2	Fixed	Delta Differential GPS Corrections
3	Fixed	Reference Station Parameters
4	Retired	Surveying
5	Tentative	Constellation Health
6	Fixed	Null Frame
7	Fixed	Beacon Almanacs
8	Tentative	Pseudolite Almanacs
9	Fixed	Partial Satellite Set Differential Corrections
10	Reserved	P(Y)-Code Differential Corrections (all)
11	Reserved	C/A-Code L1 and L2 Delta Corrections
12	Reserved	Pseudolite Station Parameters
13	Tentative	Ground Transmitter Parameters
14	Reserved	Surveying Auxiliary Message
15	Reserved	Ionosphere (Troposphere) Message
16	Fixed	Special Message
17	Tentative	Ephemeris Almanac
18	Tentative	Uncorrected Carrier Phase Measurements
19	Tentative	Uncorrected Pseudorange Measurements
20	Tentative	RTK Carrier Phase Corrections
21	Tentative	RTK Pseudorange Corrections

In addition to the more common C/A-code implementations, differential systems can also utilize the higher-resolution Y-code (see message Type 10 in Table 14-4) for even better accuracies. With *antispoofing* now in effect, however, only authorized PPS users are able to access the L2 signal, and so Y-code-DGPS systems have not been perceived as viable products by the vendor community. Prices for DGPS-capable mobile receivers run about $2K, while the reference stations cost somewhere between $10K and $20K. Magnavox is working with CUE Network Corporation (one of several companies providing DGPS correction services) to market a nationwide network to pass differential corrections over an FM link to paid subscribers (GPS Report, 1992).

Typical code-phase DGPS accuracies are around 2 to 4 meters SEP, with better performance seen as the baseline distance between the mobile unit and the fixed reference station is decreased, and the sophistication of receiver hardware and software improves with time. Many receiver manufacturers "smooth" the pseudorange measurements with the carrier, thus eliminating some of the inherent noise for improved performance (McPherson, 1991). The Coast Guard is in the process of implementing *differential GPS* in all major US harbors, with an expected accuracy of around 1 meter SEP (Getting, 1993). A differential GPS system already in operation at O'Hare International Airport in Chicago has demonstrated that aircraft and service vehicles can be located to 1 meter (Hambly, 1992). Surveyors have used differential GPS to achieve centimeter accuracy for years, but this practice involves *carrier-phase tracking* versus *code-phase tracking* (see next section), and until recently required long static dwell times plus significant postprocessing of the collected data (Byrne, 1993).

An interesting variant of kinematic DGPS is reported by Motazed (1993) in conjunction with the *Non-Line-of-Sight Leader/Follower (NLOSLF)* program underway at RedZone Robotics, Inc., Pittsburgh, PA. The NLOSLF operational scenario involves a number of vehicles in a convoy configuration that autonomously follow a lead vehicle driven by a human operator, both on-road and off-road at varying speeds and separation distances. A technique Motazed refers to as *intermittent stationary base differential GPS* is used to provide global referencing for purposes of bounding the errors of a sophisticated Kalman-filter-based GPS/INS position estimation system.

Under this innovative concept, the lead and final vehicle in the convoy alternate as fixed DGPS reference stations. As the convoy moves out from a known location, the final vehicle remains behind to provide differential corrections to the GPS receivers in the rest of the vehicles. After traversing a predetermined distance in this fashion, the convoy is halted and the lead vehicle assumes the role of differential reference station, providing enhanced accuracy to the trailing vehicle as it catches up to the pack. During this time, the lead vehicle takes advantage of on-site dwell to further improve the accuracy of its own fix. Once the last vehicle joins up with the rest, the base-station roles are reversed again, and the convoy resumes transit in "inchworm" fashion along its intended route. Disadvantages to this approach (from a robotics perspective) include the need for intermittent stops, and the accumulating ambiguity in the actual location of the appointed reference station.

Omnitech Robotics, Inc. of Englewood, CO, has developed an integrated vehicle position and orientation (pose) estimation subsystem called *COMPASS (compact outdoor multipurpose pose assessment sensing system)*. Shown in Figure 14-14, *COMPASS* uses a sophisticated suite of position and orientation sensors fused into a single statistically optimal pose estimate using Kalman filter techniques:

- Differential GPS.
- 3-axis inertial-grade accelerometers.
- 3-axis angular rate gyroscopes.
- Barometric altimeter.
- 2-axis fluxgate magnetometer.
- Optical encoder wheel odometry.
- 2-axis inclinometer.
- Magnetic landmark acquisition.

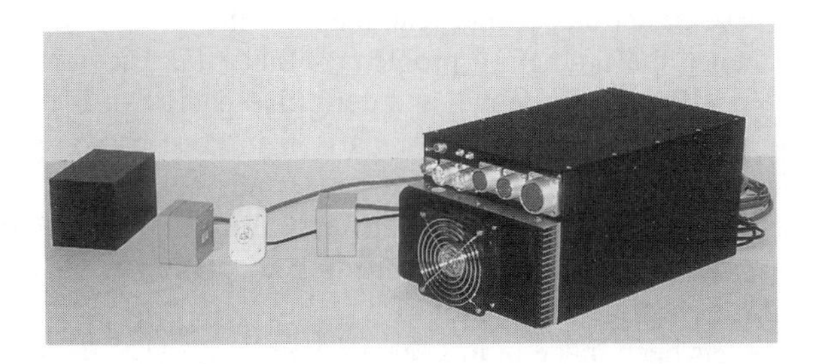

Figure 14-14. Omnitech's *COMPASS* is a commercially available navigational subsystem that uses a Kalman filter approach for fusing data from a magnetic compass, rate gyros, accelerometers, vehicle odometry, and differential GPS (courtesy Omnitech Robotics, Inc.).

The vehicle odometry and inertial navigation systems provide fast analog sensing of position, velocity, angular rate, and orientation, allowing *COMPASS* to provide three-dimensional pose information at a 32-Hz rate. The C/A-code-based DGPS and magnetic landmark sensors supply lower-update absolute position information to constrain the error accumulation to an accuracy of 10 centimeters (1-sigma 2D); in open areas where less accuracy is required, the magnetic landmarks can be omitted with a resulting accuracy of 1 meter. An optional *carrier-phase DGPS capability* (see next section) is available upon request.

Carrier-Phase Differential GPS

As previously mentioned, surveyors use a technique known as *carrier-phase differential*, also referred to (perhaps incorrectly) as *codeless differential*, to achieve improved accuracy over *code-phase differential* schemes. The L1 carrier phase is normally measured using the C/A-code, and therefore is not "codeless." The L2 phase, however, now that the P-code has been encrypted, is measured using "codeless" techniques as follows.

Recall the *Y-code* chip rate is directly equal to the satellite cesium clock rate, or 10.23 MHz. Since the L1 carrier frequency of 1575.42 MHz is generated by multiplying the clock output by 154, there are consequently 154 carrier cycles for every *Y-code* chip. This implies even higher measurement precision is possible if time-of-arrival is somehow referenced to the carrier instead of the pseudocode itself. In other words, while the *Y-code* chip wavelength is 30 meters, the L1 carrier wavelength is only 19 centimeters. Codeless interferometric differential GPS schemes can measure the phase of this carrier signal to within ±2 millimeters, and subsequently achieve 2- to 5-centimeter accuracies, but must start at a known geodetic location and typically require dwell times of several seconds or more (McPherson, 1991).

This on-station dwell time is required at start-up to resolve the unknown integer number of whole carrier cycles making up the pseudorange. (Recall from Chapter 6 that phase-detection schemes can measure the relative phase shift over only one cycle, hence the ambiguity problem.) Since the pseudo-random code generated by the satellites has a known and definitive pattern, it is relatively easy to establish a precise index for phase comparison in code-based systems. The individual cycles that make up the L-band carrier all look just alike, however, making initial synchronization much more difficult. Resolving cycle ambiguity in real time has been the principle problem impeding adaptation of the carrier-phase technique to real-time mobile applications.

Much progress has been made in this regard, however, over the past few years. The Army's Engineer Topographic Laboratories is in the process of developing a real-time *carrier-phase-differential* system that is expected to provide 1- to 3-centimeter accuracy (SEP) at a 60-Hz rate when finished sometime in 1996 (McPherson, 1991). Driven by a number of perceived application areas (i.e., automated agriculture, fleet management, robotics), many other developmental organizations and GPS equipment vendors are also pursuing viable solutions with encouraging results. The off-the-shelf Trimble *MARS* system claims a 2- to 10-centimeter real-time kinematic accuracy at slow speeds typical of the MDARS Exterior robot (i.e., less than 15 kilometers per hour). NovAtel has developed a similar system with a kinematic accuracy of less than 20 centimeters, while Premier GPS reportedly achieves 2-centimeter accuracy using proprietary software in conjunction with any of several commercially available GPS cards.

The significant engineering challenge remaining in this rapidly evolving arena is reliable avoidance, identification, and correction of *cycle slips*. A *cycle slip* is defined as a sudden gain or loss of some integer number of cycles due to receiver dynamics or the temporary occlusion of a satellite (Bock, et al., 1985). Increasing the receiver tracking bandwidth can reduce the occurrence of the former, but with an associated degradation in solution accuracy (Wong, et al., 1988). If the number of *cycle-slip-free* satellites in view ever falls below four, a new static initialization sequence must be performed to eliminate the resulting phase ambiguities (Cannon & Lachapelle, 1992). Modern receivers are addressing this vulnerability through the use of redundant satellite channels, and higher-precision C/A-code phase correlation to minimize the ambiguity interval in the first place, thus facilitating faster recovery.

GPS Summary

Conley (1993) presents an excellent assessment of stand-alone (i.e., non-differential) GPS performance and reliability with an emphasis on Control and Space Segment factors. A reasonable extraction from the open literature of achievable position accuracies for the various GPS configurations is depicted in Table 14-5. As in the case of the earlier Transit system, the Y-code has dual-frequency estimation for atmospheric refraction, and no *selective availability* error component, so accuracies are better than stand-alone single-frequency C/A-code systems. Commercial DGPS accuracy, however, exceeds stand-alone military Y-code accuracy, particularly for small-area applications such as airports. Differential Y-code is currently under consideration and may involve the use of a geosynchronous satellite to disseminate the corrections over a wide area, but the imposed *anti-spoofing* measure seriously restricts availability to potential users. Carrier-phase differential is more likely to surface as the prime contender for high-accuracy kinematic navigational systems, with several prototypes already up and running.

Table 14-5. Summary of achievable position accuracies for various implementations of GPS.

GPS Implementation Method	Position Accuracy
C/A-Code Stand Alone	100 meters 2drms
Y-Code Stand Alone	16 meters SEP
Differential C/A-Code	1 meter SEP
Differential Y-Code	TBD
Phase Differential (Codeless)	1 centimeter SEP

In summary, the fundamental problems associated with using GPS for mobile robot navigation are as follows:

- Periodic signal blockage due to foliage, structures, and hilly terrain.
- Multipath interference due to reflective structures, water, and land masses.
- Insufficient positional accuracy in stand-alone and code-phase differential mode.
- Cycle slips in carrier-phase differential mode.

GPS World provides a comprehensive listing of receiver equipment in each January issue, while Byrne, et al. (1993) present a detailed evaluation of performance for five popular models (Magnavox *6400*, Magnavox *GPS Engine*, Magellan *GPS Module,* Rockwell *NavCore V*, and Trimble *Pacer*) operating in single-point stand-alone mode. McLellan and Battie (1994) present a similar evaluation for several differential-capable receivers (Motorola *PVT6*, Magnavox *GPS Engine*, Rockwell *NavCore V*, Trimble *SVeeSix Plus*, and the Magellan *GPS Module*) operating in both static and kinematic stand-alone and differential modes.

In addition, the University of New Brunswick maintains an Internet-accessible archive of GPS information, to include receiver reviews and user feedback, in the form of the *Canadian Space Geodesy Forum (CANSPACE)*. Electronic mail messages pertaining to news, comments, questions, and answers are regularly exchanged among forum participants on topics such as Navstar GPS, Glonass, Transit, very long baseline interferometry, satellite laser ranging, etc. Although initially intended to link Canadian geodesists and geophysicists together, this valuable resource is now open to the public with a World Wide Web Universal Resource Locator of "http://www.unb.ca/Geodesy/ index.html". For more information, contact the University of New Brunswick at *lang@unb.ca*. (See Appendix.)

With a wide variety of envisioned high-volume applications and intense competition amongst a proliferation of vendors, the cost of differential GPS systems will continue to fall while reliability and accuracy further improve. Although (with the exception of Oldsmobile) the US automotive industry is taking a rather cautious wait-and-see approach, an estimated 25,000 GPS systems are sold by six major car makers in Japan each month (Berg, 1994). The Coast Guard intends to have its harbor-approach DGPS coverage completed by 1996, and the US Army Corps of Engineers is planning similar installations for the country's major navigable rivers. The FAA, meanwhile, has initiated efforts investigating the applicability of DGPS for precision aircraft approach systems at commercial airports. It seems very likely from this gathering momentum that carrier-phase DGPS, augmented by additional dead-reckoning sensors, will be the navigation system of choice for exterior robotic applications over the coming decade.

14.3 References

ARINC, "GPS NAVSTAR Global Positioning System User's Overview," Technical Report YEE-82-009D, ARINC Research Corporation, Under Contract F09603-89-G-0054/0006 to NAVSTAR GPS Joint Program Office, Los Angeles, CA, March, 1991.

Arradondo-Perry, J., "GPS World Receiver Survey," *GPS World*, pp. 46-58, January, 1992.

Berg, P., "Meanwhile, Most US Automakers are Sitting on the Fence," *Car and Driver*, pp. 113-115, May, 1994.

Blackwell, E.G., "Overview of Differential GPS Methods," Navigation: Journal of the Institute of Navigation, Vol. 32, No. 2., pp. 114-125, Summer, 1985.

Bock, Y., Abbot, R.I., Counselman, C.C., Gourevitch, S.A., King, R.W., "Establishment of Three-Dimensional Geodetic Control by Interferometry with the Global Positioning System," *Journal of Geophysical Research*, Vol. 90, No. B9, pp. 7689-7703, August, 1985.

Brown, R.G., Hwang, P.Y.C., Introduction to Random Signals and Applied Kalman Filtering, 2nd ed., New York, NY, John Wiley and Sons, p. 420, 1992.

Byrne, R.H., Klarer, P.R., Pletta, J.B., "Techniques for Autonomous Navigation," Sandia Report SAND92-0457, Sandia National Laboratories, Albuquerque, NM, March, 1992.

Byrne, R.H., "Global Positioning System Receiver Evaluation Results," Sandia Report SAND93-0827, Sandia National Laboratories, Albuquerque, NM, September, 1993.

Cannon, M.E., Lachapelle, G., "Analysis of a High-Performance C/A-Code GPS Receiver in Kinematic Mode," *Navigation: Journal of the Institute of Navigation*, Vol. 39, No. 3, pp. 285-300, Fall, 1992.

Conley, R. "GPS Performance: What is Normal?" Navigation: Journal of the Institute of Navigation, Vol. 40, No. 3, pp. 261-281, Fall, 1993.

Dodington, S.H., "Electronic Navigation Systems," *Electronic Engineer's Handbook,* D. Christiansen and D. Fink, eds., 3rd edition, New York, McGraw Hill, pp. 76-95, 1989.

Duchnowski, L.J., "Vehicle and Driver Analysis with Real-Time Precision Location Techniques," *Sensors*, pp. 40-47, May, 1992.

Ellowitz, H.I., "The Global Positioning System," *Microwave Journal*, pp. 24-33, April, 1992.

Feynman, R.P., Leighton, R.B., Sands, M., *The Feynman Lectures on Physics*, Vol. 1, Addison-Wesley, Reading, MA, 1963.

Getting, I.A., "The Global Positioning System," *IEEE Spectrum,* pp. 36-47, December, 1993.

Gothard, B.M., Etersky, R.D., Ewing, R.E., "Lessons Learned on a Low-Cost Global Navigation System for the Surrogate Semi-Autonomous Vehicle," Proceedings, SPIE Vol. 2058, Mobile Robots VIII, pp. 258-269, 1993.

GPS Report, Phillips Business Information, Potomac, MD, November, 1992.

Hambly, R.M., Chicago O'Hare International Airport Differential GNSS Trials for Airport Surface Surveillance," IEEE/AiAA 11th Digital Avionics Systems Conference, 92CH3212-8, Seattle, WA, October, 1992.

Harris, J.C., "An Infogeometric Approach to Telerobotics," Proceedings, IEEE National Telesystems Conference, San Diego, CA, pp. 153-156, May, 1994.

Hawking, S. W., *A Brief History of Time*, Bantam Books, New York, June, 1990.

Hurn, J., *GPS, A Guide to the Next Utility*, No. 16778, Trimble Navigation, Sunnyvale, CA, November, 1993.

Hurn, J., "Differential GPS Explained," No. 23036, Trimble Navigation, Sunnyvale, CA, July, 1994.

Kaman, "Threat Array Control and Tracking Information Center, Product Literature, PM1691, Kaman Sciences Corp., Colorado Springs, CO, 1991.

Kihara, M., Okada, T., "A Satellite Selection Method and Accuracy for the Global Positioning System," *Navigation: Journal of the Institute of Navigation*, Vol. 31, No. 1, pp. 8-20, Spring, 1984.

Lachapelle, G., Casey, M., Eaton, R.M., Kleusberg, A., Tranquilla, J., Wells, D., "GPS Marine Kinematic Positioning Accuracy and Reliability," *The Canadian Surveyor*, Vol. 41, No. 2, pp. 143-172, Summer, 1987.

Langley, R.B., "The Mathematics of GPS," *GPS World*, pp. 45-49, July/August, 1991.

Langley, R.B., "RTCM SC-104 DGPS Standards," *GPS World*, pp. 48-53, May, 1994.

Lyons, P., "The GPS Olds: Miss Daisy Moves to the Front Seat," *Car and Driver*, pp. 112-113, May, 1994.

Mathers, B.H., "Differential GPS Theory and Operation," Naval Command Control and Ocean Surveillance Center, San Diego, CA, July, 1994.

McLellan, J.F., Battie, J.P., "Testing and Analysis of OEM GPS Sensor Boards for AVL Applications," IEEE Position, Location, and Navigation Symposium, PLANS'94, Las Vegas, NV, April, 1994.

McPherson, J.A., "Engineering and Design Applications of Differential Global Position System (DGPS) for Hydrographic Survey and Dredge Positioning," US Army Corps of Engineers, Washington, DC, Engineering Technical Letter No. 1110-1-150, 1 July, 1991.

Motazed, B., "Measure of the Accuracy of Navigational Sensors for Autonomous Path Tracking," Proceedings, SPIE Vol. 2058, Mobile Robots VIII, pp. 240-249, 1993.

Motorola, Mini-Ranger Falcon, Product Literature, Motoroloa Government and Systems Technology Group, Scottsdale, AZ, undated.

Purkey, M., "On Target," *Golf Magazine*, pp. 120-121, May, 1994.

RTCM, "RTCM Recommended Standards for Differential NavStar GPS Service, Version 2.1," Radio Technical Commission for Maritime Services, Special Committee No. 104, January, 1994.

Schofield, J.A., "Sensors Keep Cars Purring," *Design News*, pp. 133-135, 10 October, 1994.

Spilker, J. J., Jr., "GPS Signal Structure and Performance Characteristics," *Navigation: Journal of the Institute of Navigation*, Vol. 25, No. 2, pp. 121-146, Summer, 1978.

Stansell, T., "The Navy Navigational Satellite System," *Navigation: Journal of the Institute of Navigation*, Vol. 18, No. 1, Spring, 1971.

Stokes, K.W., "Remote Control Target Vehicles for Operational Testing," Association for Unmanned Vehicles Symposium, July, 1989.

Wong, R.V.C., Schwarz, K.P., Cannon, M.E., "High-Accuracy Kinematic Positioning by GPS-INS," *Navigation: Journal of the Institute of Navigation*, Vol. 35, No. 2, pp. 275-287, Summer, 1988.

15
Ultrasonic and Optical Position-Location Systems

15.1 Ultrasonic Position-Location Systems

Ultrasonic trilateration schemes offer a medium- to high-accuracy, low-cost solution to the position location problem for mobile robots operating in relatively small work envelopes, assuming no significant obstructions are present to interfere with wave propagation. The advantages of a system of this type fall off rapidly, however, in large multiroom facilities due to the significant complexity associated with installing multiple networked beacons throughout the operating area.

Two general implementations exist: 1) a single transducer transmitting from the robot, with multiple fixed-location receivers and 2) a single receiver listening on the robot, with multiple fixed transmitters serving as beacons. The first of these categories is probably better suited to applications involving only one or at most a very small number of robots, whereas the latter case is basically unaffected by the number of passive receiver platforms involved (i.e., somewhat analogous to the *Navstar GPS* concept). Both trilateration methods will be discussed in Section 15.1.1.

An alternative ultrasonic position location scheme involves first digitizing the operating area and building a reference database of range signatures as seen by a circular ring of range sensors on the robot. This database can later be searched for an appropriate match with current sensor readings to establish the absolute position of the platform. The advantage of this approach is that no expensive beacon installations or other alterations to the environment are required, but the tradeoffs are reduced accuracy and slower update rates. This *signature-matching* technique will be discussed in Section 15.1.2.

15.1.1 Ultrasonic Transponder Trilateration

An early absolute position-location concept considered towards the end of development on ROBART I (recall the robot had no dead-reckoning capability) was an ultrasonic transponder network consisting of a master receiver unit on the robot, with three or more slaved transmitters situated around the room in known locations. The master would trigger the remote slaves via a short-range RF link, whereupon each would emit a burst of ultrasonic energy. The slaves were theoretically assigned individual operating frequencies to make them uniquely identifiable. The robot's ultrasonic receiver would begin listening for the incoming signals, timing their individual arrivals with respect to the RF trigger signal. The software could then determine the robot's position through simple trilateration.

It was further reasoned that if the master unit were equipped with two separate receiver transducers at a known orientation and lateral separation, the robot's heading could be established as well. In practice, however, a number of engineering issues come into play to complicate matters. In addition to obvious errors resulting from variations in the speed of sound, there is the inherent uncertainty associated with the finite size of the transducers themselves. For example, the original (and only) Polaroid electrostatic transducer available at that time was a full 2 inches in diameter, which differs substantially from an idealized point source or receiver. The measurement uncertainty resulting from transducer width is aggravated in this particular application by the need for wide-angle coverage, resulting in off-axis operation (i.e., the transducers are not always directly facing each other). Ranging errors induced in transmitter/receiver pairs (Figure 15-1) due to angular misalignment between transducers is extensively treated by Lamancusa and Figueroa (1990).

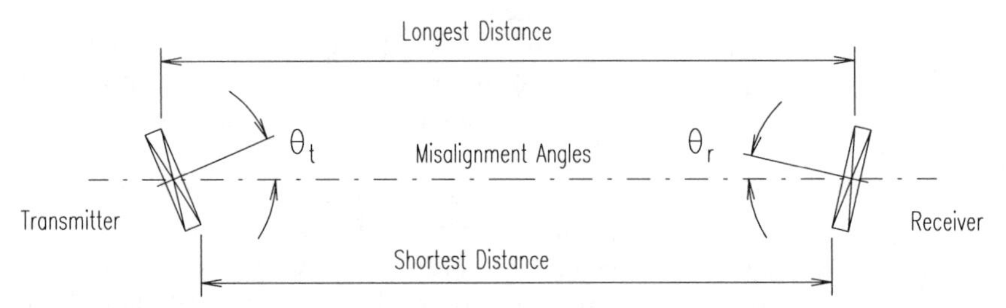

Figure 15-1. Misalignment in opposed transmitter/receiver pairs introduces a range measurement uncertainty due to the finite width of the transducers (adapted from Lamancusa and Figueroa, 1990).

On first glance it would appear that the linear relationship between effective beamwidth and transducer diameter (see Chapter 8) is advantageous in this situation, in that the necessary wide-angle coverage can be achieved in concert with an equally desirable reduction in the finite size of the sensor element.

Unfortunately, overall sensitivity falls off rapidly with any decrease in transducer diameter, significantly limiting the effective volume of coverage. An alternative strategy to achieve an omni-directional receive capability would be to employ an array of outward-facing receiver elements, monitor all transducers in the array for waveform detection, and then take into account the precise location of the actual transducer that first received the incoming acoustic wave. This approach, however, is a little less than elegant from the standpoint of increased complexity and doesn't really eliminate the range measurement errors associated with off-normal transducer alignment. IS Robotics (ISR, 1994) rather painlessly achieves the same results by situating an upward-looking sensor directly beneath a cone-shaped reflector, as will be discussed later.

I briefly toyed around at one point with the idea of using a high-voltage spark gap (Figure 15-2) as an omni-directional emitter on the robot, having read somewhere a large ultrasonic noise component could be generated in such a fashion. This approach would theoretically reduce range measurement uncertainties by more closely emulating a point source, with omni-directional propagation away from the robot to peripherally located receiver elements aimed towards the operating area. The capacitive-discharge spark generators I considered at the time, however, were even less attractive in terms of complexity, power consumption, and potential interference with other electronics, so this concept was never actually reduced to practice.

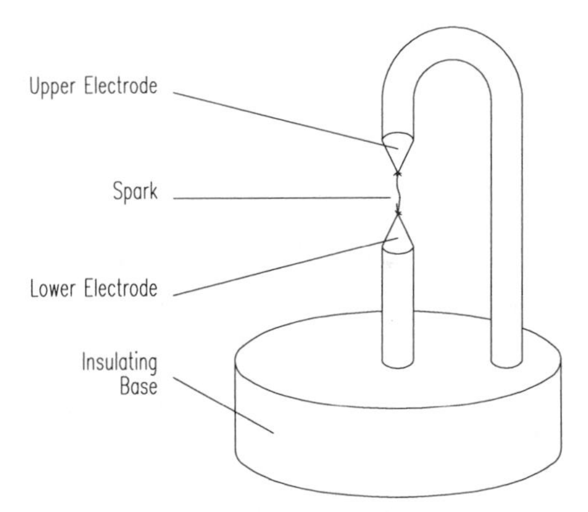

Figure 15-2. Conceptual omni-directional spark-gap emitter considered for use as an approximated point source in an ultrasonic position-location system for ROBART I.

To investigate feasibility of the ultrasonic master/slave position-location concept for possible use on ROBART II, Dunkin (1985) built and evaluated a prototype device as part of his thesis work at the Naval Postgraduate School in Monterey, CA. Although three slave units are needed for an unambiguous general solution in two dimensions, Dunkin showed a working system required only two slaves if certain conditions were taken into account. The equations describing the coordinates of the intersections of the two range arcs reduce to a pair of quadratics of the form:

$$A x^2 + B x + C = 0.$$

By solving for all known solutions, the robot's position can be determined by comparing the set of possible solutions with the estimated position, or by using various restrictions. For example, in the setup of Figure 15-3, the solution is constrained to only the positive quadrant of the Cartesian coordinate system shown, therefore all negative solutions are discarded (Dunkin, 1985).

Dunkin reported overall system accuracy to be ± 12 inches over a test area approximately 18 by 26 feet, primarily due to poor repeatability in the propagation delays associated with the RF control link which triggered the slaves. The magnitude of this inaccuracy obviously precludes any attempt to derive vehicle heading from complementary solutions for a pair of receivers mounted a known distance apart, since the ambiguity is of the same order as the maximum possible receiver separation. Significant improvement in accuracy could theoretically be obtained through optimization of the first prototype circuitry, and in fact Dunkin reported accuracies of ± 3.6 inches for the same test area when the RF link propagation uncertainties were eliminated through temporary use of hard-wired control lines.

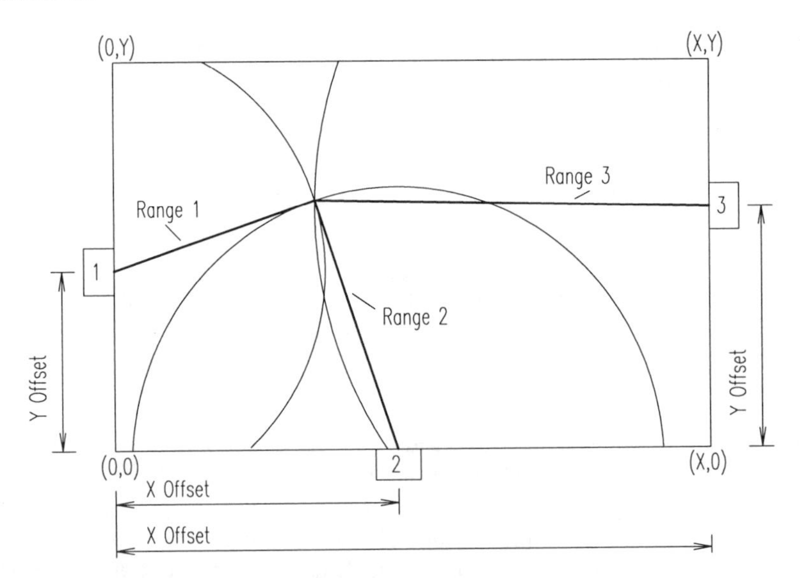

Figure 15-3. Example placement of three slave ultrasonic transmitters used to establish the position of the master receiver (robot) through trilateration (adapted from Dunkin, 1985).

IS Robotics 2-D Location System

IS Robotics, Inc. (ISR), Somerville, MA, markets an inexpensive ultrasonic trilateration system of this type that allows their *Genghis series* robots to localize position to within ±0.5 inches over a 30- by 30-foot operating area (ISR, 1994). The ISR system consists of a base-station master hard-wired to two slave ultrasonic "pingers" positioned a known distance apart (typically 90 inches) along the edge of the operating area as shown in Figure 15-4. Each robot is equipped with a receiving ultrasonic transducer situated beneath a cone-shaped reflector for omni-directional coverage. Communication between the base station and individual robots is accomplished using a Proxim spread-spectrum (902-928 MHz) RF link.

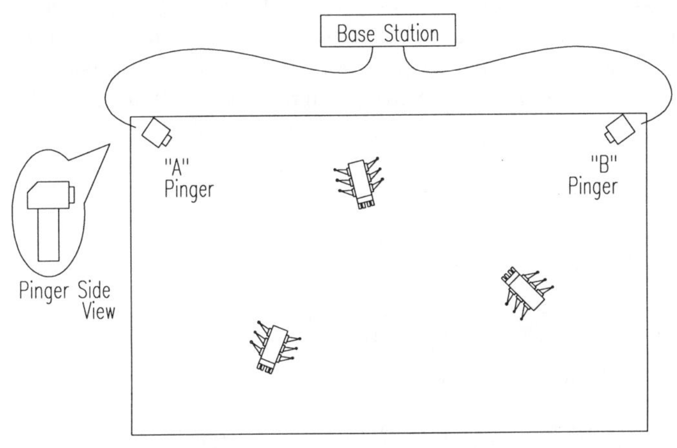

Figure 15-4. The ISR *Genghis* series of legged robots localize X-Y position with a master/slave trilateration scheme using two 40-KHz ultrasonic "pingers" synchronized via an RF spread-spectrum modem (courtesy IS Robotics, Inc.).

The base station alternately fires the two 40-KHz ultrasonic pingers every half second, each time transmitting a two-byte radio packet in broadcast mode to advise all robots of pulse emission. As with Dunkin's prototype, the elapsed time between radio packet reception and detection of the ultrasonic wavefront is used to calculate the distance between the robot's current position and the known location of the active beacon. Inter-robot communication is accomplished over the same spread-spectrum channel using a *time-division multiple-access* scheme controlled by the base station. Principle sources of error include variations in the speed of sound, the finite size of the ultrasonic transducers, non-repetitive propagation delays in the electronics, and ambiguities associated with time-of-arrival detection.

Tulane University 3-D Location System

Researchers at Tulane University in New Orleans, LA, have come up with some interesting methods for significantly improving the time-of-arrival measurement accuracy for ultrasonic transmitter/receiver configurations, as well as compensating for the varying effects of temperature and humidity. In the hybrid scheme illustrated in Figure 15-5, envelope peak detection is employed to establish the approximate time of signal arrival, and consequently eliminate ambiguity interval problems (Chapter 6) for a more precise phase-measurement technique that provides final resolution (Figueroa & Lamancusa, 1992). The desired 0.001-inch range accuracy required a time unit discrimination of ±75 nanoseconds at the receiver, which can easily be achieved using fairly simplistic phase measurement circuitry, but only within the interval of a single wavelength. The actual distance from transmitter to receiver is the summation of some integer number of wavelengths (determined by the coarse time-of-arrival measurement), plus that fractional portion of a wavelength represented by the phase measurement results.

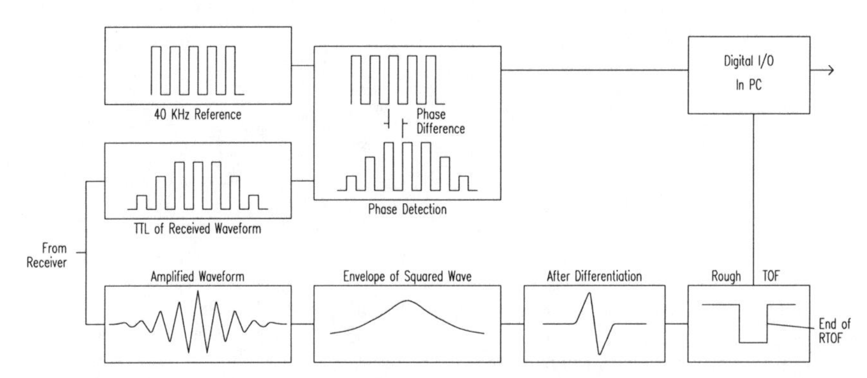

Figure 15-5. A combination of adjustable thresholding and phase detection is employed to provide higher accuracy in time-of-arrival measurements in the Tulane University ultrasonic position-location system (Figueroa & Lamancusa, 1992).

Details of this time-of-arrival detection scheme and associated error sources are presented by Figueroa and Lamancusa (1992). Range measurement accuracy of the prototype system was experimentally determined to be 0.006 inches using both adjustable thresholding (based on peak detection) and phase correction, as compared to 0.021 inches for adjustable thresholding alone. The high-accuracy requirements were necessary for an application that involved tracking the end effector of a six-DOF industrial robot (Figueroa, et al., 1992). The test setup shown in Figure 15-6 incorporates seven 90-degree Massa piezoelectric transducers operating at 40-KHz, interfaced to a 33-MHz IBM-compatible PC. The general position-location strategy was based on a trilateration method developed by Figueroa and Mahajan (1994).

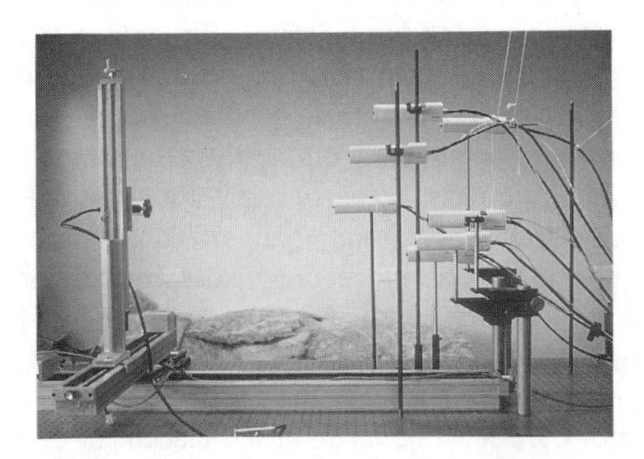

Figure 15-6. Experimental setup of the 3-D position location system showing the transmitter mounted on an X-Y-Z translation stage, and seven receivers distributed around the periphery of the work envelope (courtesy Tulane University).

The set of equations describing time-of-flight measurements for an ultrasonic pulse propagating from a mobile transmitter located at point (u, v, w) to various receivers fixed in the inertial reference frame can be listed in matrix form as follows (Figueroa & Mahajan, 1994):

$$\begin{Bmatrix} (t_1 - t_d)^2 \\ (t_2 - t_d)^2 \\ * \\ * \\ * \\ (t_n - t_d)^2 \end{Bmatrix} = \begin{bmatrix} 1 & r_1^2 & 2x_1 & 2y_1 & 2z_1 \\ 1 & r_2^2 & 2x_2 & 2y_2 & 2z_2 \\ * & & & & \\ * & & & & \\ * & & & & \\ 1 & r_n^2 & 2x_n & 2y_n & 2z_n \end{bmatrix} \begin{Bmatrix} \frac{p^2}{c^2} \\ \frac{1}{c^2} \\ -\frac{u}{c^2} \\ -\frac{v}{c^2} \\ -\frac{w}{c^2} \end{Bmatrix}$$

where:

t_i = measured time of flight for transmitted pulse to reach i^{th} receiver
t_d = system throughput delay constant
r_i^2 = sum of squares of ith receiver coordinates
(x_i, y_i, z_i) = location coordinates of ith receiver
(u, v, w) = location coordinates of mobile transmitter
c = speed of sound
p^2 = sum of squares of transmitter coordinates.

The above equation can be solved for the vector on the right to yield an estimated solution for the speed of sound c, transmitter coordinates (u, v, w), and an independent term p^2 that can be compared to the sum of the squares of the transmitter coordinates as a checksum indicator (Figueroa & Mahajan, 1994). An important feature of this representation is the use of an additional receiver (and associated equation) to enable treatment of the speed of sound itself as an unknown, thus ensuring continuous on-the-fly recalibration to account for temperature and humidity effects. (The system throughput delay constant t_d can also be automatically determined from a pair of equations for $1/c^2$ using two known transmitter positions. This procedure yields two equations with t_d and c as unknowns, assuming c remains constant during the procedure.) A minimum of five receivers is required for an unambiguous three-dimensional position solution, but more can be employed to achieve higher accuracy using a least-squares estimation approach. Care must be taken in the placement of receivers to avoid singularities as defined by Mahajan (1992).

Figure 15-7. The ceiling-mounted ultrasonic position-location system achieves 0.01-inch accuracy with an update rate of 100 Hz (courtesy Tulane University).

Figueroa and Mahajan (1994) report a follow-up version intended for mobile robot position location (Figure 15-7) that achieves 0.01-inch accuracy with an update rate of 100 Hz. The prototype system tracks a TRC *Labmate* over a 9- by 12-foot operating area with five ceiling-mounted receivers, and can be extended to larger floor plans with the addition of more receiver sets. An RF datalink will be used to provide timing information to the receivers and to transmit the subsequent X-Y position solution back to the robot. (The current prototype incorporates a hard-wire tether.) Three problem areas are being further investigated to increase the effective coverage and improve resolution:

- Actual transmission range does not match the advertised operating range for the ultrasonic transducers, probably due to a resonant frequency mismatch between the transducers and electronic circuitry.
- The resolution of the clocks (6 MHz) used to measure time of flight is insufficient for automatic compensation for variations in the speed of sound.
- The phase-detection range-measurement correction sometimes fails when there is more than one wavelength of uncertainty. This problem can likely be solved using the frequency division scheme described by Figueroa and Barbieri (1991).

15.1.2 Ultrasonic Signature Matching

An alternative technique employed on ROBART II for periodically resetting the actual *X-Y* position and orientation was ultrasonic signature matching (Everett & Bianchini, 1987). This methodology was based on previous work done at Carnegie Mellon University, and later by Harrington and Klarer (1986; 1987) at Sandia National Laboratories (Figure 15-8).

Figure 15-8. The Sandia Indoor Robot (courtesy Sandia National Laboratories).

Under this scheme, the robot must first perform a one-time "room digitizing" operation by moving in a raster-scan fashion through all unobstructed areas, stopping every 12 inches to fire the upper navigational sonar array. The ring of 24 ultrasonic sensors (spaced 15 degrees apart) thus generates a database of range returns, stored in polar coordinates, with an entry for each 12-inch square floor map unit marked as free. The database entries in effect represent unique signatures of the workspace (as seen by the ring of sensors) for each X-Y position in the map. The navigational array is placed as high as possible from the ground in order to minimize distortion due to the changing positions of transient objects.

When later performing a position estimation operation, the database of the operating area is searched for a location with a signature best matching the current set of range readings. An individual sensor return matches a database value if it falls within a specified window of acceptance, approximately 1.5 times the database resolution. Starting with the current dead- reckoned map position, the position estimator searches the database in an expanding fashion, looking for the entry (position) with the highest number of correlations matching the range values taken at the robot's present physical location.

The search algorithm also skews the current sonar data one sensor position (i.e. ±15 degrees) in each direction, in an attempt to correct for any error in current heading. If the highest number of correlations is not greater than a minimum threshold, the estimator searches a new set of neighbors farther from the original dead-reckoned position. Initial results using this technique at Sandia showed a sharp differential between the number of fits for a correct database match with respect to neighboring locations.

When a match is found with a sufficiently high correspondence, the robot's position is known to within 12 inches (the database resolution), and heading to within 15 degrees. To improve the positional accuracy, the estimator will interpolate a new location within the map, using the four sensor range values pointing 0, 90, 180, and 270 degrees relative to the robot (as long as each of these readings match their corresponding database returns within the specified tolerance).

The robot can also interpolate its heading to within about 1.5 degrees by performing several position estimations as above, rotating clockwise by 1 degree after each estimate. As long as the computed *X-Y* position and heading remain the same as the previous estimate, the robot continues to rotate and take range readings. If the estimated heading suddenly changes by 15 degrees while the estimated position remains unchanged, then it is assumed the robot has turned approximately halfway between the previous heading and the new heading. The interpolated heading can at this point be derived by subtracting 7.5 degrees (half the rotation interval) from the most recent heading estimate. If the *X-Y* position changes, or the heading changes by more than 15 degrees, then heading interpolation using this approach is discounted.

This database-search technique has proven to be fairly reliable for determining the robot's *X-Y* position during extensive testing, provided the operating environment does not change significantly. Some degradation is acceptable, as long as approximately 15 or more sensor readings of the 24 total are not affected. The number of correlations attained serves as a built- in indicator of database degradation, however, since as this number begins to approach the critical mark discussed above, the robot can simply initiate a new digitization routine to update the database. The only hitch here is some means of precisely monitoring the robot's position and orientation during this *build-database* process is required in order to ensure the database entries are themselves valid. To date, this has meant human supervision of the room digitizing operation.

15.2 Optical Position-Location Systems

Optically based position location systems typically involve some type of scanning mechanism operating in conjunction with fixed-location references strategically placed at pre-defined locations within the operating environment. A number of variations on this theme are seen in practice:

- Scanning detectors with fixed active-beacon emitters.
- Scanning emitter/detectors with passive retroreflective targets.
- Scanning emitter/detectors with active transponder targets.
- Rotating emitters with fixed detector targets.

One of the principle problems associated with optical beacon systems, aside from the obvious requirement to modify the environment, arises from the need to preserve a clear line of sight between the robot and the beacon. Preserving an unobstructed view is sometimes difficult if not impossible in certain applications such as congested warehouse environments. In the case of passive retroreflective targets, problems can sometimes arise from unwanted returns from other reflective surfaces, but a number of techniques exists for minimizing such interference.

15.2.1 CRAWLER I Homing Beacon

A very rudimentary method of beacon homing was implemented on the CRAWLER I robot introduced in Chapter 1 to facilitate automatic battery recharging. The scanning sensor was simply a cadmium-sulfide photoresistor mounted in the end of a 12-inch plastic tube (Figure 15-9). The collimating tube was rotated in the horizontal plane of the beacon by a small DC gearmotor from a sign display. A special contact sensor at the bottom of the rotating shaft was used to index the 0-degree position coinciding with the forward axis of travel.

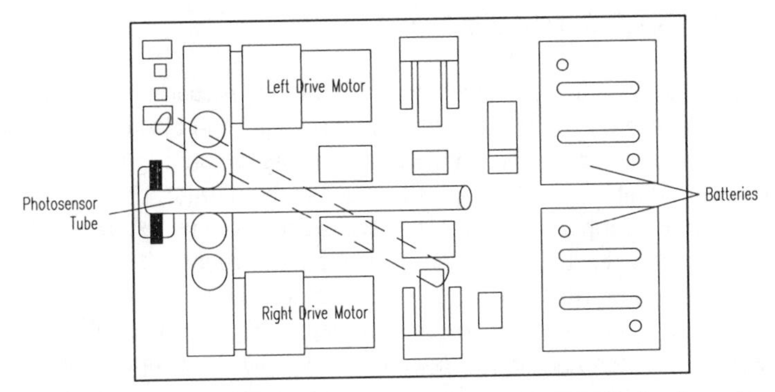

Figure 15-9. Top view of componentry layout on the CRAWLER I robot, showing the photosensor collimating tube aligned with the forward axis of travel, and displaced 30 degrees.

All of the control logic was implemented with surplus mechanical relays. The concept was simple enough; the photocell scanner rotated at a constant rate, and the turning action was started and halted by: 1) the shaft index sensor and 2) beacon detection by the photocell. The first sweep was used to determine if a beacon were present and in which direction the vehicle needed to turn. A pair of triple-pole double-throw relays were used to implement an electromechanical flip-flop to remember on which side of the vehicle the beacon had most recently been seen. (Keep in mind this was a high-school science project back in the mid-sixties).

For example, if the beacon lay off to the right side, the scanner would initiate a platform turn to the right by stopping the right drive motor when the collimating tube passed through the forward index (i.e., pointed straight ahead). As the sweep continued from left to right, the photocell would eventually detect the beacon, whereupon the control logic would restart the right drive motor. This process would repeat with each new sweep, and each time the turning action would last for a shorter period than before because the beacon would have moved closer to the 0-degree index as the robot turned toward it. When the beacon lay dead ahead, the sweep would start and stop the turn at the same time, therefore straight-line travel would not be interrupted.

A minor wrinkle in this scheme involved the fact that if the beacon were instead off to the left side, the events that initiated and terminated the turning action had to be interchanged. In other words, beacon detection would start the turn, and sweep index detection would halt it. The mechanics of this strategy were worked out in the relay logic, relying on the *photocell-left-or-right* information stored in the flip-flop to control the process. Fine tuning to achieve stability was accomplished by varying the sweep speed of the scanner with a rheostat. The results were surprisingly effective in light of the extremely crude design implementation.

15.2.2 ROBART II Recharging Beacon

Figure 15-10 shows the physical structure of the recharging system used by ROBART II. Situated above the aluminum base plate and electrically insulated by means of a Plexiglas spacer is a cylindrical housing containing the recharging and system power supplies. At the top of the unit is the homing beacon, visible from any direction and supported by a metal pipe with a coaxial extension that can be adjusted to match the height of the sensor on the robot. The homing beacon consists of a vertical-filament incandescent lamp surrounded by an optical filter that blocks the visible portion of the emitted energy (Kodak Wratten 89B or equivalent). The homing beacon is activated by the robot upon command via a dedicated radio link. Enclosed within the cylinder are the power supplies, a radio receiver, and associated decoding and control electronics.

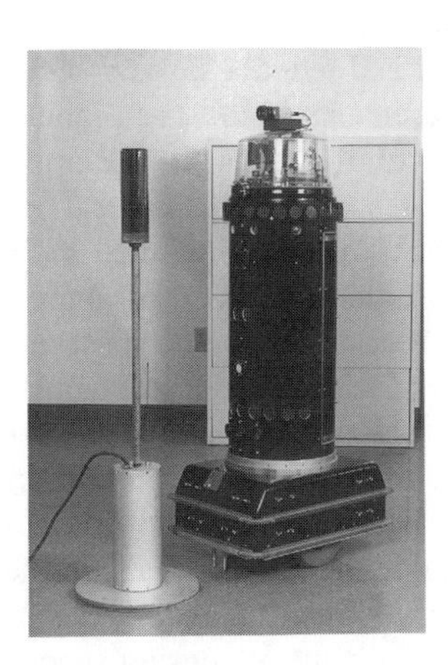

Figure 15-10. ROBART II approaches its automatic recharging station by homing in on the active near-infrared beacon shown at the top of the support pole.

There are two power supplies associated with the recharging station itself. A relatively low-current source remains energized at all times to power the radio receiver and also to energize the recharger contacts through a current-limiting resistor. This "sense" voltage (about 20 volts DC) allows the robot to know when a valid electrical connection has been established with the recharger. In addition, the sense voltage will drop to around 14 volts as soon as the battery has been connected as a load, activating the high-power battery charging supply after a 2-second delay. This second power supply furnishes the current required to recharge the battery and is automatically shut off when the robot disconnects. The delay is incorporated to allow the mating contacts to debounce before power is applied, markedly reducing contact erosion and pitting.

Once the battery monitor circuit on the robot detects a low-battery condition, the Scheduler initiates the docking sequence by requesting a path to the recharging station. The path planner first draws an imaginary circle of 24-inch radius around the last known location of the charger as encoded in the world model. A path is found to the nearest point of intersection with this circle and the robot then moves to this designated start position. The head pans from full left to full right, digitizing and storing the ambient light intensity of the room. The Scheduler next activates the homing beacon on the recharging station via the radio link and enters the beacon acquisition mode. The robot rescans the room from right to left, looking for a light source that was not present prior to beacon activation.

Positive identification is achieved by turning off the beacon while observing the selected target, after which the path planner updates the new beacon position in the model. The head pan position is servo-controlled to balance the outputs from two silicon photodetectors, and the resulting relative bearing used to calculate a steering command to cause the robot to turn in the appropriate direction. The robot relies on this optical tracking system to control heading while closing on the charger and reduces speed as a function of stand-off distance based on sonar range measurements from a head-mounted Polaroid transducer.

The task of connecting with the recharger is simplified by making the contact surfaces symmetrical with respect to the vertical pole supporting the homing beacon, so the same target is always presented to the mating contacts on the advancing robot regardless of the direction of approach. The cylindrical metal housing at the base of the beacon support pipe serves as the point of contact for the GND leg, with the respective mating surface being a metal strip attached to the front bumper of the robot. The inherent spring action of the tactile bumper keeps the conductive strip in tight contact with the housing once the two come together.

The connection for the HOT leg is made through the mating of two spring probes with the circular aluminum base plate. The probes extend vertically downward from the edge of the robot chassis. As the front of the chassis passes over the plate moving toward the metal housing supporting the beacon, the spring probes are brought into contact with the plate. Contact is maintained as motion continues toward bumper impact. The geometry of the configuration ensures the probes will be in contact with the plate as long as the front bumper contact is touching the power supply housing. Considerable margin for alignment error is allowed since the strip is 10 inches wide.

This recharging strategy was basically just a refinement and repackaging of the system developed earlier for ROBART I, which in turn was based on the original concept used by the CRAWLER robots back in 1965. In all three cases, the appropriate battery charger could be placed anywhere in the room, and the robot could quickly find it with no prior knowledge of its whereabouts. The underlying design philosophy was driven by an emphasis on making the docking operation as forgiving as possible to accommodate the known inadequacies of the navigation system. It eventually occurred to me that perhaps things should be the other way around. A fixed-location charger equipped with a homing beacon can provide a convenient mechanism for re-referencing a slightly disoriented robot. An alternative charging station incorporating this capability was developed for ROBART II and is discussed in Section 16.1.1.

15.2.3 Cybermotion Docking Beacon

The automated docking system used on the Cybermotion *Navmaster* robot incorporates a rather unique combination of a structured-light beacon to establish bearing, along with a one-way ultrasonic ranging system that determines stand-off distance. The optical portion consists of a pair of near-infrared transceiver units, one mounted on the front of the robot and the other situated in a known position and orientation within the operating environment. These two optical transceivers are capable of full-duplex data transfer between the robot and the dock at a rate of

9600 bits per second. Separate modulation frequencies of 154 and 205 KHz are employed for the uplink and downlink respectively to eliminate crosstalk. Under normal circumstances, the dock-mounted transceiver waits passively until interrogated by an active transmission from the robot. If the interrogation is specifically addressed to the assigned ID number for that particular dock, the dock control computer activates the beacon transmitter for 20 seconds. (Dock IDs are jumper selectable at time of installation.)

Figure 15-11 shows the fixed-location beacon illuminating a 90-degree field of regard broken up into two uniquely identified zones, designated for purposes of illustration here as the *Left Zone* and *Right Zone*. An array of LED emitters in the beacon head is divided by a pair of lenses and a double-sided mirror situated along the optical axis. Positive zone identification is initiated upon request from the robot in the form of a *NAV Interrogation* byte transmitted over the optical datalink. LEDs on opposite sides of the mirror respond to this *NAV Interrogation* with slightly different coded responses. The robot can thus determine its relative location with respect to the optical axis of the beacon based on the response bit pattern detected by the onboard receiver circuitry.

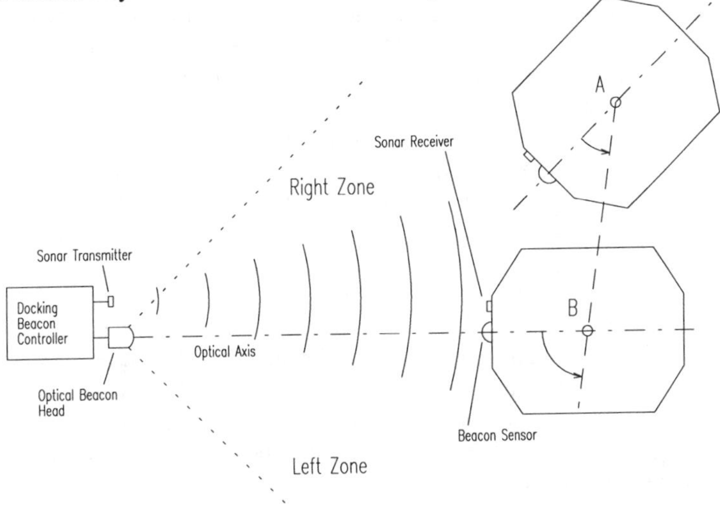

Figure 15-11. The structured-light near-infrared beacon on the Cybermotion battery recharging station defines an optimal path of approach for the *K2A Navmaster* robot.

When a docking action is requested (i.e., for recharging batteries, re-referencing position and azimuth, pick-up or delivery of material), the *Navmaster* moves into the general vicinity designated for dock approach (i.e., point *A*) and attempts to communicate with the dock via the near-infrared link. If no link is established, the robot scans first left and then right, aborting with a *Dock Failure* status if the search is unsuccessful. Assuming communications are established and the beacon is activated as discussed above, the robot sets the *Request to Dock* output bit at the dock and then waits for an acknowledgment. This handshaking procedure allows for any auxiliary equipment associated with the dock (that could conceivably interfere with the robot's approach) to report its status.

Once cleared to approach, the robot turns in the appropriate direction and executes the steepest possible (i.e., without losing sight of the beacon) intercept angle with the beacon optical axis. Crossing the optical axis at point *B* is flagged by a sudden change in the bit pattern of the *NAV Response Byte*, whereupon the robot turns inward to face the dock. The beacon optical axis establishes the nominal path of approach, and in conjunction with range offset information uniquely defines the robot's absolute location. This situation is somewhat analogous to a TACAN station (Dodington, 1989) with but a single defined radial.

Measuring the offset distance from vehicle to dock is accomplished in rather elegant fashion through use of a dedicated non-reflective ultrasonic ranging configuration. This high-frequency (>200 KHz) narrow-beam (15 degrees) sonar system consists of a piezoelectric transmitter mounted on the docking beacon head and a complimentary receiving transducer mounted on the front of the vehicle. A ranging operation is initiated upon receipt of the *NAV Interrogation Byte* from the robot; the answering *NAV Response Byte* from the docking beacon signals the simultaneous transmission of an ultrasonic pulse. The difference at the robot end between time of arrival for the *NAV Response Byte* over the optical link and subsequent ultrasonic pulse detection is used to calculate separation distance. This dual-transducer master/slave technique assures an unambiguous range determination between two well defined points and is unaffected by any projections on or around the docking beacon and/or face of the robot.

During transmission of a *NAV Interrogation Byte*, the left and right sides of the LED array located on the robot are also driven with uniquely identifiable bit patterns. This feature allows the docking beacon computer to determine the robot's actual heading with respect to the nominal path of approach. Recall the docking beacon's structured bit pattern establishes (in similar fashion) on which side of vehicle centerline the docking beacon is located. This heading information is subsequently encoded into the *NAV Response Byte* and passed to the robot to facilitate course correction. The robot closes on the beacon, halting at the defined stop range (not to exceed 8 feet) as repeatedly measured by the docking sonar. Special instructions in the path program can then be used to reset vehicle heading and/or position.

15.2.4 Hilare

Early work incorporating passive beacon tracking at the Laboratoire d'Automatique et d'Analyse des Systemes (LAAS), Toulouse, France, involved the development of a navigation subsystem for the mobile robot Hilare (Bauzil, et al., 1981). The system consisted of two near-infrared emitter/detectors mounted with a 25-cm vertical separation on a rotating mast, used in conjunction with passive reflective-beacon arrays at known locations in three corners of the room.

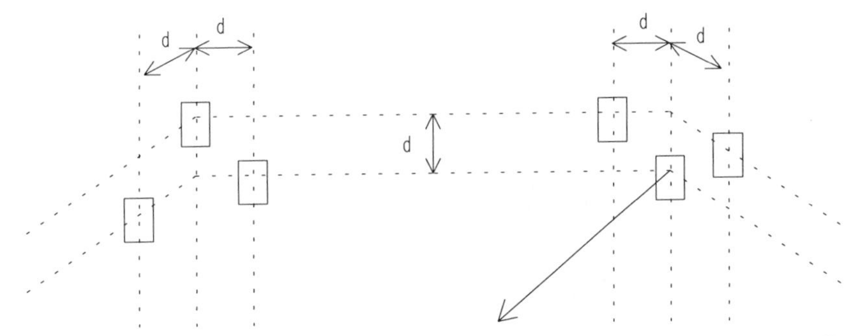

Figure 15-12. Retroreflective beacon array configuration used on the mobile robot Hilare (adapted from Bauzil, et al., 1981).

Each of these beacon arrays was constructed of retroreflective tape applied to three vertical cylinders, spaced in a recognizable configuration as shown in Figure 15-12. One of the arrays was inverted so as to be uniquely distinguishable for purposes of establishing an origin. The cylinders were vertically spaced to intersect the two planes of light generated by the rotating optical axes of the two emitters on the robot's mast. A detected reflection pattern as in Figure 15-13 confirmed beacon acquisition. Angular orientation relative to each of the retroreflective arrays was inferred from the stepper motor commands that drove the scanning mechanism; lateral position was determined through simple triangulation.

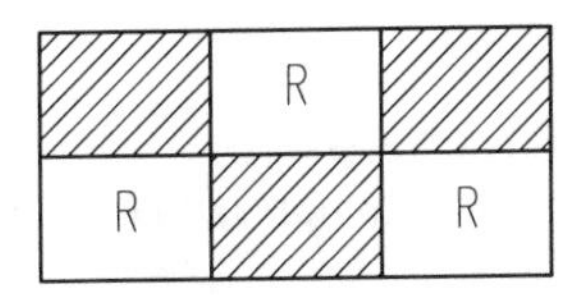

Figure 15-13. A confirmed reflection pattern as depicted above was required to eliminate potential interference from other highly specular surfaces (Bauzil, et al., 1981).

15.2.5 NAMCO *Lasernet*® *Scanning Laser Sensor*

The NAMCO *Lasernet*® *Scanning Laser Sensor* introduced in Chapter 4 employs retroreflective targets distributed throughout the operating area of an *automated guided vehicle* in order to measure range and angular orientation (Figure 15-14). A servo-controlled rotating mirror pans a near-infrared laser beam through a horizontal arc of 90 degrees at a 20-Hz update rate. When the beam sweeps across a target of known dimensions, a return signal of finite duration is sensed by the detector. Since the targets are all the same size, the signal generated by a close target will be of longer duration than that from a distant one.

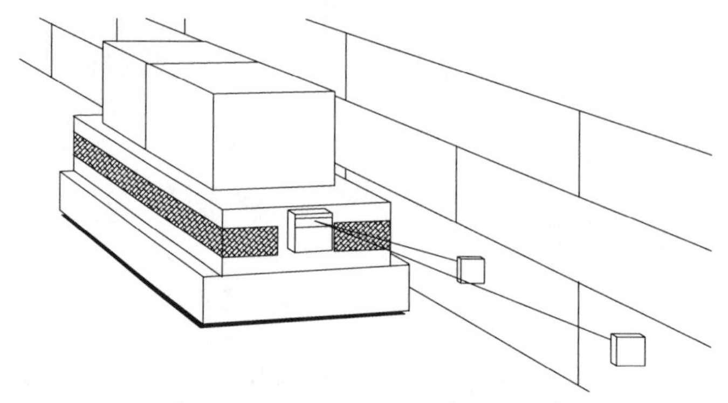

Figure 15-14. The *Lasernet*® system can be used with projecting wall-mounted targets to guide an AGV at a predetermined offset distance (courtesy NAMCO Controls).

Angle measurement is initiated when the scanner begins its sweep from right to left; the laser strikes an internal synchronization photodetector that starts a timing sequence. The beam is then panned across the scene until returned by a retroreflective target in the field of view. The reflected signal is detected by the sensor, terminating the timing sequence (Figure 15-15). The elapsed time is used to calculate the angular position of the target in the equation (NAMCO, 1989):

$$\theta = V T_b - 45$$

where:

θ = target angle
V = scan velocity (7200 degrees/sec)
T_b = interval between scan initiation and target detection.

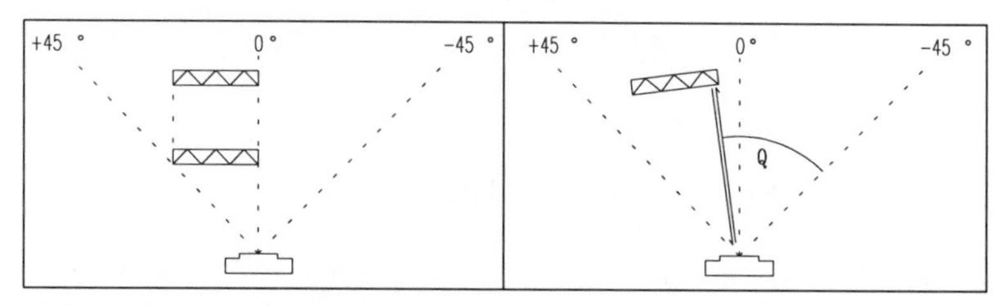

Figure 15-15. (A) Perceived width of a retroreflective target of known size is used to calculate range; (B) while the elapsed time between sweep initiation and leading edge detection yields target bearing (courtesy NAMCO Controls).

This angle calculation determines either the leading edge of the target, the trailing edge of the target, or the center of the target, depending upon the option selected within the *Lasernet*® software option list. The angular accuracy is ± 1 percent, and the angular resolution is 0.1 degrees for the analog output; accuracy is within ±0.05 percent with a resolution of 0.006 degrees when the RS-232 serial port is used. The analog output is a voltage ranging from 0 to 10 volts over the range of ±45 degrees, whereas the RS-232 serial port reports a proportional "count value" from zero to 15,360 over this same range.

15.2.6 Caterpillar *Self-Guided Vehicle*

Caterpillar Industrial, Inc., Mentor, OH, manufactures a free-ranging AGV for materials handling that relies on a scanning laser triangulation scheme to provide positional updates to the vehicle's onboard dead-reckoning system. The Class-I laser rotates at 2 rpm to illuminate passive retroreflective barcode targets affixed to walls or support columns at known locations up to 15 meters away (Gould, 1990; Byrne, et al., 1992). The barcodes serve to positively identify the reference target and eliminate ambiguities due to false returns from other specular surfaces within the operating area. An onboard computer calculates X-Y positional updates through simple triangulation to null out accumulated dead-reckoning errors (Figure 15-16).

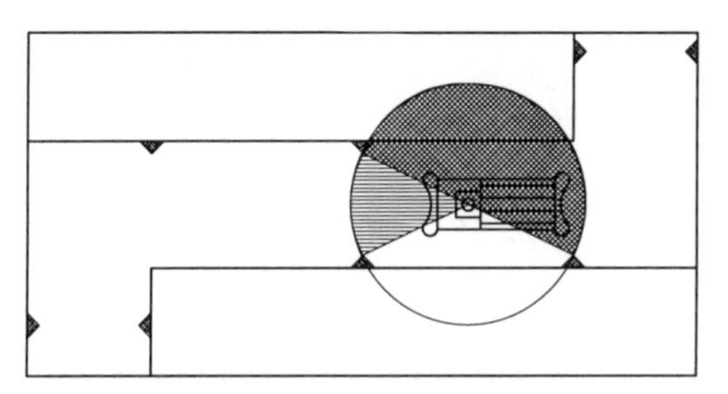

Figure 15-16. Retroreflective barcode targets spaced 10 to 15 meters apart are used by the Caterpillar SGV to triangulate position (adapted from Caterpillar, 1991a).

Some target occlusion problems have been experienced in an exterior application due to heavy fog, as would be expected, and minor difficulties have been encountered as well during periods when the sun was low on the horizon (Byrne, et al., 1992). The *Self-Guided Vehicle* relies on dead reckoning under such conditions to reliably continue its route for distances of up to 10 meters before the next valid fix.

The mobility configuration is a hybrid combination of tricycle and differential drives, employing two independent series-wound DC motors powering 18-inch rear wheels through sealed gearboxes (Caterpillar, 1991b). High-resolution resolvers (Chapter 2) attached to the single front wheel continuously monitor steering angle and distance traveled. A pair of mechanically scanned near-infrared proximity sensors sweeps the path in front of the vehicle for potential obstructions. Additional near-infrared sensors monitor the area to either side of the vehicle, while ultrasonic sensors cover the back.

15.2.7 TRC *Beacon Navigation System*

Transitions Research Corporation, Danbury, CN, has incorporated their LED-based *LightRanger* discussed in Section 6.1.7 into a compact, low-cost navigational referencing system for open-area autonomous platform control. The TRC *Beacon Navigation System* calculates vehicle position and heading at ranges up to 80 feet within a quadrilateral area defined by four passive retroreflective beacons as shown in Figure 15-17 (TRC, 1994). A static 15-second unobstructed view of all four beacons is required for initial acquisition and set-up, after which only two beacons must remain in view as the robot moves about the area. No provision is yet provided to periodically acquire new beacons along a continuous route, so operation is currently constrained to a single zone roughly the size of a small building (i.e., 80 by 80 feet).

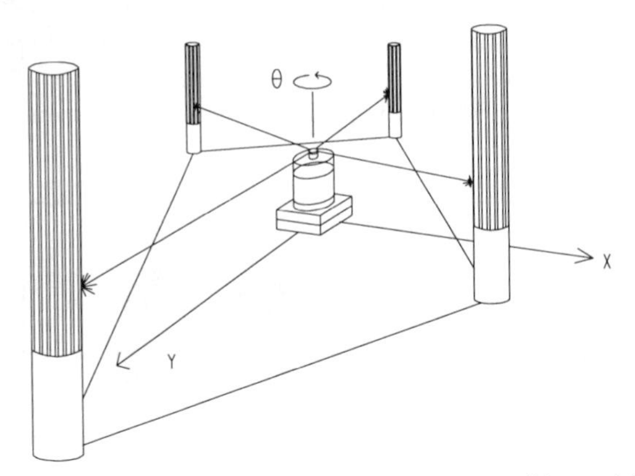

Figure 15-17. The TRC *Beacon Navigation System* calculates position and heading based on ranges and bearings to two of four passive beacons defining a quadrilateral operating area (courtesy TRC).

System resolution is 12 centimeters in range and 0.125 degrees in bearing for full 360-degree coverage in a horizontal plane. The scan unit (less processing electronics) is a cube approximately 10 centimeters on a side, with a maximum 1-Hz update rate dictated by the 60-rpm scan speed. A dedicated *68HC11* microprocessor continuously outputs navigational parameters (X, Y, θ) to the vehicle's onboard controller via an RS-232 serial port. Power requirements are 0.5 amps at 12 volts DC and 0.1 amp at 5 volts DC.

15.2.8 Intelligent Solutions *EZNav Position Sensor*

Intelligent Solutions, Inc. (ISI), Marblehead, MA, offers a laser-based scanning beacon system that computes vehicle position and heading out to 600 feet using cooperative electronic transponders (100 feet with passive reflectors). The *EZNav*

Position Sensor is a non-ranging triangulation system with an absolute bearing accuracy of ±0.03 degrees at a scan rate of 600 rpm (ISI, 1994a). The scanner mechanism consists of a rotating mirror attached at a 45-degree angle to the vertical shaft of an incremental optical encoder; for increased azimuthal accuracy, a timer interpolates between encoder counts (ISI, 1994b). Motor velocity is servoed under interrupt control every 100 milliseconds for a resolution of 0.1 percent.

The eye-safe near-infrared laser generates a 1-milliwatt output at 810 nanometers, amplitude modulated at 1.024 MHz to minimize interference from low-frequency ambient light sources. The fan-shaped beam is spread 4 degrees vertically to ensure target detection at long range while traversing irregular floor surfaces, with horizontal divergence limited to 0.3 milliradians. Each target is uniquely coded, and up to 32 separate targets can be processed in a single scan, with vehicle X-Y position calculated every 100 milliseconds (Maddox, 1994). The sensor package (Figure 15-18) weighs 10 pounds, measures 15 inches high and 12 inches in diameter, and has a power consumption of only 300 milliamps at 12 volts.

Figure 15-18. The *EZNav Position Sensor* can process up to 32 separate beacon targets per scan, with a complete X-Y position update computed at a 10-Hz rate (courtesy Intelligent Solutions, Inc.).

The *EZNav Position Sensor* is the latest development of this scanning laser technology. A similar version of this position sensor called *Lasernav* was previously developed by ISI engineers while working for Denning Mobile Robots. The *Lasernav* unit is used by the Mobile Robotics Laboratory at Georgia Tech as the primary navigational sensor on a Denning MRV-2 robotic base (Figure 15-19), in support of their work in reactive multiagent robotic systems (Balch & Arkin, 1994).

Figure 15-19. The earlier *Lasernav* prototype is used by Georgia Tech as the primary navigational referencing sensor on a Denning *MRV-2* robotic base (courtesy Georgia Institute of Technology).

15.2.9 Imperial College Beacon Navigation System

Premi and Besant (1983) of the Imperial College of Science and Technology, London, England, describe an AGV guidance system that incorporates a vehicle-mounted laser beam rotating in a horizontal plane that intersects three fixed-location reference sensors as shown in Figure 15-20. The photoelectric sensors are arranged in collinear fashion with equal separation, and are individually wired to a common FM transmitter via appropriate electronics so that the time of arrival of laser energy is relayed to a companion receiver on the vehicle. A digitally coded identifier in the data stream identifies the activated sensor that triggered the transmission, thus allowing the onboard computer to measure the separation angles α_1 and α_2. An absolute or indexed incremental position encoder monitoring laser scan azimuth is used to establish platform heading.

This technique has some inherent advantages over the use of passive retroreflective targets, in that false acquisition of specular surfaces is eliminated, and longer ranges are possible since target reflectivity is no longer a factor. More robust performance is achieved through elimination of target dependencies, allowing a more rapid scan rate to facilitate faster positional updates. The one-way nature of the optical signal significantly reduces the size, weight, and cost of the onboard scanner with respect to that required for retroreflective beacon acquisition. Tradeoffs, however, include the increased cost associated with installation of power and communications lines and the need for significantly more expensive beacons. This can be a significant drawback in very-large-area installations, or scenarios where multiple beacons must be incorporated to overcome line-of-sight limitations.

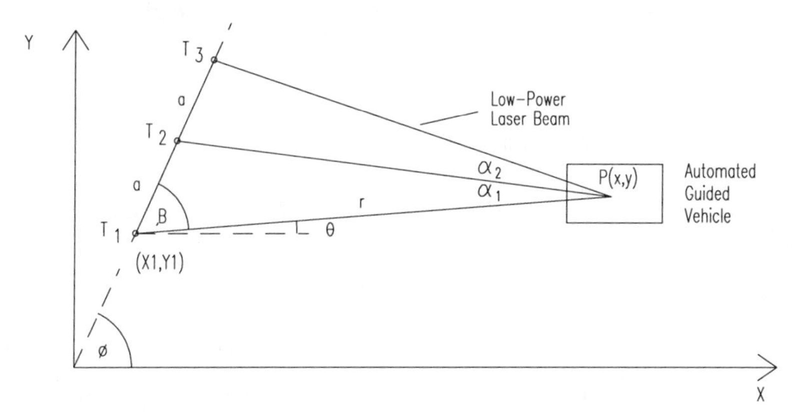

Figure 15-20. Three equidistant collinear photosensors are employed in lieu of retroreflective beacons in the Imperial College laser triangulation system for AGV guidance (adapted from Premi & Besant, 1983).

AGV position P(X,Y) is given by the equations (Premi & Besant, 1983):

$$x = x_1 + r\cos\theta$$

$$y = y_1 + r\sin\theta$$

where:

$$r = \frac{a\sin(\alpha_1 + \beta)}{\sin\alpha_1}$$

$$\beta = \arctan\left[\frac{2\tan\alpha_1\tan\alpha_2}{\tan\alpha_2 - \tan\alpha_1}\right] - \alpha_1$$

$$\theta = \phi - \beta$$

15.2.10 MTI Research *CONAC*

A similar type system using a predefined network of fixed-location detectors is currently being marketed by MTI Research, Inc., Chelmsford, MA. MTI's *CONAC (Computerized Optoelectronic Navigation and Control)* is a relatively low-cost high-performance navigational referencing system employing a vehicle-mounted laser *STROAB (STRuctured Optoelectronic Acquisition Beacon)*. The scanning laser beam is spread vertically to eliminate critical alignment, allowing the reference *NOADs (Networked Optoelectronic Acquisition Datums)* to be mounted at arbitrary heights (Figure 15-21). Detection of incident illumination by a *NOAD* triggers a response over the network to a host PC, which in turn calculates the implied angles α_1 and α_2. An index sensor built into the *STROAB* generates a special rotation reference pulse to facilitate heading measurement. Indoor accuracy is advertised at ± 0.05 inches for position and ±0.05 degrees for heading.

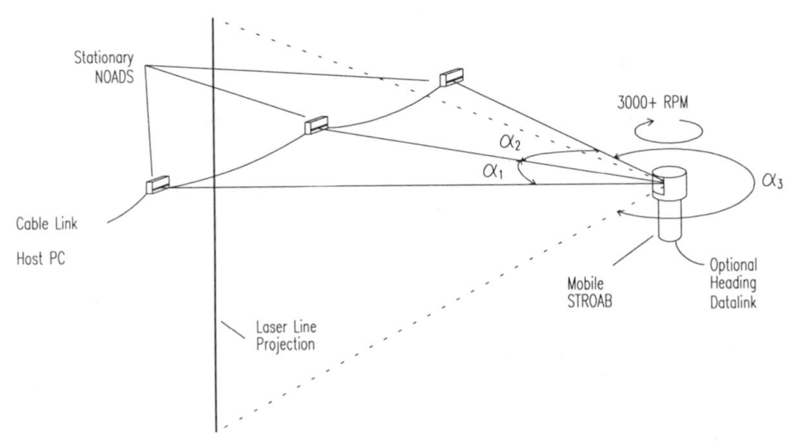

Figure 15-21. The patent-pending *Computerized Optoelectronic Navigation and Control (CONAC)* system employs a number of networked detectors tracking a rapidly scanning laser mounted on the moving vehicle (courtesy MTI Research, Inc.)

The reference *NOADs* are strategically installed at known locations throughout the area of interest, and daisy chained together with ordinary four-conductor modular telephone cable. Alternatively the *NOADs* can be radio linked to eliminate cable installation problems, as long as power is independently available to the various *NOAD* sites. (One project with MIT/Lincoln Laboratory involves a 2-kilometer system with a radio-*NOAD* spacing of 100 meters.) *STROAB* acquisition range is sufficient to where three *NOADs* can effectively cover a full acre assuming no interfering structures block the view. Additional *NOADs* are typically employed to increase fault tolerance, with the optimal set of three *NOADs* dynamically selected by the host PC, based on the current location of the robot and any predefined visual barriers. The selected *NOADs* are individually addressed over the network in accordance with assigned codes (set into DIP switches on the back of each device at time of installation).

An interesting and rather unconventional aspect of *CONAC* is that no fall-back dead reckoning capability is incorporated into the system (MacLeod & Chiarella, 1993). The 3000-rpm angular rotation speed of the laser *STROAB* facilitates rapid position updates at a 25-Hz rate, which MTI claims is sufficient for safe automated transit at highway speeds, provided line-of-sight contact is preserved with at least three fixed *NOADs*. To minimize chances of occlusion, the lightweight (less than 9 ounces) *STROAB* is generally mounted as high as possible on a supporting mast.

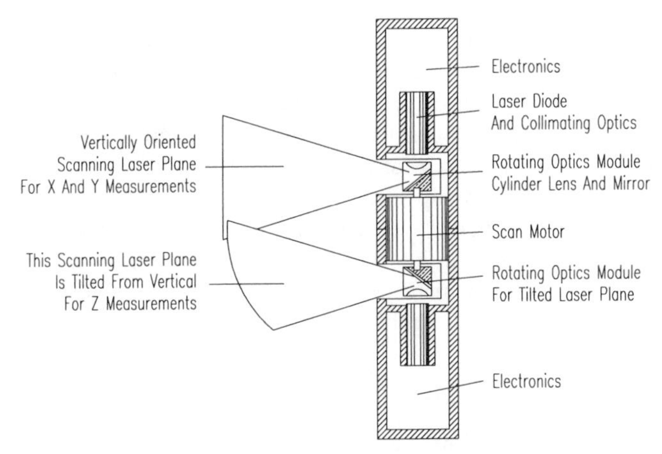

Figure 15-22. Simplified cross-sectional view of the dual-laser position location system now under development for tracking multiple mobile sensors in 3-D applications (courtesy MTI Research, Inc.).

A stationary active-beacon system that tracks an omni-directional sensor mounted on the robot is currently under development to allow for tracking multiple units. (The current *CONAC* system allows only one beacon to be tracked at a given time.) The basic system consists of two synchronized stationary beacons that provide bearings to the mobile sensor to establish its *X-Y* location. A hybrid version of this approach employs two lasers in one of the beacons as illustrated in Figure 15-22, with the lower laser plane tilted from the vertical to provide *Z*-axis resolution for three-dimensional applications.

Figure 15-23. A small model car driven under *CONAC* control executes a preprogrammed raster-scan pattern in an outdoor parking lot before hitting the jump ramp with unerring precision (courtesy MTI Research, Inc.).

Long-range exterior position accuracy for the current *CONAC* system is specified as ±0.25 inches at distances out to 600 feet, with no degradation in heading accuracy over interior (±0.05 degrees). The system was successfully demonstrated in an outdoor environment when MTI engineers outfitted a Dodge *Caravan* with electric actuators for steering, throttle, and brakes, then drove the unmanned vehicle at speeds up to 50 miles per hour (Baker, 1993). Absolute position and heading accuracies were sufficient to allow the *Caravan* to maneuver among parked vehicles and into a parking place using a simple *AutoCad* representation of the environment. Figure 15-23 shows a 1/10th-scale model race car outfitted with a *STROAB* going airborne after hitting a small ramp in an outdoor parking lot.

15.2.11 MDARS Lateral-Post Sensor

A 360-degree field-of-view beacon tracking system of the type discussed above was considered for use on the MDARS Interior robot to assist in position referencing in semi-structured warehouse operations, but never implemented for a number of reasons. For starters, the effectiveness of a multiple-beacon triangulation scheme is to a large extent dependent on the overall field of view, which suggests the ideal scanning system should be located as high as possible with 360-degree coverage. In the case of a security robot, unfortunately, these same criteria likewise influence performance of both the surveillance camera and the intrusion-detection suite. Having three such sensor systems competing for a full-circle view at the very top of the robot introduces some non-trivial design challenges, complicated further still by the fact that video and datalink antennae also work best when situated above all other componentry.

Yet another important consideration is the likelihood of line-of-sight contact with a number of widely distributed beacons. Space is generally a premium in warehouse environments, and as a consequence vision is often restricted to straight shots up and down long narrow aisles. Surrounding shelving severely limits the effective coverage area for a conventional scanning unit and necessitates additional reflective beacons closely spaced down each individual aisle, adding to the cost of an already expensive system. Protruding targets in narrow aisles are vulnerable to fork truck damage, reducing overall reliability while increasing maintenance costs. The bottom line is much of the flexibility of a 360-degree capability is lost in crowded warehouse applications.

In light of these concerns, a derivative of the rotating laser referencing technique called *lateral post detection* was incorporated on MDARS to significantly reduce costs by exploiting the forward motion of the robot for scanning purposes. Short vertical strips of 1-inch-wide retroreflective tape are placed on various immobile objects (usually structural-support posts) on either side of a *virtual path* segment (Figure 15-24). The exact *X-Y* locations of these tape markers are encoded into the *virtual path* program. Installation takes only seconds, and since the flat tape does not protrude into the aisleway at all, there is little chance of damage from a passing fork truck.

Figure 15-24. A short vertical strip of retroreflective tape can be seen affixed to the structural post in the MDARS warehouse environment (courtesy Naval Command Control and Ocean Surveillance Center).

A pair of Banner *Q85VR3LP* retroreflective sensors mounted on the turret of the *Navmaster* robot face outward to either side as shown in Figure 15-25. These inexpensive sensors respond to reflections from the tape markers along the edges of the route, triggering a "snapshot" *virtual path* instruction that records the current side-sonar range values. The longitudinal position of the platform is updated to the known marker coordinate, while lateral position is inferred from the sonar data, assuming both conditions fall within specified tolerances.

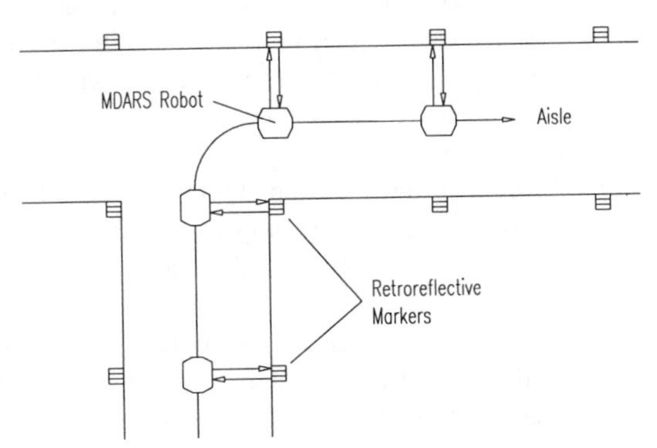

Figure 15-25. Polarized retroreflective sensors are used to locate vertical strips of reflective tape attached to shelving support posts in the Camp Elliott warehouse installation of the MDARS security robot.

The accuracy of the marker correction is much higher (and therefore assigned greater credibility) than that of the lateral sonar readings due to the markedly different uncertainties associated with the respective targets. The polarized Banner sensor responds only to the presence of a retroreflector while ignoring even highly specular surrounding surfaces, whereas the ultrasonic energy from the sonar will echo back from any reflective surface encountered by its relatively wide beam. Protruding objects in the vicinity of the tape (quite common in a warehouse environment) result in a shorter measured range value being read than the reference distance for the marker itself. The overall effect on *X*-*Y* bias is somewhat averaged out in the long run, as each time the vehicle executes a 90-degree course change the association of *X* and *Y* components with tape versus sonar updates is interchanged.

This lateral-post referencing concept was implemented on the MDARS unit in May 1994 and tested in an operational warehouse environment at Camp Elliott in San Diego, CA. The *Navmaster* robot was run continuously back and forth along a 150-foot path, with seven tape markers set on posts 20 feet apart. No other navigational referencing instructions were contained in the path program. Initial heading and location errors were quickly nulled out after the second or third post was detected, and accumulated errors remained essentially insignificant for the remaining length of the path. Each time the robot reversed course at the end of a run, some noticeable heading error was introduced on the diagnostic display but then quickly resolved as lateral-post updates were processed on the return leg.

We tried to get the system to fail by purposely injecting errors into the sonar range measurements. An increasing number of markers were corrupted throughout the course of this test by placing protruding objects (i.e., false sonar targets) immediately adjacent to the retroreflective tape. These objects were extended further and further into the aisle until at the end of the test, four of seven markers were in error with associated offsets of 7.5", 16", 10.5", and 6.5". Various combinations were tried in terms of which markers to perturb in an effort to generate the worst-case scenario (i.e., every other one, or four in a row). In general, the system remains very stable, and simply shifts the path laterally to accommodate the induced range offset. The robot's track basically follows a profile determined by the perceived sonar targets, and the platform continues to navigate in a very robust fashion without any noticeable instabilities.

An alternative triangulation configuration (Figure 15-26) can be employed to completely eliminate this sonar range ambiguity altogether by taking advantage of the excellent target discrimination feature of the Banner sensor. Two identical retroreflective units are mounted at oblique angles from the normal and with known separation baseline d_s, symmetrically oriented so their converging beams cross at point E a fixed distance x_c (about 3 to 4 feet) from the robot. The measured difference between leading-edge detection by the two sensors in conjunction with the known speed V_r of the robot determines the effective relative displacement d of the target along path CD (or path C'D') in the robot's reference frame. The sequence of detection (i.e., Sensor A followed by Sensor B, or vice versa) determines whether the tape stripe lies inside of or beyond the point of convergence of the beams, thus eliminating any ambiguity.

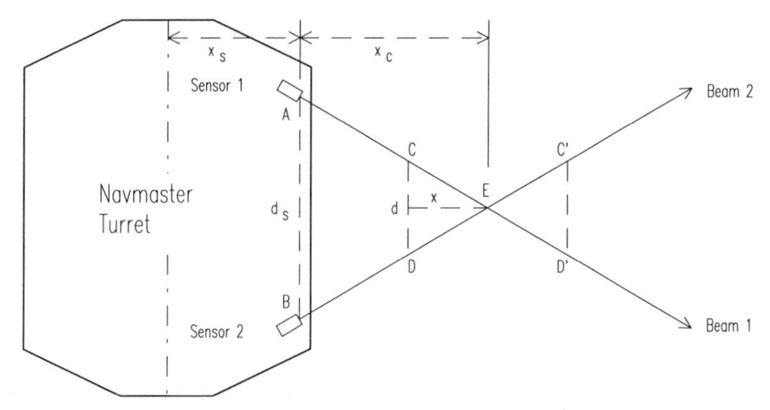

Figure 15-26. A pair of retroreflective sensors on each side of the *Navmaster* turret can be used to triangulate range to the retroreflective tape marker.

Triangles *ABE* and *CDE* are similar triangles, and so:

$$\frac{x}{x_c} = \frac{d}{d_s} \quad \text{which yields:} \quad x = \frac{d\,x_c}{d_s} = \frac{v_r\,t\,x_c}{d_s}$$

where:

x = target offset from point of beam convergence
x_c = known distance from robot to point of convergence
d = effective displacement of target due to robot motion
v_r = velocity of robot
t = measured time between target detections.

The vehicle lateral offset x_v from the known position of the tape marker is then given by:

$$x_v = x_s + x_c \pm x$$

where:

x_v = perceived lateral distance from vehicle centerline to marker position
x_s = lateral mounting offset between sensors and vehicle centerline.

The sign of x is positive if sensor 2 detects the tape marker before sensor 1.

One of the drawbacks of this method is increased possibility of target occlusion due to protruding objects on either side of the tape. Preliminary MDARS experience has shown that the highest probability of post detection exists for straight-in sighting directly normal to the path axis. An angled beam, on the other hand, is more likely to be masked as shown in Figure 15-27.

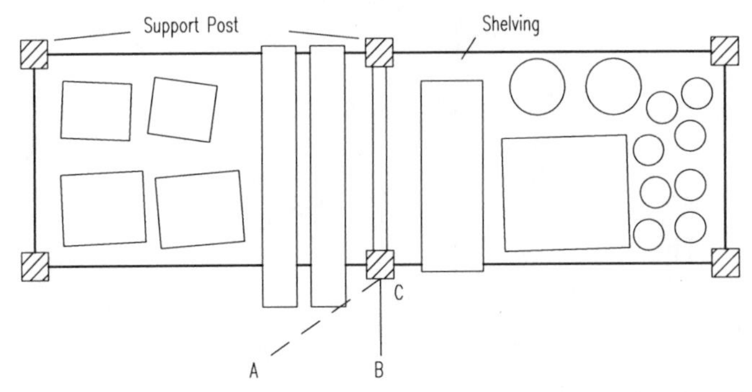

Figure 15-27. Protruding objects in the vicinity of a reflective marker will interfere with the sonar range measurements and can also occlude an off-normal optical sensor view such as along path AC.

A second problem encountered with the triangulation ranging approach in this scenario is the dependence on platform velocity. In the case of the *Navmaster,* a relatively slow update rate (10 Hz) of velocity information over the internal communications net could introduce errors if acceleration or deceleration should occur during the timing operation. Constant velocity cannot be assured in the vicinity of the markers since the collision avoidance strategy automatically servos speed of advance in relationship to perceived congestion.

15.3 References

Baker, A., "Navigation System Delivers Precision Robot Control," Design News, p. 44, December, 1993.

Balch, T., Arkin, R.C., "Communication in Reactive Multiagent Robotic Systems," *Autonomous Robots*, Vol. 1, pp. 1-25, Kluwer Academic Publishers, Boston, MA, 1994.

Bauzil, G., Briot, M., Ribes, P.., "A Navigation Subsystem Using Ultrasonic Sensors for the Mobile Robot Hilare," Proceedings of 1st Conference on Robot Vision and Sensory Control, Stratford/Avon, U.K., pp. 47-58, 13 April, 1981.

Byrne, R.H., Klarer, P.R., Pletta, J.B., "Techniques for Autonomous Navigation," Sandia Report SAND92-0457, Sandia National Laboratories, Albuquerque, NM, March, 1992.

Caterpillar, Product Literature, SGV-1092/91, Caterpillar Self Guided Vehicle Systems, Mentor, OH, 1991a.

Caterpillar, Product Literature, SGV-1106/91, Caterpillar Self Guided Vehicle Systems, Mentor, OH, 1991b.

Dodington, S.H., "Electronic Navigation Systems," *Electronic Engineer's Handbook,* D. Christiansen and D. Fink, eds., 3rd edition, New York, McGraw Hill, pp. 76-95, 1989.

Dunkin, W.M., "Ultrasonic Position Reference Systems for an Autonomous Sentry Robot and a Robot Manipulator Arm", Masters Thesis, Naval Postgraduate School, Monterey, CA, March 1985.

Everett, H.R., Bianchini, G.L., "ROBART II; An Intelligent Security Robot", Proceedings, U.S. Army Training and Doctrine Command Artificial Intelligence and Robotics Symposium, June 1987.

Figueroa, J.F., Doussis, E., Barbieri, E., "Ultrasonic Ranging System for 3-D Tracking of a Moving Target," 92-WA/DSC-3, Proceedings, Winter Annual Meeting, American Society of Mechanical Engineers, Anaheim, CA, November, 1992.

Figueroa, J.F., Lamancusa, J.S., "A Method for Accurate Detection of Time of Arrival: Analysis and Design of an Ultrasonic Ranging System," Journal of the Acoustical Society of America, Vol. 91, No. 1, pp. 486-494, January, 1992.

Figueroa, J.F., Barbieri, E., "Increased Measurement Range Via Frequency Division in Ultrasonic Phase Detection Methods," *Acustica*, Vol. 73, pp. 47-49, 1991.

Figueroa, J.F., Mahajan, A., "A Robust Navigation System for Autonomous Vehicles Using Ultrasonics," *Control Engineering Practice*, Vol. 2, No. 1, pp. 49-59, 1994.

Gould, L., "Is Off-Wire Guidance Alive or Dead?" Managing Automation, pp. 38-40, May, 1990.

Harrington, J.J., Klarer, P. R., "Development of a Self- Navigating Mobile Interior Robot Application as a Security Guard/Sentry", Sandia Report SAND86-0653, Sandia National Laboratories, July, 1986.

Harrington, J.J., Klarer, P.R., "SIR-1: An Autonomous Mobile Sentry Robot," Technical Report SAND87-1128, UC-15, Sandia National Laboratories, May, 1987.

ISI, "EZNav Position Sensor," Product Literature, Intelligent Solutions, Inc., Marblehead, MA, 1994a.

ISI, "EZNav Descriptive Manual," Product Literature, Intelligent Solutions, Inc., Marblehead, MA, April, 1994b.

ISR, *Radio Communications Option, Genghis Edition*, Product Literature, IS Robotics, Inc., Somerville, MA, May, 1994.

Lamancusa, J.S., Figueroa, J.F., "Ranging Errors Caused by Angular Misalignment Between Ultrasonic Transducer Pairs," *Journal of the Acoustical Society of America*, Vol. 87, No. 3, pp. 1327-1335, March, 1990.

MacLeod, E.N., Chiarella, M., "Navigation and Control Breakthrough for Automated Mobility," Proceedings, SPIE Mobile Robots VIII, Vol. 2058, pp. 57-68, 1993.

Maddox, J., "Smart Navigation Sensors for Automatic Guided Vehicles," *Sensors*, pp. 48-50, April, 1994.

Mahajan, A., "A Navigation System for Guidance and Control of Autonomous Vehicles Based on an Ultrasonic 3-D Location System," Master's Thesis, Mechanical Engineering Department, Tulane University, July, 1992.

NAMCO, "LNFL03-A 5M/4-90," *Lasernet* Product Bulletin, NAMCO Controls, Mentor, OH, November, 1989.

Premi, S.K., Besant, C.B., "A Review of Various Vehicle Guidance Techniques That Can Be Used by Mobile Robots or AGVs," 2nd International Conference on Automated Guided Vehicle Systems, Stuttgart, Germany, June, 1983.

TRC, *Beacon Navigation System*, Product Literature, Transitions Research Corporation, Danbury, CN, 1994.

16
Wall, Doorway, and Ceiling Referencing

For purposes of navigational referencing, indoor robotic systems can take advantage of a number of established landmarks in the form of wall structures, doorways, and ceilings or overhead beams that are not available in outdoor scenarios. (Outdoor applications, on the other hand, can take advantage of differential GPS, which is ineffective indoors due to signal blockage.)

16.1 Wall Referencing

Interior walls are probably the most commonly used structural attribute for deriving position and orientation information, with performance results determined primarily by the inherent accuracy limitations of the measurement techniques employed. Existing methodologies can be divided into four general classes:

- *Tactile* — The robotic platform aligns itself through direct physical contact with a wall of known orientation and location.
- *Non-Contact (Static)* — The platform faces off to a wall from a stationary position and determines offset and orientation from non-contact range data.
- *Non-Contact (Dynamic)* — The platform derives offset and heading from continuous real-time range data while in motion.
- Some combination of the above.

16.1.1 Tactile Wall Referencing

One obvious solution to the navigational re-referencing problem would be to bring the robot into actual contact (and alignment) with a predesignated wall section free of obstructions. The robot's heading under these conditions would be precisely orthogonal to the known wall orientation, with its positional offset from the wall equally unambiguous. While not very elegant, this method is extremely robust in the sense that range measurement inaccuracies are virtually eliminated. An added advantage is seen in the effective minimization of any backlash in the drivemotor reduction gears, since both gear trains are preloaded in the same direction. This tactile wall-referencing concept was implemented on ROBART II as an interim measure, pending the development of more sophisticated methods that did not require the robot to deviate from assigned functions solely for the purpose of resetting the navigational position and heading parameters.

To initiate this re-referencing procedure, the platform first moves to a position about 3 feet out and facing an unobstructed wall, based on the current dead-reckoning position information. The recalibration routine is then requested, whereupon the *Scheduler* computer (see Chapter 1) on board the robot assumes control. With the robot stationary, the *Scheduler* requests a sonar update from the *collision avoidance array* and checks to see that the robot is indeed within 4 feet from the wall. If the measured range exceeds 4 feet, an error message is generated. Otherwise, the ranges seen by transducer #1 (mounted on the head) and transducer #8 (center of lower array) are compared; with the robot facing an unobstructed wall, these ranges should be nearly equal. If the lower range is less than the upper range by more than a specified tolerance, some obstruction is assumed to be present between the robot and the wall (Figure 16-1), and this situation is reported to the *Planner* as an error condition.

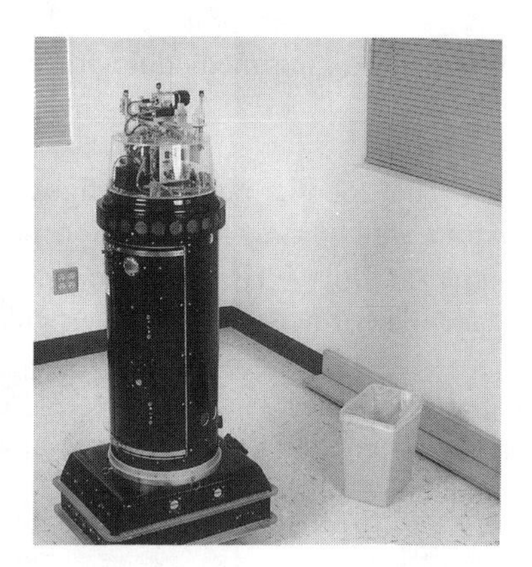

Figure 16-1. An interfering obstruction shows up in the form of conflicting range values for the upper and lower forward-looking sonar sensors.

Assuming no discrepancies are detected, the *Scheduler* requests repeated updates from sonar transducer #8 and initiates forward travel, decreasing speed as the range to the wall falls off. When the measured distance drops below 19 inches, the *Scheduler* checks the lower three *near-infrared optical proximity sensors* for wall confirmation. The sensitivities of the outer two proximity sensors are set to acquire the wall surface at a distance of 26 inches, while the center is set for a distance of 32 inches; therefore, all three should see the wall at 19 inches. If such is not the case, action is taken in accordance with the following rules:

- If none of the sensors see a target, forward motion is halted and an error message is sent to the *Planner*.
- If the center sensor only sees a target, forward motion is halted, and an error message is sent to the *Planner*.
- If the left sensor only does not see a target, the right drive motor is halted, causing the robot to turn right.
- If the right sensor only does not see a target, the left drive motor is halted, causing the robot to turn left.

The software loops in this mode until all three sensors see the wall, whereupon straight-line travel is resumed, or an error condition occurs. The last two rules above have the effect of correcting any gross misalignments with the wall surface prior to impact. Preliminary alignment could also be accomplished in the wall approach by doing a line-fitting operation on data from the lower *collision avoidance array* (see next section).

Figure 16-2. When ROBART II is firmly docked at the wall charging strip, heading and longitudinal position errors are eliminated.

At this point, the robot should be moving forward towards the wall at minimum speed (1.07 inches/second). The *Scheduler* waits for wall impact with the tactile bumper (Chapter 3) and stops each drivemotor when its associated side of the bumper indicates contact. For example, if the left side of the bumper deflects first, the *Scheduler* stops the left drivemotor, allowing the right motor to continue until such time as the right side of the bumper deflects. This reflexive action turns the robot in such a way as to square it off to the wall, whereupon forward motion stops. The *Scheduler* next backs the platform away from the wall exactly 1 inch and then resumes forward travel at minimum speed for two seconds. Wheel slippage occurs for approximately half a second as the robot's forward travel is halted upon contact with the wall (Figure 16-2), thereby symmetrically preloading the drive reduction gears to minimize backlash errors.

With the spring-loaded tactile bumper pressed firmly against the wall, and both motors stopped, the *Scheduler* requests range data from the two side-looking sensors in the upper *navigational sonar array*. One (or both) of these ranges represents the distance to a known lateral reference target, such as a bookcase (Figure 16-3) or orthogonal wall surface. This measured range is relayed to the *Planner*, completing the navigational parameter update. With this information, the robot's perceived location is updated in the model, thus eliminating any accumulated dead-reckoning errors. If the wall surface is suitably equipped with contact plates for recharging the onboard batteries (Figure 16-2), this method of recalibration becomes a little more practical than would otherwise be the case, in that the robot needs to make physical contact anyway in order to recharge.

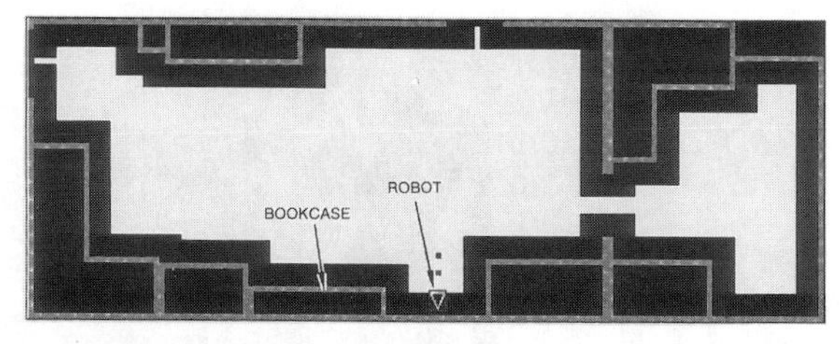

Figure 16-3. Lateral position errors are nulled out by taking a sonar range reading to the bookcase on the robot's right (Everett, et al., 1990).

16.1.2 Non-Contact Wall Referencing

Stationary walls of known orientation offer an attractive mechanism for resetting system heading as well as one component of the lateral position, even without actual physical contact. The concept as implemented on ROBART II calls for positioning the robot near an unobstructed wall surface, and then sequentially firing the 11 transducers in the lower *collision avoidance sonar array*. A line-fitting operation can then be performed on the subsequent data (Table 16-1) from the five transducers in the vicinity of the minimum range value. (The minimum range theoretically should represent the reading from that transducer whose axis was most nearly orthogonal to the wall surface.) The angle of the resulting line with respect to the robot is used to adjust the robot's perceived heading based on the known orientation of the wall (Figure 16-4). In addition, the robot's longitudinal offset from this wall is made available as well.

Table 16-1. Measured sonar data for angular orientation of -7.5 degrees (see Figure 16-4).

Range (inches)	Bearing (degrees)	X (inches)	Y (inches)
35.55	36	20.87	28.72
30.75	18	9.50	29.24
30.75	0	0.00	30.75
35.55	-18	-10.97	33.76
41.95	-36	-24.66	33.94

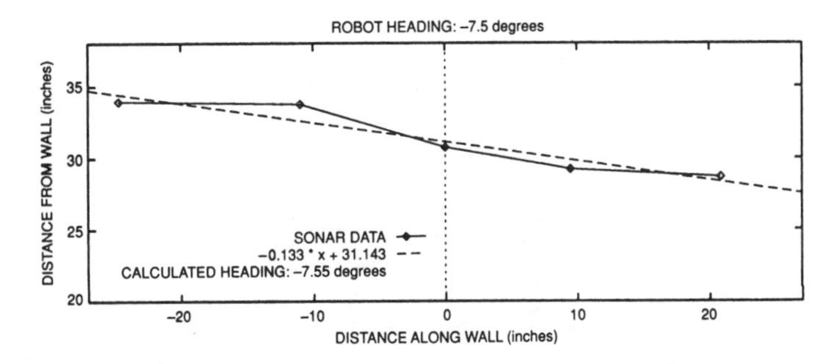

Figure 16-4. A calculated angular offset of -7.55 degrees is obtained for an actual orientation of -7.5 degrees (adapted from Everett, et al., 1990).

A complete navigational update (*X-Y* position and heading) could theoretically be obtained in a corner situation at the intersection of two orthogonal walls. Attempts to determine platform heading using this concept on ROBART II, however, met with only limited success due to specular reflection and beam divergence (see Figure 16-5). These problems were aggravated by the physical orientation of the ranging sensors, which fanned out radially from the cylindrical housing. This arrangement works to your advantage when trying to detect an obstacle for collision avoidance purposes, as the odds of a single transducer being nearly normal to the target surface are greatly increased. On the other hand, the odds of two or more transducers in the radial array being normal to a planar wall surface are likewise inherently low. The range values associated with those sensors which are not normal to the wall surface, obviously, are going to be adversely affected (Table 16-2), as accuracy falls off when the angle of incidence varies from the perpendicular. Since fairly accurate data from at least three transducers is required for the wall referencing algorithm to function properly, this represents a fundamental problem.

Table 16-2. Measured sonar ranges for angular orientation of 7.5 degrees (see Figure 16-5).

Range (inches)	Bearing (degrees)	X (inches)	Y (inches)
43.55	36	25.60	35.23
35.55	18	10.99	33.81
35.55	0	0.00	35.55
33.95	-18	-10.49	32.29
41.95	-36	-24.66	33.94

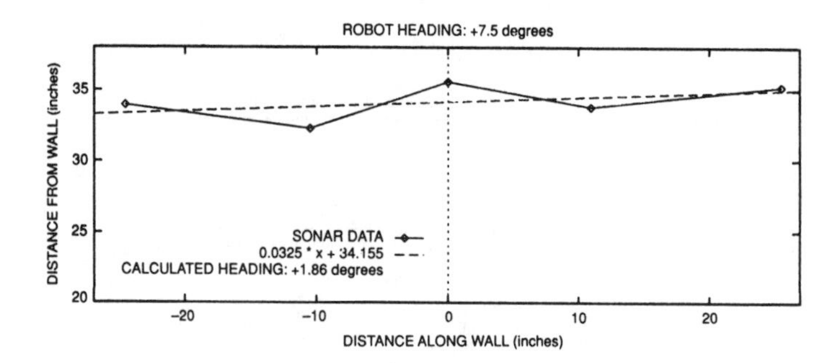

Figure 16-5. A calculated offset of 1.86 degrees is obtained for an actual orientation of 7.5 degrees.

One possible solution that was considered called for placing two or more additional ultrasonic ranging transducers along the front panel of the robot's base, which was a planar as opposed to cylindrical surface as shown in Figure 16-6. The platform would first rotate in place to turn to the heading indicated by the axis of the minimum range value discussed above, and then fire the front panel sensors; the resulting range values should be close in value if in fact the front panel were aligned parallel to the wall. If the difference was not within a specified tolerance, the robot would rotate slightly to correct the discrepancy. Once roughly aligned in this fashion, the front panel sensors, all normal to the target surface, would provide the highly accurate range data needed by the line-fit algorithm, which would subsequently determine the robot's precise angular orientation with respect to the wall. Alternatively, inexpensive short-range (5 to 6 feet) optical ranging systems with tightly focused beams and less susceptibility to problems associated with specular reflection could be employed for this application in place of the ultrasonic rangefinders.

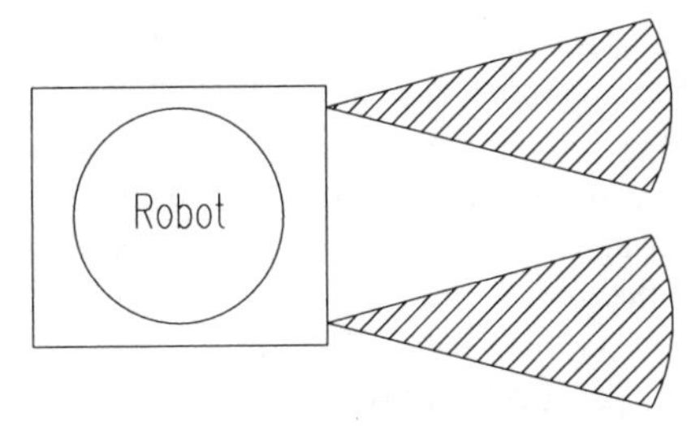

Figure 16-6. Potential mounting configuration for two additional sonar transducers to facilitate wall referencing on ROBART II (adapted from Everett, et al., 1990).

The Cybermotion *Navmaster* robot employs a *virtual path instruction* known as "wall approach" to reset the longitudinal displacement parameter only (i.e., no attempt is made to determine platform heading). The robot is known to be approaching a wall directly ahead and has been told that at the end of the current path segment it should be a certain distance from this wall. Knowing the absolute position of the wall, the robot can then update the unknown coordinate using the measured range in the forward direction upon completion of the move. Cybermotion's "wall-approach" instructions are generally used in conjunction with "wall-following" instructions (see next section) for a complete navigational solution (Holland, et al., 1990).

16.1.3 Wall Following

Wall following is another type of referencing technique, similar to the method described above, except that it takes place while the robot is traveling along a wall rather than facing it. Both the TRC *HelpMate* (Krishnamurthy, et al., 1988) and the Cybermotion *Navmaster* (Holland, et al., 1990) robots rely extensively on *wall-following* strategies. The nomenclature is somewhat misleading in the sense that the robot does not actually follow the wall in a servo-controlled fashion, but rather obtains a navigational reference from it. Only a single side-looking transducer is required, as the forward motion of the robot is exploited to provide the baseline separation required for heading calculation. While *wall following* can be used to effectively reset the robot's heading and lateral (either the *X* or *Y*) position coordinate, the longitudinal displacement along the path of travel remains unknown.

The basic procedure is described by Kadonoff (1990) and typically applied where the robot is traveling parallel to a wall of known position and orientation with a specified lateral separation. During the execution of this path segment, the robot repetitively fires a non-contact ranging sensor that is perpendicular to and facing the wall. Over a period of time the system thus accumulates several data points, each consisting of the measured range to the wall and the associated longitudinal position of the robot along the path of travel. A straight-line fit can be made to these data points using standard linear regression techniques (Devore, 1982). If a "good" fit is obtained (i.e., the data do not deviate significantly from a straight line), the line is accepted and the lateral offset from the wall as well as the current heading of the robot can be calculated as described below. With this information, the robot can adjust course to correct its offset and heading, turning toward or away from the wall as appropriate.

A simple example is illustrated in Figure 16-7. The robot begins the *wall-following* maneuver at point A and proceeds to point B, with the measured sonar ranges indicated in the figure by lines emanating from the robot and terminating somewhere near the wall shown at the bottom. Table 16-3 lists the range data collected by the side-looking sonar as a function of longitudinal displacement along path segment AB.

Table 16-3. Sonar versus actual (measured) range readings along path segment AB of Figure 16-7.

Path Position	Sonar Range	Actual Range	Path Position	Sonar Range	Actual Range	Units
0.0	33.2	33.7	59.8	36.2	36.6	inches
5.4	33.7	33.7	66.8	36.4	36.6	inches
11.2	33.8	34.2	72.2	36.5	37.1	inches
19.3	34.2	34.7	78.4	36.6	37.1	inches
23.9	34.6	34.7	84.8	36.8	37.4	inches
29.4	34.8	35.2	90.4	37.2	37.4	inches
36.0	35.3	35.6	96.6	37.6	37.9	inches
42.7	35.5	36.1	102.5	37.9	37.9	inches
49.9	35.9	36.1	108.0	38.0	38.4	inches
55.9	36.1	36.6	114.0	38.2	38.4	inches

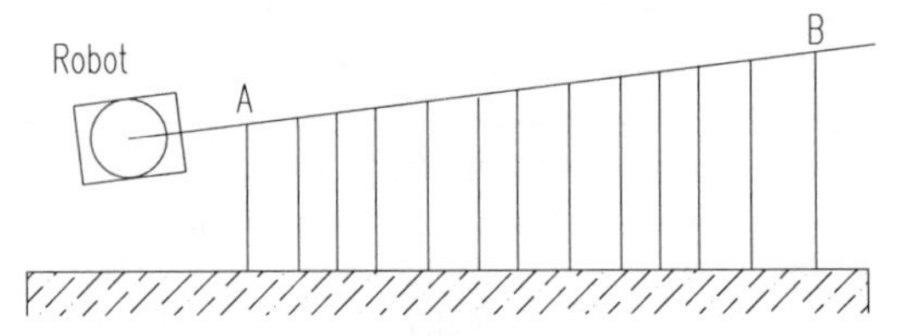

Figure 16-7. A line-fit operation is performed on several sonar range readings taken while the robot is in motion to establish relative heading and lateral offset with respect to the wall.

The linear regression equations used to calculate the slope, intercept, and estimated variance are as follows (Everett, et al., 1990):

$$m = \frac{n\sum(x_i y_i) - \sum x_i \sum y_i}{n\sum x_i^2 - (\sum x_i)^2}$$

$$Y_I = \frac{\sum y_i - m\sum x_i}{n}$$

$$\sigma^2 = \frac{\sum y_i^2 - Y_I \sum y_i - m\sum(x_i y_i)}{n-2}$$

where:

m = slope
n = number of sonar readings taken (20 in this example)
σ^2 = variance
Y_I = intercept.

Using these formulas, the equation of the line (Figure 16-8) resulting from the use of the sonar range values is:

$$y = 0.0416\,x + 33.885 \qquad \sigma^2 = 0.0530.$$

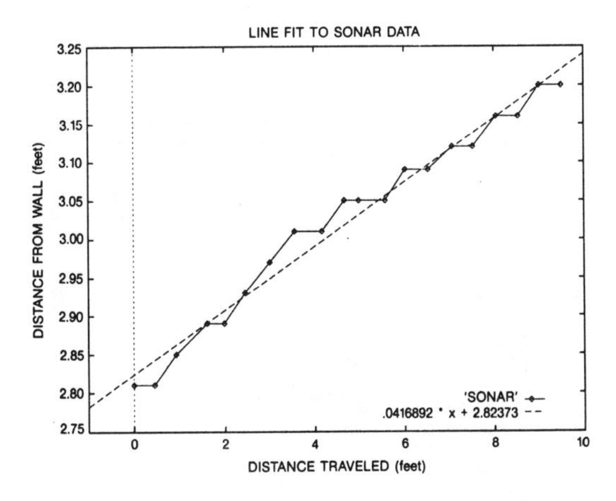

Figure 16-8. Plot of actual sonar data from Table 16-3 and resulting least-squares fit (Everett, et al., 1990).

Similarly, the equation of the line (Figure 16-9) using the robot's measured position from the wall is:

$$y = 0.0420\,x + 33.517 \qquad \sigma^2 = 0.0335.$$

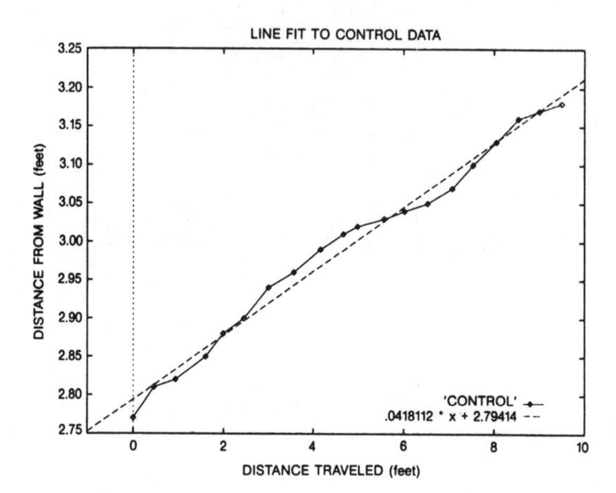

Figure 16-9. Plot of measured (reference) data from Table 16-3 and resulting least-squares fit. Undulations in the data are caused by imperfections in the wall itself (Everett, et al., 1990).

Figure 16-10 below shows a comparison of the two lines: the slopes are extremely close and the sonar data is offset from the measured (reference) data by only 0.03 feet (0.36 inch).

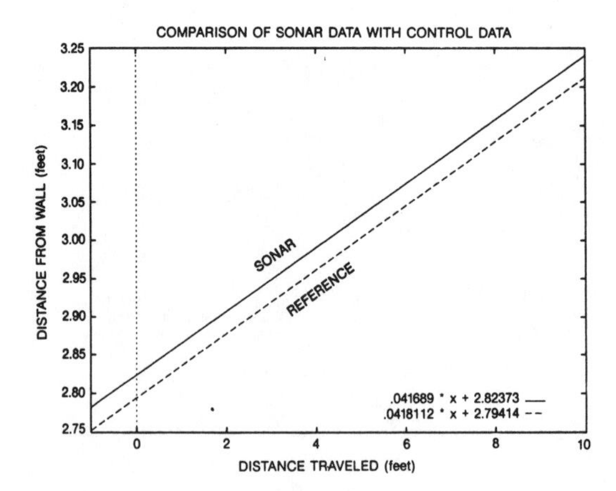

Figure 16-10. A comparison of the sonar and reference line-fit operations depicted in Figure 16-8 and Figure 16-9 above depict a lateral offset of about 0.3 inches (Everett, et al., 1990).

The robot's heading with respect to the wall can be calculated by taking the arctangent of the slope. For the sonar data, this yields:

$$\theta = \arctan(0.0416) = 2.382 \text{ degrees}$$

while for the measured (reference) data:

$$\theta = \arctan(0.0420) = 2.405 \text{ degrees.}$$

In this particular example, the sonar-estimated heading of 2.382 degrees varies by only 0.023 degrees in comparison to the actual measured results.

16.2 Doorway Transit Referencing

The concept of using existing interior doorways as navigational landmarks has always been appealing, in that no modifications to the surrounding environment are required. In certain indoor environments, the robot by necessity must travel through a doorway to enter an adjoining space. If in so doing the system could obtain an accurate positional update, then such would indeed represent an elegant solution to the problem of cumulative dead-reckoning errors. The doorway penetration approach employed on ROBART II can be decomposed into the following tasks:

- Finding the doorway.
- Entering the doorway.
- Verifying the doorway.
- Determining longitudinal position relative to doorway.
- Determining lateral position relative to doorway.
- Determining heading (angular orientation) relative to doorway.

The first of these tasks is addressed through use of a combination of ultrasonic ranging sensors, which have good distance measurement capability but poor angular resolution, and optical proximity sensors, which typically have superior angular resolution but little or no ranging capability. The problem is greatly simplified by virtue of the fact that the *Planner* knows where the door is located within the map structure and can direct the robot reasonably well to the vicinity of this position. In addition, the *Planner* always orients the path segment that actually penetrates the door opening to be orthogonal to the associated wall. With such *a priori* information, the task of finding the doorway's actual position with respect to the robot is greatly simplified.

To accomplish this task, the *Planner* informs the *Scheduler* that the current path segment penetrates a door opening and provides the estimated bearing and distance to the door. The *Scheduler* rotates the head to this bearing (typically straight ahead), thus pointing the long-range near-infrared proximity sensor (Figure 16-11) at the center of the anticipated passage. Unless the robot is significantly misaligned due to accumulated dead-reckoning errors, the proximity sensor will return a "no target" condition, as it should be looking through the open doorway. If this is not the case, the head begins scanning 15 degrees either side of centerline in an attempt to find the opening. If this search fails to locate the doorway, an error condition is returned informing the *Planner* that the robot is either significantly lost to where the door penetration routine won't work, or the door is closed.

Figure 16-11. ROBART II searches for the door opening using the head-mounted programmable near-infrared proximity detector discussed in Chapter 7 (courtesy Naval Command Control and Ocean Surveillance Center).

Assuming the opening is detected, the *Scheduler* next attempts to locate the left and right edges by panning the head and watching the proximity sensor output for a "target" condition, indicative of energy being reflected from the door casings (see doorway detail, Figure 16-12) and adjacent wall areas to either side. Head position angles corresponding to the left and right boundaries are then averaged to yield a relative bearing to the actual center of the doorway.

The *Scheduler* alters the robot's heading to be coincident with this bearing and begins looking at the sonar data from the center five transducers in the *collision avoidance array* for range confirmation. Measured distance to the door should be within a specified tolerance of the estimated range provided earlier by the *Planner*, less distance traveled in the interim, otherwise another error condition is returned. If the robot is more than 5 feet from the doorway, the center three transducers should all indicate ranges within this window of acceptance.

As the robot closes on the doorway, the beam from the center transducer should eventually break through the opening, with a corresponding increase in range to target. This occurs at the point where the effective beamwidth at the indicated distance becomes less than the width of the doorway, assuming the robot is perfectly aligned with the center of the opening. (Perfect alignment is typically not the case, however, resulting in a slight delay as the beam narrows further on approach, before the jump in range is observed.)

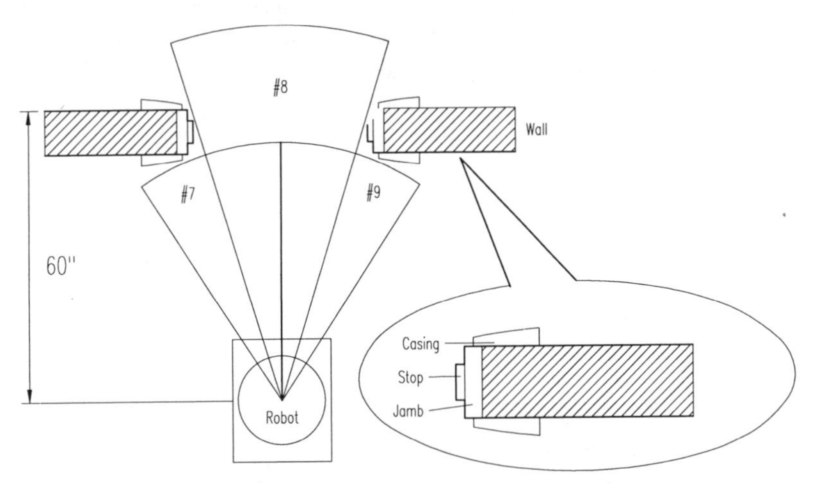

Figure 16-12. Energy is reflected from the left and right door casings, but the center beam (#8) penetrates the door opening at a distance of about 5 feet.

It may appear the robot's alignment with the doorway could be calculated in advance at a distance of approximately 4 feet by simply comparing the range returns from transducers #7 and #9, but this method turns out to be unreliable due to the possibility of furniture or other objects on either side of the door interfering with the beam. In addition, doorways are sometimes placed in the corner of a room in close proximity to an adjoining wall, which will interfere with the sonar readings on that particular side. For this reason, it was found necessary to let the robot get very close to the opening as discussed above before assessing alignment.

At the instant the center beam penetrates the opening, the two adjoining beams from transducers #7 and #9 should by virtue of their orientation in the array be directed at the left and right door casings, as shown in Figure 16-12. The respective range readings from these two transducers at this point should again be consistent with the previously estimated range to the doorway, until such time as the indicated ranges decrease to around 36 inches, whereupon these beams should break through the opening, as shown in Figure 16-13. If either of these ranges decreases below 12 inches prior to penetration, the robot is likely to impact the side of the door, and the *Scheduler* will have to execute a corrective maneuver to attain better alignment.

The next step in the procedure calls for deriving *X-Y* positional data while passing through the door opening. The most obvious solution for the transverse fix is to ping the two side-looking transducers in the upper *navigational sonar array* at the left and right door jambs; the only difficulty here would be knowing exactly when to ping. One solution might be to ping continuously during transit, and then use the minimum range value thus obtained. An alternative approach would be to estimate the distance to the center of the opening from the last set of range values measured by *collision avoidance* transducers #7 and #9 just prior to penetration, and then ping the door jambs after traversing that amount of distance. In either case, the left and right range readings thus obtained specify the robot's lateral position, and for purposes of verification should add together to yield the width of the door passage, typically 36 inches.

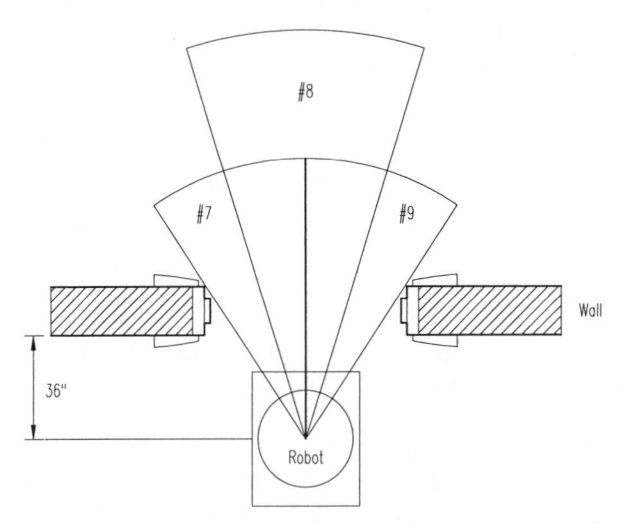

Figure 16-13. As the robot closes on a 36-inch doorway, all three sonar beams should penetrate the opening at approximately 3 feet.

The task of obtaining a longitudinal fix during doorway transit is a little more difficult. The longitudinal fix could be derived from the last set of readings obtained by the forward-looking sonar transducers mentioned above, but the accuracy would be somewhat suspect. Alternatively, if the transverse fix discussed above is obtained by successive pinging of the casings, then postanalysis of the data should yield a door edge profile in the sense that ranges to either side will decrease to some minimum upon entry, remain at that minimum plus or minus some tolerance value for a finite length of time proportional to the width of the jamb (thickness of the wall), and then increase. The midpoint of this period of minimum ranges would then correspond to the midpoint of the door jamb width (centerline of the wall), which is of course the desired longitudinal fix.

Both of the above solutions, however, assume an ideal door opening in the center of an unobstructed wall and will suffer significantly from the presence of objects near the open doorway, not the least of which might be the door itself. (When in the open position, the door folds back to one side, adding several inches in projected target surface that will interfere with the ranging process.) This is primarily due to problems associated with specular reflection and beam divergence in the ultrasonic rangefinders employed. Diffuse-mode near-infrared proximity sensors are often employed in an effort to compensate for some of the limitations in ultrasonic systems, in that the beams can be tightly focused, and specular reflection is less significant due to the shorter wavelengths involved (Banner Engineering, 1993a, 1993b). This type of proximity sensor provides no range measurement capability, however, other than that which can be inferred from the strength of returning energy, which varies as a function of target reflectivity (Chapter 3).

If the sensors are mounted on the robot so as to be orthogonal to the direction of travel, however, they could be used to detect the leading edge of the door casing as the robot passed through the opening. As shown in Figure 16-14, the elapsed time between target detection by sensors mounted on either side of the robot also could be used to calculate the angular orientation of the robot with respect to the doorway, in accordance with the following formula:

$$\sin\theta = \frac{vt}{d}$$

where:

θ = angular orientation
v = velocity of robot
t = elapsed time between detections
d = target separation distance.

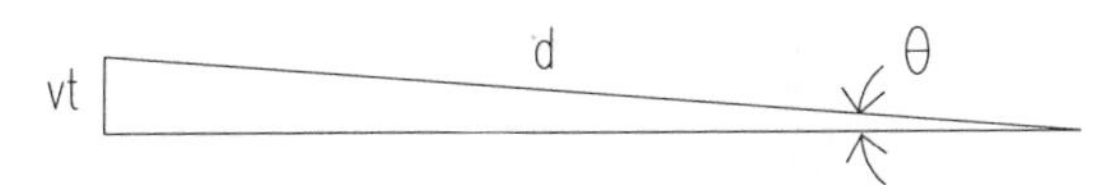

Figure 16-14. Elapsed time between doorway-overhead detection by left and right proximity sensor pairs can be used to calculate heading.

To achieve any useful accuracy in deriving the heading of the robot in this fashion, however, the following conditions must apply:

- The sensors must be rigidly mounted to retain their orthogonal relationship to the robot.
- The sensors must have well defined narrow beams.
- The excess gain must be sufficiently high to ensure rapid detection as the targets move into view.
- The time between left and right target detection must be accurately measured.
- The robot's heading must remain constant for this period.

- The distance traveled by the robot during this period must be accurately measured.
- The targets must stand out clearly from their respective backgrounds with no ambiguity.
- The lateral separation *d* between targets must be known in advance or measurable during transit.

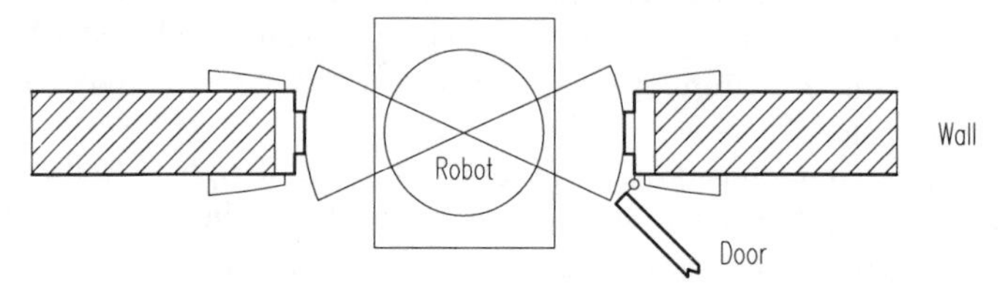

Figure 16-15. The door stops as well as the actual door can interfere with the ranging process (Everett, et al., 1990).

The first six conditions outlined above are easily met, but the latter two pose a problem. As previously discussed, objects on either side of the doorway can effectively mask the location of the door casing to make accurate leading-edge detection impossible. One way around this would be to apply strips of retroreflective tape to the door casings to create cooperative targets, and reduce the gain of the proximity sensors to where only these strips triggered a detection.

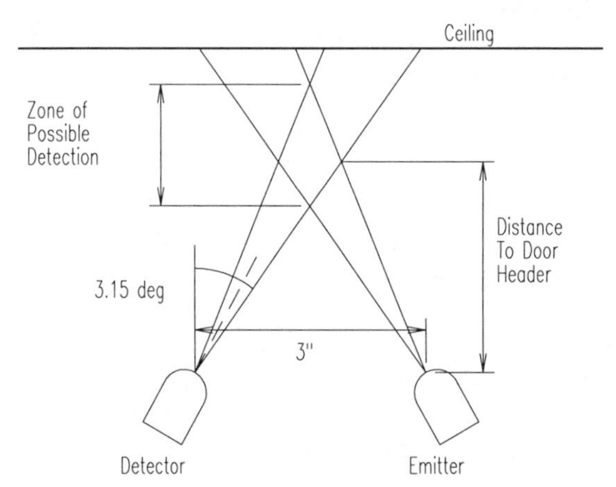

Figure 16-16. Objects (such as the ceiling) outside the zone of possible detection will be ignored by an emitter/detector pair configured in convergent mode.

The use of retroreflective tape, however, requires the environment be modified to accommodate the robot, which is not in keeping with the objective of using existing (unmodified) doorways as navigational aids. Such strips are somewhat obtrusive and distracting to humans and can be accidentally removed or painted over by maintenance crews. In addition, setting the detection threshold of the sensors to respond only to the retroreflective strips violates the requirement for high excess gain. In reality, the critical threshold setting required is likely to be impossible to achieve under dynamic real-world conditions. The robot may pass through the opening closer to one side than the other, and the distances involved can vary as well due to different doorway widths ranging anywhere from 30 inches to 72 inches or more.

Even if the leading edges could be precisely detected, ambiguities arise in measuring the distance between the actual locations of the left and right targets using ultrasonic ranging techniques. Referring to Figure 16-15, we see that both the door stops as well as the actual door itself can interfere with the ranging process. The resulting measurement accuracy, although acceptable for determining the lateral position of the robot in the doorway, would be insufficient for the desired final resolution in heading.

One solution to these problems is to reorient the proximity sensors to where the beams are vertical as opposed to horizontal, yet still orthogonal to the direction of robot motion. The target separation distance is thus a constant, precisely determined by and equal to the sensor displacement d on board the robot, eliminating one of the above concerns altogether. The upper door casing now becomes the target, where there is much less possibility of obstructions being present that might interfere with leading-edge detection. To further address this issue, the proximity sensors can be configured in the *convergent mode* as opposed to *diffuse mode*, taking advantage of the fact that the distance to the overhead casing will be fairly constant, regardless of the path followed by the robot through the doorway. (Standard door height is 80 inches.) This means objects (such as a ceiling or overhead light fixture) outside of the zone of potential detection will be ignored as illustrated in Figure 16-16, allowing for even greater excess gain to be employed.

Figure 16-17. Two Banner emitter/detector pairs arranged on either side of the robot's head are used to detect passage under an overhead door frame.

A photo of this overhead doorway-detection configuration as installed on ROBART II is provided in Figure 16-17. The emitter and detector units are Banner models *SM31EL* and *SM31RL*, respectively (Banner, 1993a). These near-infrared sensors are normally intended to be operated in the *break-beam mode* (see Chapter 3) at distances out to 100 feet but were found to effectively detect diffuse target surfaces at a nominal sense range of 40 inches when arranged in the *convergent mode*.

16.3 Ceiling Referencing

One obvious disadvantage of a navigational strategy that depends primarily on dynamic wall referencing to control accumulated errors is the inability of the system to operate in unstructured scenarios that do not provide sufficient access to appropriate wall structures. A good example is seen in the case of the MDARS Interior robot that patrols primarily in warehouse environments, where walls exist only along the outer perimeter of the building and are typically obscured even there by stored materials (Gage, et al., 1995). Similarly, warehouse environments are generally devoid of internal doorway structures, at least to the extent such would be found in an office building of equivalent square footage. Almost all buildings, however, are fairly consistent from the standpoint of unobstructed ceiling access, opening up distinct possibilities for a number of overhead referencing strategies.

16.3.1 Polarized Optical Heading Reference

One of the initial concepts considered in the development of ROBART II as a heading update mechanism called for placement of a number of active beacons on the ceiling above the robot's operating area. These modulated near-infrared sources were to be fitted with polarizing filters of known angular origin (i.e., referenced to building north). The modulated output of any one of these sources would automatically trigger a special head-mounted receiver whenever the robot traversed within the associated footprint of illumination of the source. An optical filter that passed only the near-infrared component (Kodak Wratten 89B) of the incoming energy would be placed in front of the detector to minimize the effects of ambient lighting.

Upon detection of a beacon, a second polarizing filter mounted on the robot (just above the receiver's upward-looking PIN photodiode detector) would be rotated under servo control to ascertain the null point for which the minimum receiver output signal was observed. This null point, of course, would be directly related to the pre-established orientation of the polarizing filter on the source. There is somewhat of an ambiguity in this proposed scheme since there would exist two null positions, 180 degrees apart, corresponding to when the polarizing gratings of the two filters were made orthogonal to one another. This ambiguity was to be resolved by choosing that null position most closely in agreement with the robot's fluxgate compass heading.

It was speculated the ceiling mounted sources could be modulated in such a way as to be uniquely identifiable to the robot, thus allowing them to serve double duty as lateral position markers. This enhancement, however, would require the PIN photodiode to be replaced by a suitable two-axis position sensitive detector or CCD array. Rathbone, et al. (1986) proposed a similar system for AGV guidance that would employ an upward-looking imaging sensor able to positively identify and track a number of specifically coded near-infrared LED beacons mounted on the ceiling. This polarized heading reference concept was never seriously pursued due to the desire to avoid modifying the robot's environment to accommodate the needs of the navigation system, and the less than elegant requirement to mechanically servo the angular orientation of the polarizing filter at the detector.

16.3.2 Georgia Tech Ceiling Referencing System

A more robust ceiling referencing scheme employing an upward-looking digital camera was implemented about this same time frame by researchers at the Material Handling Research Center at Georgia Tech, for use in free-ranging AGV navigation (Holcombe, et al., 1988; Bohlander, et al., 1989). In addition, the Georgia Tech approach significantly reduced required installation costs through the use of passive landmarks instead of active-emitter targets (Figure 16-18).

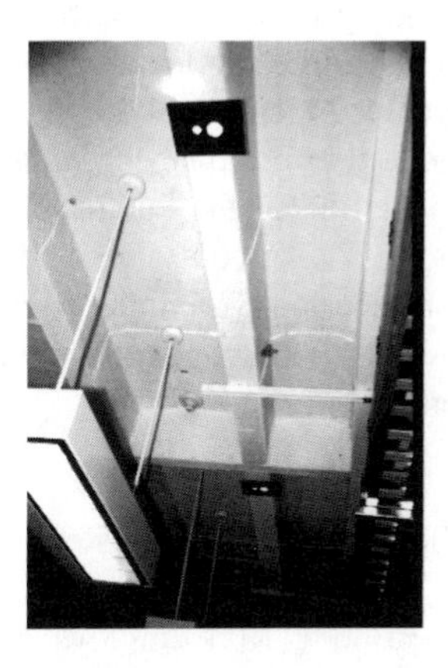

Figure 16-18. Asymmetrical retroreflective targets of known orientation and location serve as passive overhead markers for the *Landmark Tracking Camera System* (courtesy Georgia Institute of Technology).

A number of identical retroreflectors were mounted at known locations and orientations on the ceiling of the laboratory environment, approximately 5 meters apart. The *Landmark Tracking Camera System* (Figure 16-19) employed a xenon strobe to illuminate the overhead targets for subsequent detection by an 8192-pixel digital-RAM camera interfaced directly to an 8-bit *6500*-series microprocessor. The asymmetrical targets consisted of two different-diameter circles made of retroreflective paper affixed to a piece of black cardboard. Since the cooperative targets had a significantly higher reflectivity than the surrounding background, the camera threshold could be adjusted to eliminate everything except the desired circular targets from the binary image, thereby substantially reducing the amount of required processing (Holcombe, et al., 1988).

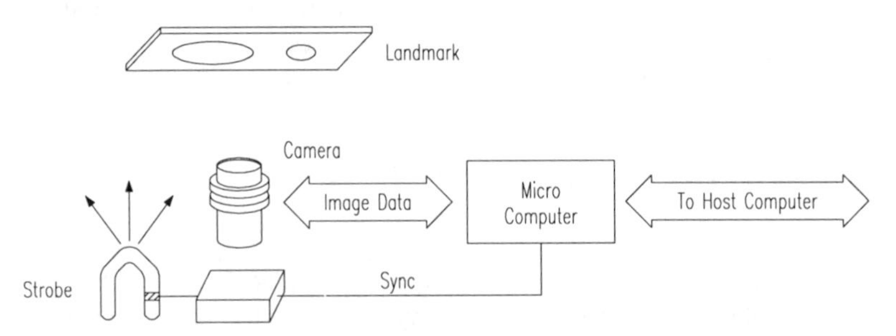

Figure 16-19. A xenon flash was used to illuminate the retroreflective landmark for subsequent detection by a binary digital-RAM camera (adapted from Holcombe, et al., 1988).

The *first moments* were calculated from the captured video data to find the centroid of the landmark, which was intentionally located within the diameter of the larger circle. An edge-finding operation was then performed (starting at the centroid location) to bound the large circle, after which the circle centerpoint could be calculated. The perceived location of the landmark centroid with respect to the center of the larger circle thus established the landmark's orientation. The relative *X-Y* location and orientation of the landmark as computed by the vision system was then passed to the AGV controller and integrated with odometry position estimates using an aperiodic Kalman filter. Experimental testing involving the collection of over 12,000 images showed the 3-σ position error to be less than 0.09 inches (Holcombe, et al., 1988).

16.3.3 TRC *HelpMate* Ceiling Referencing System

Transitions Research Corporation (TRC) employs a vision-based navigational referencing system on their *HelpMate* robot to determine platform lateral offset and heading from the known orientation of overhead lights (Krishnamurthy, et al., 1988; King & Weiman, 1990). In office buildings and hospitals, for example, such lighting fixtures are generally rectangular in nature and aligned parallel to the longitudinal axis of hallways and corridors. By thresholding the image and performing a series of edge-detect operations, the left and right boundaries of a light fixture can be reliably established. After first locating the light closest to the robot (i.e., at the top of the video image), the operation is repeated to identify a second light further down the hall. The left and right boundaries of the two lights are then extended to find their point of intersection on the horizon, known as the *vanishing point*.

The *HelpMate* vision system consists of an *80386*-based PC/AT computer, an associated frame grabber, and a forward-looking CCD camera tilted upward from the horizontal at some preset angle of pitch ϕ. The origin of the sensor's Cartesian coordinate system (depicted in Figure 16-20) is the midpoint of the scan line associated with the optical axis of the camera. The *vanishing point* is determined to be at pixel coordinates (i, j), where the extended centerline of the perceived light pattern also intersects the horizon. From this pixel information, the tangent components (u, v) of the vanishing point can be calculated, given the physical dimensions of the camera sensor and the focal length of the lens. The robot's heading θ (with respect to the perceived longitudinal axis of the light fixtures) can then be calculated as follows (King & Weiman, 1990):

$$\theta = \arctan(u \cos\phi)$$

where:

θ = relative heading of the robot
u = measured lateral offset to vanishing point
ϕ = camera pitch angle (fixed) with respect to horizontal.

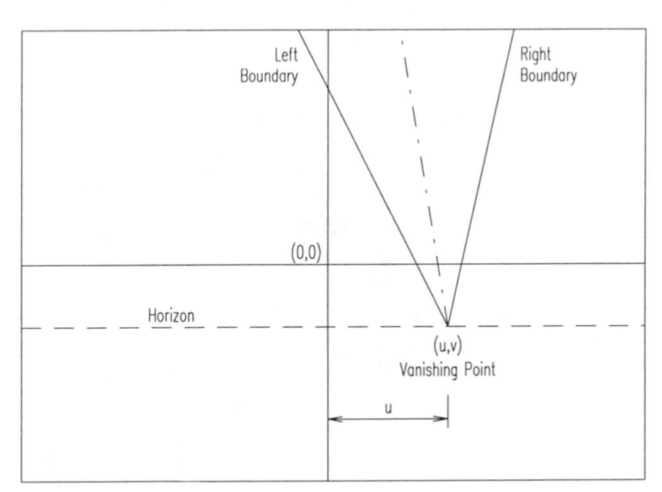

Figure 16-20. The lateral displacement u from the origin (in camera coordinates) of the *vanishing point* defined by the intersection of the perceived light centerline and the horizon determines the heading of the robot (adapted from King & Weiman, 1990).

If the vertical distance between the camera and the overhead lights is known in advance (or measured by sonar), the robot's lateral offset can also be calculated as follows (King & Weiman, 1990):

$$x_o = -\frac{c\cos\theta}{m\cos\phi} - \frac{c\sin\phi\sin\theta}{\cos\phi}$$

where:

x_o = lateral displacement of the robot from centerline of ceiling lights
c = vertical distance from camera to lights
m = slope of perceived centerline in camera coordinates.

16.3.4 MDARS Overhead-Beam Referencing System

An overhead vision system that combines elements of the Georgia Tech prototype and the doorway penetration system used on ROBART II is under consideration to support autonomous operation of the MDARS Interior robot in completely unstructured warehouse scenarios (Figure 16-21). The lack of definitive rack structures coupled with constantly changing load-out conditions encountered in bulk-storage facilities of this type poses a significant navigational challenge. While the intended aisleways may be predefined and in some cases even marked with paint stripes on the floor, there is no repeatability to the sonar profile created by the stored items. In fact, when portions of the warehouse are temporarily depleted of stock, there very likely is no target surface at all within the sensor's effective range. *Wall-following* and *wall-approach* re-referencing techniques clearly do not apply under these circumstances.

Figure 16-21. Lack of definitive walls or rack structures in bulk-storage facilities pose a significant navigational challenge (courtesy Naval Command Control and Ocean Surveillance Center).

As a consequence, an overhead optical referencing system is thought to be the only practical near-term solution to the needs of unstructured warehouse navigation. Since ceiling heights may vary anywhere from 10 to 60 feet or more, the higher resolution of a CCD-array camera was chosen over a more simplistic active near-infrared retroreflective proximity detector. To simplify the image-processing requirements, the overhead targets would consist of identical sections of 1-inch-wide retroreflective tape several feet in length. The tape sections would be attached to the underside of the roof-support rafters running perpendicular to the path of travel and actively illuminated by a strobe when the robot was in approximate position directly below. As in the case of the Georgia Tech system, the camera iris would be stopped down to eliminate all background returns other than the reflective tape itself.

The length of the installed tape segments would be chosen in conjunction with the fixed field of view of the camera and the local ceiling height to meet the following criteria:

- At least 50 percent of the tape segment is within the camera's horizontal field of view from any possible lateral position of the robot on the path below.
- The total length of the tape segment is slightly shorter than the full horizontal field of view of the camera.

The first requirement ensures there is sufficient length to the detected target image to adequately determine the slope of the line, and hence the platform heading. The second criteria guarantees detection of a tape end point, from which the robot's lateral position can be calculated.

16.4 References

Banner, *Photoelectric Controls*, Product Catalog, Banner Engineering Corp., Minneapolis, MN, 1993a.

Banner, *Handbook of Photoelectric Sensing*, Banner Engineering Corp., Minneapolis, MN, 1993b.

Bohlander, R.A., Holcombe, W.D., Larsen, J.W., "An Advanced AGVS Control System: An Example of Integrated Design and Control," Material Handling Research Center, Georgia Institute of Technology, Atlanta, GA, 1989.

Devore, J. L., *Probability & Statistices for Engineering and the Sciences*, pp. 422-436, Brooks/Cole Publishing Company, 1982.

Everett, H.R., "Survey of Collision Avoidance and Ranging Sensors for Mobile Robots," Technical Report No. 1194, Naval Command Control and Ocean Surveillance Center, San Diego, CA, 29 March, 1988.

Everett, H.R., Gilbreath, G.A., Tran, T., Nieusma, J.M., "Modeling the Environment of a Mobile Security Robot," Technical Document 1835, Naval

Command Control and Ocean Surveillance Center, San Diego, CA, June, 1990.

Gage, D.W., Everett, H.R., Laird, R.T., Heath-Pastore, T.A., "Navigating Multiple Robots in Semi-Structured Environments," ANS 6th Topical Meeting on Robotics and Remote Systems, Monterey, CA, February, 1995.

Holcombe, W.D., Dickerson, S.L., Larsen, J.W., Bohlander, R.A., "Advances in Guidance Systems for Industrial Automated Guided Vehicles," SPIE Vol. 1007, Mobile Robots III, Cambridge, MA, November, 1988.

Holland, J.M., Everett, H.R., Gilbreath, G.A., "Hybrid Navigational Control Scheme, SPIE Vol. 1388, Mobile Robots V, Boston, MA, November, 1990.

Holland, J.M., "An Army of Robots Roams the Night," International Robot and Vision Automation Show and Conference, Detroit, MI, pp. 17.1-17.12, April, 1993.

Kadonoff, Mark B., "Ultrasonic Wall-Following Controller For Mobile Robots", Mobile Robots IV, W.J. Wolfe, W.H. Chun, Editors, Proc. SPIE 1195, pp. 391-401, 1990.

King, S.J., Weiman, C.F.R., "HelpMate Autonomous Mobile Robot Navigation System," SPIE Vol. 1388, Mobile Robots V, Boston, MA, pp. 190-198, November, 1990.

Krishnamurthy, B., Barrows, B., King, S., Skewis, T., Pong, W., Weiman, C., "HelpMate: a Mobile Robot for Transport Applications," SPIE Vol. 1007, Mobile Robots III, Cambridge, MA, pp. 314-320, November, 1988.

Rathbone, R.R., Valley, R.A., Kindlmann, P.J., "Beacon-Referenced Dead Reckoning: A Versatile Guidance System", *Robotics Engineering*, December, 1986.

17
Application-Specific Mission Sensors

The ultimate goal of an autonomous robotic system is, of course, to perform some useful function in place of its human counterpart. Some of the more common applications currently being pursued include:

- Material handling (King & Weiman, 1990; MacLeod & Chiarella, 1993; Mattaboni, 1994).
- Floor cleaning (Bancroft, 1994).
- Physical security (Everett, 1988; George, 1992; Holland, 1993; Gage, et al., 1995).
- Inventory management (ISRA, 1994; Gage, et al., 1995).
- Nuclear and hazardous waste inspection (Byler, 1993; Heckendorn, et al., 1993).

The last three of these categories are probably the more interesting from the standpoint of their mission-specific sensing needs. To reasonably bound the discussion in keeping with the illustrative intent of this chapter, we will examine in detail only those two scenarios being addressed under the ongoing MDARS program: 1) physical security and 2) automated inventory assessment.

17.1 The Security Application

One of the earliest perceived applications for an autonomous mobile robot was acting as an intelligent sentry or security guard. Numerous sensors are readily available to support the detection functions (i.e., fire, smoke, intrusion, toxic gas, flooding, radiation). The ability to maintain an effective security presence under adverse (severe weather, degraded visibility) or even hazardous (nuclear, chemical, and biological) conditions is important, and therefore appropriately addressed by robotic systems. Reliable detection of intruders involves discrimination from background conditions of some property or properties unique to the presence or motion of a human, with sufficient signal-to-noise ratio to minimize the occurrence of nuisance alarms.

Security sensors of this type generally are classified either as *presence sensors*, which can detect a motionless body, or *motion sensors*, which require the intruder to move before detection is possible. A robust solution generally involves evaluation of more than just a single attribute, such as for example:

- Target motion.
- Thermal signature.
- Aspect ratio.
- Temporal history.

Passive detectors for the most part sense a change in ambient conditions due to movement or presence of an intruder within their field of view. This change could be associated with the observed level of illumination, thermal energy, noise, or even vibration normally present in an unoccupied space. *Active detectors* provide a controlled energy input into the observed environment, reacting to changes with respect to a monitored reference as caused by perturbations within the area of coverage. For this reason, active detectors can sometimes be tailored to provide more sensitivity or selectivity in a specific situation.

17.1.1 Acoustical Detection

A simple form of passive detection capability intended primarily for indoor scenarios can be illustrated by the use of a microphone that allows the system to "listen" for sounds in the protected area. Figure 17-1 shows the circuitry employed on ROBART II; an automatic gain control feature in the amplifier stage adjusts to ambient conditions, and any subsequent increase in signal level is detected by the *LM-339* comparator.

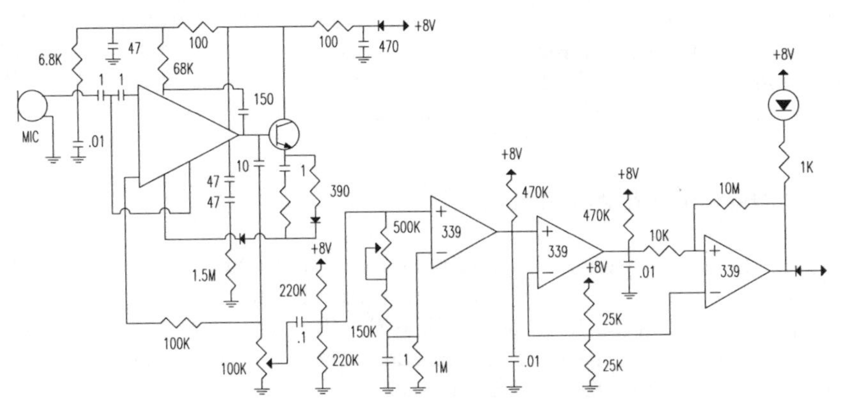

Figure 17-1. Schematic diagram of the acoustical amplifier circuitry used on ROBART II (Everett, et al., 1990).

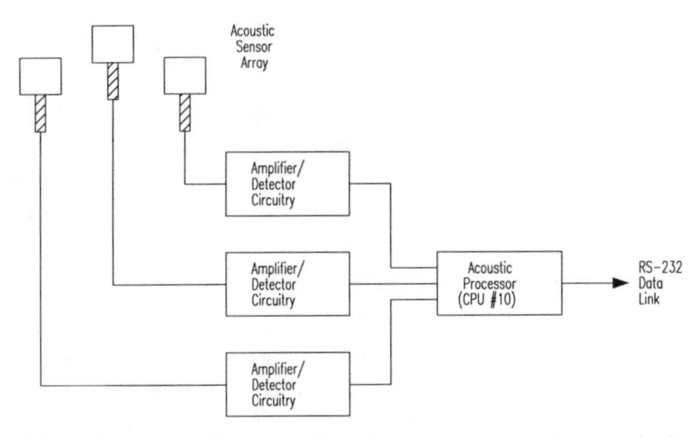

Figure 17-2. Block diagram of the *Acoustic Detection Array* used to passively determine a relative bearing to the source of a detected disturbance.

A three-channel *acoustic detection array* intended to provide bearing information to the source of detected noise was developed using the circuitry presented in Figure 17-1. The sensor array consists of three omni-directional microphones symmetrically oriented 120 degrees apart and separated by a distance *d*. The system will calculate a bearing to a sudden acoustical disturbance when the sound travels across the array and triggers the microphones in a specific sequence, the exact order being dependent on the relative position of the source. A block diagram of the system is presented in Figure 17-2.

Figure 17-3. Three omni-directional microphones situated 120 degrees apart form a passive *acoustic detection array* that can localize the source of a perceived disturbance (courtesy Naval Command Control and Ocean Surveillance Center).

The array is mounted on top of ROBART II as shown in Figure 17-3, with the three transducers individually supported by coil springs. The springs provide some degree of acoustical isolation, while raising the transducers to yield a clear path for wavefront propagation without any blockage by the video camera. Because of the symmetrical orientation, the direction of the disturbance can be classified as being in one of six sectors by examining the detection sequence of the comparators associated with each of the three audio channels.

Each sector is bounded by two lines, the first extending from the array center *O* through the first sensor S_i, to be triggered, and a second originating at *O* and passing through a point B_{xy} midway between the first two sensors detecting the incoming noise. The subscripts *x* and *y* are taken from the first and second sensors to trigger, as depicted in Figure 17-4. Assuming the intruder (source of sound) is some distance away from the robot when initially detected, we can neglect the difference between the robot's height and that of the source with little adverse effect on resultant accuracy, and consider all derivations in a two-dimensional plane. Table 17-1 lists the detection-sequence information for each sector.

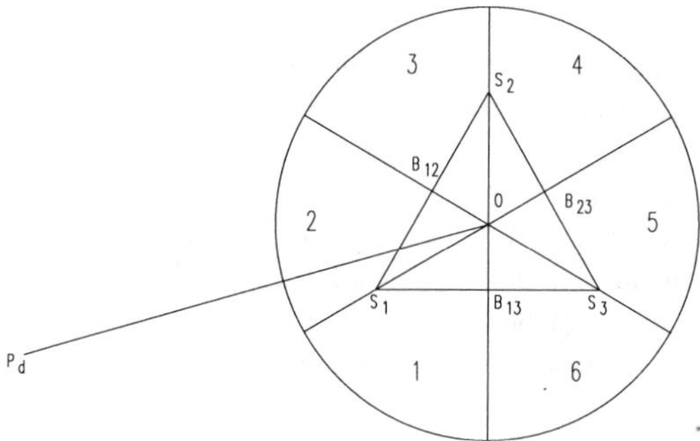

Figure 17-4. Diagram illustrating the relationship in the horizontal plane for acoustic sensors S_1, S_2, and S_3, and the corresponding sectors 1 through 6.

Table 17-1. Sensor firing sequence for the six potential sectors.

Sector #	1st detection	2nd detection
1	sensor #1	sensor #3
2	sensor #1	sensor #2
3	sensor #2	sensor #1
4	sensor #2	sensor #3
5	sensor #3	sensor #2
6	sensor #3	sensor #1

The individual sensor outputs are active low, with a negative transition triggered by the arrival of an incoming noise. Referring to Figure 17-5, the time delay T_1 (between first and second firings) and delay T_2 (between second and third firings) can be measured in order to determine the angle to the source. In keeping with the previously stated assumption that the intruder is some distance away from the sensors when first detected, wavefront propagation can be simplistically modeled as parallel lines perpendicular to a line extending from the array center O to the source of the detected disturbance at P_d. In addition, the speed of sound in air is assumed to be constant over the region of travel involved. For each sector, therefore, it is possible to calculate a bearing to the perceived location of the source, relative to the line segments $OB_{xy}(O_i)$ and to $OS_{xy}(O_i)$. Details of the derivation are provided by Tran (Everett, et al., 1990).

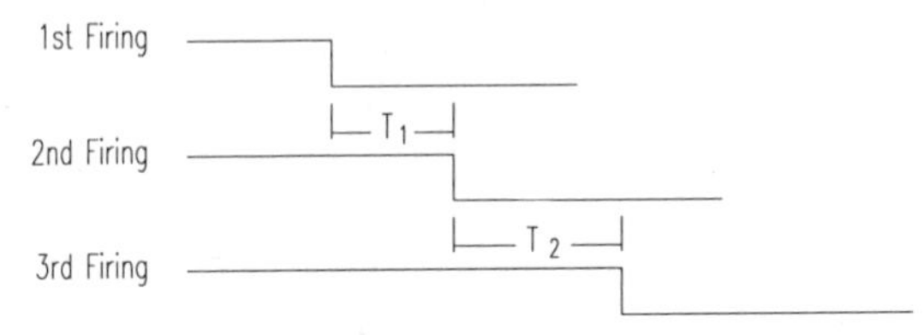

Figure 17-5. The relationship between time delays T_1 and T_2 is used to calculate the bearing to the perceived source.

While the *acoustic detection array* used on ROBART II was very effective in detecting impulse-type disturbances and involved less than $100 worth of hardware, it was completely insensitive to gradually increasing or steady-state noise. A much more sophisticated capability is seen in the *Integrated Acoustic Sensor* system employed on the Surrogate Teleoperated Vehicle (STV) presented in Chapter 1. Developed by SAIC Bio-Dynamics, Eugene, OR, the passive system is designed to alert the remote STV operator to approaching vehicles by detecting the low-frequency sounds of the engine (RST, 1993). The software first establishes a background acoustical signature during power-on initialization and then signals an alarm if the monitored noise level increases above the background threshold in four frequency bands (1-2 KHz, 2-4 KHz, 4-8 KHz, and 8-16 KHz).

Figure 17-6. The *Surrogate Teleoperated Vehicle (STV)* developed by Robotic Systems Technologies employs a 360-degree acoustical sensing array (top-left of mast) manufactured by SAIC Bio-Dynamics (courtesy Naval Command Control and Ocean Surveillance Center).

17.1.2 Vibration Sensors

Vibration sensors are commonly employed as motion detectors in automotive security systems and as window-breakage sensors in fixed-installation alarms. When deployed on a mobile security platform, such devices are usually mechanically coupled to the floor through wheel contact to detect structural vibrations due to footsteps or even earthquakes. ROBART I was equipped with a sensor of this nature made from a 12-inch length of piano wire enclosed in a vertically oriented quarter-inch-diameter section of brass tubing (Everett, 1982a; 1982b). Any vibration of the robot's frame was directly coupled to the tubing, causing the piano wire to jiggle back and forth, generating an electrical signal in a piezoelectric sense element supporting the wire at the bottom of the tube.

Mims (1987) describes a very simple and much more elegant fiber-optic vibration sensor, consisting of a phototransistor that monitors the light output from a short cantilevered length of plastic fiber attached to an LED emitter (Figure 17-7). Any transverse vibration of the housing assembly causes the free end of the fiber to be displaced in alignment from the optical axis of the phototransistor, with a subsequent reduction in light-coupling efficiency. As a consequence, the output signal from the detector is modulated in amplitude at the frequency of the applied vibration. A variation of this vibration sensing technique was incorporated on ROBART II for static security monitoring, in view of its inherent simplicity, low cost, and high output-signal amplitude.

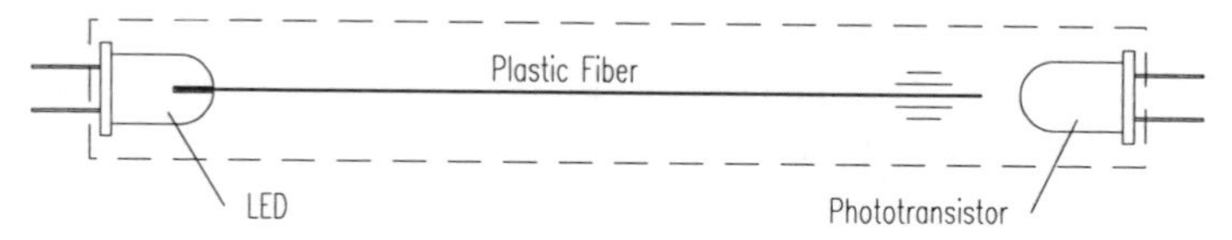

Figure 17-7. The active element in this inexpensive vibration sensor is a cantilevered length of plastic fiber cemented into a small hole in the epoxy housing of an ordinary LED (adapted from Mims, 1987).

17.1.3 Ultrasonic Presence Sensors

The *ultrasonic presence detection* system used on both ROBART II and the ModBot identifies a potential intrusion through changes in measured target distances as seen by one or more sensors in the 24-element *navigational sonar array*. The system creates a reference template consisting of the two most frequently observed range values for each of the individual transducers in the array and then compares subsequent readings to this template. The presence of an intruder within the field of view results in a range value that does not agree with the two possibilities recorded earlier in the reference template. The new range reading corresponds to the distance to the intruder, and the index (position) of the affected sensor within the 360-degree array provides a relative bearing, both of which are used by the host computer to plot the position of the suspected intruder on the map display.

17.1.4 Optical Motion Detection

The Sprague *D-1072* optical motion detector used on ROBART I responds to changes in perceived light level, incorporating a built-in lens to create a cone-shaped detection field (Weiss, 1979; Gontowski, 1983). After a brief settling period upon power-up, the circuit adjusts itself to ambient conditions, and any subsequent deviations from that setpoint will result in an alarm output. The low cost and directional nature of the device allow for several to be used collectively in an array to establish unique detection zones that help locate the relative position of the suspected security violation. The ability to provide geometric resolution of the intruder's position can be invaluable in tailoring an appropriate response in minimal time.

The *D-1072* optical motion detector suffered from three significant drawbacks that limited its utility and contributed to eventual discontinuation: 1) the current consumption of the device was fairly large, 2) it was susceptible to nuisance alarms, and 3) it responded only to visible light. The fact that the chip was incapable of sensing in the near-infrared region of the optical spectrum meant an intruder using an active night-vision device would not trigger an alarm even if the high-power source were pointed directly at the sensor. Interestingly, there are no systems in place even today at most high-security facilities employing elaborate automated equipment to warn guards the area is being illuminated by near-infrared energy.

For this reason, ROBART II was equipped with a dual-element optical motion detector designed specifically for scenarios in which the guarded installation could be under observation by potential intruders armed with a night-vision device employing a near-infrared source. The output of a cadmium-sulfide photosensor (sensitive only to visible light) is compared to an integrated lagging reference voltage derived from the same sensor, such that any change in scene intensity above a specified threshold will be detected. An identical circuit monitors the output of a silicon photosensor shielded by a near-infrared optical filter.

Simultaneous examination of the output states of both circuits reveals the type of lighting involved when motion is sensed (i.e., near-infrared, fluorescent, or incandescent). Fluorescent and incandescent light both produce energy in the visible-light portion of the energy spectrum and will activate the cadmium-sulfide detector, which is not sensitive to near-infrared. Incandescent and near-infrared sources will penetrate the optical filter to activate the broadband silicon detector, but the fluorescent source will be blocked. The following truth table applies:

Table 17-2. Sensed energy derived from detector status.

Activated detector	Sensed energy
Cadmium-sulfide	Fluorescent
Silicon	Near-infrared
Both	Incandescent

17.1.5 Passive Infrared Motion Detection

A significant development in security sensor technology is seen in the *passive infrared (PIR)* motion detector. Originally designed for both indoor and outdoor fixed-installation security systems, this type of *pyroelectric sensor* quickly found application on mobile robots due to its small size, low power consumption, and excellent performance and reliability characteristics (Everett, 1982a; 1982b; Quick, 1984). *PIRs* routinely exhibit remarkably low nuisance-alarm rates in indoor environments but can sometimes be triggered by gusty wind conditions when employed outdoors. The principle of operation as a motion detector is similar to the optical sensor described in the previous section, except a different range of wavelengths (7-16 micrometers) in the energy spectrum is being sensed.

Recall from Chapter 9 all objects with an absolute temperature above 0°K emit radiant energy in accordance with the Stephan-Boltzman equation (Buschling, 1994):

$$W = \varepsilon \sigma T^4$$

where:

W = emitted energy
ε = emissivity
σ = Stephan-Boltzman constant (5.67×10^{-12} watts/cm^{2}K^4)
T = absolute temperature of object in degrees Kelvin.

The *emissivity* of human skin is very close to unity (0.98) and the same for all races (Cima, 1984). A typical human gives off somewhere between 80 and 100 watts of radiant energy with a peak wavelength around 10 micrometers (Cima, 1990), thus producing a distinctive thermal signature.

Ordinary glass is 100-percent opaque at wavelengths longer than 5 micrometers, and therefore useless as a lens or window material for this application (Barron, 1992). Early *PIR* detectors (such as the unit used on ROBART I) employed fairly expensive germanium or zinc-selendide lenses, but more recent devices take advantage of high-density polyethylene Fresnel lenses, which are 60- to 80-percent transmissive at the wavelengths of interest. Fresnel lenses provide an equivalent degree of refraction in a much thinner package than conventional lenses, due to the discretized nature of their construction, and therefore absorb much less infrared radiation (Viggh & Flynn, 1988). An optical filter between the lens and sensing element restricts the admitted wavelengths to the region associated with human emission (i.e., 7-16 micrometers).

Typical *pyroelectric sensing elements* are thin-wafer slices of either lithium tantalate (Cima, 1984), ceramic (Philips, 1985), or polymer film (Tom, 1994) materials, with metallic electrodes deposited on opposite faces. The so-called *pyroelectric effect* arises due to thermally induced changes in polarization of the wafer or film (Philips, 1985). Incident photons absorbed by an exposed face heat the sandwiched material, generating a small charge that is in turn collected by the electrodes (Cima, 1984). The sense elements are made as thin as possible to minimize their thermal inertia for improved dynamic response.

Pyroelectric detectors can be thought of as "self-generating capacitors" (Eltec, 1993), with the voltage differential across the sense element given by (Philips, 1985):

$$v = \frac{Q}{C}$$

where:

v = voltage developed across the electrodes
Q = induced charge due to pyroelectric effect
C = effective capacitance of the detector.

As discussed in Chapter 11, current leakage paths in the sensor and associated electronics cause the voltage generated across the capacitive element to decay with time (Russell, 1993), and so the output is really proportional to the *change* in incident radiation. The pyroelectric sensor thus functions normally as a *motion detector* and not a *presence detector*.

In conventional security applications, the device is mounted such that the sense element "stares" at a stable thermal field of view and responds only when a moving entity disturbs the magnitude and distribution of incident photons (Cima, 1984). Most commercially available systems incorporate opposed-output dual-element detectors that provide common-mode rejection of global disturbances (Figure 17-8A), as was briefly discussed in Chapter 11. When a human target moves left to right through the sensor's field of view, the focused concentration of photons in the image plane moves right-to-left across the two detector elements. The lens geometry is such that the incident radiation falls almost exclusively at first on the right-hand detector, reaches a balance between the two as the intruder crosses the optical axis, and then shifts with continued motion to where the left sense element dominates. The resulting output signal is plotted as a function of time in Figure 17-8B. If the "intruder" stops moving at any point while still in view, the detector will settle out to equilibrium with no appreciable output signal.

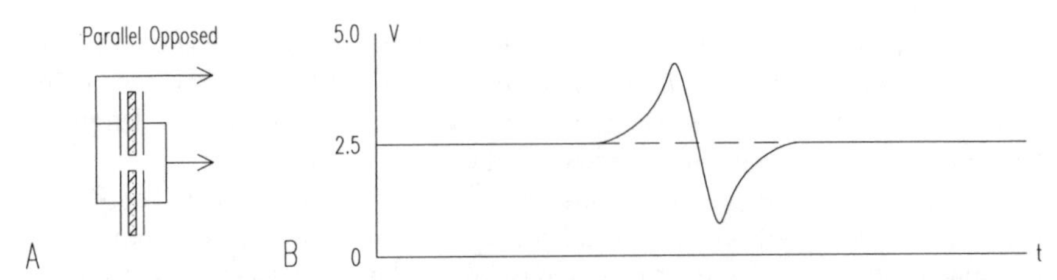

Figure 17-8. A typical output signal (B) of a dual-element detector (A) showing the characteristic rise and fall signature relative to the 2.5-volt equilibrium (adapted from Jones & Flynn, 1993).

Overall device sensitivity can be improved through use of multifaceted lenses that create a number of radial detection zones separated by blind alleys as shown in Figure 17-9A. An intruder moving laterally across these zones will inherently generate a sharp differential signal in the detector when going from a monitored to an unmonitored area, and vice versa (Philips, 1986). Vertical "stacking" of lens sections is often employed to break the foreground and background into different zones (Figure 17-9B), thereby enhancing the probability of detection for an intruder moving radially towards or away from the sensor (Cima, 1984).

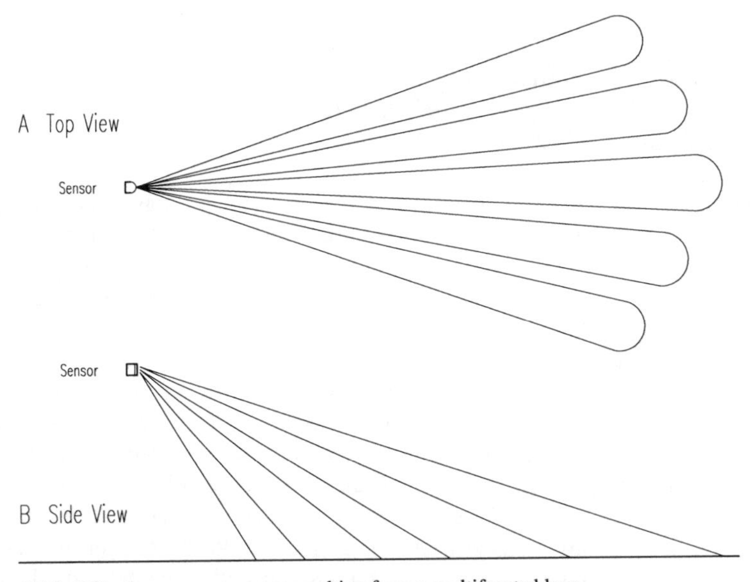

Figure 17-9. Effective coverage areas resulting from a multifaceted lens.

Some security-sensor manufacturers supply special lenses that create a so-called "pet-alley" dead-zone for blocking detection near the floor, in order to minimize nuisance alarms due to the harmless movement of indoor pets. While this approach may work fairly well for dogs, cats routinely aspire to lofty perches from which to survey their domain, and probably become even more adventuresome when their owners are away. "Quad-element" detectors have been introduced in attempts to distinguish among different sizes of stimuli (i.e., small animals versus humans), as part of a growing trend to achieve higher-resolution discrimination at the smart-sensor level (Ademco, 1989; Nippon, undated).

Eltec Instruments, Daytona Beach, FL, produces a 32-element pyroelectric array (Eltec, 1993). Tom (1994) describes a 16-element sensor design based on the *ferroelectric polymer* polyvinylidene fluoride, manufactured by AMP, Inc., Valley Forge, PA. The company's *PIRL180-100* detector incorporates a six-element polymer-film detector with a custom Fresnel lens providing 180-degree coverage divided into 20 discrete beams.

The relative performance of different types of pyroelectric detectors can be reasonably compared using a number of industry-established parameters. The *responsivity* of an IR sensor is defined as the resultant signal voltage per watt of incident radiation (Nippon, undated; Eltec, 1984):

$$R = \frac{V_s}{HA} = \frac{\Gamma}{d\omega \, C_d}$$

where:

R = *responsivity*
V_s = rms value of signal voltage
H = rms value of incident radiation (W/cm^2)
A = active sensor area
Γ = material parameter
d = electrode separation
$\omega = 2\pi f$ (chopping frequency of interrupted input radiation)
C_d = effective capacitance.

Noise equivalent power (NEP) is the radiant flux required to produce an output signal equal in magnitude to the sensor noise (Nippon, undated), or in other words, the *noise* divided by the *responsivity* (Eltec, 1984):

$$NEP = \frac{V_n \, HA}{V_s}$$

where:

NEP = *noise equivalent power* (watts)
V_n = rms value of noise signal.

The ultimate sensitivity of an IR detector is determined by its signal-to-noise ratio (Eltec, 1984). *Detectivity-star (D*)* is a term used to denote the sensor's signal-to-noise ratio when 1 watt of incident radiation falls on a 1-cm^2 sensor, and the noise is measured with a 1-Hz electrical bandwidth (Nippon, undated):

$$D^* = \frac{\sqrt{A\,\Delta f}}{NEP}$$

where:

D^* = detectivity-star
Δf = electrical bandwidth of measuring circuit.

The D^* parameter in effect normalizes the *NEP* to a given constant area for more equivalent comparison of different types of detectors (Eltec, 1984). The absolute temperature T of the blackbody radiation source, chopping frequency f, and electrical bandwidth Δf must be specified for meaningful results. (Standard values are 420°K, 1 Hz, and 1 Hz.) The larger the value of D^* the better.

Eltec *Model 442 IR-EYE* Integrated Sensor

The *Model 442 IR-Eye* pyroelectric sensor manufactured by Eltec Instruments, Inc., Daytona Beach, FL, is a parallel-opposed dual-detector configuration with integral analog signal processing (Eltec, 1991). Lithium tantalate ($LiTaO_3$), a non-hydroscopic single-crystal material that maintains its pyroelectric properties to a Curie point of 610°C, was chosen for its demonstrated sensitivity and stability (Eltec, 1993). The *Model 442* is the only current-mode (transimpedance) dual-element detector commercially available. Selected specifications for the basic sensor are provided in Table 17-3. For long-range operation in exterior settings (Cima, 1992), the sensor is incorporated into the *Model 862 Passive Infrared Telescope*, with narrow field-of-view ranges out to 500 feet.

Table 17-3. Selected specifications for the *Model 442 IR-Eye* detector (courtesy Eltec Instruments, Inc.).

Parameter	Value	Units
Spectral response	8-14	micrometers
NEP	1.1×10^{-9}	watts
D*	2.2×10^{8}	√Hz/watt
Responsivity	3.7×10^{5}	volts/watt
CMR (Minimum)	5/1	
(Maximum)	15/1	
Noise	0.36	millivolts/√Hz
Power	5-15	volts DC
	2	milliamps
Housing	TO-5	
Size (diameter)	.360	inches
(height)	.190	inches

Nippon Ceramic *Model SEA02-54* Pyroelectric Sensor

The *Model SEA02-54* pyroelectric sensor manufactured by Nippon Ceramic Co., Ltd., Tottori, Japan, is a series-opposed dual-detector configuration based on a ceramic ferroelectric material (Nippon, 1991). Typical motion-detection applications include occupancy sensing for lighting and air conditioning control, visitor enunciation, and security systems. The *Model SEA02-54* sensor and a lower-cost version (*Model RE200B*) are distributed in the United States by McGee Components, Inc., North Attleboro, MA. Selected specifications are listed in Table 17-4 below.

Table 17-4. Selected specifications for the *Model SEA02-54* and *Model RE200B* pyroelectric sensors (courtesy McGee Components, Inc.).

Parameter	*SEA02-54*	*RE200B*	Units
Spectral response	7-14	5-14	micrometers
NEP	8.8×10^{-10}	9.6×10^{-10}	watts
D*	1.6×10^{8}	1.5×10^{8}	√Hz/watt
Responsivity	3.2×10^{3}	3.3×10^{3}	volts/watt
Noise	70	80	millivolts/√Hz
Power	2.2-15	2.2-15	volts DC
	12	12	microamps
Housing	TO-5	TO-5	
Size (diameter)	9.2	9.2	millimeters
(height)	4.8	4.5	millimeters

Scanning PIR Configurations

The fact that pyroelectric sensors respond only to changes in thermal energy is actually an advantage in fixed-installation security systems, since an intruder must at some point move into the sensor's field of view to be considered a potential threat. In the case of a mobile security robot, however, this is not always the case. An intruder can easily enter a secured zone before the robot arrives on scene, and simply has to remain motionless (or hidden) when the area comes under temporary surveillance. Once the robot departs, the intruder is free to resume his or her clandestine activities until such time as the robot makes another patrol.

For this reason, investigations have been conducted into potential ways of operating pyroelectric sensors as *human presence detectors*. For example, a number of researchers have attempted to accommodate the differential nature of the pyroelectric sensor by scanning the device about the vertical axis, thus enabling detection of stationary thermal sources. The single PIR sensor incorporated on ROBART I was mounted on the robot's head to facilitate panning slowly back and forth in search of non-moving intruders (Everett, 1982a).

A commercially available scanning configuration is seen in Cybermotion's *Security Patrol Instrumentation (SPI)* module employed on their *SR2* robot (Holland, 1993). The *SPI* incorporates a scanning sensor array rotated by a small DC motor at about 60 rpm. Slip-ring connections are provided for four sensor modules spaced 90 degrees apart on the rotor assembly:

- Passive infrared vertical array.
- Visible-light vertical array.
- Continuous-wave K-Band microwave motion detector.
- Ultraviolet flame detector.

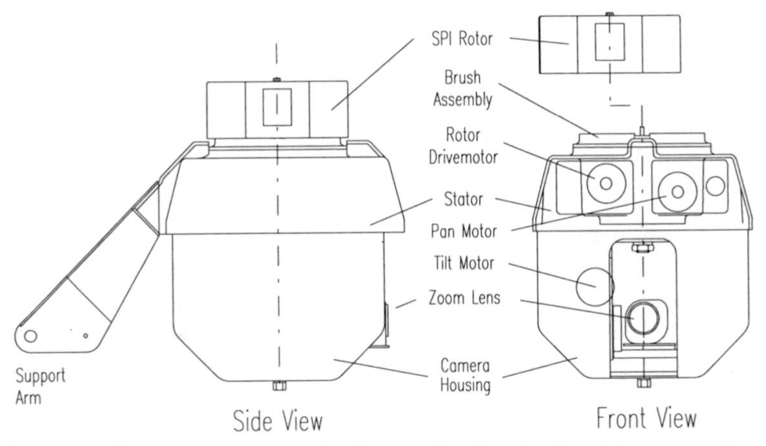

Figure 17-10. Schematic drawing of the Cybermotion *Security Patrol Instrumentation (SPI)* module with integrated surveillance camera pan-and-tilt (courtesy Cybermotion, Inc.).

The passive infrared vertical array consists of four Eltec *Model 442* sensors stacked to achieve an instantaneous field of view of 5.6 degrees horizontal and 31.6 degrees vertical. A *target identification algorithm* extracts perceived movement from scan-to-scan thermal differences and passes the results to a *target tracking algorithm*, where the highest probability targets are closely monitored. The company describes this hardware/software combination as neither a *motion detector* nor a *presence detector* in the strictest sense, but more of a "change-of-presence detector."

Viggh & Flynn (1988) describe a continuously rotating pyroelectric detector implementation on the MIT robot *Seymour* that incorporates a pair of narrow field-of-view sensors, also based on the Eltec *Model 442* detector. A *synthetic field of view* defined by the leading-edge detections associated with each of the two sensors is created by the scanning motion as illustrated in Figure 17-11. An output signal similar to that shown earlier in Figure 17-8 will be generated each time a *442* sensor sweeps across a high-contrast thermal source. The elapsed time interval between detections by the individual sensors is a function of the separation distance between the robot and the perceived hot spot, due to the diverging nature of the synthetic field of view, and thus indicative of range to a suspected human target. (The MIT system was expressly developed to support a human-following behavior routine.)

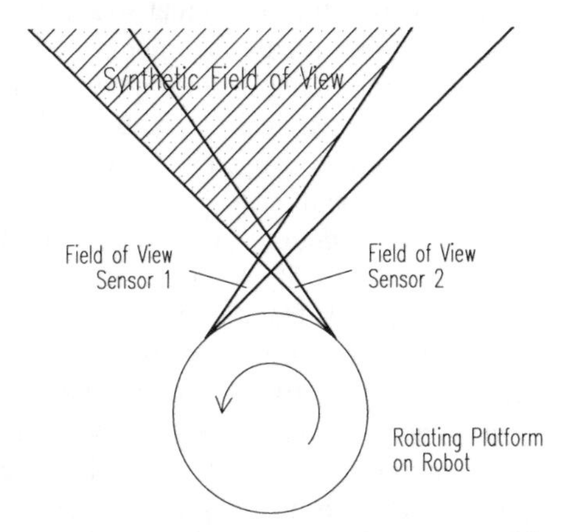

Figure 17-11. A rotating synthetic field-of-view created by two revolving *PIR* sensors yields a varying time delay between leading-edge detection of a stationary person that is proportional to range (adapted from Viggh & Flynn, 1988).

17.1.6 Microwave Motion Detection

Microwave motion detectors operating at radio frequencies rely on the *Doppler shift* introduced by a moving target (Chapters 8 and 9) to sense the relative motion of an intruder. Commercially available systems are generally of two types: 1) *continuous wave* and 2) *pulsed.* Most *continuous-wave* systems employ a *Gunn diode* serving as both transmitter and local oscillator, and a *Schottky-barrier mixer-diode* receiver feeding a low-noise intermediate-frequency (IF) amplifier (Alpha, 1987). *Pulsed* systems incorporate an additional square-wave driver as shown in Figure 17-12 to regulate the duty cycle (i.e., 10 percent typical) to conserve power.

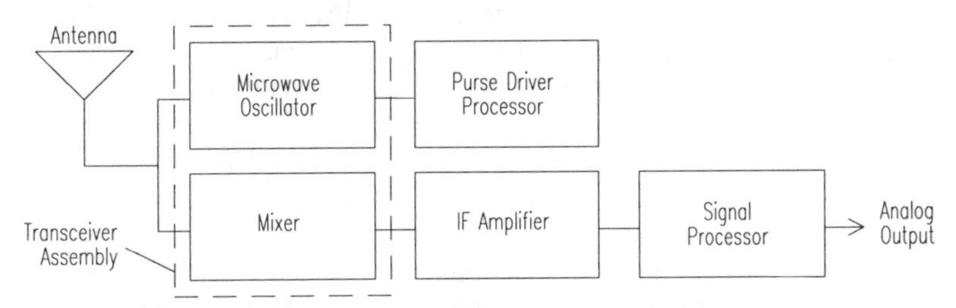

Figure 17-12. Block diagram of a typical pulsed Doppler microwave motion detector (adapted from Williams, 1989).

Both *continuous-wave* and *pulsed* configurations generally employ a common transmit and receive antenna. The *mixer diode* combines reflected energy from the target with a reference signal from the *Gunn oscillator*, generating an output signal with an amplitude that varies as a function of the phase difference between its two inputs, and with a frequency proportional to the relative radial velocity (Williams, 1991). The IF amplifier usually incorporates 60- and 120-Hertz notch filters to eliminate potential interference from power lines, as well as the fluctuations of ionized gas in fluorescent light fixtures (Williams, 1989; 1991). Some form of threshold detection is employed at the analog output to signal radial motion above a pre-established setpoint.

The electromagnetic energy emitted by microwave motion sensors can penetrate hollow walls and doorways, allowing the sensor to "see" into adjoining rooms in certain circumstances. This feature can be used to advantage by a robot patrolling a hallway to check locked office spaces and storerooms without need for entry. There is some inherent sensitivity to nuisance alarms, particularly in the presence of rotating machinery, or in scenarios where building walls and/or windows are subject to vibration from overflying aircraft or vehicular traffic. The effective detection of intruders is dependent on the degree of radial motion, target range, background interference, and the target's *effective cross-section* σ (Chapter 9). Human targets typically have a value of σ between 0.2 and 2.0 meters^2 (Alpha, 1987).

17.1.7 Video Motion Detection

Vision systems offer a sophisticated method of sensing intrusion in outdoor as well as indoor applications, with the added benefits of excellent resolution in the precise angular location of the intruder. A surveillance camera is used to digitize a scene for comparison with a previously stored image pattern representing the same region, and significant deviations between the two can be attributed to motion within the FOV. "Windowing" techniques can be employed on most systems to selectively designate certain portions of the image to be ignored (such as a tree blowing in the wind), resulting in a significant reduction in nuisance alarms. Simple algorithms that distinguish purposeful from random motion can further refine this discrimination feature at the intelligent-sensor level. Calculated boundaries of the perceived disturbance within the image frame can be used to automatically reposition the camera in closed-loop systems to keep it centered on the activity of interest.

The first step in the implementation of a video motion detection capability is of course to acquire the image. Traditional digital approaches employ a "frame grabber" to convert an entire two dimensional image into a corresponding digitized array that can be stored in computer memory. For every pixel there exists an associated RAM location describing scene intensity (gray levels) at that particular location. For a conventional 525-line television image with 512 pixels-per-line horizontal resolution, this equates to 268,800 memory locations (bytes). In other words, over a quarter megabyte of memory is required to store a single frame of video. The second step involves processing the data that has just been acquired. Due to the array size, even the simplest of operations, such as thresholding to convert to a binary image, is time consuming. More elaborate operations simply cannot be done using simplistic hardware before the next image arrives at the NTSC rate of 60 frames per second.

Experiments with the linear CCD-array cameras employed on ROBART II (see Chapter 4) showed it was possible to detect motion by examining only one horizontal line cutting through the region of interest in the scene. If several horizontal lines equally spaced throughout the scene are acquired, effective full-screen coverage could be achieved without the need to "grab" the entire frame. The image processing needs would be greatly reduced, and in most cases could be performed during the wait period between lines of interest when the acquisition system was idle. As an example, if only a single line is sufficient, the memory requirement would be reduced to half a kilobyte, with 16.4 milliseconds available for processing before the next line must be digitized.

The *Reconfigurable Video Line Digitizer* was developed for just this purpose, and consists of a high-speed (100-nanosecond conversion time) analog-to-digital (A/D) converter which samples the composite video signal of a conventional NTSC-format camera output. The composite video is also fed to a sync separator that splits off the horizontal and vertical sync pulses and provides a frame index (Figure 17-13). (Note: the single-chip *AD9502BM* video digitizer available from Analog Devices (Hansford, 1987) contains a flash A/D converter and integrated sync strippers.) The horizontal sync pulses drive a counter that identifies the scan line of interest in the scene, whereupon line digitizing is performed. The digital output of the A/D converter is written directly into dual-buffered high-speed (35-nanosecond) video RAM, in order that it might be accessed later by the microprocessor when the A/D is idle.

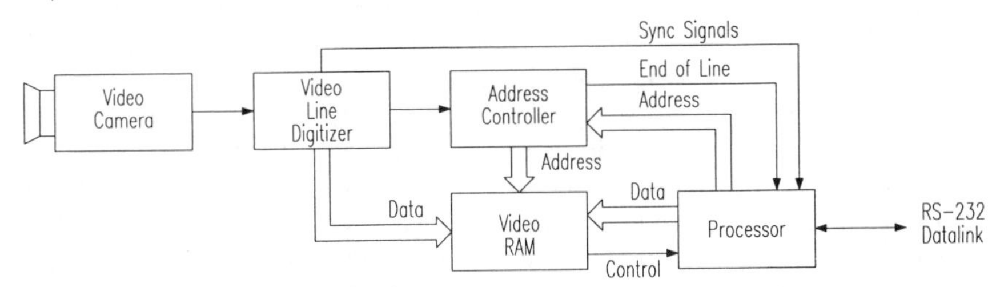

Figure 17-13. Block diagram of the video line grabber developed for use on ROBART II.

The most simplistic motion detection algorithm involves subtracting the latest intensity array from a previously acquired array, and reacting to any significant discrepancies indicative of changes in the scene under observation. In reality, some software filtering is required to eliminate noise and reduce the occurrence of nuisance alarms, but this is easily accomplished on a 512-element linear data array in the time available. (For simple motion detection schemes, 256 elements of horizontal resolution are more than adequate, further reducing required system complexity.)

Assuming full 512-pixel coverage, only 2K bytes of RAM are sufficient to support the microcomputer operating system and to save three select lines of video data, which normally would be equally spaced across the region of interest. Once motion is detected in any of the three lines, it is possible to select new lines for the next motion analysis operation. If these lines are chosen in such a fashion around the vicinity of the initially detected disturbance, it is possible over successive frames to converge on and effectively bound the area perturbed by the intrusion. In this fashion, the system can detect and output information describing the geometric area involved so as to provide servo-control inputs for camera positioning or robot motion algorithms.

17.1.8 Intrusion Detection on the Move

Most commercially available security sensors operate through detection of relative motion, and must be attached to a stable (i.e., non-mobile) mounting. In the early 1980s, Denning Mobile Robots experimented with a specially configured *microwave motion detector* developed by Alpha Industries in an attempt to address this problem (Everett, 1988). The Alpha sensor employed a programmable notch filter that theoretically could be set to filter out the Doppler component introduced by forward travel of the platform itself. In reality, however, this simplistic notch filter approach proved to be a bit idealistic, due to the off-axis velocity components arising from the cone-shaped nature of the beam and the presence of multiple harmonics. An alternative (but more costly) strategy may be to process the analog output signal with a pattern-matching algorithm trained to distinguish the less-structured Doppler components generated by a moving intruder from the more predictable results attributed to platform motion.

A more realistic possibility in the meantime would be to consider a much cheaper alternative that could potentially provide partial coverage using off-the-shelf components. *Microwave motion detectors* that discriminate direction of motion employ two receiving diodes spaced a quarter wavelength apart (Figure 17-14), so the Doppler outputs from the two mixers are 90 degrees out of phase (Williams, 1989). Conventional phase-quadrature techniques such as used in incremental encoders and interferometers can then determine direction of target motion. A common application for these direction-discriminating microwave devices is automatic traffic control, where it is necessary to distinguish between cars approaching a light and those moving away. For example, the AM Sensors *Model MSM10502* can be preset to respond only to objects moving toward the sensor, or alternatively away from the sensor.

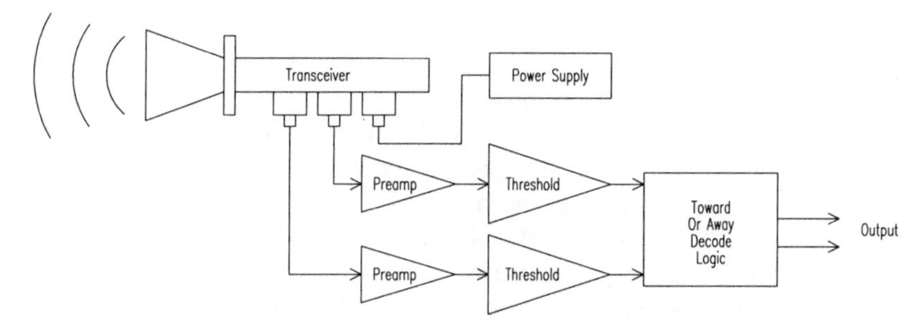

Figure 17-14. The use of two mixer diodes spaced a quarter-wavelength apart (90 degrees) allows for determination of the direction of relative motion (adapted from Williams, 1989).

Assuming stationary-object Doppler components due to forward motion of the platform (while admittedly consisting of a multitude of frequencies) are all indicative of relative movement towards the platform, the possibility for detecting intruder motion away from the platform seems obvious. A forward-looking direction-discrimination sensor, preset for receding objects only, should theoretically ignore the effects of platform motion altogether, yet respond to any mobile targets moving away from the vehicle. Similarly, a rear-facing sensor preset for approaching objects would likewise detect any mobile target gaining on the vehicle from behind. While there are obvious limitations to this detection strategy, they fortunately are somewhat aligned with the least vulnerable of potential intruder response scenarios. For example, an intruder in front of an approaching MDARS platform is most likely going to retreat rather than advance toward the vehicle, and in so doing becomes susceptible to detection.

The above approach seems worthy of further investigation as a low-risk interim solution for motion detection on the move. The most likely longer term candidate technology for truly solving the problem is probably image processing, based on video obtained from a *FLIR (forward-looking infrared)*, or a low-light-level or image-intensified CCTV camera. Conventional *FLIRs* and image-intensified cameras are not cheap, however, and the required image processing hardware is expensive as well. Recent developments in *uncooled FLIR* technology may hold promise for significantly reduced costs in the near future.

Texas Instruments Ferroelectric Focal Plane Array

Texas Instruments has been a principle contender in the recent development of a new generation of FLIRs that do not require cryogenic cooling of the detector element. Cryogenic cooling of conventional FLIRs adds to system complexity and cost, with a significant decrease in reliability, as most coolers have a *mean time between failure* of around 1500 hours. In addition, the initial cool-down period required after system start-up (before the detector becomes operational) can significantly hamper quick response in security applications.

The principle of operation for the Texas Instruments' *focal plane array* is based upon the induced pyroelectric effect in the ferroelectric-ceramic *barium-strontium titanate (BST)* near its phase transition (Hanson & Beratan, 1994). While the sensor is billed as an "uncooled" FLIR, a solid-state thermoelectric cooler is employed to keep the material stabilized at the 22°C transition temperature. This requirement is much less of a burden than the need to cool down to around 75°K, however, and solid-state coolers are considerably cheaper and more reliable than cryogenic systems.

The 328- by 245-element *detector* consists of an array of ceramic capacitors bump-bonded to CMOS VLSI readout circuitry containing a dedicated preamplifier, noise filter, buffer, and switch for each pixel (Hanson, et al., 1993). A row-address shift register and a column multiplexer are included into the IC design to generate a serial output format that can be easily processed to yield a standard composite-video signal. The *IR absorber* (Figure 17-15) consists of a three-layer resonant cavity, where the bottom layer is an opaque metal film that also serves as the common upper plate for the pyroelectric capacitors. A 1.25-micrometer-thick intermediate layer of *parylene* is used to tune the cavity for an absorption peak of approximately 10 micrometers (Hanson & Beratan, 1994). The detector face is a thin semi-transparent metal layer that matches the optical impedance of the cavity to free space.

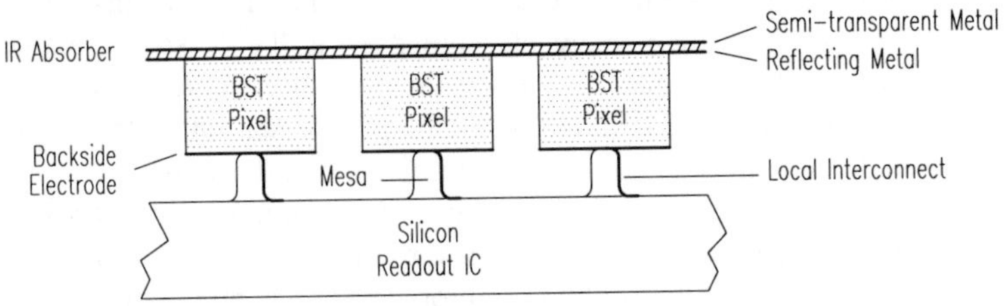

Figure 17-15. The Texas Instruments *focal plane array* is reticulated to reduce thermal crosstalk between pixels, and thermally isolated from the readout electronics by polyimide mesas (adapted from Hanson, et al., 1993).

Since pyroelectric sensors do not have a true DC response as previously discussed, a mechanical chopper is employed in the current design, with successive frames differenced to produce a final image. This technique produces a sort of halo effect around the edges of high-contrast (i.e., significant temperature differential) image features. While this artifact is of minimal concern in conventional surveillance scenarios, it can potentially interfere with automated motion-detection functions associated with robotic security equipment.

Texas Instruments received a Phase I *Low Cost Uncooled Sensor Program (LOCUSP)* contract award in 1990 to produce prototype weapons sights for the US Army, resulting in a demonstrated average *noise equivalent temperature difference (NETD)* of 0.08°C with *f*/1.0 optics (Hanson, et al., 1992). A *LOCUSP* Phase II contract was awarded in 1994 to further improve performance (Hanson &

Beratan, 1994). In an effort to transition this military-oriented technology into civilian law enforcement applications, the *Nightsight Thermal Vision System* (Figure 17-16) was jointly developed by Texas Instruments and Hughes Aircraft Company (TI, 1994). The goal was to produce a low-cost (approximately $6,000) night-vision system for installation in police cruisers.

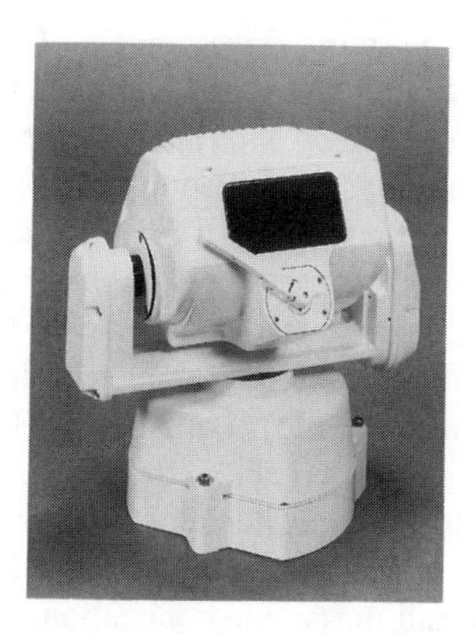

Figure 17-16. The *Nightsight Thermal Vision System* with integrated pan-and-tilt unit is available for crossbar installation on police cruisers for under $10,000 (courtesy Texas Instruments, Inc.).

Table 17-5. Selected specifications for the *Nightsight* camera (TI, 1994).

Parameter	Value	Units
Pixels	80,360	
Format	328 x 245	
Field-of-View	27 (V) by 18 (H)	degrees
NETD	< 0.12	degrees C
Detection Range (human)	20 - 900	feet
Video Interface	RS-170/NTSC	
Update rate	60	Hz
Image Polarity	selectable	
Warm up time	< 30	seconds
Power	9 - 16	volts DC
	6	watts
Size (length)	8	inches
(width)	6	inches
(height)	6	inches
Weight	< 8	pounds

The *Nightsight* kit consists of an uncooled IR sensor, pan-and-tilt unit, video display, control console, and the required interface hardware. The front window of the environmentally sealed sensor housing is equipped with an automatic defroster and wiper blade. Selected specifications are listed in Table 17-5.

Alliant Techsystems Microbolometer Focal Plane Array

Honeywell's Sensor and System Development Center, Bloomington, MN, developed in the early 1980s a revolutionary approach to high-resolution uncooled focal plane arrays, based on *silicon microbolometer* technology. In 1990, the company's former Defense Systems Group was spun off to form Alliant Techsystems and received a Phase I *LOCUSP* contract award to develop a low-cost battlefield surveillance sensor prototype (Gallo, et al., 1993).

Silicon micromachining techniques are employed to fabricate large arrays of individual detectors operating as microbolometers in the 8 to 12 micrometer region of the infrared spectrum (Gallo, et al., 1993). The 336- by 240-pixel array is made up of tiny masses suspended above the IC face by two support struts that provide excellent thermal isolation as depicted in Figure 17-17. Temperature rise due to incident radiation is quantified by measuring the resistance of a region of vanadium oxide on each small mass; the connections run down the support struts to the integrated readout electronics for each pixel. No cooler is required for operation, but a simple thermoelectric device is used to minimize the effect of thermal drift. The entire device is fabricated as a monolithic silicon IC, with a final etch step creating the gap under the masses.

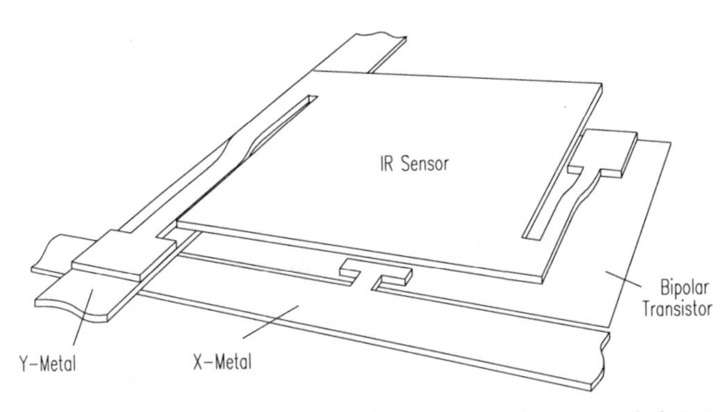

Figure 17-17. Typical microbridge detector element employed in the *microbolometer focal plane array* (adapted from Gallo, et al., 1993).

A *NETD* well under 0.1°C is achievable, along with a pixel-to-pixel thermal isolation of -142 dB, making "blooming" virtually non existent (Gallo, et al., 1993). The sensor's true DC response eliminates the need for a mechanical chopper, for improved reliability with reduced size, weight, and power consumption. In addition, the achievable *NETD* relative to comparable ferroelectric designs employing choppers is improved by a factor of two for the following reasons (Gallo, et al., 1993):

- The incident radiation is not blocked every other frame, for an improvement factor of 1.4.
- The sensor frame rate can be reduced from 60 Hz to 30 Hz, for an additional improvement factor of 1.4.

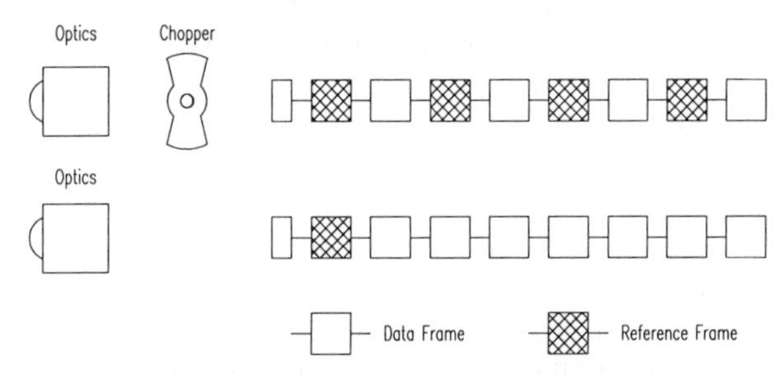

Figure 17-18. Chopperless operation provides twice the *noise equivalent temperature difference (NETD)* over comparable ferroelectric designs (adapted from Gallo, et al., 1993).

The basic *LOCUSP* sensor head comes with a dual-mode optics assembly that can be remotely operated to select either a wide (15- x 9-degree) or narrow (5- x 3-degree) field of view. Selected specifications are provided in Table 17-6 below.

Table 17-6. Selected specifications for the Alliant Techsystems uncooled FLIR (Gallo, et al., 1993).

Parameter		Value	Units
Pixels		82,320	
Format		240 x 336	
Field-of-View	(wide)	15 (V) by 9 (H)	degrees
	(zoom)	5 (V) by 3 (H)	degrees
NETD		< 0.1	degrees C
Detection Range (human)		4,920	feet
Video Interface		RS-170/NTSC	
Update rate		30	Hz
Image Polarity		selectable	
Warm up time		< 10	seconds
Power		24	volts DC
		12	watts
Size	(length less optics)	5.1	inches
	(width)	5.3	inches
	(height)	7.1	inches
Weight (less optics)		< 7.3	pounds

17.1.9 Verification and Assessment

Potential security functions assigned to a mobile sentry robot can be categorized into four general areas: (1) *detection*, (2) *verification*, (3) *assessment*, and (4) *response* (Everett, 1988). *Detection* is readily addressable by a multitude of commercially available sensors of the type presented above. *Verification* involves cross-checking with other sensors to lessen the chances of a nuisance alarm and depends heavily upon both the types of detectors employed and the operating environment. The *assessment* task acts upon the data collected to ascertain the nature of the disturbance, usually to determine if a response is necessary. The *response* itself must be tailored to the application, the operating scenario, and the nature of the situation.

The traditional problem encountered in applying off-the-shelf intrusion sensors in an automated security system has been as the detector sensitivity is raised to provide the necessary high probability of detection, there is a corresponding unacceptable increase in the nuisance alarm rate. Operators quickly lose confidence in such a system where sensors are prone to false activation. For example, passive infrared motion detectors can be falsely triggered by any occurrence that causes a localized and sudden change in ambient temperature within the sensor's coverage area. This false triggering can sometimes occur naturally, as in the case of an exterior sensor viewing trees blowing in the wind. Optical motion detectors can be activated by any change in ambient light level, as could be caused by passing automobile headlights or lightning flashes. Discriminatory hearing sensors could be triggered by loud noises originating outside the protected area, such as thunder, passing traffic, or overflying aircraft. Microwave motion detectors can respond to rotating or vibrating equipment, and so forth.

A truly robust robotic or automated security system must employ a variety of intrusion detection sensors and not rely on any single method. This redundancy thwarts attempts to defeat the system, due to the higher probability of detection with multiple sensors of different types. Equally as important, such redundancy provides a means of verification to reduce the occurrence of nuisance alarms (i.e., redundant intrusion detection schemes operating on different principles will not all respond to the same spurious interference). The strategy employed on ROBART II involves using numerous types of broad coverage sensors (Figure 17-19) as primary detection devices, and higher-resolution units in a secondary confirmation mode to verify and more clearly characterize a suspected disturbance. The robot is alert at all times, but its acuity can be enhanced by self-generated actions that activate these additional systems when needed to better discriminate among stimuli.

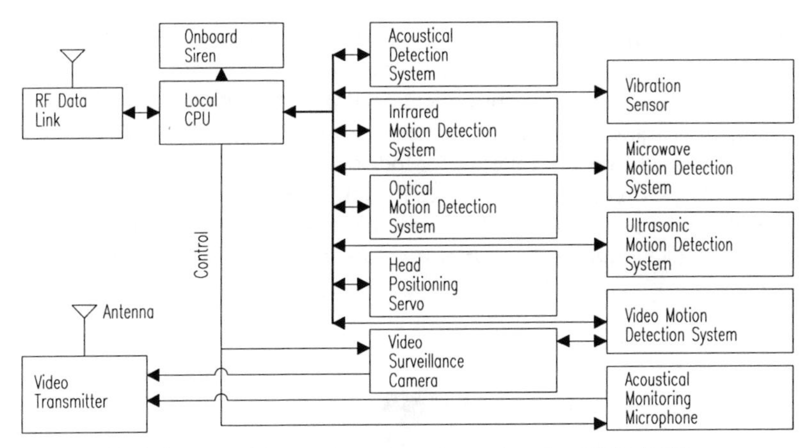

Figure 17-19. Block diagram of the security sensor suite employed on ROBART II.

The field of view is divided into four discrete zones (Figure 17-20), with different types of redundant motion detection schemes assigned to each zone. An array of 24 ultrasonic ranging units with 360-degree coverage can be activated to establish the position of a moving intruder with respect to the robot. A miniature high-resolution CCD surveillance camera is deployed on a panning mechanism for specific direction at areas of suspected disturbance. Assessment of the results is performed by appropriate software that cross-correlates among redundant primary sensors within a specific detection zone, and schedules and interprets subsequent verification by the secondary high-resolution sensors (Smurlo & Everett, 1993). The goal of the intelligent assessment software is to make the robot sensitive enough to detect any intrusion, yet smart enough to filter out nuisance alarms (Everett, et. al., 1988).

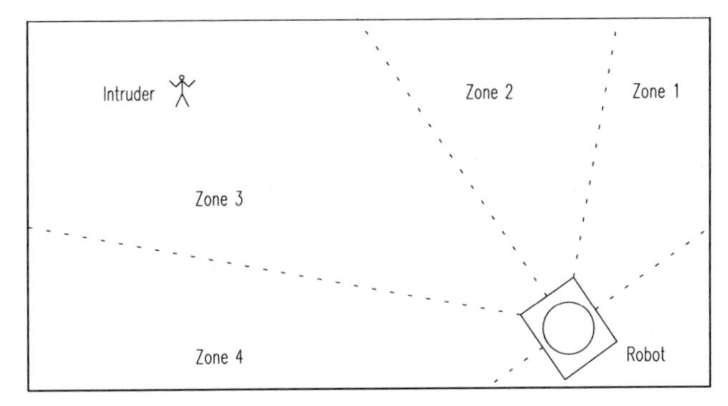

Figure 17-20. The six groups of intrusion detection sensors employed on ROBART II are arranged with full 180-degree coverage divided into four fan-shaped zones.

The field of view employed on ROBART II was purposely limited to 180 degrees for two reasons: 1) I didn't have the budget to procure the additional sensors and 2) it provided a convenient dead zone behind the robot for observers during demonstrations. Full 360-degree coverage divided into 24 discrete zones was incorporated into the follow-on ModBot design, and later carried over into the first MDARS prototype (Smurlo & Everett, 1993). The MDARS staring array was replaced in early 1994 with a commercially developed improvement in the form of the Cybermotion *Security Patrol Instrumentation (SPI)* module (Figure 17-21).

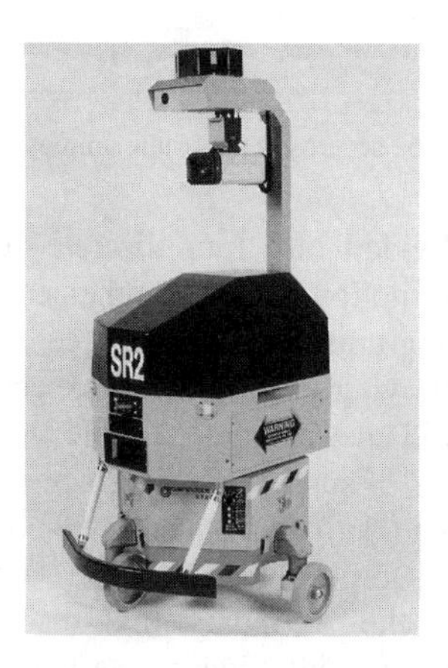

Figure 17-21. An early prototype of the *Security Patrol Instrumentation (SPI)* module with an underhung pan-and-tilt unit for surveillance camera (courtesy Cybermotion, Inc.).

17.2 Automated Inventory Assessment

The Microcircuit Technology in Logistics Applications (MITLA) Program Management Office at Wright-Patterson Air Force Base, Dayton, OH, is the focal point for *radio frequency identification* within the Department of Defense. As such, this agency is responsible for maintaining in-depth knowledge of the state of the art, and making that knowledge available to DoD customers. Ongoing developments in over 20 coordinated efforts are underway at a number of military installations in the United States (Lawlor, 1993).

The Physical Security Equipment Management Office submitted an informal request for an RF-tag market investigation to MITLA in early December 1993, as well as a list of potential suppliers previously compiled by the MDARS developers. A review of the stated MDARS needs as compared to existing capabilities within the industry subsequently conducted, with the more difficult core requirements for a long-range omni-directional system that could read/write to at least 10 or 12 feet addressed first. MITLA reported that to the best of their knowledge, only Savi Technology, Mountain View, CA, had a system (at the time) that could perform remote read/write operations in an omni-directional pattern at distances greater than 50 feet.

17.2.1 MDARS Product Assessment System

The MDARS *Product Assessment System* is physically separated into two groups of components respectively located at the host console and on the robotic platforms as depicted in Figure 17-22. The *Product Assessment Computer* collects tag data (when available) from multiple robots, storing the information in the *Product Database Computer*. The *Product Database Computer*, as the name implies, is a database that keeps track of all tags read in by the robots as well as those entered manually by the user. The *Database Access Computer* is the user interface to the *Product Database Computer*, allowing the entry of manual information, editing of existing tag information, as well as generation and viewing of various tag reports.

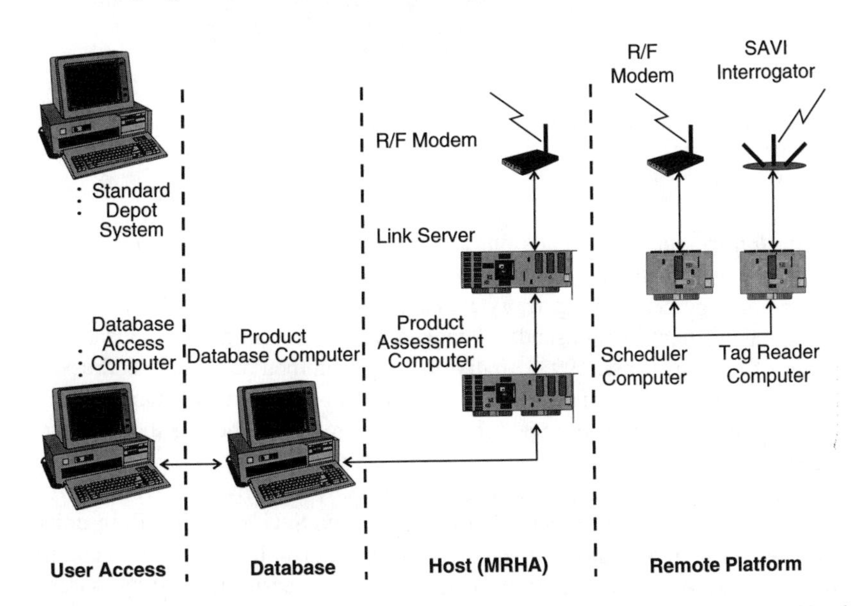

Figure 17-22. Block diagram of the MDARS Product Assessment System (courtesy Naval Command Control and Ocean Surveillance Center).

The hardware resident on each mobile robot consists of a Savi *Interrogator* for bidirectional communication with interactive RF transponder tags attached to high-value inventory items (Savi, 1994a), and a controlling *Tag Reader Computer*. When commanded by a *virtual-path program instruction*, the *Tag*

Reader Computer collects all tag information from the *Interrogator* and buffers it in internal blackboard memory for later transfer to the *Product Assessment Computer* when requested. The *Interrogator* is an off-the-shelf unit designed for unlicensed operation (below FCC Part 15 power levels) at either 315 or 433.92 MHz (Lawlor, 1993). Early models employed three 12-inch stub antennae mounted external to the half-spherical housing, 120 degrees apart for full omni-directional coverage, while the most recent version (Figure 17-23) incorporates a pair of antennae inside the housing for a rugged, less-obtrusive profile.

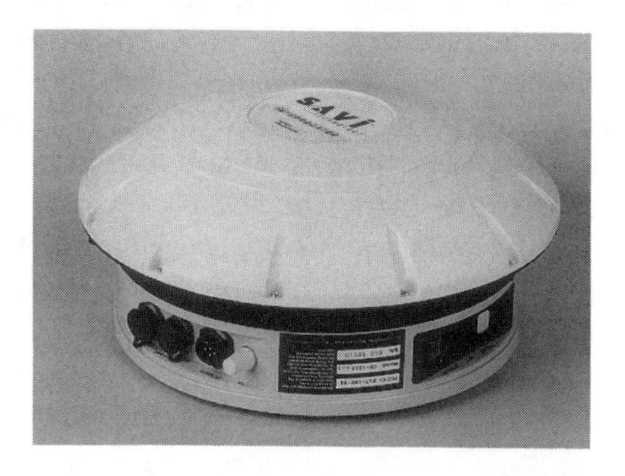

Figure 17-23. The Savi *Interrogator* is a microprocessor-controlled RF transceiver capable of omni-directional read/write operations to transponder tags located up to 150 feet away (courtesy Savi Technology, Inc.).

The *Interrogator* first sends out a wakeup signal consisting of a 3.49-second duration pulse modulated at 30-KHz, and uploads 10 bytes of data from each responding tag. Savi's proprietary *Batch Collection* algorithm allows the system to accurately identify thousands of tagged assets at a single read location in a matter of minutes (Savi, 1993). Individual tags can then be directly addressed for more complex data transfers, such as storing item-unique maintenance or special handling instructions in tag memory for future reference during the product life cycle.

Two types of RF transponder tags are currently used by the *MDARS Product Assessment System*: 1) the Savi *TyTag* and 2) the Savi *SealTag*. Both units are equipped with an onboard piezoelectric beeper that can be activated on command from an *Interrogator* to allow individual tags to be easily located by warehouse personnel (Savi, 1994b). The *TyTag* (Figure 17-24) operates on a 6-volt 600-mAh Lithium flat-pack battery and will automatically issue a low-battery warning (i.e., set a status bit in the tag's data stream) at 5.16 volts. The minimum operating voltage required to achieve a 25-foot line-of-sight range is 4.16 volts, and typical battery life is two years with two data collections per day. *TyTags* are normally intended for indoor operation only and are available with 128 or 256 bytes of non-volatile memory.

Table 17-7. Selected specifications for the Savi *Model CLIN 0003AA Interrogator.*

Parameter	Value	Units
Frequency	315 or 433	MHz
Range	300	feet
RF pattern	360	degrees
Memory	64K	bytes
Data rate (RF)	9600	baud
RS-232	9600	baud
RS-485	38.4K	baud
Power	6-15	volts DC
	100	milliamps
Size (diameter)	12	inches
(height)	4.5	inches
Weight	6.5	pounds

The *SealTag* (Figure 17-24) is enclosed in a rugged environmental package suitable for exposed outdoor operation and is available with an extended non-volatile memory of up to 128 kilobytes for mass storage of information such as product history or container manifests (Savi, 1994b). A 6-volt 1400-mAh lithium battery provides an expected service life of four years assuming two data collections per day, and battery status is automatically monitored as in the case of the *TyTag*. A real-time clock is incorporated into the SealTag design to facilitate time-stamping data or event occurrences.

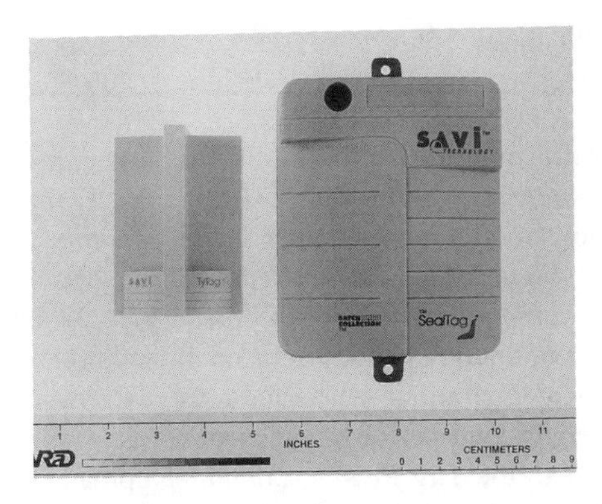

Figure 17-24. The Savi *TyTag* (left) is an interactive RF transponder with up to 256 bytes of read/write memory storage; the *SealTag* (right) can have up to 128 kilobytes and is equipped with four binary input lines that monitor external events (courtesy Savi Technology, Inc.).

An inverted-TTL RS-232 serial interface and four binary I/O lines are provide on the *SealTag* to communicate with auxiliary equipment and/or monitor external events. A change in logic level of an input line will toggle the state of an associated bit in the data stream read by the *Interrogator*, greatly expanding the versatility of the system. For example, an input line on a *SealTag* will be used in the MDARS Exterior program to monitor the physical status (i.e., open or shut) of high-security locks and will upload this information along with the lock serial number to the patrolling MDARS vehicle upon request. In this fashion, the same hardware used to verify inventory inside a locked space can also be used to collect binary type information describing related conditions (i.e., flooding, fire, smoke).

Table 17-8. Selected specifications for the *TyTag* and *SealTag*.

Parameter	*TyTag*	*SealTag*	Units
Frequency	314.975	433.92	MHz
Transmit power	< 10	< 10	microwatts
Range	200	300	feet
Memory	128 or 256	256, 8K, 128K	bytes
Environment	indoor	indoor/outdoor	
Audible beeper	yes	yes	
Real-time clock	no	yes	
Power	6	6	volts DC
(standby)	4	10	microamps
(active)	25	25	milliamps
Battery life	2	4	years
Size (length)	3.6	5.5	inches
(width)	2.35	4.5	inches
(height)	1.6	2	inches
Weight	5	10	ounces

The MDARS platform software to control the Savi *Interrogator* runs on the *Tag Reader Computer*, a Motorola *M68HC11*-based single-board computer developed exclusively to fit in the *Turret Interface Panel* of the Cybermotion platform (see again Figure 17-22). Communication between the *Tag Reader Computer* and the *Interrogator* is via a 9600-baud RS-232 serial link. The main loop of the software continuously monitors a command register awaiting direction to perform a *tag-read* operation, whereupon the *Interrogator* is instructed to transmit a *wakeup signal* and perform a subsequent tag collection. After the tag collection is completed, the *Tag Reader Computer* uploads the collected tag IDs from the *Interrogator*, and packetizes the data into its onboard memory for later collection by the MDARS *Product Assessment Computer* at the host console.

Listed below are some of the additional capabilities of the tag-reader software.

- Write to EEPROM of *Interrogator* to adjust power level of wakeup signal.

- Instruct *Interrogator* to send out a *tag-wakeup* signal.
- Determine battery status of each tag read.
- Determine signal strength of received signal from each tag.
- Read in the 24-hour clock of *Interrogator*.
- Store current X and Y coordinates of the platform.
- Instruct a specific tag to activate or deactivate its beeper.
- Perform a search to find a specific tag.

In January 1995, extensive testing was conducted by the MDARS development team at the Camp Elliott warehouse facility in San Diego (Figure 17-25) to assess the accuracy of several tag-position-estimation algorithms (Smurlo, et al., 1995). The test was also designed to determine the impact of performing tag-read operations at two different stop intervals (37.5 and 75 feet) along the route, using 173 Savi *TyTags* placed at known locations throughout the warehouse. For survey intervals of 37.5 feet, the best performing algorithm achieved an average of approximately 15 feet positional uncertainty (i.e., the difference between estimated and actual tag locations), while for survey intervals of 75 feet the uncertainty was increased to approximately 20 feet.

Figure 17-25. An earlier model of the Savi *Interrogator* mounted on top of the MDARS Interior robot undergoing feasibility testing at the Camp Elliott warehouse facility in San Diego, CA (courtesy Naval Command Control and Ocean Surveillance Center).

17.3 References

Ademco, "Quad Passive Infrared Motion Detector," *1989 Security Sourcebook: The Ademco Catalog of Products*, P6715, Ademco Alarm Device Manufacturing Company, Syosset, NY, May, 1989.

Alpha, "Theory, Operation, and Application of Microwave Motion Sensing Modules," *Sensors*, pp. 29-36, December, 1987.

Bancroft, A.J., "The First Commercial Floor Care Company that Ventured into the Production of Robotics," Conference on Intelligent Robotics in Field, Factory, Service, and Space, CIRFFSS '94, Houston, TX, pp. 669-674, March, 1994.

Barron, W.R., "The Principles of Infrared Thermometry," *Sensors*, pp. 10-19, Decmber, 1992.

Buschling, R., "Understanding and Applying IR Temperature Sensors," *Sensors*, pp. 32-37, October, 1994.

Byler, E., "Intelligent Mobile Sensor System for Drum Inspection and Monitoring," Phase I Topical Report, DOE Contract DE-AC21-92MC29112, Martin Marietta Astronautics Group, Littleton, CO, June, 1993.

Cima, D., "Using Lithium Tantalate Pyroelectric Detectors in Robotics Applications," Eltecdata #112, Eltec Instruments, Inc., Daytona Beach, FL, 1984.

Cima, D., "Using Optical Radiation for Security," Eltecdata #124, Eltec Instruments, Inc., Daytona Beach, FL, December, 1990.

Cima, D., "Surveillance Applications of the Eltec Model 862 Passive Infrared Telescope," Eltecdata #128, Eltec Instruments, Inc., Daytona Beach, FL, June, 1992.

Eltec, "Introduction to Infrared Pyroelectric Detectors," Eltecdata #100, Eltec Instruments, Inc., Daytona Beach, FL, 1984.

Eltec, "Model 442 IR-Eye Integrated Sensor," Preliminary Product Literature, Eltec Instruments, Inc., Daytona Beach, FL, April, 1991.

Eltec, "Model AR170, 32 Element Pyroelectric Array," Product Literature, Eltec Instruments, Inc., Daytona Beach, FL, December, 1993.

Everett, H.R., "A Computer Controlled Sentry Robot," *Robotics Age*, March/April, 1982a.

Everett, H.R., "A Microprocessor Controlled Autonomous Sentry Robot", Masters Thesis, Naval Postgraduate School, Monterey, CA, October 1982b.

Everett, H.R., "Security and Sentry Robots", *International Encyclopedia of Robotics Applications and Automation*, R.C. Dorf, ed., John Wiley, pp. 1462-1476, March, 1988.

Everett, H.R., Gilbreath, G.A., Alderson, S.L., Priebe, C., Marchette, D., "Intelligent Security Assessment for a Mobile Sentry Robot", Proceedings, 29th Annual Meeting, Institute for Nuclear Materials Management, Las Vegas, NV, June, 1988.

Everett, H.R., Gilbreath, G.A., Tran, T., Nieusma, J.M., "Modeling the Environment of a Mobile Security Robot," Technical Document 1835, Naval Command Control and Ocean Surveillance Center, San Diego, CA, June, 1990.

Gage, D.W., Everett, H.R., Laird, R.T., Heath-Pastore, T.A., "Navigating Multiple Robots in Semi-Structured Environments," ANS 6th Topical Meeting on Robotics and Remote Systems, Monterey, CA, February, 1995.

Gallo, M.A., Willits, D.S., Lubke, R.A., Thiede, E.C., "Low Cost Uncooled IR Sensor for Battlefield Surveillance," SPIE Infrared Technology XIX, Vol. 2020, San Diego, CA, July, 1993.

George, S.C., "Robot Revival," *Security*, pp. 12-13, June, 1992.

Gontowski, W., "Build a Motion Detector Alarm," *Electronic Experimenter's Handbook*, pp. 56-64, 1983.

Hansford, A., "The AD9502 Video Signal Digitizer and its Application", Analog Devices Application Note C1100-9-7/87, Norwood, MA, July, 1987.

Hansen, C., Beratan, H., Owen, R., Corbin, M., McKenney, S., "Uncooled Thermal Imaging at Texas Instruments," SPIE Infrared Technology XVIII, Vol. 1735, San Diego, CA, pp. 17-26, July, 1992.

Hanson, C., Beratan, H., Owen, R., Sweetser, K., "Low-Cost Uncooled Focal Plane Array Technology," Detector IRIS Meeting, Bedford, MA, August, 1993.

Hanson, C., Beratan, H. "Uncooled Pyroelectric Thermal Imaging," International Symposium on Applications of Ferroelectrics, 1994.

Heckendorn, F.M., Ward, C.W., Wagner, D.G., "Remote Radioactive Waste Drum Inspection with an Autonomous Mobile Robot," ANS Fifth Topical Meeting on Robotics and Remote Systems, American Nuclear Society, Knoxville, TN, pp. 487-492, April, 1993.

Holland, J.M., "An Army of Robots Roams the Night," International Robot and Vision Automation Show and Conference, Detroit, MI, pp. 17.1-17.12, April, 1993.

ISRA, "Military Finds Big Cost Savings from Mobile Robotics," *ISRA News,* Newsletter of the International Service Robot Association, Ann Arbor, MI, Fall, 1994.

Jones, J.L., Flynn, A.M., *Mobile Robots: Inspiration to Implementation*, AK Peters, Ltd., Wellesley, MA, p. 113, 1993.

King, S.J., Weiman, C.F.R., "HelpMate Autonomous Mobile Robot Navigation System," SPIE Vol. 1388, Mobile Robots V, Boston, MA, pp. 190-198, November, 1990.

Lawlor, M., "Microcircuit Technology Improves Readiness, Saves Resources," *Signal*, Armed Forces Communications and Electronics Association, August, 1993.

MacLeod, E.N., Chiarella, M., "Navigation and Control Breakthrough for Automated Mobility," Proceedings, SPIE Mobile Robots VIII, Vol. 2058, pp. 57-68, 1993.

Mattaboni, P., "An Update on Lab Rover: A Hospital Material Transporter," Conference on Intelligent Robotics in Field, Factory, Service, and Space, CIRFFSS '94, Houston, TX, pp. 405-406, March, 1994.

Mims, F.M., *Forrest Mims' Circuit Scrapbook II*, Howard W. Sams, Indianapolis, IN, pp. 170-171, 1987.

Nippon, "Pyroelectric Infrared Sensor," Nippon Ceramics Technical Information TI-101, McGee Components, Inc, North Attleboro, MA, undated.

Nippon, "Pyrosensor," Nippon Ceramics Product Literature PE 1001-1091, McGee Components, Inc, North Attleboro, MA, October, 1991.

Philips, "Ceramic Pyroelectric Infrared Sensors and Their Applications," Philips Technical Publication 163, Philips Semiconductors, Slatersville Division, Smithfield, RI, 1985.

Philips, "Movement Sensing Using a Multi-Element Fresnel Lens," Philips Semiconductors, Slatersville Division, Smithfield, RI, April, 1986.

Quick, C., "Animate vs. Inanimate," *Robotics Age*, Vol. 6, No. 9, August, 1984.

RST, "Surrogate Teleoperated Vehicle (STV) Technical Manual," Robotic Systems Technology, Westminster, MD, Contract No. N66001-91-C-60007, CDRL Item B001, Final Issue, 13 September, 1993.

Russell, R.A., "Mobile Robot Guidance Using a Short-Lived Heat Trail," *Robotica*, Vol. 11, Cambridge Press, pp. 427-431, 1993.

Savi, "The Savi Asset Management System," Product Brochure, Savi Technology, Inc., Mountain View, CA, 1993.

Savi, "System Components," Product Literature, Savi Technology, Inc., Mountain View, CA, April, 1994a.

Savi, "Savi Technology Ordering Guide," First Edition, Radio Frequency Identification Equipment Contract No. F33600-94-D-0077, Savi Technology, Inc., Mountain View, CA, November, 1994b.

Smurlo, R.P., Everett, H.R., "Intelligent Sensor Fusion for a Mobile Security Robot," *Sensors*, pp. 18-28, June, 1993.

Smurlo, R.P., Laird, R.T., Elaine, S., Jaffee, D.M., "The MDARS Product Assessment System," Association of Unmanned Vehicle Systems, 22nd Annual Technical Symposium and Exhibition, Washington, DC, July, 1995.

TI, "*Nightsight* Thermal Vision System," Product Literature, Texas Instruments, Inc., Attleboro, MA, November, 1994.

Tom, E., "Polymer Film Arrays in Pyroelectric Applications," *Sensors*, pp. 75-77, September, 1994.

Viggh, H.E.M., Flynn, A.M., "Infrared People Sensors for Mobile Robots," SPIE Vol. 1007, Mobile Robots III, Cambridge, MA, pp. 391-398, November, 1988.

Weiss, M., "Protect Your Valuables - Light Sensitive Security Alert," *Radio Electronics*, April, 1979.

Williams, H., "Proximity Sensing with Microwave Technology," *Sensors*, pp. 6-15, June, 1989.

Williams, H., "The Basic Principles of Microwave Proximity Sensing," *Sensors*, pp. 26-28, May, 1991.

Appendix
Alphabetical Listing of Cited Organizations

3M Traffic Control Materials
3M Center, Bldg 553-1A-01
St. Paul, MN 55144
POC: Tom Dahlin
Tel: 612-736-7505
Fax: 612-733-2227

Acuity Research, Inc.
3475P Edison Way
Menlo Park, CA 94025
POC: Robert Clark
Tel: 415-369-6782
Fax: 415-369-6785

AGV Products, Inc.
9307-E Monroe Road
Charlotte, NC 28270
POC: Mats Herrstromer
Tel: 704-845-1110
Fax: 704-845-1111

Alliant Techsystems, Inc.
600 Second Street NE
Hopkins, MN 55343-8384
POC: Mike Gallo
Tel: 612-931-6873
Fax: 612-931-4305

Alpha Industries, Inc.
20 Sylvan Road
Woburn, MA 01801
POC: Bill Sherman
Tel: 617-935-5150
Fax: 617-933-5582

AMP, Inc.
950 Forge Avenue
Valley Forge, PA 19482
POC: Edward Tom
Tel: 610-666-3500
Fax: 610-666-3509

AM Sensors, Inc.
(See Monitor Manufacturing, Inc.)

Analog Devices
One Technology Way
PO Box 9106
Norwood, MA 02062-9106
Tel: 617-329-4700
Fax: 617-326-8903

Apogee Robotics, Inc.
(Last known address)
2643 Midpoint Drive
Fort Collins, CO 80525
POC: Mike Henningsen

Applied Physics Systems
897 Independence Ave.
Suite 1C
Mountain View, CA 94043
POC: Robert Goodman
Tel: 415-965-0500
Fax: 415-965-0404

Applied Research Lab
PO Box 30
Penn State University
State College, PA 16804
POC: Henry Watson
Tel: 814-865-6345
Fax: 814-863-1183

Arizona State University
Chemistry Department
Tempe, AZ 85287-1604
POC: Prof. Neal Woodbury
Tel: 602-965-3294
Fax: 602-965-2747

Arnold Engineering, Inc.
PO Box 1567
Norfolk, NB 68072-1567
POC: Kent Liesemeyer
Tel: 402-371-6100
Fax: 402-371-1994

Ashtec, Inc.
1170 Kifer Road
Sunnyvale, CA 94086
POC: Sue MacLean
Tel: 408-524-1400
Fax: 408-524-1500

Associates and Ferren
Box 609 Wainscott-NW Road
Wainscott, NY 11975
POC: Bran Ferren
Tel: 516-537-7800
Fax: 516-537-4343

Banner Engineering Corp.
9714 10th Ave N.
Minneapolis, MN 55441
POC: Bob Garwood
Tel: 612-544-3164
Fax: 612-544-3213

Barnes and Reineke Corp.
425 East Algonquin
Arlington Heights, IL 60005
POC: Mike Fitzgerald
Tel: 708-640-3740
Fax: 708-640-0354

Barrier Systems, Inc.
1100 E. William Street, Suite 206
Carson City, NV 89701-3104
POC: Jay Ciccotti
Tel: 702-885-2500
Fax: 702-885-2598

Bell and Howell Mailmobile Co.
411 East Roosevelt Ave.
Zeeland, MI 49464-1395
POC: Rick Paske
Tel: 800-325-7400
Fax: 616-772-6380

Blue Road Research
2555 NE 205th Street
Troutdale, OR 97060
POC: Eric Udd
Tel: 503-667-7772
Fax: 503-667-7880

Bonneville Scientific
918 East 900 South
Salt Lake City, UT 84105
POC: Josephine Grahn
Tel: 801-359-0402
Fax: 801-359-0416

CANSPACE
(See University of New Brunswick)

Carnegie Mellon University
Robotics Institute
Pittsburgh, PA 15213
POC: Chuck Thorpe
Tel: 412-268-3612
Fax: 412-268-5571

Caterpillar Industrial, Inc.
Automated Vehicle Systems
5960 Heisley Road
Mentor, OH 44060-1881
POC: David Heinz
Tel: 216-357-2246
Fax: 216-357-4410

Charles Stark Draper Laboratory
555 Technology Square, MS 27
Cambridge, MA 02139
POC: Bill Kaliardos
Tel: 617-258-1989
Fax: 617-258-2121

Chesapeake Laser Systems, Inc.
222 Gale Lane
Kennett Square, PA 19348-1734
POC: Larry Brown
Tel: 610-444-4253
Fax: 610-444-2321

Computer Sciences Corp.
7405 Alban Station Court
Suite B-206
Springfield, VA 22150
POC: Susan Hower
Tel: 703-912-7880
Fax: 703-912-6082

Control Engineering
8212 Harbor Springs Road
Harbor Springs, MI 49740
POC: Bruce Lindsay
Tel: 616-347-3931
Fax: 616-347-3342

Corning, Inc.
Telecommunications Products Division
35 W. Market Street
Corning, NY 14831
POC: Vincent P. Martinelli
Tel: 607-974-3539
Fax: 607-974-3975

Cybermotion, Inc.
115 Sheraton Drive
Salem, VA 24153
POC: John Holland
Tel: 703-562-7626
Fax: 703-562-7632

David Sarnoff Research Center
201 Washington Road
Princeton, NJ 08540-6449
POC: Dr. Peter Burt
Tel: 609-734-2451
Fax: 609-734-2662

Denning Branch International Robotics
1401 Ridge Avenue
Pittsburgh, PA 15233
POC: Alan Branch
Tel: 412-322-4412
Fax: 412-322-2040

Denning Mobile Robots, Inc.
(See Denning Branch International Robotics)

Dinsmore Instrument Company
1814 Remell Street
Flint, MI 48503
POC: R.C. Dinsmore
Tel: 313-744-1330
Fax: 313-744-1790

Eltec Instruments, Inc.
PO Box 9610
Central Business Park
Daytona Beach, FL 32020-9610
POC: David Cima
Tel: 800-874-7780
Fax: 904-258-3791

Environmental Research Institute of Michigan
Box 8618
Ann Arbor, MI 48107
POC: Frank Pont
Tel: 313-994-1200
Fax: 313-994-3890

ESP Technologies, Inc.
21 LeParc Drive
Lawrenceville, NJ 06848
POC: Susan Cox
Tel: 609-275-0356
Fax: 609-275-0356

General Microwave
5500 New Horizons Blvd.
Amityville, NY 11701
POC: Mathew Jacobs
Tel: 516-226-8900, X304
Fax: 516-226-8966

Georgia Institute of Technology
Mobile Robot Laboratory
College of Computing
Atlanta, GA 30332
POC: Prof. Ronald Arkin
Tel: 404-894-8209
Fax: 404-853-0957

Hamamatsu Corp.
360 Foothill Rd.
Bridgewater, NJ 08807
POC: Norman H. Schiller
Tel: 908-231-0960
Fax: 908-231-1218

Harris Technologies, Inc.
PO Box 6
Clifton, VA 22024
POC: Jim Harris
Tel: 703-266-0900
Fax: 703-968-8827

Hewlett-Packard Components
Customer Information Center
Building 49 AV
19310 Pruneridge Avenue
Cupertino, CA 95014
Tel: 800-752-9000

Hitachi Cable America, Inc.
50 Main Street
White Plains, NY 10606-1920
POC: Ray Ikeda
Tel: 914-993-0990
Fax: 914-993-0997

Honeywell, Inc.
Microswitch Division
11 West Spring Street
Freeport, IL 61032
POC: John Mitchell
Tel: 800-537-6945
Fax: 815-235-6545

Honeywell, Inc.
Military Avionics Division
11601 Roosevelt Boulevard
St. Petersburg, FL 33716-2202
POC: Jody Wilkerson
Tel: 813-579-6473
Fax: 813-579-6832

Honeywell, Inc.
Solid State Electronics Center
12001 Highway 55
Plymouth, MN 55441
POC: Tamara Bratland
Tel: 612-954-2992
Fax: 612-954-2051

Honeywell, Inc.
Technology Center
3660 Technology Drive
Minneapolis, MN 55418
POC: James E. Lenz
Tel: 612-951-7715
Fax: 612-951-7438

Honeywell Visitronics
(See Honeywell, Micro Switch Division)

IBM Research
30 Saw Mill River Road
Hawthorne, NY 10532
POC: Jonathan Connell
Tel: 914-784-7853
Fax: 914-784-6307

Intelligent Solutions, Inc.
1 Endicott Avenue
Marblehead, MA 01945
POC: Jim Maddox
Tel: 617-639-8144
Fax: 617-639-8144

IS Robotics
Twin City Office Center, Suite #6
22 McGrath Highway
Somerville, MA 02143
POC: Colin Angle
Tel: 617-629-0055
Fax: 617-629-0126

Kaman Sciences
Remote Control Program
1500 Garden of the Gods Road
Colorado Springs, CO 80933-7463
POC: Doug Caldwell
Tel: 719-599-1285
Fax: 719-599-1942

Kearfott Guidance and Navigation
1150 McBride Avenue
Little Falls, NJ 07424
POC: James G. Koper
Tel: 201-785-6000, X5492
Fax: 201-785-5555

KVH Industries, Inc.
110 Enterprise Center
Middletown, RI 02842
POC: Sandy Oxx
Tel: 401-847-3327
Fax: 401-849-0045

LAAS - CNRS
7, avenue du Colonel Roche
31077 Toulouse Cedex, France
POC: Raja Chatila
Tel: (33) 61 33 63 28
Fax: (33) 61 33 64 55

Litton Industrial Automation
(See Saurer Automation Systems)

Macome Corp.
7-32-6 Nishikamata Ohta-ku
Tokyo 144 Japan
POC: S. Kamewaka

Magellan Systems Corp.
960 Overland Court
San Dimas, CA 91773
POC: Emile Yakoup
Tel: 909-394-6062
Fax: 909-394-7050

Magnavox Advanced Products and Systems
2829 Maricopa Street
Torrance, CA 90503
POC: Eric Furlong
Tel: 310-618-1200
Fax: 310-618-7074

Martin Marietta Aerospace Corp.
Space Systems Division
PO Box 179
Denver, CO 80201
POC: Wendell Chun
Tel: 303-971-7945
Fax: 303-971-4093

Massa Products Corp.
280 Lincoln Street
Hingham, MA 02043
POC: Paul Shirley
Tel: 617-749-4800
Fax: 617-740-2045

Massachusetts Institute of Technology
Artificial Intelligence Lab
545 Technology Square
Cambridge, MA 02139
POC: Anita Flynn
Tel. 617-253-3531
Fax: 617-253-0039

Merritt Systems, Inc.
2425 N. Courtenay Parkway
Suite 5
Merritt Island, FL 32953
POC: Daniel Wegerif
Tel: 407-452-7828
Fax: 407-452-3698

Microswitch Division
(See Honeywell, Inc.)

Millitech Corp.
PO Box 109
Deerfield, MA 01373-0109
POC: Ken Wood
Tel: 413-665-8551
Fax: 413-665-2536

Monash University
Department of Electrical and Computer Systems Engineering
Clayton, Australia VIC 3168
POC: Andrew Russell

Monitor Manufacturing, Inc.
44W320 Keslinger Road
PO Box 8048
Elburn, IL 60119-8048
POC: Thomas F. Meagher
Tel: 708-365-9403
Fax: 708-365-5646

Motorola, Inc.
Government and Systems Technology Group
8220 E. Roosevelt Road
PO Box 9040
Scottsdale, AZ 85252-9040
POC: Burt Woelkers
Tel: 602-441-7685
Fax: 602-441-7677

MTI Research, Inc.
313 Littleton Road
Chelmsford, MA 01824
POC: Edward N. MacLeod
Tel: 508-250-4949
Fax: 508-250-4605

Murata Electronics North America
2200 Lake Park Drive
Smyrna, GA 30080
POC: Satoshi Ishino
Tel: 404-436-1300
Fax: 404-436-3030

NAMCO Controls
5335 Avion Park Drive
Highland Heights, OH 44143
POC: Greg Miller
Tel: 800-NAM-TECH
Fax: 216-946-1228

NASA Goddard Space Flight Center
Robotics Branch, Code 714.1
Greenbelt, MD 20771
POC: John Vranish
Tel: 301-286-4031
Fax: 301-286-1613

NASA Jet Propulsion Laboratory
4800 Oak Grove Drive
Pasadena, CA 91109
POC: Larry Mathies
Tel: 818-354-3722
Fax: 818-354-8172

National Institute for Standards and Technology
Building 200, Room B124
Gaithersburg, MD 20899
POC: Marty Herman
Tel: 301-975-2000
Fax: 301-990-9688

National Research Council of Canada
Institute for Information Technology
Ottawa, Ontario, Canada, K1A 0R6
POC: Francois Blais
Tel: 613-993-7892
Fax: 613-952-0215

National Semiconductor Corp.
2900 Semiconductor Drive
P.O. Box 58090
Santa Clara, CA 95052-8090
Tel: 408-721-5000
Fax: 408-739-9803

Naval Command Control and Ocean Surveillance Center (NCCOSC)
RDT&E Division 5303
San Diego, CA 92152-7383
POC: Bart Everett
Tel: 619-553-3672
Fax: 619-553-6188

Naval Ocean Systems Center (NOSC)
(See Naval Command Control and Ocean Surveillance Center)

Naval Postgraduate School
Department of Computer Science
Monterey, CA 93940
POC: Prof. Bob McGhee
Tel: 408-656-2026
Fax: 408-656-2814

Naval Research Laboratory
Chemistry Division
Code 6177
Washington, DC 20375-5000
POC: Richard J. Colton
Tel: 202-767-0801
Fax: 202-767-3321

Navigation Technologies Corp.
740 Arques Avenue
Sunnyvale, CA 94086
POC: Daniel Udoutch
Tel: 408-737-3200
Fax: 408-737-3280

Nonvolatile Electronics, Inc.
11409 Valley View Road
Eden Prairie, MN 55344
POC: Jay Brown
Tel: 612-829-9217
Fax: 612-829-9241

NovAtel Communications Ltd.
6732 8 Street N.E.
Calgary, Alberta, Canada T2E 8M4
POC: Bryan R. Townsend
Tel: 403-295-4500
Fax: 403-295-0230

Oak Ridge National Laboratory
PO Box 2008
Oak Ridge, TN 37831-6304
POC: William R. Hamel
Tel: 615-574-5691
Fax: 615-576-2081

Odetics, Inc.
1515 South Manchester Highway
Anaheim, CA 92802-2907
POC: Tom Bartholet
Tel: 714-758-0300
Fax: 714-774-9452

OmniTech Robotics
2640 Raritan Circle
Englewood, CO 80110
POC: Dave Parish
Tel: 303-922-7773
Fax: 303-922-7775

Perceptron, Inc.
23855 Research Drive
Farmington Hills, MI 48335
POC: Dave Zuk
Tel: 810-478-7710
Fax: 810-478-7059

Phase Laser Systems, Inc.
14255 N. 79th Street
Suite 6
Scottsdale, AZ 85260
POC: Michael Brubacher
Tel: 602-998-4828
Fax: 602-998-5586

Philips Semiconductors
100 Providence Pike
Slatersville, RI 02876
POC: Ed Martins
Tel: 401-767-4458
Fax: 401-767-4403

Physical Security Equipment
Management Office
10101 Gridley Road
Suite 104
Fort Belvoir, VA 22060-5818
POC: Jerry Edwards
Tel: 703-704-2412
Fax: 703-704-2495

Polaroid Corp.
784 Memorial Drive
Cambridge, MA 02139
POC: Phil Jackman
Tel: 617-386-3964
Fax: 617-386-3966

Precision Navigation, Inc.
1235 Pear Avenue
Suite 111
Mountain View, CA 94043
POC: Mark Moran
Tel: 415-962-8777
Fax: 415-962-8776

Precision Technology, Inc.
4000 Chesswood Drive
Downsview, Ontario
Canada M3J 2B9
POC: Bruce Buck
Tel: 416-630-0200
Fax: 416-630-4414

Quantic Industries, Inc.
990 Commercial Street
San Carlos, CA 94070
Tel: 408-867-4074
Fax: same as above

Redzone Robotics
2425 Liberty Avenue
Pittsburgh, PA 15222
POC: Jeff Callen
Tel: 412-765-3064
Fax: 412-765-3069

RIEGL USA
Laser Measurement Systems
8516 Old Winter Garden Road #101
Orlando, FL 32835-4410
POC: Ted Knaak
Tel: 407-294-2799
Fax: 407-294-3215

Robotic Systems Technology
1110 Business Parkway
Westminster, MD 21157
POC: Scott Myers
Tel: 410-876-9200
Fax: 410-876-9470

Robotic Vision Systems, Inc.
425 Rabro Drive East
Hauppauge, NY 11788
POC: Howard Stern
Tel: 516-273-9700
Fax: 516-273-1167

Rockwell International
3200 East Renner Road
Richardson, TX 75082
POC: Larry Creech
Tel: 214-705-1704
Fax: 214-705-3284

Safety First Systems, Ltd.
42 Santa Barbara Drive
Plainview, NY 11803
POC: Alan Hersch
Tel: 516-681-3653
Fax: 516-938-6558

SAIC Bio-Dynamics
(See Robotic Systems Technology)

Sandia National Labs
PO Box 5800
Albuquerque, NM 87185-0860
POC: Paul Klarer
Tel: 505-844-2900
Fax: 505-844-5946

Sandia National Labs
Organization 9122
PO Box 5800
Albuquerque, NM 87185-0860
POC: John Sackos
Tel: 505-844-3033
Fax: 505-844-7020

SatCon Technology Corp.
161 First Street
Cambridge, MA 02142-1221
POC: Ralph Fenn
Tel: 617-349-0815
Fax: 617-661-3373

Saurer Automation Systems
11818 James Street
Holland, MI 49424
POC: Dwight Williams
Tel: 616-393-0101
Fax: 616-393-0331

Savi Technology, Inc.
450 National Avenue
Mountain View, CA 94043-2238
POC: Alan Bien
Tel: 415-428-0550
Fax: 415-428-0444

Schwartz Electro-Optics, Inc.
3404 N. Orange Blossom Trail
Orlando, FL 32804
POC: Robert Gustavson
Tel: 407-298-1802
Fax: 407-297-1794

Space Electronics, Inc.
4031 Sorrento Valley Blvd.
San Diego, CA 92121
POC: David Czajkowski
Tel: 619-452-4166
Fax: 619-452-5499

Sperry Marine, Inc.
Seminole Trail
Charlottesville, VA 22901
POC: Peter Arnold
Tel: 804-974-2000
Fax: 804-974-2259

Systron Donner Inertial Division
BEI Electronics
2700 Systron Drive
Concord, CA 94518-1399
POC: Scott Orlosky
Tel: 510-682-6161
Fax: 510-671-6590

Texas Instruments, Inc.
Defense Systems and Equipment
PO Box 655474, MS-37
Dallas, TX 75265
POC: Charles Hanson
Tel: 214-995-0874
Fax: 214-995-2231

Texas Instruments, Inc.
Nightsight
34 Forest Street
Attleboro, MA 02703
POC: Stan Kummer
Tel: 508-236-1396
Fax: 508-699-3242

Transitions Research Corp.
Shelter Rock Lane
Danbury, CT 06810
POC: John Evans
Tel: 203-798-8988
Fax: 203-791-1082

Trimble Navigation
P.O. Box 3642
Sunnyvale, CA 94088-3642
POC: Joel Avery
Tel: 408-481-8927
Fax: 408-481-2000

Tulane University
Mechanical Engineering Department
New Orleans, LA 70118-5674
POC: Prof. Fernando Figueroa
Tel: 504-865-5775
Fax: 504-865-5345

University of Michigan
1101 Beal Avenue
Ann Arbor, MI 48109-2110
POC: Dr. Johann Borenstein
Tel: 313-763-1560
Fax: 313-944-1113

University of Minnesota
111 Church Street SE
Minneapolis, MN 55455
POC: Prof. Max Donath
Tel: 612-625-2304
Fax: 612-625-8884

University of Nebraska
Department of Mathematics and Computer Science
Omaha, NE 68182
Tel: 402-554-2800
Fax: 402-554-2975

University of New Brunswick
Geodetic Research Laboratory
Department of Geodesy and Geomatics Engineering
Fredericton, N.B., Canada E3B 5A3
POC. Prof. Richard Langley
Tel: 506-453-5142
Fax: 506-453-4943

University of South Carolina
Swearingen Engineering Center
Columbia, SC 29208
POC: Prof. Joe Byrd
Tel: 803-777-9569
Fax: 803-777-8045

Unmanned Ground Vehicle Joint Program Office
(See US Army Missile Command)

US Army Engineering Topographical Laboratory (USAETL)
ATTN: CEETL-TL-SP
Building 2592
Fort Belvoir, VA 22060-5546
POC: Stephen DeLoach
Tel: 703-355-3026
Fax: 703-355-3176

US Army Missile Command
Unmanned Ground Vehicle Joint Program Office, Building 5410
Redstone Arsenal, AL 35896-8060
POC: COL Jeff Kotora
Tel: 205-876-3988
Fax: 205-842-0947

Vehicle Radar Safety Systems, Inc.
10 South Gratiot
Mt. Clemens, MI 48043
POC: Charles Rashid

VORAD Safety Systems, Inc.
10802 Willow Court
San Diego, CA 92127
POC: Kevin Wixom
Tel: 619-674-1450
Fax: 619-674-1355

Watson Industries, Inc.
3041 Melby Road
Eau Claire, WI 54703
POC: William Watson
Tel: 715-839-0628
Fax: 715-839-8248

Wright Laboratories
Robotics Lab, Building 9738
Tyndall AFB, FL 32403-5319
POC: Ed Brown
Tel: 904-283-3725
Fax: 904-283-9710

Zemco Group, Inc.
(Last known address)
3401 Crow Canyon Road
Suite 201
San Ramon, CA 94583
POC: Peter Blaney

Zexel Corp.
37735 Enterprise Court
Suite 600
Farmington Hills, MI 48331
POC: Mike Rice
Tel: 810-553-9930
Fax: 810-553-9931

Index

A

B

C

D

E

F

G

H

N

O

P

Q

R

S

T

U

V

W

Y